AF838691

ION EXCHANGE TECHNOLOGIES

Edited by **Ayben Kilislioğlu**

Ion Exchange Technologies
http://dx.doi.org/10.5772/2925
Edited by Ayben Kilislioğlu

Contributors

Ayben Kilislioğlu, Yoshihiro Tsujimoto, Hiroshi Kageyama, Berta Domènech, Julio Bastos-Arrieta, Amanda Alonso, Jorge Macanás, Maria Muñoz, Dmitri N. Muraviev, Yoshiro Kaneko, Kriveshini Pillay, Dorota Kołodyńska, Zbigniew Hubicki, Roman Rogoziński, Kyeong-Won Park, Aiymgul M. Akimkhan, Marisa Nascimento, Patrícia F. Prado, Paulo Sérgio M. Soares, Vicente P. de Souza, Vincenzo M. Sglavo, Tevfik Ünaldı, Selahattin Kadir, Hidayat Ullah Khan, Xingcheng Ding, Xunyue Liu

Published by InTech

Janeza Trdine 9, 51000 Rijeka, Croatia

Copyright © 2012 InTech

All chapters are distributed under the Creative Commons Attribution 3.0 license, which allows users to download, copy and build upon published articles even for commercial purposes, as long as the author and publisher are properly credited, which ensures maximum dissemination and a wider impact of our publications. However, users who aim to disseminate and distribute copies of this book as a whole must not seek monetary compensation for such service (excluded InTech representatives and agreed collaborations). After this work has been published by InTech, authors have the right to republish it, in whole or part, in any publication of which they are the author, and to make other personal use of the work. Any republication, referencing or personal use of the work must explicitly identify the original source.

Notice

Statements and opinions expressed in the chapters are these of the individual contributors and not necessarily those of the editors or publisher. No responsibility is accepted for the accuracy of information contained in the published chapters. The publisher assumes no responsibility for any damage or injury to persons or property arising out of the use of any materials, instructions, methods or ideas contained in the book.

Publishing Process Manager Iva Lipovic
Technical Editor InTech DTP team
Cover InTech Design team

Additional hard copies can be obtained from orders@intechopen.com

Ion Exchange Technologies, Edited by Ayben Kilislioğlu
p. cm.
ISBN 978-953-51-0836-8

We are IntechOpen, the world's largest scientific publisher of Open Access books.

3,250+
Open access books available

106,000+
International authors and editors

112M+
Downloads

151
Countries delivered to

Our authors are among the
Top 1%
most cited scientists

12.2%
Contributors from top 500 universities

WEB OF SCIENCE™

Selection of our books indexed in the Book Citation Index
in Web of Science™ Core Collection (BKCI)

Interested in publishing with us?
Contact book.department@intechopen.com

Numbers displayed above are based on latest data collected.
For more information visit www.intechopen.com

For EU product safety concerns:
IN TECH d.o.o., Prolaz Marije Krucifikse Kozulić 3, 51000 Rijeka, Croatia,
info@intechopen.com or visit our website at intechopen.com.

Meet the editor

Professor Ayben Kilislioğlu is currently working in the department of chemistry, Istanbul University (IU), Turkey. She received her master of science degree in physical chemistry from IU in 1994. She received her doctor of philosophy degree in physical chemistry from IU in 2000. She worked as visiting research assistant professor at the University of Illinois, Chicago, department of chemistry, between 2005-2006. She also worked at University of Chicago in Dr. Graeme Bell's Lab in 2007. She has research experience in adsorption, surface characterization and ion exchange. She worked on different projects funded by Istanbul University Grant Commission. She has published several research articles and a book chapter in this area.

Contents

Preface

With a history dating back to 1850, ion exchange is not a new phenomenon. But it become more and more important and widely used for some applications like selective ion removal, product recovery and purification.

This book presents innovative theories, structural and computational models, experimental laboratory work and practical industrial applications about ion exchange. Original research will report scientifically valuable information useful to academicians pursuing new synthesis and characterizations for ion exchangers and practitioners using ion exchange science and technology in industry.

The book consists of fifteen chapters about the recent improvements with very important applications of the ion-exchange reactions. It also contains useful information about the reasons why ion exchange can be extremely important for technology. Some of the different types of methods, like cation exchange resin with complex agent, using anion exchange resin, and/or methods using impregnating resin (extraction chromatgraphy), are explained.

It is very difficult to cover all areas of ion exchange in one book. However, we hope that this book will help scientists with ideas for future inventions in the field of ion exchange.

Professor Ayben Kilislioğlu
Department of Chemistry,
Istanbul University
Turkey

Section 1

Ion Exchange Equilibria

Chapter 1

Thermodynamics of Ion Exchange

Ayben Kilislioğlu

Additional information is available at the end of the chapter

http://dx.doi.org/10.5772/53558

1. Introduction

1.1. Ion exchange equilibria

During an ion exchange process, ions are essentially stepped from the solvent phase to the solid surface. As the binding of an ion takes place at the solid surface, the rotational and translational freedom of the solute are reduced. Therefore, the entropy change (ΔS) during ion exchange is negative. For ion exchange to be convenient, Gibbs free energy change (ΔG) must be negative, which in turn requires the enthalpy change to be negative because ΔG = ΔH - TΔS. Both enthalpic (ΔH°) and entropic (ΔS°) changes help decide the overall selectivity of the ion-exchange process [Marcus Y., SenGupta A. K. 2004]. Thermodynamics have great efficiency on the impulsion of ion exchange. It also sets the equilibrium distribution of ions between the solution and the solid. A discussion about the role of thermodynamics relevant to both of these phenomena was done by researchers [Araujo R., 2004].

As the basic rule of ion exchange, one type of a free mobile ion of a solution become fixed on the solid surface by releasing a different kind of an ion from the solid surface. It is a reversible process which means that there is no permanent change on the solid surface by the process. Ion exchange has many applications in different fields like enviromental, medical, technological,.. etc. To evaluate the properties and efficiency of the ion exchange one must determine the equilibrium conditions. At equilibrium conditions no mass diffusion and concentration gradients occur through the surface of ion exchanger. Ion Exchange equilibria can be explained by plotting the concentration($C^s_{A^+}$) or equivalent fraction(x^{s+}_A) of A^+ at the surface versus the concentration (C_{A^+}) or equivalent fraction(x^+_A) of this ion in the equilibrium solution. For monovalent ions the equivalent fractions are given by [1];

$$x_A^{s+} = \frac{m_A^{s+}}{m_A^{s+} + m_B^{s+}}$$

INTECH
open science | open minds

© 2012 Kilislioglu, licensee InTech. This is a paper distributed under the terms of the Creative Commons Attribution License (http://creativecommons.org/licenses/by/3.0), which permits unrestricted use, distribution, and reproduction in any medium, provided the original work is properly cited.

$$x_B^{s+} = \frac{m_B^{s+}}{m_A^{s+} + m_B^{s+}} \quad (1)$$

Where m denotes the concentration of the specified ions at the surface. Ion exchange reactions may be described by the law of mass action using the activities of the ions, rather than their concentrations [Bergaya et al., 2006];

$$K = \frac{a_B^{s+} a_A^{+}}{a_A^{s+} a_B^{+}} \quad (2)$$

where a_A^+ and a_B^+ are the activities of ions A^+ or B^+ in solution a_A^{s+} and a_B^{s+} the activities of these ions at the surface of the exchanger. K represents the thermodynamic equilibrium constant. To explain ion exchange equilibria some common models are used. First type of this model is based on the mass action law [Gorka et al., 2008; de Lucas et al., 1992; Melis et al., 1995; Valverde et al., 1999]. In the second group ion exchange is treated as an adsorption process [Gorka et al., 2008] and in the third, the solid phase is considered to provide sites with fixed charges for ion exchange, as well as sites on which molecular adsorption takes place[Gorka et al., 2008; Novosad & Myers,1982; Myers & Byington, 1986; Ioannidis & Anderko, 2001].

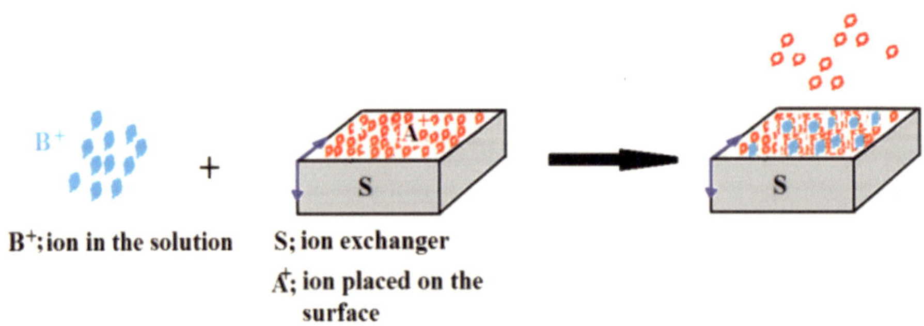

Figure 1.

$$B^+ + AS \rightarrow BS + A^+$$

1.2. Ion exchange isotherm

The ion exchange equilibrium can be qualified by suitable equilibrium isotherms. These isotherms are a graphical representation of the correlation between the equilibrium and experimental terms at constant temperature. The concentration of an ion in the solid expressed as a function of its concentration in the solution under specified conditions and at constant temperature. The most common ion exchange isotherm represents solidarity between ionic compositions of two phases: the ion exchange material and solution [Zagorodni A. A., 2007]

The selectivity is a widely used characteristic of ion exchange systems. It shows the choice of the material to one ion in comparison with another ion. The selectivity is a comparative value. The easiest definition of selectivity can be done by comparing the equivalent fractions of the ion in two phases. The exchanger is considered as selective towards one ion, if (at the equilibrium state) the equivalent fraction of this ion in the exchanger phase is higher than in the surrounding solution. Such a special selection characteristic usually depends on the problem to be solved and on experimental data available [Zagorodni A. A., 2007].

In the presence of confounding processes (such as chemisorption, microbial degradation, precipitation, and steric effects), selectivity coefficients can be used to generate exchange isotherms utilizing a reverse order to the usual procedure [Bloom S. A. and Mansell R. S., 2001].

Author details

Ayben Kilislioğlu
Department of Chemistry, Istanbul University, Turkey

2. References

Araujo R., (2004). Thermodynamics of ion exchange. Journal of Non-Crystalline Solids 349 (2004) 230–233 www.elsevier.com/locate/jnoncrysol.

Bergaya F., Lagaly G., and Vayer M. (2006). Cation and Anion Exchange, In: *Handbook of Clay Science*, Elsevier Ltd.

Bloom S. A. and Mansell R. S. (2001) An algorithm for generating cation exchange isotherms from binary selectivity coefficients SSSAJ Vol. 65 No. 5, p. 1426-1429 https://www.soils.org/publications/sssaj/articles/65/5/1426

De Lucas,A., Zarca,J., Canizzares, P.(1992). Ion-exchange equilibrium of Ca^{2+}, Mg^{2+}, $K^{+}$and H^{+} ions on amberlite IR-120:experimental determination and theoretical predictions of the ternary and quaternary equilibrium data. *Separation Science and Technology,* Vol.27,No.6, pp.823-841.

Gorka, A., Bochenek, R., Warchol, J., Kaczmarski, K., Antos D. (2008). Ion exchange kinetics in removal of small ions. Effect of salt concentration on inter-and intraparticle diffusion. *Chemical Engineering Science,* Vol.63,pp.637-650.

Ioannidis,S., Anderko, A. (2001). Equilibrium modeling of comoined ion-exchange and molecular adsorption phenomena. *Industrial and Engineering Chemistry Research*, Vol.40, No.2, pp. 714-720

Marcus Y., SenGupta A. K. (2004). Ion Exchange and Solvent Extraction A Series of Advances Volume 16. ISBN: 0-8247-5489-1

Melis, S., Cao, G., Morbidelli, M. (1995). A new model fort he simulation of ion exchange equilibria. *Industrial and Engineering Chemistry Research,* Vol.34, No.11, pp. 3916-3924.

Myers, A.L., Byington, S. (1986). Thermodynamics of ion exchange. Prediction of multicomponent equilibria from binary data, In: *Rodrigues,A.E.(Ed),Ion Exchange: Science and Technology.* NATO ASI Series E, No.107. Martinus Nijhoff, Dordrecht.

Novasad, J., Myers, A.L. (1982). Thermodynamics of ion exchange as an adsorption process. *The Canadian Journal of Chemical Engineering.* Vol.60, No. 4, pp.500-503.

Valverde, J.L., De Lucas, A., Rodrigues, J.F.(1999). Comparison between heterogeneous and homogeneous MASS action models in the prediction of ternary ion exchange equilibria. Industrial and Engineering Chemistry Research. Vol.38, No.1, pp.251-259.

Zagorodni A. A., (2007). Physico-chemical Description of Ion Exchange Processes Ion Exchange Materials, Pages 169-198

Section 2

Ion Exchange Technologies

Chapter 2

Ion-Exchange Reactions for Two-Dimensional Quantum Antiferromagnetism

Yoshihiro Tsujimoto and Hiroshi Kageyama

Additional information is available at the end of the chapter

http://dx.doi.org/10.5772/52111

1. Introduction

1.1. Insertion of metal halide array

Topotactic low-temperature reactions such as intercalation, deintercalation and ion-exchange reactions provide a rational design of new structures of non-molecular extended solids which are otherwise not accessible by conventional high-temperature solid-state reactions [1-4]. Among candidate oxide materials used as a precursor of such reactions, the most intensively studied system is the Dion-Jacobson (DJ) type layered perovskite (Figure 1). The chemical formula of the DJ phase is expressed as $A'[A_{n-1}B_nO_{3n+1}]$, where A' is an alkali metal (Na, Rb, ...), A is an alkaline earth or rare earth metal (Ca, Sr, La, ...), B is a d^0 transition metal (Ti, Nb, Ta, ...), and n is the number of perovskite layers (2, 3, 4, ...) [5, 6]. Here, alkali metal ions at the A' site are highly reactive because of ionic (i.e., weak) A'-O bonding, while the perovskite unit $[A_{n-1}B_nO_{3n+1}]$ is strongly bonded and is chemically inert. By exploiting ion-exchange reactions with various reagents, a wide variety of new or improved chemical and physical functionalities including (photo)catalysis [7, 8], ionic conductivity [9] and superconductivity [10] have been developed. However, ion-exchange reactions are rarely employed for the purpose of materials design toward low-dimensional magnetism, because those who work in this filed (mostly physicists) are not familiar with such chemical processing. In addition, compounds obtained by soft-chemical approaches involve negative effects on the crystal structure, for example, poor crystallinity, non-stoichiometry and defects. In particular, it is known that a tiny defect can easily destroy or mask intrinsic magnetic properties of low-dimensional quantum systems.

In 1999, John B. Wiley *et al.* in the University of New Orleans reported a new type of ion-exchange reactions involving the simultaneous co-exchange of transition-metal cation and chloride anion [11]. As shown in Figure 1(a) and 1(b), the reactions of the n = 2 and 3 DJ

INTECH
open science | open minds

© 2012 Tsujimoto and Kageyama, licensee InTech. This is a paper distributed under the terms of the Creative Commons Attribution License (http://creativecommons.org/licenses/by/3.0), which permits unrestricted use, distribution, and reproduction in any medium, provided the original work is properly cited.

phases using CuX_2 (X = Cl, Br) result in the Rb^+-to-$(CuX)^+$ exchange, yielding new metastable compounds $(CuX)^+[A_{n-1}B_nO_{3n+1}]^-$ [11, 12]. Here, the Cu^{2+} ions are octahedrally coordinated, bridging between two apical oxide ions from the perovskite blocks and also surrounded by four halogen ions. The CuO_2X_4 octahedra share edges to form a square lattice network in the *ab* plane. $(CuX)A_{n-1}B_nO_{3n+1}$ has two interesting features in terms of magnetic interactions. First, the chemical bonds between the transition-metal cations and halide anions are covalent, thereby providing sizable Cu-*X*-Cu superexchange interactions within the Cu*X* layer. Second, the two-dimensional (2D) Cu*X* layers are well-separated by the non-magnetic perovskite blocks $A_{n-1}B_nO_{3n+1}$ along the *c*-axis, with the interlayer distance of 12 Å (n = 2) and 16 Å (n = 3). Thus, $(CuX)A_{n-1}B_nO_{3n+1}$ is regarded as an ideal candidate of S = 1/2 2D square lattice systems (Figure 1(c)). What is also important is the fact that this ion-exchange reaction proceeds stoichiometrically, making it possible in principle to observe intrinsic quantum phenomena.

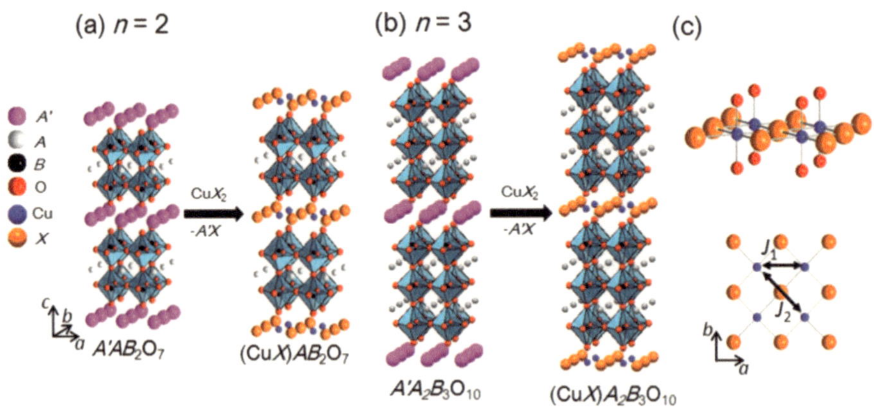

Figure 1. Ion-exchange reaction involving the insertion of copper halide layer into the interlayer spaces of the Dion-Jacobson type layered perovskites $A'[A_{n-1}B_nO_{3n+1}]$ for (a) n = 2 and (b) n = 3. (c) Local coordination environment around the copper atom. J_1 and J_2 denote nearest-neighbor and 2nd nearest-neighbor exchange constants, respectively.

1.2. A new family of two-dimensional quantum antiferromagets

A simple geometrical consideration reveals that the magnetism of $(CuX)A_{n-1}B_nO_{3n+1}$ is described by a square-lattice J_1-J_2 model, where J_1 and J_2 represent the nearest neighbor (NN) interaction and the 2nd NN interaction (Figure 1(c)). According to the Goodenough-Kanamori rule, J_1 and J_2 are expected to be ferromagnetic (FM) and antiferromagnetic (AFM), respectively, giving a situation where geometrical frustration is present in the lattice. The J_1-J_2 model was originally proposed to explain the origin of high-T_c superconductivity in carrier-doped cuprates with S = 1/2 square lattice [13]. Although the detail of theoretical phase diagram is still controversial [14-18], regardless of the approaches employed, several interesting phases appear as a function of $\alpha = J_2/J_1$ (see Figure 2). Let us define hereafter positive and negative J as AFM and FM interaction, respectively. The most interesting phase predicted is a spin liquid phase with a spin-singlet ground state, being located at around $\alpha \sim |0.5|$ (in the regions of both $J_1 < 0$, $J_2 > 0$ and J_1, $J_2 > 0$), where frustration effect is most prominent. So far, several prototypes of the J_1-J_2 model were reported, for example, $Li_2VO(Si,Ge)O_4$ [19] and $Pb_2VO(PO_4)_2$ [20]. Unfortunately, all of them undergo magnetic long-range ordering at low temperatures and the estimated α values are far from $\alpha \sim |0.5|$. In contrast, the magnetic study for $(CuCl)LaNb_2O_7$ revealed that this material does not show long-range magnetic order down to low temperatures [21]. $(CuCl)LaNb_2O_7$ and related compounds obtained by ion-exchange process exhibit a rich variety of quantum phenomena due to geometrical frustration. Hereafter, we will overview recent studies of the ion-exchanged $(CuX)A_{n-1}B_nO_{3n+1}$ compounds, which allows a systematic understanding and tuning of magnetic properties.

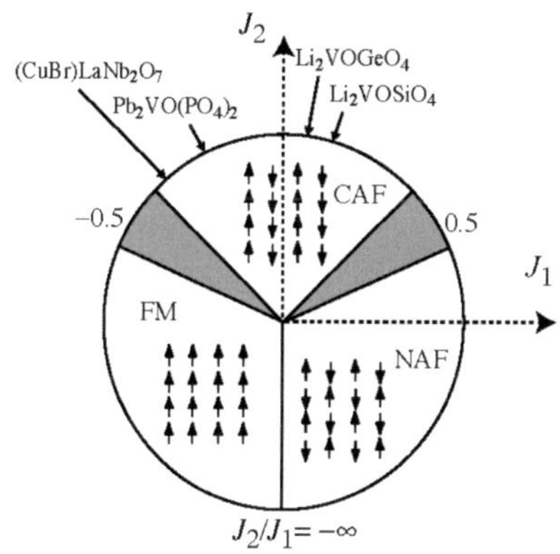

Figure 2. Phase diagram of the J_1-J_2 model on the square lattice where FM, NAF, CAF represent a (0 0) ferromagnetic state, a (π π) Néel antiferromagnetic state, and a (π 0) collinear antiferromagnetic state.

2. Crystal structures and magnetic properties

2.1. Double layered system

2.1.1. $(CuCl)LaNb_2O_7$

A polycrystalline sample of $(CuCl)LaNb_2O_7$ is typically synthesized by heating a mixture of $RbLaNb_2O_7$ and anhydrous $CuCl_2$ at 320 °C for 1 weak. The white specimen becomes light green after the ion-exchange reaction. This color change is due to insertion of Cu^{2+}. The crystal structure of $(CuCl)LaNb_2O_7$ was originally assigned as tetragonal (space group *P4/mmm*), with one copper ion at (1/2, 1/2, 1/2) and one chlorine ion at (0, 0, 1/2) per unit cell, forming the ideal $S = 1/2$ square lattice. Shortly later, the neutron powder diffraction suggested the same space group but a random distribution of chlorine atoms at more general site (x, 0, 1/2) [22]. The tetragonal lattice constants are a = 3.88 Å, c = 11.73 Å. Compared with $RbLaNb_2O_7$ (a = 3.90 Å, c = 11.03 Å), the c-axis of the ion-exchanged product significantly increases by ~ 0.7 Å, while the a-axis remains almost unchanged. This is consistent with cupper occupation between two terminal apical oxide ions from NbO_6 octahedra.

The absence of long-range magnetic order in $(CuCl)LaNb_2O_7$ down to low temperatures was demonstrated first by means of magnetic susceptibility and inelastic neutron scattering (INS) experiments [21]. On cooling, the magnetic susceptibility χ (= M/T) (Figure 3) exhibits a Curie-Weiss (CW) behavior followed by a broad peak at around 16.5 K, which is characteristic of low-dimensional AFM system. The χ decreases rapidly with decreasing temperature further. The Weiss temperature θ estimated from the CW fitting is –9.6 K. The magnitude of θ is smaller than the temperature at the broad maximum, indicating the presence of competing FM and AFM interactions. The obtained value of C (= 0.385 emu K/mol) agrees well with that expected theoretically for 1 mol of Cu^{2+}, ensuring (almost) completion of the ion-exchange reaction. The upturn at 3.5 K in the raw data is ascribed to a tiny amount of Cu-containing impurities or defects in the CuCl layers. When the impurity contribution was subtracted, the intrinsic susceptibility approaches to zero for $T \rightarrow 0$. As shown in Figure 4(a), the INS measurements (constant-Q scan in zero field), carried out at JRR-3, Tokai, Japan, provide direct evidence for the spin-singlet ground state with a one-triplet excitation gap of Δ = 2.3 meV. The full widths at half maximum (FWHM) for the 2.3 meV excitation is 1.3 meV, being close to the instrumental resolution. The scattering intensity of the one-triplet excitation decreases with increasing T and levels off above approximately 20 K, a temperature comparable to the spin gap. There is another excitation centered at 5.0 meV, whose T-dependence shows the same tendency as that of the one-triplet transition. Accordingly, the second mode is the collective bound state excitations of several elementary triplets.

In general, magnetic excitations in 2D systems with a spin-singlet ground state are highly dispersive along the magnetic plane. However, the spectrum shape in constant Q scans for the one-triplet mode in powder $(CuCl)LaNb_2O_7$ specimen is completely symmetric and almost independent of Q (ΔE ~ 0.2 meV), indicating an extremely localized nature of the excited triplets. Such a flat dispersion is characteristic of isolated spin dimer systems such as $CaCuGe_2O_6$ [23]. Since the Cu spins in $(CuCl)LaNb_2O_7$ are not isolated, the observation of the dispersionless spin spectrum is a consequence of geometrical frustration. The same behavior has been observed in the 2D Shastry-Sutherland system $SrCu_2(BO_3)_2$, where compared to in-plane exchange constants (5 – 8 meV) the dispersion of the single-triplet excitation is much suppressed (~ 0.02 meV) [24]. As shown in Figure 4(b), the Q dependence of the scattering intensity of the singlet-triplet excitation, I(1.7 K) – I(20 K), exhibits a fast oscillation. The geometrical frustration in $(CuCl)LaNb_2O_7$ gives a situation that the spin-singlet pairs are *effectively* isolated, which may allow us to use the isolated dimer model:

$$I(Q) \propto F^2(Q)\left(1-\frac{\sin QR}{QR}\right) \tag{1}$$

where the second term $\sin QR/(QR)$ is the interference term reflecting the intradimer distance R. Unexpectedly, the estimated value of R is 8.8 Å, which is far longer than the NN Cu-Cu distance (~ 3.9 Å), and is rather close to the 4th NN distance (8.67 Å). This result contradicts the J_1-J_2 model, where the NN bonds are responsible for the spin-singlet formation.

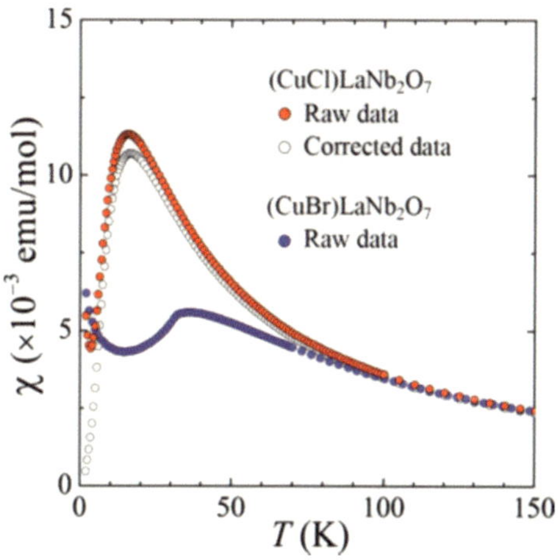

Figure 3. Temperature dependence of the magnetic susceptibility in $(CuX)LaNb_2O_7$ (X = Cl, Br) [21, 37].

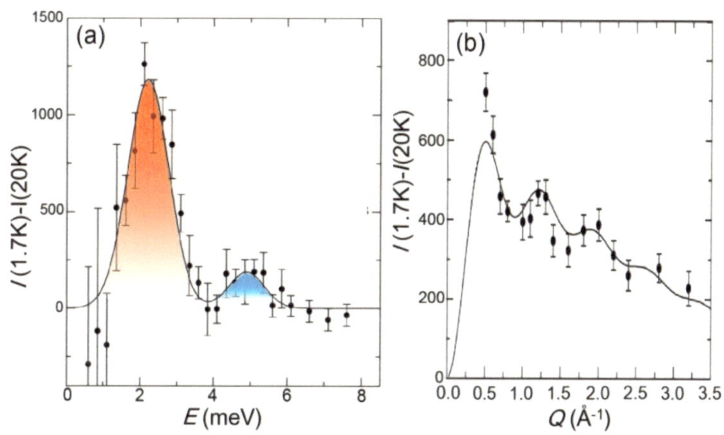

Figure 4. (a) Constant-Q scan at 0.8 Å^{-1} in $(CuCl)LaNb_2O_7$. The solid circles are the difference, I(1.7 K) – I(20 K) [21]. (b) The Q dependence of the transition at 2.3 meV. The solid line indicates the fit according to eq. (1), which resulted in R = 8.8 Å [21].

Figure 5 shows the field dependence of magnetization M-H and the differential magnetization dM/dH at 1.3 K [25], demonstrating field-induced phase transitions at H_{c1} = 10.3 T and H_{c2} = 30.1 T. Reflecting the spin-singlet ground state with a finite energy gap, the initial magnetization slope is very small. Note that the small increase for $H < H_{c1}$ comes from a tiny amount of impurities or defects as also seen in the magnetic susceptibility for T < 5 K. The intrinsic magnetization curve is obtained by subtracting the extrinsic component (2.0% of free magnetic ions of S = 1/2) from the raw magnetization. The field-induced transition at H_{c1} should be caused by the level crossing of the singlet and one of triplet states, following the relation $H_{c1} = \Delta/g\mu_B$, where Δ is the gap in zero field. However, the observed value of H_{c1} is much smaller than 18.5 T calculated from the neutron data (Δ = 2.3 meV). This observation indicates that some other states like a two-triplet bound state causes the level crossing at H_{c1}.

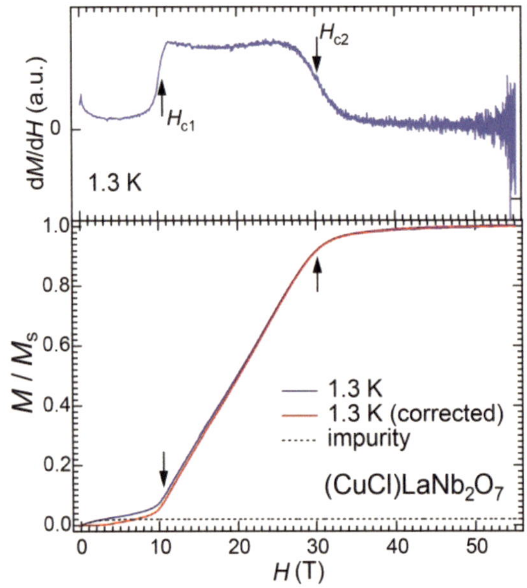

Figure 5. Magnetization curve measured at $T = 1.3$ K and its derivative [25]. Arrows point to the transition fields.

For $H_{c1} < H$, the magnetization gradually increases up to the saturated magnetization. This intermediate state is stable over a wide field range ($H_{c2} - H_{c1} = 19.8$ T), which is due to substantial interdimer interactions. No magnetization plateaus are observed, unlike $SrCu_2(BO_3)_2$ with 1/8, 1/4, and 1/3 plateaus [26, 27] and NH_4CuCl_3 with 1/4 and 3/4 plateaus [28]. Such a difference is understood in terms of competition between localization and delocalization among excited triplets. The magnetization is correlated with the total number of magnons N through the relation $M = g\mu_B N$. The chemical potential of the $S^z = +1$ magnons is described as $\mu = g\mu_B(H - H_{c1})$. The transverse components of the exchange interactions generate hopping of the magnons, while the longitudinal components generate repulsion of the magnons. When the delocalization term is dominant as in the present material, the Bose-Einstein condensation (BEC) of excited triplets, or magnons, occurs. On the other hand, when the localization term is dominant, magnons are crystallized in an ordered fashion to minimize the repulsive interactions, leading to a plateau in the magnetization curve.

Figure 6(a) shows the temperature dependence of total specific heat C_p (in the inset) and C_p/T under applied magnetic fields ranging from 0 to 14 T [29]. The zero field data reveal no anomaly associated with a long-range magnetic ordering, in consistent with magnetic susceptibility and INS experiments. The Schottky anomaly is seen at around 7 K. After subtracting the lattice contribution βT^3 ($\beta = 0.717$ mJ/K^4 mol), the magnetic contribution C_m is obtained, and the subsequent T-integration of C_m/T gives the magnetic entropy S_m of 1.1 J/mol K, which is 13% smaller than the total magnetic entropy ($R\ln2$) for 1mol of $S = 1/2$.

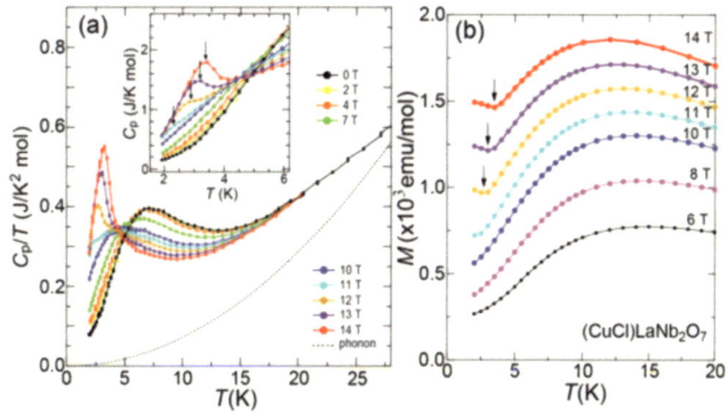

Figure 6. (a) C_P/T for (CuCl)LaNb$_2$O$_7$ collected under several fields. The dotted curve represents the phonon contribution βT^3. The inset shows the C_P vs T plots. Arrows denote the transition temperature [29]. (b) Low-temperature magnetizations measured at various external magnetic fields up to 14 T [29].

When the magnetic field is increased up to 7 T, the broad maximum of C_P/T shifts gradually to lower temperature, indicating reduction of the spin gap owing to Zeeman splitting of the excited triplet states (Figure 6(a)). The broad maximum in $C_P(T)/T$ changes into a cusp at and above 11 T (H_{c1}), corresponding to the field-induced transition to the BEC state. When the magnetic field is further increased, the peak associated with the phase transition develops and shifts systematically to higher temperatures. The transition temperatures T_N at 11 T and 14 T are 2.3 K and 3.3 K, respectively. The growth of the peak area at T_N is consistent with the BEC scenario. Figure 6(b) shows the temperature dependence of magnetization at various constant magnetic fields. A cusp-like anomaly above 10 T shows the occurrence of the BEC transition. The increase in magnetization with decreasing temperature below T_N corresponds to the increased magnon density on the basis of the BEC scenario. Shown in Figure 7 is a T vs H phase diagram. All $C_P(T)$, $M(T)$ and $M(H)$ data agree well with one another. A theoretical curve based on the Hartree-Fock approximation gives the power low behavior $[H_c(T) - H_c(0)] \propto T^{\phi}$ with the critical exponent $\phi = 3/2$ [30].

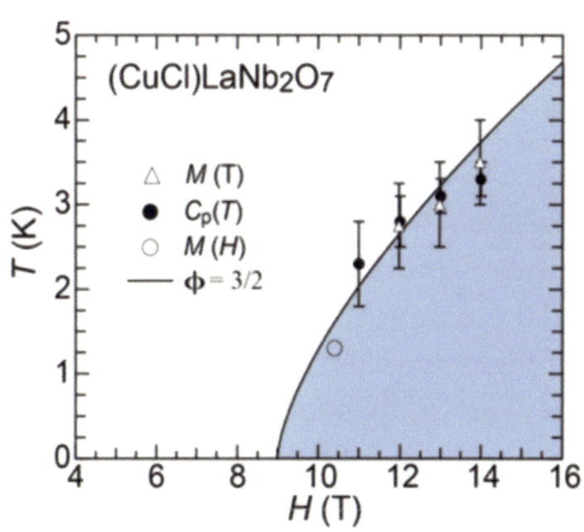

Figure 7. Phase boundary between spin-singlet state and the BEC state in (CuCl)LaNb$_2$O$_7$ determined from the results of $M(T)$, $M(H)$, and $C_P(T)$. The solid line represents a theoretical curve described in the text [29].

Nuclear magnetic resonance (NMR) is a useful tool to clarify the local symmetry. It is revealed that none of the Cu, Cl, and La sites have the tetragonal symmetry [31], which is incompatible with the initially reported tetragonal structure with *P4/mmm* [11]. For example, the Cu- and Cl-NMR spectra could not be accounted for by C_4 symmetry along the *c* direction. Furthermore, transmission electron microscopy (TEM) measurements demonstrated superlattice reflections indicating the doubling along *a*- and *b*-axes ($2a \times 2b$). From the obtained spectra, a single-site occupation without site disorder is indicated for both Cu and Cl, in contradiction to the early neutron diffraction study [22]. In addition, the Nb hyperfine coupling constant for $H \parallel c$ is as large as 2.84(8) T/μ_B, suggesting a strong Cu-O-Nb hybridization. Thus, the in-plane magnetic interactions come not only by the Cu-Cl-Cl superexchange pathways but also by the Cu-O-O-Cu super-superexchange pathways. Figure 8 shows two possible superstructures of the CuCl layer, proposed on the basis of the NMR and TEM results [31].

A single crystal X-ray diffraction is essential to determine the *real* crystal structure. However, the difficulty is that one cannot directly obtain the single crystal of $(CuCl)LaNb_2O_7$ by a conventional high-temperature flux method, because it is a metastable phase. Nevertheless, the $(CuCl)LaNb_2O_7$ single crystals could be obtained topochemically through the ion-exchange reaction of the $CsLaNb_2O_7$ single crystals [32, 33]. The $CsLaNb_2O_7$ crystal was grown by a self-flux method [34]. As shown in Figure 9, transparent, plate-shaped crystals with a typical size of $0.5 \times 0.5 \times 0.1$ mm^3 are obtained. The crystals of $CsLaNb_2O_7$ were embedded in $CuCl_2$ powder, for the ion exchange at 340 °C for 1 week. This yielded dark-green single crystals of $(CuCl)LaNb_2O_7$. Energy dispersive spectroscopy (EDS) shows the absence of Cs in the crystal after the reaction. The time-dependent study revealed that, despite the use of large single crystal and low temperature for reaction, the diffusion occurs surprisingly fast, at a rate of 20 μm h^{-1}.

The crystal structure of $(CuCl)LaNb_2O_7$ was refined using single crystal X-ray diffraction and high- resolution powder neutron diffraction data, and Figure 9(b) shows the determined structure [32]. $(CuCl)LaNb_2O_7$ adopts the orthorhombic $2a \times 2b \times c$ superstructure with the *Pbam* space group. The lattice constants at 2 K are a = 7.7556 Å, b = 7.7507(5) Å, and c = 11.7142(4) Å. The closeness of the in-plane constants is the reason why earlier low-resolution X-ray/neutron studies gave a tetragonal cell. In the new structure, the NbO_6 octahedra are strongly tilted around the *a*-axis in a staggered manner. The tilting pattern of the NbO_6 is correlated with the positions of both the Cu and Cl atoms. In particular, the Cl atoms move significantly along the *b* direction and slightly along *a* direction from their tetragonal position. The Cu ions occupy the 4*h* site are displaced along the *a*- and *b*-axis, yielding different Cu-Cu distances of 3.626 Å and 4.129 Å along the *b*-axis and 3.885 Å along the *a*-axis. The Cu ions take octahedral coordination against two oxide ions with a distance of 1.865 Å and against four chlorine ions with two shorter bonds (2.386 Å and 2.389 Å) and two longer bonds (3.136 Å and 3.188 Å). When the local *z* and *x* axes for each Cu^{2+} are taken along the Cu-O and the short Cu-Cl bonds, respectively, the overall symmetry of the magnetic orbital including the ligand *p* orbitals has the $z^2–x^2$ character, making the exchange interactions in the CuCl layer highly anisotropic.

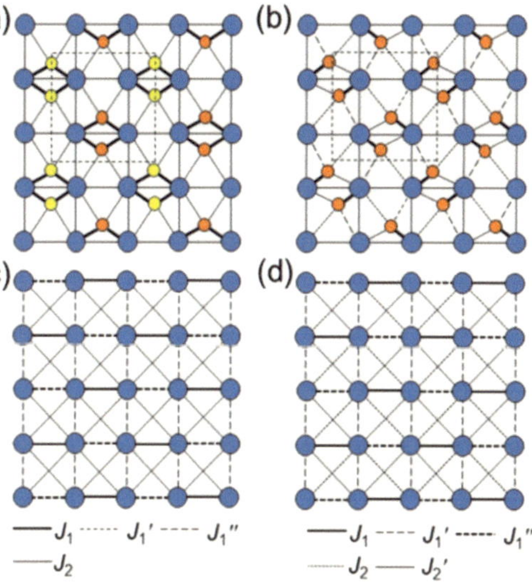

Figure 8. (a, b) Superstructures and (c, d) their corresponding exchange paths for the CuCl layer, suggested by NMR/NQR and TEM experiments. The yellow/orange and blue circles represent the Cl and Cu ions, respectively [31]. The dashed lines in (a) and (b) indicate the unit cell for the superstructure. In the structure (a), the yellow and orange circles represent Cl atoms above and below the Cu layer, respectively.

The first principles calculations based on the new structure [32] yielded the superexchange interactions (see Figure 10(a) and Table 1). Interestingly, the 4th NN interaction, J_4, of the Cu-Cl-Cl-Cu exchange path is the strongest and is AFM. The other 4th NN interaction, J_4' is also AFM, but is much weaker than J_4 ($J_4'/J_4 = 0.18$). This is understood as J_4 has a larger Cu-Cl-Cl angle and a shorter Cl-Cl distance than J_4' (164.9° and 3.835 Å for J_4 vs 156.0° and 4.231 Å for J_4'). The other major J's are FM. The relative strength of the six exchange interactions is $J_4 > J_{1a} > J_{2a} > J_4' > J_{2b} > J_{1b}$.

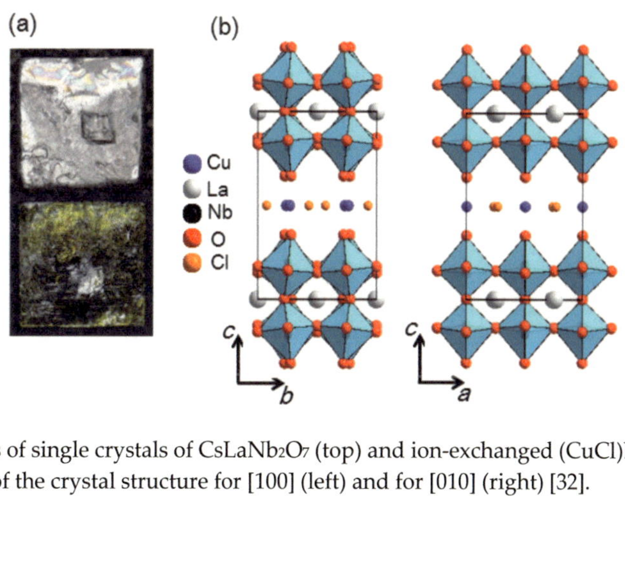

Figure 9. (a) Photos of single crystals of $CsLaNb_2O_7$ (top) and ion-exchanged $(CuCl)LaNb_2O_7$ (bottom) [33]. (b) Projection of the crystal structure for [100] (left) and for [010] (right) [32].

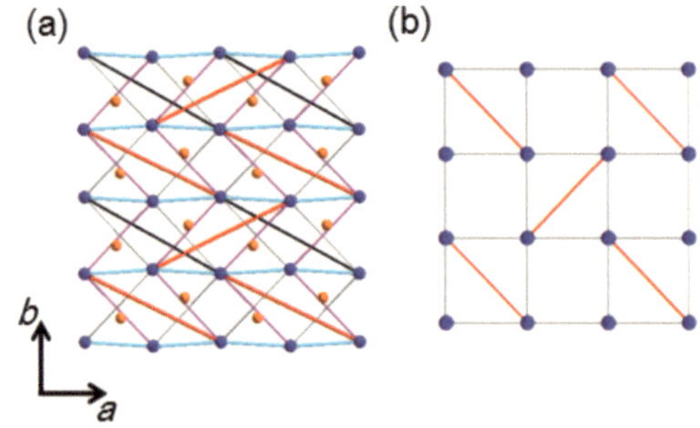

Figure 10. (a) Exchange interactions in the CuCl layer in (CuCl)$LaNb_2O_7$ based on the revised structural model. The lines connecting Cu atoms represent exchange interactions: J_{1a} (light blue), J_{2a} (pink), J_{2b} (gray), J_4 (red), and J_4' (black). J_{1b} is not plotted for clarity [32]. (b) Shastry-Sutherland model where spin dimers form the orthogonal array [35].

J_s	path	d (Å)	angle (°)	J_s/J_4
J_{1a}	Cu-Cl-Cu	3.88548	108.9, 75.8	–0.39
J_{1b}	Cu-Cl-Cu	3.88548	80.9	–0.04
J_{2a}	Cu-Cl-Cu	5.46148	156.7	–0.38
J_{2b}	Cu-Cl-Cu	5.5053	170.2	–0.14
J_4	Cu-Cl-Cl-Cu	8.81262	164.9	1
J_4'	Cu-Cl-Cl-Cu	8.53250	150.0	0.18

Table 1. Exchange paths for up to the fourth NN interactions in (CuCl)$LaNb_2O_7$ and the corresponding coupling constants relative to the strongest AFM interaction J_4 [32].

The strongest AFM J_4 gives the spin-singlet dimers in (CuCl)$LaNb_2O_7$. The distance of Cu-Cl-Cl-Cu (J_4 bond) is 8.533 Å, which is in good agreement with the intradimer distance obtained from the INS analysis (Figure 4b). Using the set of J values, the INS data was analyzed again and the better agreement was obtained [32]. Most remarkably, the 4th NN Cu ions form spin singlet arranged in a staggered manner in the *ab* plane (Figure 10a) and the coupling between them is primarily ferromagnetic. Therefore, the spin lattice is best described as ferromagnetically coupled Shastry-Sutherland quantum spin singlet [32, 35]. In otehr words, (CuCl)$LaNb_2O_7$ is a ferromagnetic counterpart of $SrCu_2(BO_3)_2$.

2.1.2. *(CuBr)$LaNb_2O_7$*

(CuBr)$LaNb_2O_7$ is synthesized in a way similar to (CuCl)$LaNb_2O_7$, but using $CuBr_2$ [11]. The color of the cupric bromide is brown. The structure of (CuBr)$LaNb_2O_7$ has the tetragonal *P*4/*mmm* space group (a = 3.90 Å and c = 11.70 Å), without any superstructure. When this structure is assumed when refined, the Br atoms exhibits an extraordinarily large displacement parameter U_{iso} = 0.087 Å^2. As in the case of (CuCl)$LaNb_2O_7$, a deviation from the tetragonal *P*4/*mmm* structure is inferred from the NMR/NQR experiments [36]. The Cl-to-Br replacement resulted in the magnetic order. Figure 3 shows $\chi(T)$ of this material [37]. The CW fitting gives $\theta = -5.1$ K. The $\chi(T)$ shows a round maximum at T_{max} = 36 K, which is much larger than θ in magnitude, suggesting competing AFM and FM interactions. A tiny kink is seen at 30 K, right below T_{max}, which is ascribed to magnetic order. Consistently, the specific heat measurements show a peak at 32 K. The neutron diffraction experiments demonstrated a collinear antiferromagnetic (CAF) order with a modulation vector of (1/2 0 1/2). The estimated magnetic moment is 0.60(11) μ_B. The reduced magnetic moment probably results from quantum fluctuations. Figure 11 shows the high-field magnetization $M(H)$ measured at 1.3 K. For $0 < H < 30$ T, $M(H)$ increases linearly with magnetic fields, which is followed by non-linear growth. The upward curvature should be caused by the gradual suppression of zero-point oscillations by the external field and is expected to be observed in the framework of spin-wave theory. On the assumption of the J_1-J_2 model [17], J_1/k_B and J_2/k_B are estimated as –35.6 K and 41.3 K, giving $\alpha = |J_2/J_1| = 1.10$, which is located within the range of the CAF phase in the J_1-J_2 model (Figure 2).

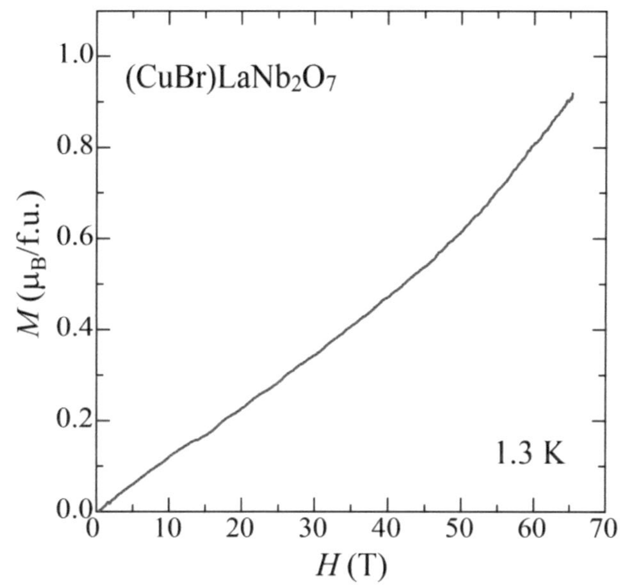

Figure 11. High-field magnetization for $(CuBr)LaNb_2O_7$ at 1.3 K [37].

2.1.3. Cu(Cl, Br)La(Nb, Ta)$_2$O$_7$

Two kinds of solid solutions have been investigated, against Br-for-Cl [38, 39] and Ta-for-Nb [39, 40] substitution. For $(CuCl_{1-x}Br_x)LaNb_2O_7$ ($0 \le x \le 1$), the *a*-axis linearly increases with *x*, while the *c*-axis decreases. The temperature dependence of the magnetic susceptibility (Figure 12(a)) shows the reduction of Néel temperature from T_N =32 K (x = 1), 25 K (x = 0.67), 21 K (x = 0.5), 17 K (x = 0.33) (Figure 12(b)). The magnetization curves for $x \ge 0.33$ show a linear increase over a wide field range, and the saturation field decreases with decreasing the Br concentrations, implying a weaker superexchange constant for Cu-Cl-Cu than Cu-Br-Cu.

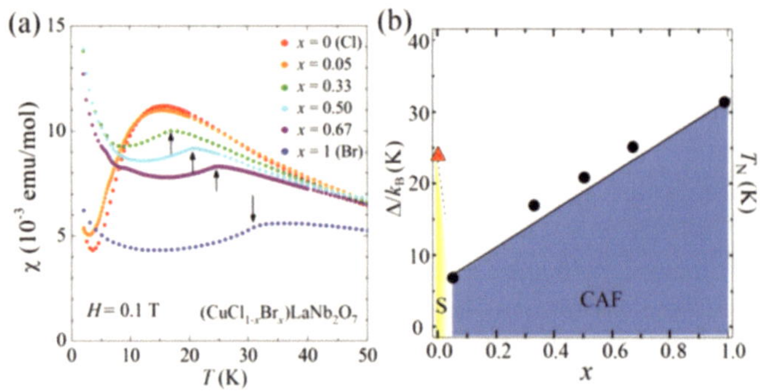

Figure 12. (a) Magnetic susceptibilities of $(CuCl_{1-x}Br_x)LaNb_2O_7$ [38]. (b) x dependence of T_N. S and CAF stand for the spin-singlet state and the collinear antiferromagnetic state [38].

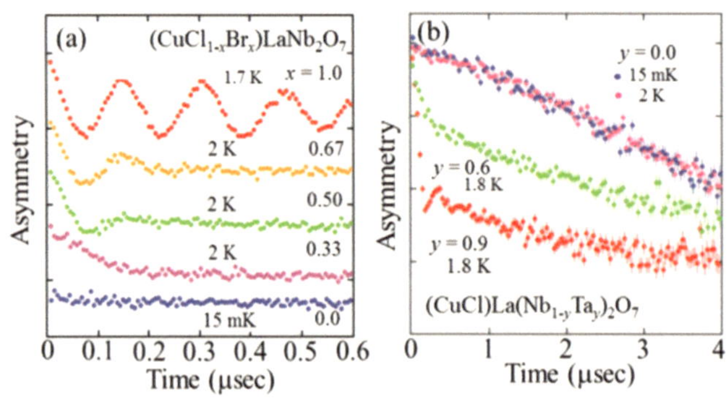

Figure 13. ZF-μSR time spectra in $(CuCl_{1-x}Br_x)LaNb_2O_7$ and $(CuCl)La(Nb_{1-y}Ta_y)_2O_7$ [39].

Muon spin relaxation (μSR) offers a unique opportunity in detecting a static magnetic order at high sensitivity. Figure 13 shows the Zero-field (ZF) μSR spectra of $(CuCl_{1-x}Br_x)LaNb_2O_7$ [39]. The $x = 1$ spectrum exhibits a long-lived precession below T_N, indicating the existence of a well-defined local field at the muon site expected for homogeneous long-range order. With decreasing x, the internal field at ~ 2 K becomes increasingly inhomogeneous as shown by the damping of the oscillation. The $x = 0.05$ also exhibits a long-range magnetic order at 7 K, in spite of the spin–gapped behavior in magnetic susceptibility and magnetization. The inhomogeneous static local field was observed in the low x range, a characteristic behavior often seen in dilute alloy spin-glass systems [41]. The change in the initial damping rate between $x = 0.33$ and 0.05 indicates that the size and/or spatial uniformity of the ordered moment changes between these two Br concentrations. In fact, the neutron scattering experiments for $x = 0.05$ reveals the CAF ordered structure with a magnetic moment being significantly reduced to 0.2(1) μ_B. The long-range magnetic order induced by only 5%-Br doping and the significantly reduced ordered moment indicates that $(CuCl)LaNb_2O_7$ is located in the vicinity of the quantum phase boundary adjacent to the ordered state.

Contrasting behaviors are observed in the Nb-site substituted system, $(CuCl)La(Nb_{1-y}Ta_y)_2O_7$ ($0 \leq y \leq 1$) [40]. Reflecting similar ionic radii between Nb^{5+} and Ta^{5+}, the lattice parameters are almost constant with y. However, the magnetic ground state is affected significantly by the

Nb-to-Ta substitution. The magnetic susceptibility of $(CuCl)LaTa_2O_7$ exhibits a broad maximum at T_{max} = 11.5 K, which is close to 16.5 K in $(CuCl)LaNb_2O_7$. However, it exhibits only a slight decrease below T_{max}, indicating a magnetic ground state. The neutron diffraction and μSR experiments (Figure 13(b)) evidence a magnetic order at 7 K, with the reduced magnetic moment of 0.69(1) μ_B and the propagation vector of (1/2 0 1/2). Unlike the (Cl, Br) solid solution, the spin-singlet state is quite robust against the Ta substitution. Substantial substitution up to y ~ 0.4 is necessary to induce the magnetic order. It is remarkable that the spin disordered phase persists up to y = 0.9. This indicates the occurrence of the phase separation. In this range, T_N remains almost unchanged against the Ta concentrations (Figure 14). The volume fraction of the spin-disordered phase increases with decreasing y. It should be noted that the observed phase separation is not caused by chemical inhomogeneity because of uniform distribution of the Nb and Ta atoms as revealed by TEM/EDS images.

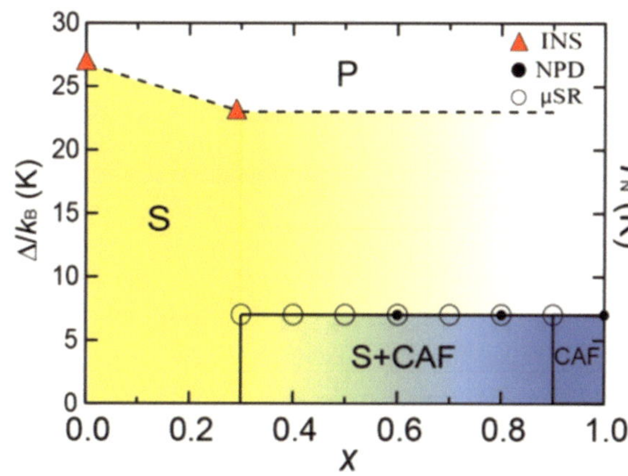

Figure 14. Magnetic phase diagram as a function of temperature and concentration. P, S and CAF stand for the paramagnetic state, the spin-singlet state, and the collinear type AFM state, respectively [40].

The difference in magnetic behaviors between *A*- and *B*-site solid solutions is accounted for by the chemical randomness; the superexchange interactions mediated by halogen atoms should be larger than those mediated the non-magnetic perovskite block. Some theories on 1D systems showed the staggered spin-spin correlations in the vicinity of spin vacancies, which results in the emergence of the ordered phase as observed in $CuGeO_3$ [42] and $SrCu_2O_3$ [43]. By analogy, the induced magnetic order by a small amount of Br substitution in $(CuCl)LaNb_2O_7$ may be associated with staggered correlations induced around Br ions.

2.2. Triple layered system

2.2.1. $(CuBr)Sr_2Nb_3O_{10}$

$(CuBr)Sr_2Nb_3O_{10}$ is a triple-layered compound, prepared by the ion-exchange reaction of $RbSr_2Nb_3O_{10}$ with $CuBr_2$ at 350 °C for 1 week. The crystal structure of $(CuBr)Sr_2Nb_3O_{10}$ at room temperature is tetragonal (a = 3.91069(4) Å, c = 16.0207(3) Å) with the space group

P4/mmm [44, 45]. As in the double layer compounds, a random displacement of the Br atoms from the ideal site is pointed out. In comparison with the double-layered system, the *c* axis is expanded by ~ 4 Å corresponding to the one-perovskite unit. Thus, the enhanced two-dimensionality is expected in magnetism.

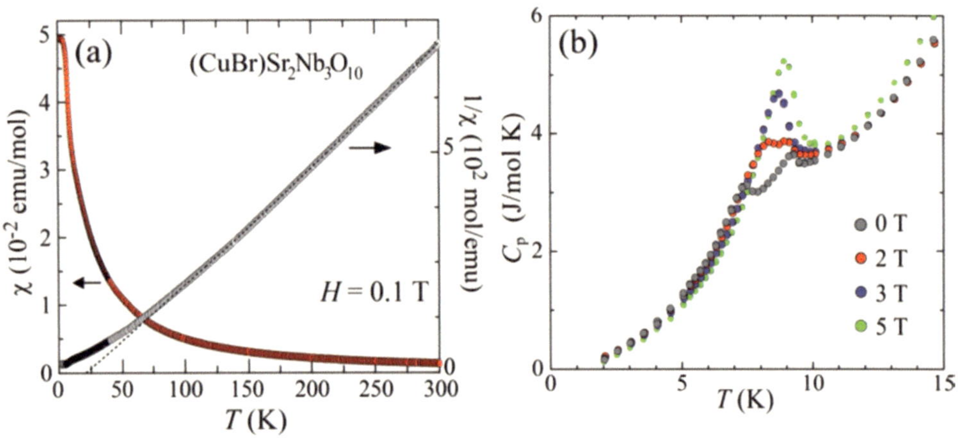

Figure 15. (a) Temperature dependence of the (inverse) magnetic susceptibility for $(CuBr)Sr_2Nb_3O_{10}$ measured at 0.1 T [44]. (b) C_P vs *T* measured at 0, 2, 3, and 5 T [44].

Figure 15(a) shows the temperature dependence of the magnetic susceptibility for $(CuBr)Sr_2Nb_3O_{10}$ measured at $H = 0.1$ T. The CW fitting gives $\theta = 20.9(3)$ K. The positive θ implies that FM interactions are dominant over the AFM ones, which is in contrast to the negative θ for the double layered compounds: –9.6 K for $(CuCl)LaNb_2O_7$ [21] and –5.1 K for $(CuBr)LaNb_2O_7$ [37]. Instead of a broad maximum in $\chi(T)$ typically seen for low-dimensional spin systems, $\chi(T)$ increases with reducing *T* until it flattens out below 5 K. A noticeable anomaly associated with a long-range magnetic order is not seen. However, specific heat measurements in Figure 15(b) show successive phase transitions at 9.3 K (T_{c1}) and 7.5 K (T_{c2}). Non-discontinuous character and the absence of *T* hysteresis in $C_P(T)$ as well as $\chi(T)$ indicate the 2nd-order phase transitions. Successive phase transitions are characteristic phenomena observed in spin frustrated systems such as $CsNiCl_3$ and $CsCoCl_3$ [46]. Application of magnetic fields finally merges two transitions together at around 3 T. At higher field, the transition temperature has a dome-shaped boundary peaking at 5 T, as shown in Figure 16. Zero-field μSR spectra demonstrate a long-range magnetic order below T_{c2}, but a paramagnetic state for $T_{c2} < T < T_{c1}$ [45]. Thus, the transitions at T_{c1} and T_{c2} are, respectively, structural and magnetic.

The most prominent observation in $(CuBr)Sr_2Nb_3O_{10}$ is a 1/3 magnetization plateau in the magnetization curve (see Figure 17). The plateau becomes obscured with increasing *T* and vanishes at 9 K, in agreement with the phase boundary determined by heat capacity measurements. Since the magnetization curves at 4.2 K and 1.3 K are nearly identical, non-flat plateau is not due to thermal effects. Dzyaloshinskii-Moriya interactions and/or staggered *g* tensors are a possible origin as discussed in $SrCu_2(BO_3)_2$ [47].

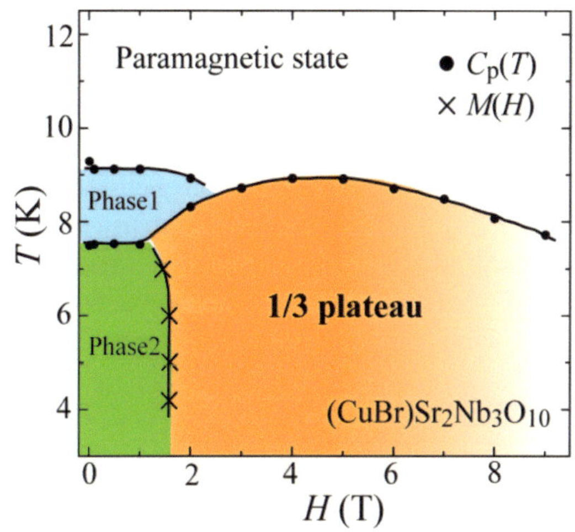

Figure 16. *T-H* phase diagram for $(CuBr)Sr_2Nb_3O_{10}$, determined by $C_p(T)$ and $M(H)$ [44].

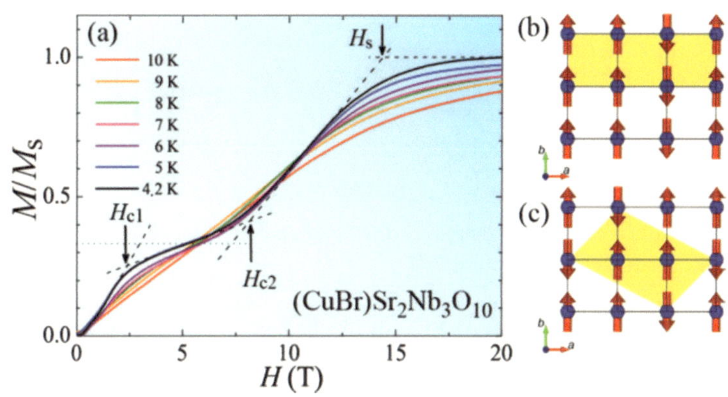

Figure 17. (a) High-field magnetization curves for $(CuBr)Sr_2Nb_3O_{10}$ [44]. (b, c) Originally proposed magnetic structures in the 1/3 plateau state. Only Cu ions are depicted. Magnetic unit cells are represented in yellow.

The 1/3 magnetization plateau has been theoretically predicted for various triangle-based lattices. Experimentally, triangular- and diamond-lattice antiferromagnets such as Cs_2CuBr_4 [48] and $Cu_3(CO_3)_2(OH)_2$ [49] exhibit the 1/3 plateau. However, for commensurability reasons, 1/2 and 1/4 plateaus are naturally expected for the square-based systems [50]. The exception is found in $SrCu_2(BO_3)_2$ with the Shastry-Sutherland lattice, which is due to the stronger 2nd NN interdimer interaction than the NN interdimer one [27]. Oshikawa *et al.* formulated the quantization condition [51]; $p(S - m)$ = integer, where p and m are the period of the spin state, the magnetization per site, respectively. For $S = 1/2$, the minimal necessary condition of the 1/3 plateau ($m = 1/6$) is $p = 3$. Since $(CuBr)Sr_2Nb_3O_{10}$ has one Cu^{2+} ion in its chemical unit cell ($p = 1$), the breaking of translational symmetry is needed to satisfy the quantization conditions. Based on these considerations, we initially proposed two possible in-plane magnetic structures with a propagation vector of (1/3 0) and (1/3 1/6) as shown in Figure 17(b) and 17(c), respectively [44].

In order to observe the magnetic structure at zero field and at the 1/3 plateau, the neutron powder diffraction experiments were carried out at ILL in Grenoble [52]. Figure 18 shows the magnetic diffraction patterns without magnetic field at 8 K and 2 K after subtraction of the 26 K nuclear data. The magnetic Bragg reflections were discernible below T_{c2}, while no magnetic reflections were detected above T_{c2}. The magnetic peaks can be indexed with the propagation vector of (0 3/8 1/2), and the magnetic structure is shown in Figure 19. The Cu moments rotate around the *c*-axis within the *ab* plane, and the helix propagates along the *b*-axis with a rotation of 135° between adjacent moments. The helical chains are coupled ferromagnetically along the *a*-axis. The ordered moment at 2 K is 0.79(7) μ_B/Cu^{2+}.

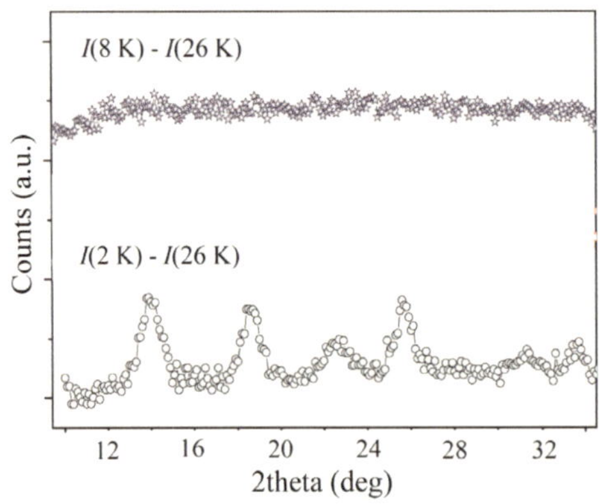

Figure 18. The low angle region of the magnetic diffraction patterns for $(CuBr)Sr_2Nb_3O_{10}$ at 8 and 2 K after subtraction of the 26 K nuclear pattern [52].

The observed helical AFM structure is incompatible with those expected from the simple J_1-J_2 model in the absence of magnetic field. It can only be explained when an additional magnetic interaction J_3 is introduced. In the J_1-J_2-J_3 model, several magnetically ordered states appear depending on J_1, J_2, J_3 [53-55]: (0 0) FM phase, (π π) NAF phase, (π 0) CAF phase, (0 q) spiral phase with $q = \cos^{-1}[-(J_1 + 2J_2)/4J_3]$, ($q$ q) spiral phase with $q = \cos^{-1}[-J_1/(2J_2 + 4J_3)]$. The propagation vector of (0 3/8) in $(CuBr)Sr_2Nb_3O_{10}$ corresponds to the third case. The magnetic Bragg reflections at 4.5 T (in the 1/3 plateaus region) can be indexed with an incommensurate propagation vector (0 1/3 0.446), which corresponds to the magnetic structure in Figure 17(c). Recently, Furukawa *et al.* have found theoretically helical AFM ordered states in the ferromagnetically coupled Shastry-Sutherland model [56]. Whether this model can explain the 1/3 plateau or not is interesting to check.

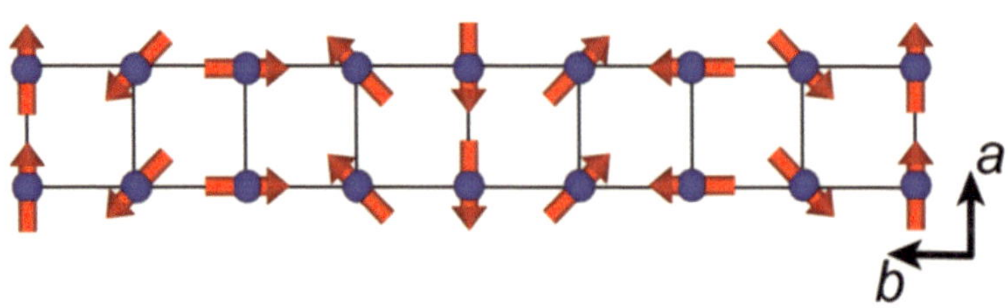

Figure 19. The magnetic structure of $(CuBr)Sr_2Nb_3O_{10}$ in zero field [52].

2.2.2. (CuBr)$A_2B_3O_{10}$ (A = Ca, Sr, Ba, Pb; B = Nb, Ta)

A series of triple layered copper bromides, (CuBr)$A_2B_3O_{10}$, was synthesized [45]. Since the Rb compounds could not be formed for A = Pb and Ba, Cs$A_2B_3O_{10}$ was prepared as a precursor. $A'Pb_2Ta_3O_{10}$ (A' = Rb, Cs) could not be isolated because of formation of $Pb_3Ta_4O_{13}$ as a main phase. The ion-exchange reaction using $CuBr_2$ successfully proceeded at 320 – 350 °C. From the XRD patterns at room temperature, all the (CuBr)$A_2B_3O_{10}$ compounds are found to be isostructural with (CuBr)$Sr_2Nb_3O_{10}$. Both a- and c- axes show a linear increase with the ionic radius of A^{2+}, except Pb. A deviation from the linear relation for A = Pb is due to the steric effect of the $6s^2$ lone pair in Pb [57]. There is no B-site dependence because of the similar ionic radii.

The χ-T curves of (CuBr)$Ca_2B_3O_{10}$ exhibit a broad maximum. T_{max}, is 15 K for Nb and 22 K for Ta. The Weiss temperature θ is positive for all the compounds. (CuBr)$Ba_2Nb_3O_{10}$ shows a drastic drop at 5 K, suggesting an antiferromagnetic transition, while its Ta counterpart shows a monotonous temperature dependence, which is similar to that of (CuBr)$Sr_2Nb_3O_{10}$. Figure 20 shows the magnetizations normalized by the saturation magnetizations. The 1/3 plateau state is remarkably robust against the chemical substitution. As seen in Figure 20(c), (CuBr)$Pb_2Nb_3O_{10}$ exhibits a 1/3 plateau between H_{c1} = 1.7 T and H_{c2} = 5.8 T. The narrowed plateau region of the plateau (4.1 T) indicates that the A-site replacement by the larger cation destabilizes the plateau phase. Further increasing the size of the A-site cation leads to the disappearance of the plateau; no plateau was observed in (CuBr)$Ba_2Nb_3O_{10}$ (Figure 20(d)). Decreasing the size of the A-site (vs. Sr) produces the same results; no plateau was observed in (CuBr)$Ca_2Nb_3O_{10}$ (Figure 20(a)). As shown in Fig. 20(e), the $H_{c2} - H_{c1}$ vs. $r_{A^{2+}}$ plot clearly demonstrates an intimate correlation between $r_{A^{2+}}$ and the stability of the plateau phase.

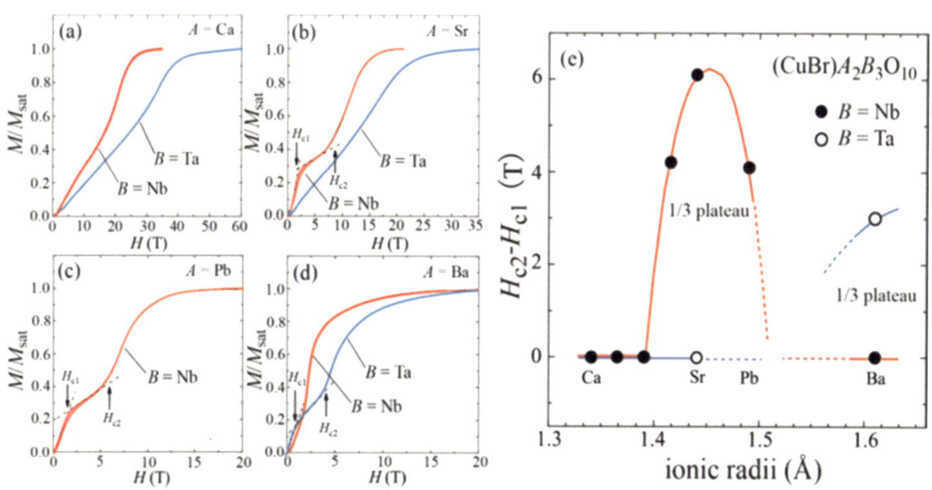

Figure 20. (a)-(d) Magnetization curves measured for (CuBr)$A_2B_3O_{10}$. (e) The compositional dependence of the plateau width $H_{c2} - H_{c1}$ as a function of the ionic radius of the A site [45].

The 1/3 plateau phase can be tuned not only by the *A* site but also by the *B* site. In the case of *B* = Ta, only $(CuBr)Ba_2Ta_3O_{10}$ showed a 1/3 plateau with H_{c1} = 1.0 T and H_{c2} = 4.0 T. Although the data available are limited, it is anticipated, in analogy to the Nb case, that the plateau-stabilizing region is shifted to the right in Figure 20(e). Conditions for the appearance of the plateau would come from a subtle balance between the magnetic interactions mediated by Br and those mediated by BO_6–BO_6.

2.3. *Quadruple layered system*

2.3.1. *$(CuCl)Ca_2NaNb_4O_{13}$*

A quadruple layered compound, $(CuCl)Ca_2NaNb_4O_{13}$ [58], was prepared by the reaction of $RbCa_2NaNb_4O_{13}$ [59] and $CuCl_2$ at 320 °C for 1 week. Laboratory XRD patterns of both $RbCa_2NaNb_4O_{13}$ and $(CuCl)Ca_2NaNb_4O_{13}$ at room temperature could be indexed in the tetragonal cell with the lattice parameters a = 3.869 Å and c = 18.859 Å, and a = 3.866 Å and c = 19.608 Å, respectively. The TEM and synchrotron powder diffraction experiments revealed superlattice reflections that are correlated with the original cell by $2a \times 2a \times c$ (Figure 21). Such a structural deviation from the ideal structure is reasonable in light of the tolerance factor ($t = [r_A + r_O]/\sqrt{2}[r_B + r_O]$). Perovskite compounds with the ideal cubic structure such as $SrTiO_3$ have t = 1. When t is smaller than unity, coherent octahedral tilting takes place. Indeed, t = 0.955 for $(CuCl)Ca_2NaNb_4O_{13}$, implying the octahedral tilting in the perovskite blocks. However, the precursor exhibited very weak and diffuse superlattice reflections. The difference in the TEM patterns would be related to the difference in the bonding nature connecting adjacent perovskite blocks; in $RbCa_2NaNb_4O_{13}$, the adjacent $Ca_2NaNb_4O_{13}$ blocks are weakly bound via the Rb cations, while in $(CuCl)Ca_2NaNb_4O_{13}$, the adjacent perovskite blocks are connected covalently via the edge-shared $CuCl_4O_2$ octahedra, which results in a long-ranged coherent octahedral tilting. In order to find the most reasonable space group for $(CuCl)Ca_2NaNb_4O_{13}$, Aleksandrov's analysis of symmetry reduction in response to octahedral tilting in layered perovskites was employed [60]. Among possible space groups, only *I*4/*mmm* with (++0) tilt met the extinction conditions derived from the TEM and XRD results. Here, + and 0 denote, respectively, in-phase tilt and no tilt in Glazer's notation [61].

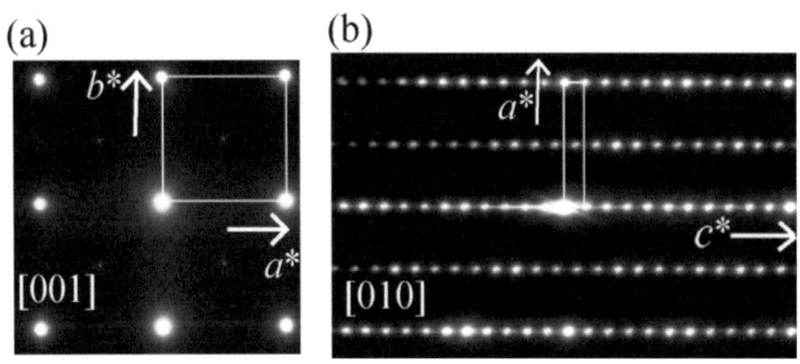

Figure 21. Electron diffraction patterns of $(CuCl)Ca_2NaNb_4O_{13}$ at room temperature obtained along the [001]- and [010]-zone axes [58].

The temperature dependence of magnetic susceptibility for (CuCl)$Ca_2NaNb_4O_{13}$ did not show an anomaly associated with magnetic ordering, which is also supported by μSR and specific heat measurements. FM interactions are dominant over AFM interactions, similar to the triple-layered CuBr compounds [45]. Figure 22 shows the magnetization curve measured at 1.3 K. The $M(H)$ does not show either a fractional magnetization plateau or other field-induced phase transitions, but a rather slow increase. The $M(H)$ does not saturate even at 57 T, implying that the in-plane interaction is fairly strong and that there is a strong competition between FM and AFM interactions.

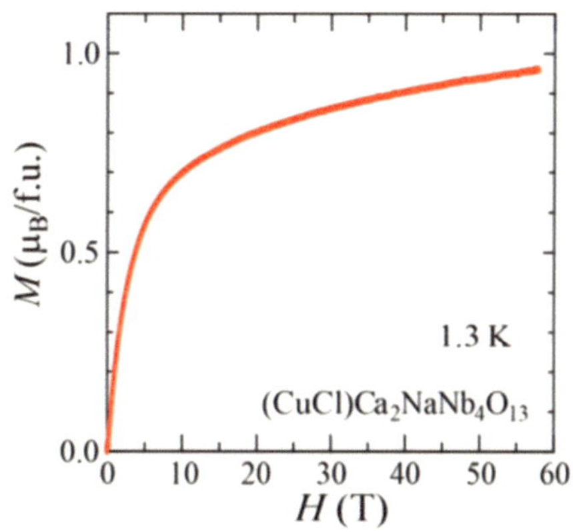

Figure 22. Magnetization curve of (CuCl)$Ca_2NaNb_4O_{13}$ at 1.3 K [58].

n	compounds	space group	magnetic features	T_N (K)	Δ/k_B (K)	θ (K)	Ref.
2	(CuCl)$LaNb_2O_7$	*Pbam* ($2a\times2b\times c$)	Spin-singlet state		26.7	–9.6	32
		P4/mmm					11
	($CuCl_{1-x}Br_x$)$LaNb_2O_7$ ($0.05 \le x \le 1.0$)	*P4/mmm*	CAF order	7 (x = 0.05) 17 (x = 0.33) 21 (x = 0.50) 25 (x = 0.67) 32 (x = 1.0)		–	38, 39
	(CuCl)La$(Nb_{1-y}Ta_y)_2O_7$ ($0.2 \le y \le 1.0$)	*P4/mmm*	Coexistence of spin ordered and disordered states	7 ($0.2 \le y \le 0.9$)	23.2 (y = 0.2)	–5.6 (x = 0.20) –3.8 (x = 0.40) –3.1 (x = 0.60) –2.5 (x = 0.80)	39, 40
	(CuCl)$LaTa_2O_7$	*P4/mmm*	CAF order	7		–1.2	39, 40
3	(CuBr)$(Ca_{1-x}Sr_x)_2Nb_3O_{10}$ ($0 \le x \le 0.50$)	*P4/mmm*	AFM order	13 (x = 0)		4.6 (x = 0)	45
	(CuBr)$(Ca_{1-x}Sr_x)_2Nb_3O_{10}$ (x = 0.75, 1.0)	*P4/mmm*	Helical AFM order, 1/3 magnetization plateau	7.5 (x = 1.0)		20.9 (x = 1.0)	44, 45
	(CuBr)$Pb_2Nb_3O_{10}$	*P4/mmm*	AFM order, 1/3 magnetization plateau	~ 6		17.4	45
	(CuBr)$Ba_2Nb_3O_{10}$	*P4/mmm*	AFM order	5		14.9	45
	(CuBr)$Ca_2Ta_3O_{10}$	*P4/mmm*	AFM order	16.6		3.2	45
	(CuBr)$Sr_2Ta_3O_{10}$	*P4/mmm*	AFM order	11		13.2	45
	(CuBr)$Ba_2Ta_3O_{10}$	*P4/mmm*	AFM order, 1/3 magnetization plateau	5		14.7	45
4	(CuCl)$Ca_2NaNb_4O_{13}$	*I4/mmm* ($2a\times2a\times c$)	Paramagnetic state	–		22.4	58

Table 2. Space group, magnetic features, Neel temperature, T_N, spin-gap energy, Δ/k_B and Weiss temperature, θ for $(CuX)A_{n-1}B_nO_{3n+1}$ (X = Cl, Br; A = Na, La, Ca, Sr, Ba, Pb; B = Nb, Ta; n = 2, 3, and 4).

3. Conclusion

We have demonstrated that the ion-exchange reaction using the DJ phase that involves the simultaneous co-exchange of metal cations and halide anions is effective approach to design a new class of two-dimensional quantum spin antiferromagnets, with tuned in-plane magnetic interactions and a variety of quantum phases. A series of $(CuX)A_{n-1}B_nO_{3n+1}$ with the $S = 1/2$ square lattice exhibits novel quantum spin phenomena such as spin-liquid state, quantized magnetization plateau, quantum phase separation, and field-induced BEC state. They are caused by the strong competition between FM and AFM interactions. Table 2 summarizes the structural and magnetic properties in $(CuX)A_{n-1}B_nO_{3n+1}$. Note that most studies on spin frustrated magnetism has been focused so far on the systems composed of only AFM interactions. It is only recently that attentions have been paid to frustrated systems including FM interactions from both experimentally and theoretically points of view. We hope that the $(CuX)A_{n-1}B_nO_{3n+1}$ system will continue to contribute to the development of new physics of the frustrated spin systems in the future. The ion-exchange reaction can be extended to other compounds with transition metals with different spin quantum number, for example, $(MCl)LaNb_2O_7$ (M = Mn^{2+}, Fe^{2+}, etc.) [62–65] and $(NiCl)Sr_2Ta_3O_{10}$ [66]. Reflecting the classical nature of magnetic moment, $(MCl)LaNb_2O_7$ (M = Co, Fe, Mn) exhibit antiferromagnetic order, but several anomalous behaviors have been still observed due to geometrical frustration. The nickelic chloride with $S = 1$ undergoes the successive magnetic phase transitions with an intermediate phase characterized by a partial magnetic order. The advantage of the ion-exchange reactions is that, once one finds an appropriate precursor or hosts, any desired spin lattices can be constructed in principle. Using this topochemical strategy, various magnetic lattices (triangle, kagomé, pyrochlore, etc.) might be prepared.

Author details

Yoshihiro Tsujimoto
National Institute for Materials Science, Japan

Hiroshi Kageyama
Graduate School of Engineering, Kyoto University, Japan

Acknowledgement

The authors thank Prof. John B. Wiley, Prof. Yoshitami Ajiro, Prof. Kazuyoshi Yoshimura, Dr. Atsushi Kitada, Prof. Kazuma Hirota, Dr. Masakazu Nishi, Prof. Keisuke Totsuka, Prof. Masayuki Hagiwara, Prof. Yasuo Narumi, Prof. Koichi Kindo, Prof. Hiroi Zenji, Dr. Makoto Yoshida, Prof. Masashi Takigawa, Dr. Masaki Ichihara, Dr. Cédric Tassel, Prof. Yasutomo J. Uemura, Prof. Graeme M. Luke, Dr. Tatsuo Goko, Prof. Seung-Hun Lee, Prof. Bella Lake, Prof. Mike H. Whangbo, Prof. Werner Paulus, Dr. Olivier J. Hernandez, Dr. Clemens Ritter, Dr. Kunihiro Nakano, Prof. Yutaka Ueda, Dr. Sk Mohammad Yusuf, Prof. John P. Attfield, Dr. Tsutomu Momoi, Prof. Nic Shannon, and

Prof. Mikio Takano for their fruitful discussion and collaborations. This work was supported by the Japan Society for the Promotion of Science (JSPS) through its "Funding Program for World–Leading Innovative R&D on Science and Technology (FIRST) Program", Grants-in-Aid for Science Research in the Priority Areas "Novel States of Matter Induced by Frustration" (No. 19052004) and Grant-in-Aid for Scientific Research (A) (No. 22245009) from the Ministry of Education, Culture, Sports, Science and Technology of Japan, and CREST.

4. References

[1] Gopalakrishnan J. Chimie Douce Approaches to the Synthesis of Metastable Oxide Materials. Chem. Mater. 1995;7 1265-1275.

[2] Takada K, Sakurai H, Takayama-Muromachi E, Izumi F, Dilanian A, Sasaki T. Superconductivity in Two-Dimensional CoO_2 Layers. Nature 2003;422 53-55.

[3] Choy J-H, Kwon S-J, Park G-S. High T_c Superconductors in the Two-Dimensional Limit: $[(Py\text{-}C_nH_{2n+1})_2HgI_4]\text{-}BiSr_2Ca_{m-1}Cu_mO_y$ (m = 1 and 2). Science 1998;280 1589-1592.

[4] Tsujimoto Y, Tassel C, Hayashi N, Watanabe T, Kageyama H, Yoshimura K, Takano M, Ceretti M, Ritter C, Paulus W. Infinite-Layer Iron Oxide with a Square-Planar Coordination. 2007;450 1062-1066.

[5] Schaak R, Mallouk T. E. Perovskites by Design: A Toolbox of Solid-State Reactions. Chem. Mater. 2002;14 1455-1471.

[6] Sanjaya Ranmohotti K. G, Josepha E, Choi J, Zhang J, Wiley J. B. Topochemical Manipulation of Perovskites: Low-Temperature Reaction Strategies for Directing Structure and Properties. Adv. Mater. 2011;23 442-460.

[7] Ebina Y, Sasaki T, Harada M, Watanabe M. Restacked Perovskite Nanosheets and Their Pt-loaded Materials as Photocatalysis. Chem. Mater. 2002;14 4390-4395.

[8] Machida M, Mitsuyama T, Ikeue K, Matsushima, S, Arai M. Photocatalytic Property and Electronic Structure of Triple-Layered Perovskite Tantalates, $MCa_2Ta_3O_{10}$ (M = Cs, Na, H, and $C_6H_{13}NH_3$). J. Phys. Chem. B 2005;109 7801-7806.

[9] Toda K, Watanabe J, Sato M. Synthesis and Ionic Conductivity of New Layered Perovskite Compound, $Ag_2La_2Ti_3O_{10}$. Solid State Ionics 1996;90 15-19.

[10] Nagai I, Abe Y, Kato M, Koike Y, Kakihana M. Synthesis and Superconducting Properties of Li-Intercalated Niobium Oxide $Li_xAB_2Nb_3O_{10}$. Solid State Ionics 2002;151 265-268.

[11] Kodenkandath T. A, Lalena J. N, Zhou W. L, Capenter E. E, Sangregorio C, Falster A. U, Simmons W. B, O'Connor C. J, Wiley J. B. Assembly of Metal-Anion Arrays within a Perovskite Host. Low-Temperature Synthesis of New Layered Copper-Oxyhalides, $(CuX)LaNb_2O_7$, X = Cl, Br. J. Am. Chem. Soc. 1999;121 10743-10746.

[12] Kodenkandath T. A, Kumbhar A, Zhou W. L, Wiley J. B. Construction of Copper Halide Networks within Layered Perovskites. Synthesis and Characterization of New-Temperature Copper Oxyhalides. Inorg. Chem. 2001;40 710-714.

[13] Anderson P. W. The Resonating Valence Bond state in La_2CuO_4 and Superconductivity. Science 2012;235 1196-1198.
[14] Read N, Sachdev S. Valence-Bond and Spin-Peierls Ground States of Low-Dimensional Quantum Antiferromagnets. Phys. Rev. Lett. 1989;62 1694-1697.
[15] Gelfand M, Singh R. R. P, Huse D. A. Zero-Temperature Ordering in Two-Dimensional Frustrated Quantum Heisenberg Antiferromagnets. Phys. Rev. B 1989;40 10801-10809.
[16] Becca F, Mila F. Peiers-Like Transition Induced by Frustration in a Two-Dimensional Antiferromagnet. Phys. Rev. Lett. 2002;89 037204-1-4.
[17] Shannon N, Schmidt B, Penc K, Thalmeier P. Finite Temperature Properties and Frustrated Ferromagnetism in a Square Lattice Heisenberg Model. Eur. Phys. J. B 2004;38 599-616.
[18] Shannon N, Momoi T, Sindzingre P. Nematic Order in Square Lattice Frustrated Ferromagnets. Phys. Rev. Lett. 2006;96 027213-1-4.
[19] Melzi R, Carretta P, Lascialfari A, Mambrini M, Troyer M, Millet P, Mila F. $Li_2VO(Si,Ge)O_4$, a Prototype of a Two-Dimensional Frustrated Quantum Heisenberg Antiferromaget. Phys. Rev. Lett. 2000;85 1318-1321.
[20] Nath R, Furukawa Y, Borsa F, Kaul E. E, Baenitz M, Geibel C, Johnston D. C. Single-Crystal ^{31}P NMR Studies of the Frustrated Square-Lattice Compound $Pb_2VO(PO_4)_2$. Phys. Rev. B 2009;80 214430-1-10.
[21] Kageyama H, Kitano T, Oba N, Nishi M, Nagai S, Hirota K, Viciu L, Wiley J. B, Yasuda J, Baba Y, Ajiro Y, Yoshimura K. Spin-Singlet Ground State in Two-Dimensional $S = 1/2$ Frustrated Square Lattice: $(CuCl)LaNb_2O_7$. J. Phys. Soc. Jpn. 2005;74 1702-1705.
[22] Caruntu G, Kodenkandath T. A, Wiley J. B. Neutron Diffraction Study of the Oxychloride Layered Perovskite, $(CuCl)LaNb_2O_7$. Mat. Res. Bull. 2002;37 593-598.
[23] Zheludev A, Shirane G, Sasago Y, Hase M, Uchinokura K. Dimerized Ground State and Magnetic Excitations in $CaCuGe_2O_6$. Phys. Rev. B 1996;53 11642-11646.
[24] Kageyama H, Nishi M, Aso N, Onizuka K, Yoshihama T, Nukui K, Kodama K, Kakurai K, Ueda Y. Direct Evidence for the Localized Single-Triplet Excitations and the Dispersive Multitriplet Excitations in $SrCu_2(BO_3)_2$. Phys. Rev. Lett. 2000;84 5876-5879.
[25] Kageyama H, Yasuda J, Kitano T, Totsuka K, Narumi Y, Hagiwara M, Kindo K, Baba Y, Ajiro Y, Yoshimura K. Anomalous Magnetization of Two-Dimensional $S = 1/2$ Frustrated Square-Lattice Antiferromagnet $(CuCl)LaNb_2O_7$. J. Phys. Soc. Jpn. 2005;74 3155-3158
[26] Kageyama H, Yoshimura K, Stern R, Mushnikov N. V, Onizuka K, Kato M, Kosuge K, Slichter C. P, Goto T, Ueda Y. Exact Dimer Ground State and Quantized Magnetization Plateaus in the Two-Dimensional Spin Sytem $SrCu_2(BO_3)_2$. Phys. Rev. Lett. 1999;82 3168-3171.

[27] Onizuka K, Kageyama H, Narumi Y, Kindo K, Ueda Y, Goto T. 1/3 Magnetization Plateau in $SrCu_2(BO_3)_2$ - Stripe Order of Excited Triplets -. J. Phys. Soc. Jpn. 2000;64 1016-1018.

[28] Shiramura W, Takatsu K, Kurniawan B, Tanaka H, Uekusa H, Ohashi Y, Takizawa K, Mitamura H, Goto T. Magnetization Plateaus in NH_4CuCl_3. J. Phys. Soc. Jpn. 1998;67 1548-1551.

[29] Kitada A, Hiroi Z, Tsujimoto Y. Kitano T, Kageyama H, Ajiro Y, Yoshimura K. Bose-Einstein Condensation of Quasi-Two-Dimensional Frustrated Quantum Magnet $(CuCl)LaNb_2O_7$. J. Phys. Soc. Jpn. 2007;76 093706-1-4.

[30] Nikuni T, Oshikawa M, Oosawa A, Tanaka H. Bose-Einstein Condensation of Dilute Magnons in $TlCuCl_3$. Phys. Rev. Lett. 2000;84 5868-5871.

[31] Yoshida M, Ogata N, Takigawa M, Yamaura J, Ichihara M, Kitano T, Kageyama H, Ajiro Y, Yoshimura K. Magnetic and Structural Studies of the Quasi-Two-Dimensional Spin-Gap System $(CuCl)LaNb_2O_7$. J. Phys. Soc. Jpn. 2007;76 104703-1-9.

[32] Tassel C, Kang J, Lee C, Hernandez O, Qiu Y, Paulus W, Collet E, Lake B, Guidi T, Whangbo M.-H, Ritter C, Kageyama H, Lee S.-H. Ferromagnetically Coupled Shastry-Sutherland Quantum Spin Singlets in $(CuCl)LaNb_2O_7$. Phys. Rev. Lett. 2010;105 167205-1-4.

[33] Hernandez O. J, Tassel C, Nakano K, Paulus W, Ritter C, Collet E, Kitada A, Yoshimura K, Kageyama H. First Single-Crystal Synthesis and Low-Temperature Structural Determination of the Quasi-2D Quantum Spin Compound $(CuCl)LaNb_2O_7$. Dalton Trans. 2011;40 4605-4613.

[34] Kumada N, Kinomura N, Sleight A. W. $CsLaNb_2O_7$, Acta Crystallogr., Sect. C: Cryst. Struct. Commun. 1996;C52 1063-1065.

[35] Shastry B. S, Sutherland B. Exact Ground State of a Quantum Mechanical Antiferromagnet. Physica B 1981;108 1069-1070.

[36] Yoshida M, Ogata N, Takigawa M, Kitano T, Kageyama H, Ajiro Y, Yoshimura K. Antiferromagnetic Nuclear Resonance in the Quasi-Two-Dimensional $(CuBr)LaNb_2O_7$. J. Phys. Soc. Jpn. 2008;77 104705-1-7.

[37] Oba N, Kageyama H, Kitano T, Yasuda J, Baba Y, Nishi M, Hirota K, Narumi Y, Hagiwara M, Kindo K, Saito T, Ajiro Y, Yoshimura K. Collinear Order in Frustrated Quantum Antiferromagnet on Square Lattice $(CuBr)LaNb_2O_7$. J. Phys. Soc. Jpn. 2006;75 113601-1-4.

[38] Tsujimoto Y, Kitada A, Kageyama H, Nishi M, Narumi Y, Kindo K, Kiuchi Y, Ueda Y, Uemura Y. J, Ajiro Y, Yoshimura K. Synthesis Structural and Magnetic Properties of the Soild Solution $(CuCl_{1-x}Br_x)LaNb_2O_7$. J. Phys. Soc. Jpn. 2010;79 014709-1-4.

[39] Uemura Y. J, Aczel A. A, Ajiro. Y, Carlo J. P, Goko T, Goldfield D. A, Kitada A, Luku G. M, MacDougall G. J, Mihailescu I. G, Rodriquez J. A, Russo P. L, Tsujimoto Y, Wiebe C. R, Williams T. J, Yamamoto T, Yoshimura K, Kageyama H. Muon Spin Relaxation of the Frustrated Quasi-Two-Dimensional Square-Lattice Spin System Cu(Cl, Br)La(Nb,

$Ta)_2O_7$: Evolution From Spin-Gap to Antiferromagnetic State. Phys. Rev. B 2009;80 174408-1-9.

[40] Kitada A, Tsujimoto Y, Kageyama H, Ajiro Y, Nishi M, Narumi Y, Kindo K, Ichihara M, Ueda Y, Uemura Y. J, Yoshimura K. Quantum Phase Transition in (CuCl)La($Nb_{1-x}Ta_x)_2O_7$. Phys. Rev. B 2009;80 174409-1-5.

[41] Uemura Y. J, Yamazaki T, Harshman D. R, Senba M, Ansaldo E. J. Muon-Spin Relaxation in AuFe and CuMn Spin Glasses. Phys. Rev. B 1985;31 546-563.

[42] Kojima K. M, Fudamoto Y, Larkin M, Luke G. M, Merrin J, Nachumi B, Uemura Y. J, Hase M, Sasago Y, Uchinokura K, Ajiro Y, Revcolevschi A, Renard J. P. Antiferromagnetic Order with Spatially Inhomogeneous Ordered Moment Size of Zn- and Si-Doped $CuGeO_3$. Phys. Rev. Lett. 1997;79 503-506.

[43] Ohsugi S, Tokunaga Y, Ishida K, Kitaoka Y, Azuma M, Fujishiro Y, Takano M. Impurity-Induced Staggered Polarization and Antiferromagnetic Order in Spin-1/2 Heisenberg Two-Leg Ladder Compound $SrCu_2O_3$: Extensive Cu NMR and NQR Studies. Phys. Rev. B 1999;60 4181-4190.

[44] Tsujimoto Y, Baba Y, Oba N, Kageyama H, Fukui T, Narumi Y, Kindo K, Saito T, Takano M, Ajiro Y, Yoshimura K. 1/3 Magnetization Plateau in Spin-1/2 Square Lattice Antiferromagnet (CuBr)$Sr_2Nb_3O_{10}$. J. Phys. Soc. Jpn. 2007;76 063711-1-4.

[45] Tsujimoto Y, Kageyama H, Baba Y, Kitada A, Yamamoto T, Narumi Y, Kindo K, Nishi M, Carlo J. P, Aczel A. A, Williams T. J, Goko T, Luke G. M, Uemura Y. J, Ueda Y, Yoshimura K. Synthesis Structure and Magnetic Properties of the Two-Dimensional Quantum Antiferromagnets (CuBr)$A_2B_3O_{10}$ (A = Ca, Sr, Ba, Pb; B = Nb, Ta) with the 1/3 Magnetization Plateau. Phys. Rev. B 2008;78 214410-1-10.

[46] Collins M. F, Petrenko O. A. Triangular Antiferromagnets. Can. J. Phys. 1997;75 605-655.

[47] Kodama K, Miyahara S, Takigawa M, Horvatic M, Berthier C, Mila F, Kageyama H, Ueda Y. Field-Induced Effects of Anisotropic Magnetic Interactions in $SrCu_2(BO_3)_2$. J. Phys.: Condens. Matter 2005;17 L61-L68.

[48] Ono T, Tanaka H, Aruga Katori H, Ishikawa F, Mitamura H, Goto T. Magnetization Plateau in the Frustrated Quantum Spin System Cs_2CuBr_4. Phys. Rev. B 2003;67 104431-1-7.

[49] Kikuchi H, Fujii Y, Chiba M, Mitsudo S, Idehara T, Tonegawa T, Okamoto K, Sakai T, Kuwai T, Ohta H. Experimental Observation of the 1/3 Magnetization Plateau in the Diamond-Chain Compound $Cu_3(CO_3)_2(OH)_2$. Phys. Rev. Lett. 2005;94 227201-1-4.

[50] Honecker A. Lanczos Study of the S = 1/2 Frustrated Square-Lattice Antiferromagnet in a Magnetic Field. Can. J. Phys. 2001;79 1557-1563.

[51] Oshikawa M, Yamanaka M, Affleck I. Magnetization Plateaus in Spin Chains: "Haldane Gap" for Half-Integer Spins. Phys. Rev. Lett. 1997;78 1984-1987.

[52] Yusuf S. M, Bera A. K, Ritter C, Tsujimoto Y, Ajiro Y, Kageyama H, Attfield J. P. Magnetic Correlation in the Square-Lattice Spin System $(CuBr)Sr_2Nb_3O_{10}$: A Neutron Diffraction Study. Phys. Rev. B 2011;84 064407-1-6.

[53] Feldner, H, Cabra D. C, Rossini G. L. Ferromagnetic Frustrated Spin Systems on the Square Lattice: Schwinger Boson Study. Phys. Rev. B 2011;84 214406-1-7.

[54] Sindzingre P. Spin-1/2 Frustrated Antiferromagnet on a Spatially Anisotropic Square Lattice Contribution of Exact Diagonalization. Phys. Rev. B 2004;69 094418-1-14.

[55] Sindzingre P, Seabra L, Shannon, Momoi T. Phase Diagram of the Spin-1/2 J_1-J_2-J_3 Heisenberg Model on the Square Lattice with Ferromagnetic J_1. J. Phys.: Conf. Ser. 2009;145 012048-1-6.

[56] Furukawa S, Dodds T, Kim Y. B. Ferromagnetically Coupled Dimers on the Distorted Shastry-Sutherland Lattice: Application to $(CuCl)LaNb_2O_7$. Phys. Rev. B 2011;84 054432-1-19.

[57] Belik A. A, Azuma M, Saito T, Shimakawa Y, Takano M. Crystallographic Features and Tetragonal Phase Stability of $PbVO_3$, a New Member of $PbTiO_3$ Family. Chem. Mater. 2005;17 269-273.

[58] Kitada A, Tsujimoto Y, Yamamoto T, Kobayashi Y, Narumi Y, Kindo K, Aczel A. A, Luke G. M, Uemura Y. J, Kiuchi Y, Ueda Y, Yoshimura K, Ajiro Y, Kageyama H. Quadruple-Layered Perovskite $(CuCl)Ca_2NaNb_4O_{13}$. J. Solid State Chem. 2012;185 10-17.

[59] Sugimoto W, Ohkawa H, Naito M, Sugawara Y, Kuroda K. Synthesis and Structures of Reduced Niobates with Four Perovskite-like Layers and Their Semiconducting Properties. J. Solid State Chem. 1999;148 508-512.

[60] Aleksandrov K. S. Structural Phase Transitions in Layered Perovskitelike Crystals. Crystallography Reports 1995;40 251-272.

[61] Glazer A. M. The Classification of Tilted Octahedra in Pervskites. Acta Crystallogr. B 1972;28 3384-3392.

[62] Viciu L, Caruntu G, Royant N, Koenig J, Zhou W. L, Kodenkandath T. A, Wiley J. B. Formation of Metal-Anion Arrays within Layered Perovsktie Hosts. Preparation of a Series of New Metastable Transition-Metal Oxyhalides, $(M\mathrm{Cl})LaNb_2O_7$ (M = Cr, Mn, Fe, Co). Inorg. Chem. 2002;41 3385-3388.

[63] Viciu L, Golub V. O, Wiley J. B. Structural, Thermal and Magnetic Characterization of the Manganese Oxyhalide Layered Perovskite, $(MnCl)LaNb_2O_7$. J. Solid State Chem. 2003;175 88-93.

[64] Viciu L, Koenig J, Spinu L, Zhou W. L, Wiely J. B. Insertion of a Two-Dimensional Iron-Chloride Network between Perovskite Blocks. Synthesis and Characterization of the Layered Oxyhalide, $(FeCl)LaNb_2O_7$. Chem. Mater. 2003;15 1480-1485.

[65] Kitada A, Tsujimoto Y, Yajima T, Yoshimura K, Ajiro Y, Kobayashi Y, Kageyama H. Two-dimensional Frustrated Antiferromagnets $(M\mathrm{Cl})LaNb_2O_7$ (M = Mn, Co, Cr). J. Phys.: Conf. Series 2011;320 012035-1-6.

[66] Tsujimoto Y, Kitada A, Uemura Y. J, Goko T, Aczel A. A, Williams T. J, Luke G. M, Narumi Y, Kindo K, Nishi M, Ajiro Y, Yoshimura K, Kageyama H. Two-Dimensional *S* = 1 Quantum Antiferromagnet $(NiCl)Sr_2Ta_3O_{10}$. Chem. Mater. 2010;22 4625-4631.

Chapter 3

Bifunctional Polymer-Metal Nanocomposite Ion Exchange Materials

Berta Domènech, Julio Bastos-Arrieta, Amanda Alonso, Jorge Macanás, Maria Muñoz and Dmitri N. Muraviev

Additional information is available at the end of the chapter

http://dx.doi.org/10.5772/51579

1. Introduction

The unusual electrical, optical, magnetic, and chemical properties of metal colloids (better known in nowadays as metal nanoparticles, MNPs) have attracted increasing interest of scientists and technologists during the last decade. In fact, although Nanoscience and Nanotechnology are quite recent disciplines, there have already been a high number of publications that discuss these topics. [1-11] What is more, there are quite new high impact peer-reviewed journals especially devoted to these research fields and there is also a particular subject category "Nanoscience & Nanotechnology" in the Journal Citation Reports from Thomson Reuters.

MNPs can be obtained by various synthetic routes, such as electrochemical methods, decomposition of organometallic precursors, reduction of metal salts in the presence of suitable (monomeric or polymeric) stabilizers, or vapour deposition methods. Sometimes, the presence of stabilizers is required to prevent the agglomeration of nanoclusters by providing a steric and/or electrostatic barrier between particles and, in addition, the stabilizers play a crucial role in controlling both the size and shape of nanoparticles.

In this sense, the development of polymer-stabilized MNPs (PSMNPs) is considered to be one of the most promising solutions to the MNPs stability problem. For this reason, the incorporation of MNPs into polymeric matrices has drawn a great deal of attention within the last decade as polymer-metal nanocomposites have already demonstrated unusual and valuable properties in many practical applications.

The modification of commercially available ion exchange resins and the development of suitable polymeric membranes with metal nanoparticles (MNPs) having certain functionality, such as for example, biocide or catalytic activity has proved to be a theme of

INTECH
open science | open minds

© 2012 Muraviev et al., licensee InTech. This is a paper distributed under the terms of the Creative Commons Attribution License (http://creativecommons.org/licenses/by/3.0), which permits unrestricted use, distribution, and reproduction in any medium, provided the original work is properly cited.

great interest. The main advantage of the nanocomposite ion exchange materials is the location of metal nanoparticles near the surface of the polymer what substantially enhances the efficiency of their biocide and catalytic application.

1.1. Metal nanoparticles

The most commonly accepted definition for a nanomaterial is "a material that has a structure in which at least one of its phases has a nanometer size in at least one dimension." [12] Regarding this definition, it is possible to classify nanoobjects in three groups:

i. 1D nanometer-size objects (e.g., thin films)
ii. 2D nanometer-size objects (e.g., nanowires, nanorods and nanotubes)
iii. 3D nanometer-size objects (e.g. nanoparticles and/or nanoclusters)

Such materials include porous materials (with porous sizes in the nanometer range), polycrystalline materials (with nanometer-sized crystallites), materials with surface protrusions separated by nanometric distances, or nanometer-sized metallic clusters. Among all of these materials, metal nanoparticles (MNPs) have attracted increasing interest of scientists and technologists during the last decade, due to their unique electrical, optical, magnetic, and chemical properties.

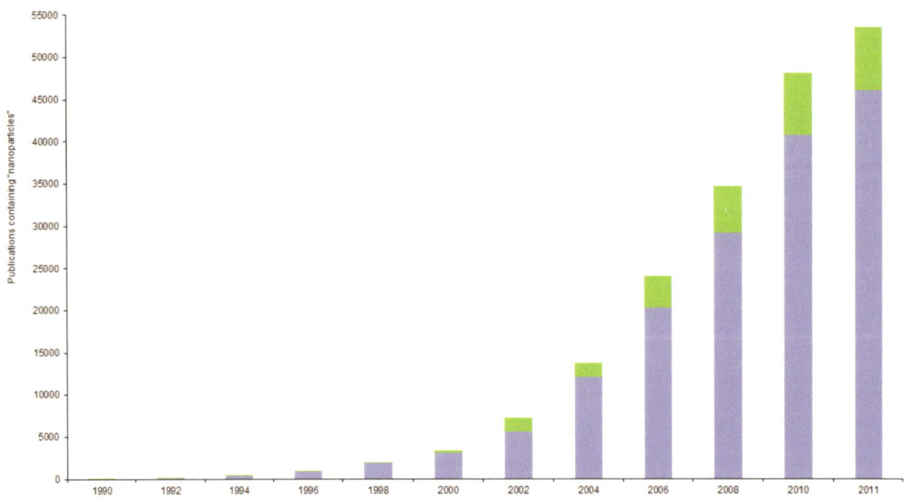

Figure 1. Bibliographic analysis based on the search of "nanoparticles" in Scifinder Scholar. Blue represents scientific publications (e.g. books, articles, reviews) and green represents patents.

Within the last two decades a new focus has been initiated to control and to better understand the nanometer-size objects due to the appearance of a new interdisciplinary field, which is known now as Nanoscience and Nanotechnology. This has stimulated a new wave of intensive and more detailed studies of MNPs and various nanocomposites on their base. Figure 1 shows the tendency in publication about this issue, where it is shown that the number of publications has increased exponentially due to their wide applications in different fields such as Medicine, Chemistry, and Physics and so on. Moreover, not only scientific publications have been growing in the last decades but also a huge number of patents have been issued in the last decade.

The main goal of Nanoscience and Nanotechnology is the creation of useful/functional materials, devices and systems through the control of matter on the nanometer length scale and exploitation of novel phenomena and properties (physical, chemical and biological) at that scale. To achieve such goal it is necessary to use a multidisciplinary approach: inputs from physicists, biologists, chemists and engineers are required for the advancement of the understanding in the preparation, application and impact of new nanotechnologies.

1.2. General properties

Because of the decrease in the scale of the materials, their behaviour changes in a remarkable form. In fact, the reduction of the bulk materials to a nanometric size induces size-dependant effects resultant from:

i. An increase of the ratio surface-volume, what gives to an increase in the total surface area and in the fraction of the entities (e.g. atoms) in the surface of the material, as shown Figure 2.
ii. Changes in the electronic structure of the entities forming the nanoparticles and in the nanoparticles as whole.
iii. Changes in the associations (e.g. interatomic distances) of the entities forming the nanoparticle and presence of defects.
iv. Confinement and quantic effects (due to the confinement of the charge carriers in a particle of size comparable to the wavelength of the electron).

Figure 2. Schematic representation of the change in the ratio surface/volume between a bulk microsphere and the same microsphere composed by NPs.

This can be illustrated, for example, by the dependence of gold melting point on the size of gold nanoparticles, or by suspensions of Ag nanoparticles with sizes ranging from 40 to 100 nm which show different colours. [13] In addition, there are physical phenomena that do not exist in materials with larger grain sizes, as the general quantum-size effect for optical transitions in semiconductor nanocrystals which occurs in very small nanoparticles (<10 nm) due to the quantum confinement effects inherent in particles of that size.

Is due to all of these new properties that, indeed, research centred on nanoscopic materials have a large field of application which extends from the semiconductor industry, where the ability to produce nanometer-scale features leads to faster and less expensive transistors [6], to biotechnology, where luminescent nanoparticles are extremely interesting as bioprobes. [14] Some other particular examples are catalyst for fuel cells [15] or electrocatalysts used in sensing devices with enhanced properties. But, as a rule of thumb, nanoparticles, due to the large percentage of surface atoms [12, 16], have already made a major impact on the field of surface science, as Catalysis or Biocide treatment.

1.3. MNPs Preparation: Stability challenges and stabilization mechanisms

In general there are two routes for the preparation of MNPs (see Figure 3):-Down and Bottom-Up. The top-down methods are those that reduce the macroscopic particles to the nanoscale. This route is not very suitable to prepare uniform particles of very small sizes. In contrast, with the bottom-up methods it is possible to obtain uniform particles (usually of different shapes and structures). These routes start from atoms that can be added (either in solution or gas phase) to form larger particles.

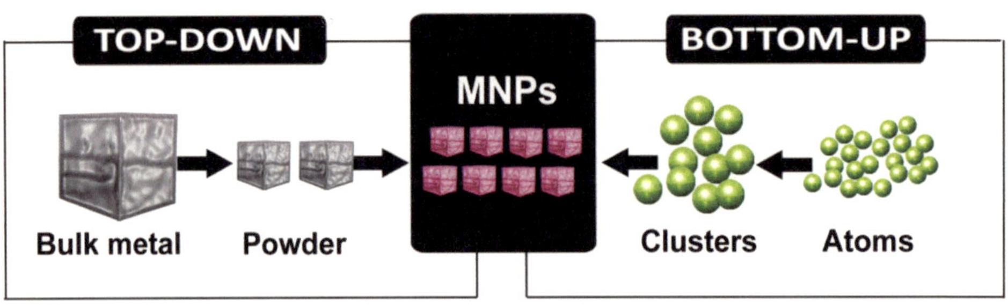

Figure 3. Scheme of Top-Down and Bottom-Up approaches to the synthesis of MNPs.

Overall, a good method to classify the different methods of synthesis of MNPs is by Physical, Physicochemical and Chemical routes (See Figure 4). [13] Many synthetic pathways can be used but the chemical ones are generally cheaper and do not require equipment or instruments as specific as in the case of physical methods the physical methods.

However, the main drawback which still limits the wide application of MNPs is their insufficient stability dealing with their high tendency to self-aggregate. [17] MNPs are so reactive that when they touch each other, they surfaces fuse, what results in a loss of the nanometric size and in their special properties. These features of nanoparticles, in part determined by the conditions of synthesis, create enormous difficulties in their fabrication and application. [18]

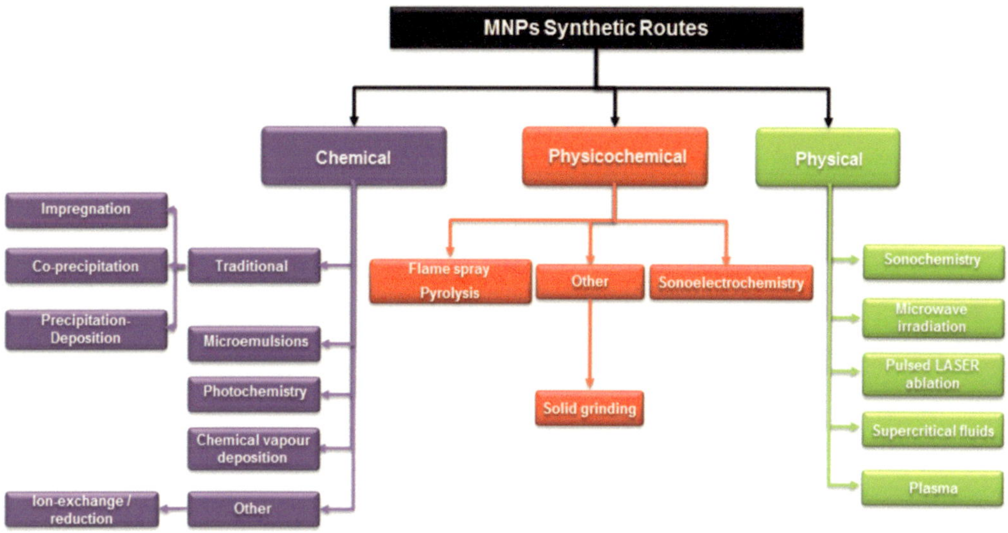

Figure 4. Physical, Physicochemical and Chemical routes for the preparation of MNPs.

It is noteworthy that NPs can aggregate not only as a result of a further manipulation but also during their growth. A typical mechanism of aggregation is the Ostwald ripening which is a growth mechanism where small particles dissolve, and are consumed by larger particles. [19] So, the average nanoparticle size increases with time, the particle concentration decreases and their solubility diminishes.

Therefore, the stabilization of MNPs is specifically required to:

i. prevent their uncontrollable growth
ii. prevent particle aggregation
iii. control their final shape and size
iv. allow particle solubility in various solvents
v. terminate the particle growth reaction

The successful synthesis of nanoparticles usually involves three steps: nucleation, growth, and termination by a capping agent or ligand (or stabilizing agent) through colloidal forces.[20] These colloidal forces can be classified in three main types as follows: Van der Waals interactions, electrical double-layer interactions, and steric interactions. In addition, hydrophobic and solvation forces may be important. [21]

Some of the mechanisms regarding the stabilization forces have been thoroughly revised in the literature.[13, 22]

Specially the use polymer-assisted fabrication of inorganic nanoparticles is probably one of the most efficient and universal ways to overcome the stability problem of MNPs and to save their properties. Metal nanoparticles synthesized by this approach exhibit long-time stability against aggregation and oxidation while nanoparticles prepared in the absence of polymers are prone to quick aggregation and oxidation.[18, 23]

In this sense, stabilization of MNPs can be done by different strategies. In the ex-situ synthesis, NPs are dispersed after their synthesis in a solid or liquid medium by using different mechanochemical approaches. The problem is that in these cases, the success of the stabilization is limited by the possibility of re-aggregation of the MNPs along the time. On the opposite hand, by the in-situ synthesis, MNPs are grown directly in the stabilizer medium yielding a material that can be directly used for a foreseen purpose. For this reason, in-situ approaches are getting much attention, because of their technological advantages (Figure 5).

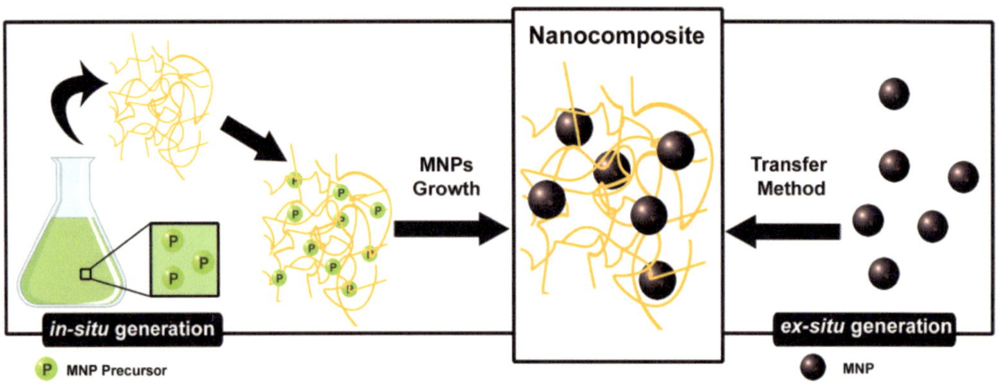

Figure 5. Schematic comparison between in-situ and ex-situ methodologies.

Despite the methodology employed, it is of crucial importance to understand the processes occurring in polymer interactions with nanoparticles. In this regard, the mechanism of MNP stabilization with polymers can be explained by two approaches which run simultaneously in the system and influence one another: the substantial increase of viscosity of the immobilizing media (the polymer matrix), and the decrease of the energy of particle-particle interaction in PSMNP systems versus non-stabilized MNP dispersions. [24]

In the first approach, the substantial increase of viscosity of the immobilizing media (the polymer matrix), the Coagulation velocity depends on factors as the range of attraction forces, Brownian motion velocity, concentration of colloidal solution, presence of electrolytes… As follows from the Smoluchowsky equation [25], the rate constant of particle coagulation, k_c, is inversely proportional to the viscosity of the media, η, (here k stands for the Boltzman constant, and T is the temperature):

$$k_c = \frac{8kT}{\eta} \quad (1)$$

The second approach is the decrease of the energy of particle-particle interaction in PSMNP systems versus non-stabilized MNP dispersions. The potential energy of attraction U_r between two spherical particles of radius r and minimum distance l_o between their surfaces can be given by the following equation:

$$U_r \approx \frac{Ar}{12\,l_o} \text{ at } r >> l_o \tag{2}$$

where A is the effective Hamaker's constant with dimensions of energy. The value of A is known to be close to *kT* for polymer particles (e.g 6.3×10^{-20}J for polystyrene), while for the metal dispersions it is far higher (40×10^{-20} J for silver). [24]

2. Inter-Matrix synthesis in ion exchange matrices

The ion-exchange synthesis of Metal Nanoparticles (MNPs) refers to a group of methods which can be generally classified as Inter-Matrix Synthesis (IMS) technique. The main feature of IMS is the dual function of the matrix, which allows the stabilization of the MNPs to prevent their uncontrollable growth and aggregation and provides a medium for the synthesis.

It is noteworthy that IMS was essentially the first method employed by the humans to incorporate nanoparticles inside inorganic materials. In this sense, one of the oldest nanocomposite materials found is the "Lycurgus cup" which dates back to the late 4th century B.C.[26] and it is made of a sort of glass that changes its colours depending on the incident light: in reflected light the glass turns green, but when the light is shone directly through it, it turns red due to the presence of small amount of Ag-Au-MNPs with the diameter of approximately 70 nm. It is remarkable that Greco-Roman techniques have been used up to modern times: related recipes were described by Arabian authors during the medieval period [27], during the Renaissance as practical application of alchemical knowledge[28], and by modern chemists, from the Encyclopedie of Diderot and d'Alembert [29] through to the present day.

Another example is the "lustre pottery"[30] employed at the same time in Asian an European countries by a simple two-step procedure:

1. the immobilization of metal cations (MNP precursors) inside the ceramic matrix.
2. the reduction of metal ions to the zero-valent state with carbon monoxide leading to the formation of MNPs.

Examples of this ceramics are showed in Figure 6.

Although, as it has been shown above, humanity had used the properties of the nanoscale materials for a long time, the fundamentals of the scientific Nanoscience and Nanotechnology studies did not appear since the middle of the XIXth. As an example, in 1949 the first communication of the IMS is published by Mills and Dickinson.[31] In this pioneered publication it is described the preparation of the anionic resin containing Cu-MNP ("colloid copper") and the use of this nanocomposite material for the removal of oxygen from water due to its interaction with copper MNPs. Since that, a great number of researchers focused their efforts to the development of a new class of ion-exchange materials, combining ion-exchange and redox properties (known also as "redoxites" or "electron-ion exchangers").

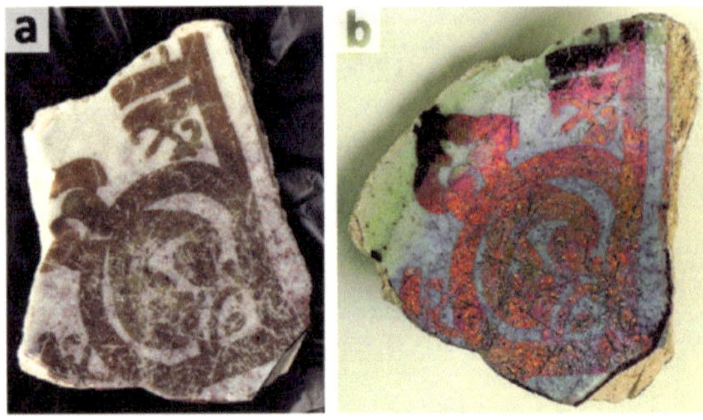

Figure 6. Lustre Pottery. Fatimid-sherds excavated from Fustât; (a) the sample and (b) sample with an orientation that corresponds to the diffraction angle and lustre shining is observed.

2.1. General principles

This section describes the principles of the IMS in ionic exchange matrices. This technique takes advantatges of the in-situ approach (above mentioned), and has a wide range of application because of the multiple metal-polymer existing possibilities. In this sense, even if the number of polymers is reduced to those with ion exchange capacities, the multiple possibilities remain and a different number of polymer-nanocomposites can be obtained. Figure 7 shows the multiple possibilities of the IMS methodology.

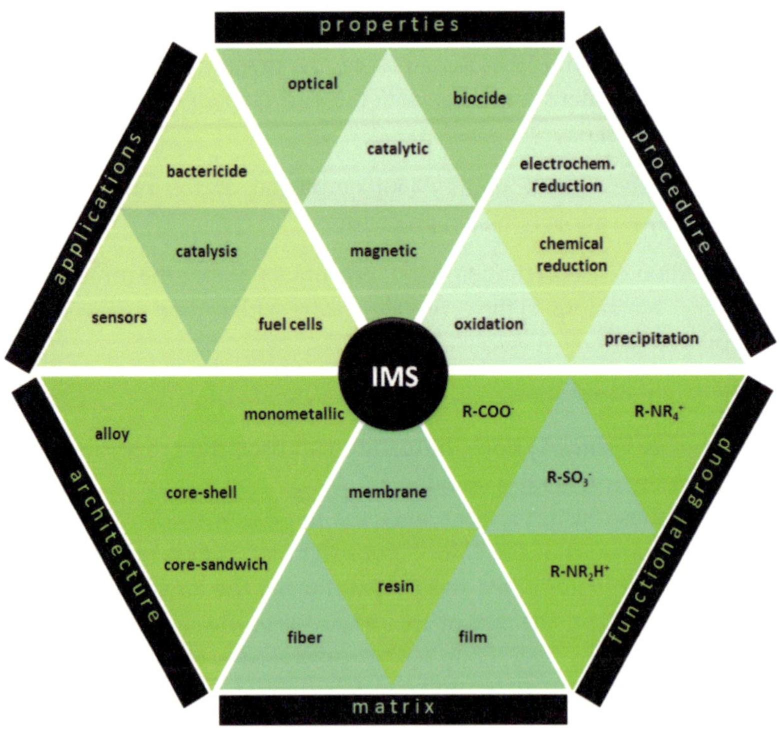

Figure 7. Scheme of the multiple possibilities of Ion-Matrix Synthesis.

The general principles of IMS, valid for any kind of polymer matrix and type of nanoparticle, are based on:

1. The nanoreactor effect: the confinement of the particles by the polymer molecules which allows limiting the size and the particle size distribution; and,
2. The barrier effect: the polymer molecules locally isolate the formation of each single NP preventing the contact between their surfaces and therefore their aggregation.

These guidelines are only achievable if NPs precursors can properly be immobilized in the polymeric matrix. In this sense, Ion Exchange matrices are the perfect template to retain the ionic species, either metal cations, anions or any kind of coordination compound.

To illustrate this approach, two anion charged groups as sulfonic group (SO_3^-) can efficiently interact with the metal cation (M_1^{2+}) which afterwards can undergo a chemical reaction (precipitation, reduction, etc.) which finally will yield to the formation of the MNP.

The same can be done for an anion exchange matrix bearing a functional group such as a quaternary ammonium ($-NR_4^+$), capable to immobilize metal complexes (i.e. $[CoCl_4]^{2-}$)or other anions (i.e. BH_4^-).

Figure 8 illustrate the two main consecutive stages which rule the IMS technique: (i) the immobilization of the metal ion or complex (in a Cation Exchange Matrix, CEM) or immobilization of the reductant (in an Anion Exchange Matrix, AEM) and (ii) the reduction of the metal ion inside the matrix.

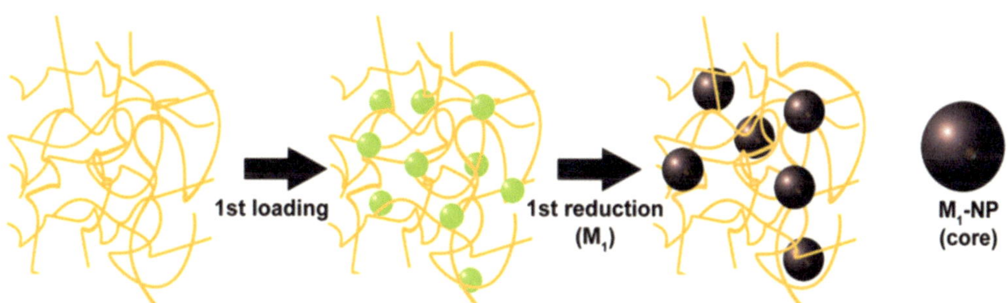

Figure 8. Monometallic MNPs preparation inside a polymeric ion-exchange matrix by IMS. Green spheres represent the M_1^{2+} cation, and the black ones the MNPs obtained after the reduction.

These two stages may be described by the following equations (equations 3-4 and 5-6) considering the presence of strong acid groups ($2R\text{-}SO_3^-$) in the CEMs and the presence of strong basic groups ($R\text{-}R_3N^+$) in the AEMs. M_1 represents a divalent metal and R the organic radical.

For CEMs:

$$2\left(R-SO_3^-Na^+\right) + M_1^{2+} \leftrightarrow \left(R-SO_3^-\right)_2 M_1^{2+} + 2\,Na^+ \quad (3)$$

$$\left(R-SO_3^-\right)_2 M_1^{2+} + 2\,NaBH_4 + 6H_2O \rightarrow 2\left(R-SO_3^-Na^+\right) + 7H_2 + 2B\left(OH\right)_3 + M_1^0 \quad (4)$$

For AEMs:

$$R-\left(R_3N\right)^+ Cl^- + NaBH_4 \leftrightarrow R-\left(R_3N\right)^+ BH_4^- + NaCl \quad (5)$$

$$2\left(R-\left(R_3N\right)^+ BH_4^-\right) + \left[M_1\right]^{2+} Cl_2 + 6H_2O \rightarrow 2\left(R-\left(R_3N\right)^+ Cl^-\right) + \\ + 7H_2 + 2B\left(OH\right)_3 + M_1^0 \quad (6)$$

A deeper look in the second stage reveals that the MNPs formation (equations 4 and 6) is, indeed, a combination of an ion-exchange reaction and a reduction reaction, accordingly the reduction of the metal ions to the zero-valent metal takes place in the solution boundary, close to the ion exchange groups:

For CEMs:

$$\left(R-SO_3^-\right)_2 M_1^{2+} + 2\,Na^+ \leftrightarrow 2\left(R-SO_3^-Na^+\right) + M_1^{2+} \quad (7)$$

$$M_1^{2+} + 2\,BH_4^- + 6H_2O \rightarrow 7H_2 + 2B\left(OH\right)_3 + M_1^0 \quad (8)$$

For AEMs:

$$R-\left(R_3N\right)^+ BH_4^- + Cl^- \leftrightarrow R-\left(R_3N\right)^+ Cl^- + BH_4^- \quad (9)$$

$$M_1^{2+} + 2\,BH_4^- + 6H_2O \rightarrow 7H_2 + 2B\left(OH\right)_3 + M_1^0 \quad (10)$$

Additionally, as it can be seen from equations 7 and 9 in both cases the matrix is regenerated after the second stage of the IMS, so it is possible to apply consecutive IMS cycles to increase the total loaded metal [32] or to obtain bimetallic nanoparticles (core-shell, alloys or core-sandwich).

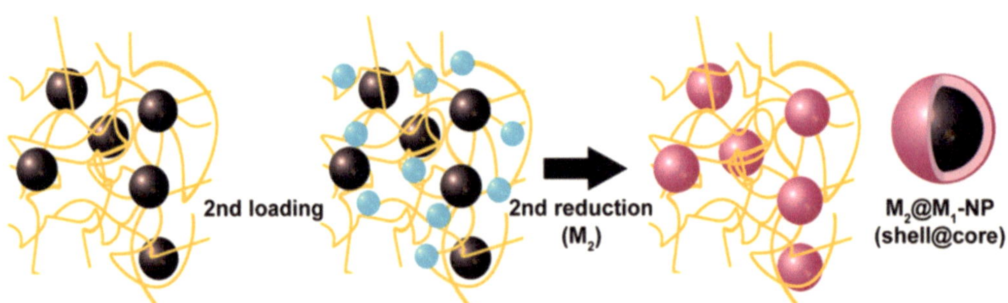

Figure 9. Bimetallic core-shell MNPs preparation. Black spheres represent the MNPs obtained after he first loaing-reduction cycle, blue spheres the M_2^{2+} cations and the pink ones are the final core-shell MNPs.

In Figure 9, there is an schematic representation for the preparation of core-shell NPs by coating the monometallic MNPs obtained after the first cycle with a secondary functional metal shell. As follows, the formation of core-shell MNPs (M_1-M_2, represented as $M_2@M_1$ where M_2 is the coating and M_1 is the core) allows modification of charge and functionality, improves the stability or combines the properties of both metals to make their future applications more efficient. The final activity of the nanocomposite is determined by the properties of the shell metal, although in some cases the properties of the metal core may add an additional advantage to the final nanocomposite (e.g. magnetic core).

To better understand the procedure shown in Figure 9, let us consider the reactions corresponding to the IMS of PSMNPs inside the parent polymeric matrix after the first loading-reduction cycle for example in the CEM:

$$2\left(R-SO_3^-Na^+\right) + M_1^{\,0} + M_2^{\,2+} \leftrightarrow \left(R-SO_3^-\right)_2 M_2^{\,2+} + M_1^{\,0} + 2\,Na^+ \tag{11}$$

$$\left(R-SO_3^-\right)_2 M_2^{\,2+} + M_1^{\,0} + 2NaBH_4 + 6H_2O \rightarrow 2\left(R-SO_3^-Na^+\right) + \\ + 7H_2 + 2B\left(OH\right)_3 + M_2@M_1 \tag{12}$$

According to some authors [33], the second metal ion can act as an oxidizing agent towards the core-metal (M_1^0) resulting in the oxidation of the first metal by the following transmetallation reaction:

$$M_2^{\,2+} + 2M_1^{\,0} \rightarrow M_1^{\,2+} + M_2^{\,0}@M_1^{\,0} \tag{13}$$

As it can be seen, in the general IMS procedure described before, there are always two species bearing the same charge: the matrix and the reducing agent (in CEMs) or the metal ion (in AEMs). This means that there is an electrostatic repulsion between the matrix and one of the species mentioned that impede the penetration inside the polymeric matrix, referred to as Donnan-exclusion effect [34, 35].

The Donnan-exclusion effect is based on the exclusion (inability to deeply penetrate inside the polymer) of co-ions when the sign of their charge coincides with that of the polymer

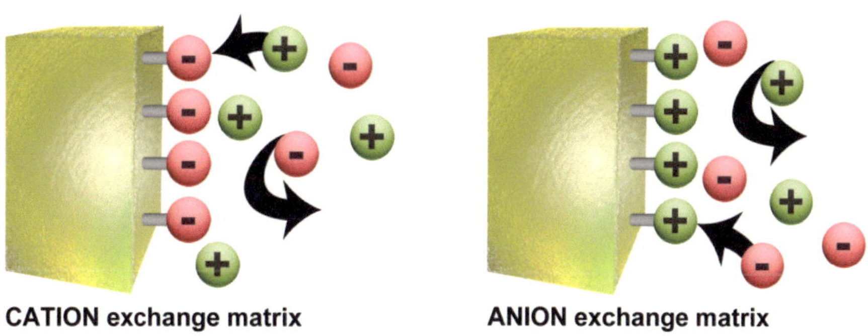

Figure 10. Donnan Exclusion Effect.

functional groups. Consequently, ion penetration inside the matrix is balanced by the sum of two driving forces acting in opposite directions: the gradient of the ion concentration and the Donnan-effect itself. The result of these two driving forces is the formation of the MNPs mainly near the surface of the polymer matrix (see Figure 10).

Regarding the final application of the nanocomposite, this is a really suitable distribution, since MNPs remain maximally accessible for substrates of interest such as chemical reagents or bacteria.

2.2. Requirements for parent polymers

Many materials which contain functional groups can be used as supports for the IMS disregarding their form or shape: granulated form, fibrous or membranes. In all cases, when using the IMS technique it is important to take into account both the polymer properties and the final application of the nanocomposite since as both points dictate certain necessary requirements to the parent matrix (inside which the MNPs must be synthesized).

For example, when using a nanocomposite for making a sensor or a biosensor to be applied in aqueous solutions, the polymer must be insoluble in water. But, at the same time, the polymer has to provide sufficient permeability towards the analyte under study (ions or molecules). Besides, the polymer must either slightly swell in water or at least be hydrophilic to enhance both the sensor and the response rate.

Similarly, the solubility of the nanocomposite in some organic solvents allows for the preparation of homogeneous Polymer Stabilized MNPs solutions (PSMNPs "inks") that can be deposited onto the desired surfaces (e.g. electrodes) to modify their properties. This solubility would also allow the characterization via microscopic analysis, electrochemical techniques and others.[36]

The main requirements for a polymer to be used as a matrix for the IMS technique are the following:

- The polymer must be chemically compatible with the MNPs surface.
- The polymer must bear functional groups which would act as nanoreactors.
- Appropriate distances between the coordinating centers to insure the hopping of charge carriers.
- Sufficient flexibility of the polymer chain segments to facilitate movements of ionic carriers.
- Appropriate swelling ratio of the matrix.
- Adequate hydrophilicity.

In polymer functional matrices, ionic transport occurs in a highly amorphous, viscoelastic (solid) state. In this sense, the most intensively studied polymers are based on poly(oxa alkanes), poly(aza alkanes), or poly(thia alkanes).

In general, it is possible to state that functional groups define the chemical properties of the polymer matrix, by bearing on the surface a negative or positive charge. Due to this fact different dissociation properties of group lead to strong and weak exchangers (which are named similar to that of strong and weak electrolytes). Based on these functional groups, classification of Ion Exchange matrices involves four main groups:

1. Cation exchangers (with anionic functionalities and positively charged mobile ions)
   a. strong acid exchangers (e.g., containing sulfonic acid groups or the corresponding salts)
   b. weak acid exchangers (e.g., containing carboxylic acid groups or the corresponding salts)
2. Anion exchangers (with cationic functionalities)
   a. strong base exchangers (e.g., containing quaternary ammonium groups)
   b. weak base exchangers (e.g., containing amine groups)

The Ion exchange capacity (IEC) is the main feature of ion exchange materials. Taking into account that an ion exchanger can be considered as a "reservoir" containing exchangeable counterions, the counterion content in a given amount of material is defined essentially by the amount of fixed charges which must be compensated by the counterions, and thus is essentially constant.

Above all mentioned polymer matrices, this chapter is mainly focused in the use of cross-linked polymers in the form of resins and in commercial or tailored polymeric membranes.

2.2.1. Resin beads

As IMS is based on the feasibility of the polymeric support used, which must contain ionic functional groups, one of the typical matrices that accomplish this requirement are Ion-exchange resins, also known as granulated polymers.[34, 37]

Ion-exchange resins are commercial products commonly available and their shape and size allow these materials to be easily and quantitatively recovered by simple filtration or decantation. Ion exchange resins are usually used in water treatment processes (e.g. water softening) but have many other applications in chemical production. For instance, several common applications include immobilization of biological an inorganic catalysts, extraction procedures, metal recovery and separation and acid-base catalysis.[38, 39]

The first effort to obtain more stable synthetic resins for ion exchange reactions is attributed to B.A. Adams and E.L. Holmes, who in 1935 published the condensation polymerization of methanal (formaldehyde) with phenol or polysubstituted benzene compounds to give reversible exchange resins. [40] Based on the same concept Adams and Holmes quickly developed anion exchange resins, obtained by the condensation between methanal and phenylamines giving directly a copolymer matrix bearing weak basic secondary amine groups, that in presence of strong acid solutions result in acid amine salts (anion exchangers). Although nowadays the polymerization mechanism is entirely different, based on the so called addition or vinyl polymerization (first applied by D'Alelio in 1944) commercial ion exchange resins production uses the same principles as their predecessors, and the two kind of resin explained before are still the most commonly used resins.[41]

Regarding their chemical composition, most ion-exchange resins are based on cross-linked polystyrene- divinylbenzene (DVB) copolymers bearing ion-exchanging functional groups. Besides, from the morphological point of view, the following types can be considered:

- Gel type: resins with a macroscopic homogeneous and elastic framework which contains the solvent employed in its synthesis. What characterizes the resin is the presence of channels (instead of pores) in the matrix, with a size depending on the proportion of DVB employed in the synthesis. Is the size of these channels what determines the size of the specie, ion or molecule allowed to pass though it and the velocity and diffusion.
- Macroporous type: resins with a higher cross-linking level than gel type. Inside their structure macropores and micropores coexist as a result of the elimination of the solvent, what gives to these kinds of resins a high inner surface area. In this sense, it is necessary a high amount of DVB to maintain the resin structure. The high level of porosity affect to the swelling properties, making the inner structure heterogeneous, thus their behaviour in to polar solvents or to nonpolar solvents solvents is really different. As advantages it is important to say that they are easily sulfonated and they are really resistant.
- Isoporous type: Their structure is modified during their synthesis, thus polymers with a relative uniform pore size are obtained. These resins present a low sensibility to the fouling, have a higher capacity, high regenerating effectiveness and low production costs.
- Film type resins: They are an special type of ion exchangers that consist in an exchange material in a film shape deposited above an inert support. These resins present quick kinetics and allow working in high pressure conditions (i.e. chromatography).

Considering the feasibility of this matrix type for the synthesis of MNPs, it is noteworthy to mention that they lead to obtain a stable support for the embedded MNPs, are insoluble in water and offer different types of functionalities (e.g. sulfonic, carboxylic, quaternary ammonium), as well as different distributions of functional groups. Some common important parameters, from chemical and physical points of view, are listed in the following table (Table 1).

Chemical Parameters	Physical Parameters
Polymer structure, Ion conductivity	Grading
Functional group type	Particle Size
Ion Exchange Capacity	Pore size and morphology
pH working range	Form
Chemical stability	Density
Water absorption (swelling)	Shipping weight

Table 1. Common important parameters in Ion Exchange Resins.

From those parameters listed in Table 1, the functional groups nature is one of the key ones since it defines the chemical properties and applicability of ion-exchange resins and, what is crucial for IMS, the sign of the matrix charge (either positive or negative). Moreover, the different dissociation properties of the functional groups leads to the distinction between strong and weak exchangers, which have to be considered separately since they have a remarkably different chemical behaviour. In this sense, in Figure 11 are shown some typical resin beads with their polymeric structure formula.

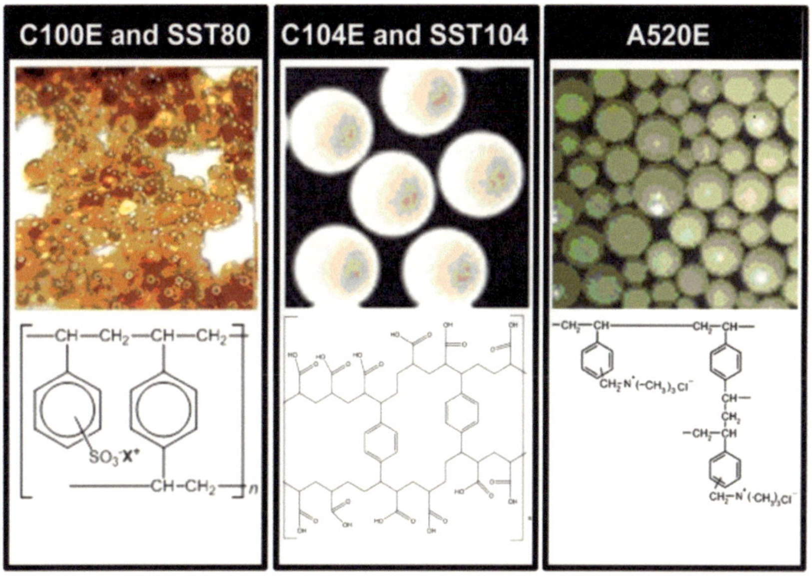

Figure 11. Physical and chemical features of some ion exchange resins.

Another very important parameter to be taken into account is porosity, which affects some bulk properties of the resins, and which have important consequences on their catalytic applications through direct influence on swelling capacity, equilibration rate, and selectivity. In this sense, swelling is an important to be taken account on resin behaviour, because depending on the nature of the ion-exchange resin, this interaction with the solvent may lead to a volume increase (swelling volume) up to 800% and a decrease around 90% of the cross-linking percentage, Hence, gel-type resins are generally preferred over macroporous ones due to enhanced mass transfer inside the polymer beads, resulting in good active-sites accessibility to all soluble reactants.

2.2.2. *Membranes*

The separation of substances by membranes has been (and still is) essential in the industry development and in the human life. Among various separation membranes, the ion exchange membranes, membranes with ionic groups permeable to electrolytes in an aqueous solution, are widely used in different fields: dialysis, solid polymer electrolyte of batteries, analytical chemistry, etc.

In its origins, the ion exchange membrane was developed from two different sources: the finding of ion exchange phenomena in soil and biological phenomena in cell membranes. In 1939 many researchers focused to the establishment of the basis of the studies on electrochemical ion exchange membrane. For example, in 1939 K.H. Meyer, J.F. Sievers and T. Teorell obtained the first artificial charged membrane with the aim of developing a theory of membrane potential; in 1949 Sollner published a paper concerning bi-ionic potential (a measure of permselectivity between ions with the same charge through the membrane); etc. But it is not until 1950, with the work of M.R.J. Wyllie, W. Juda and M.R.C. McRae, when the first ion exchange membranes synthesis is published.

After these works, studies on ion exchange membranes, their synthetic methods, modifications, theorical explanations and applications in industry became very active. But what really made the researchers focus in the development of this kind of membrane is that the charged groups of the membrane act as a fixed carrier for various ionic materials and provide new applications of the membrane.[42, 43] (See Table 2).

Characteristcs	Application	Example
Ion conductivity	Electrodialysis	Separation between electrolyte and non-electrolyte
	Separator for electrolysis	Synthesis of H_2O_2
	Diffusion Dialysis	Acid or alkali recovery from waste
	Neutralization Dialysis	Desalination of water
	Donnan Dialysis	Recovery of precious metals
	Up-hill transport	Separation and recovery of ions
	Piezodialysis	Desalination
	Thermo-dialysis	Desalination
	Battery	Concentration cell
	Fuel cell	H_2-O_2
	Actuator	Catheter for medical use
Hydrophilicity	Pervaporation	Dehydration of water miscible organic solvents
	Dehumidification	Dehumidification of air
	Sensors	Gas sensor
Fixed Carrier (Ion exchange groups)	Facilitated transport	Separation of sugars
	Modified Electrodes	-

Table 2. Principal applications of Ion Exchange membranes

Though ion exchange membranes can be used in many fields, most are used in electrochemical processes such as electrodialysis, separation of electrolytes and solid polymer electrolytes for fuel cells (which really boosted the development of these membranes).

The properties required basically depend on their final application, but generally they can be summarized as:

1. low electrical resistance,
2. high transport number of counterions,
3. low diffusion coefficient of salt,
4. low osmotic water and low electroosmotic water,
5. permselectivity for specific ions with the same charge,
6. antiorganic properties,
7. mechanical strength,
8. dimensional stability,
9. high chemical stability and durability,
10. low cost.

Ion Exchange membranes can be classified in various ways: by their structure and microstructure, by their functionality, materials, etc. But maybe one of the simplest classifications is the morphology which, in first instance, will determine their preparation methodology. In this sense, it is possible to classify ion exchange membranes in two main types: Heterogeneous and Homogeneous membranes.[43]

In an initial stage of membrane development, heterogeneous ion exchange membranes were actively developed by blending finely powdered ion- exchange materials and a binder. In a general procedure, cation or anion exchange resins are homogenously blended and heated with a thermoplastic polymer (i.e. polyethylene, polypropylene, etc.) and the mixture is formed as a membrane by pressing or heating.

Although these types of membranes are easily prepared and have a great mechanical strength, their electrochemical properties are lower than the ones of homogenous ones in which the fixed charged groups are evenly distributed over the entire membrane polymer matrix. This homogeneity in the homogeneous membranes structure is due to the fact that they can be produced, e.g. by polymerization or polycondensation of functional monomers such as phenylsulfonic acid with formaldehyde, or by functionalizing a polymer such as polysulfone dissolved in an appropriate solvent.

But the completely homogeneous and the macroscopically heterogeneous ion-exchange membranes are extreme structures. Most ion-exchange membranes show a certain degree of heterogeneity on the microscopic scale. Thus, other properties may be considered to classify them. In this regard, according to the distribution and species of the fixed charge (ion exchange groups) it is possible to difference between:

i. cation exchange membranes (with anionic charged groups)

ii. anion exchange membranes (with cationic charged groups)
iii. amphoteric ion exchange membranes (with both cation and anion exchange groups at random throughout the membrane
iv. bipolar exchange membranes (bilayer membranes composed by a cation exchange membrane layer and an anion exchange membrane layer)
v. mosaic ion exchange membranes (which have certain domains that may be separated with an isolator of cation-exchange groups and also domains with anion-exchange groups).

On the other hand, a classification based on the constituent materials allows grouping such membranes as:

1. membranes composed of hydrocarbons or partially halogenated hydrocarbons
2. perfluorocarbon membranes
3. inorganic membranes
4. composite membranes of inorganic ion exchanger and organic polymer (e.g., hybrids).

Nowadays, one of the most employed commercial ion exchange membranes is Nafion, a cation exchange homogenous perfluorinated membrane. It is an excellent proton conductor: it has excellent chemical stability, high ionic conductivity, good mechanical strength, good thermal stability, etc. ideal for performance in fuel cells. The main drawback of these membranes and of those containing Fluor in their structure (i.e. Selemion) are their high cost and, specially, the absence of pores that limits their application to the transport of ions in solution or vapour (pervaporation). Thus, the search of new homogeneous cation exchange membranes has been focussing much of efforts. In these sense, sulfonated polymers open a new window to the ion exchange membranes field. Some typical sulfonated polymers are shown in Figure 12.

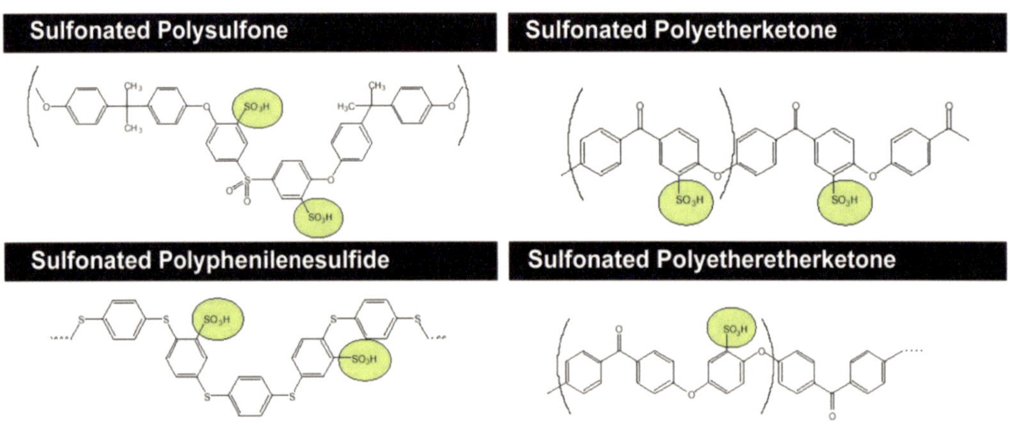

Figure 12. Typical Sulfonated polymers.

One of the polymers that fulfils the properties to be casted as an ion exchange membrane is sulfonated poly(ether-ether ketone) (SPEEK) which is nowadays attracting great interest regarding the fabrication of membranes for fuel cells, due to its thermoplastic properties, its high chemical strength, its high stability towards oxidation and its low cost.[44, 45]

However, regarding this last option it is important to take into account that an excess of ionic groups (in this case, sulfonic groups) could cause the dissolution of the polymeric material in water since an increase of the ionic groups in the polymer directly increases the hydrophilicity of the material.

To cope with this limitation, one good option is the sulfonation of a polymer with a very hydrophobic group so as to reduce the hydrophilicity of the final polymer, the polyethersulphone with Cardo group (PES-C) [46-48] which bears a five-member lactone ring and whose sulfonation can be done in a simple way. Among all the stability properties mentioned, this polymer can be casted by wet phase inversion methodology to obtain porous membranes be applied in filtration. By controlling the ratio of sulfonic groups in the polymer, different porosity can be obtained.

3. Environmental and safety concerns

Perception and knowledge are important parts of public understanding of nanotechnology. They can be influential, for an achievable benefit obtained and the possible risks and hazard which it could imply.

The use of engineered nanoparticles (ENPs) in the environment, as a consequence of the development of nanotechnology, is a serious case of worldwide concern. However, a few studies have already demonstrated the toxic effects of nanoparticles on various organisms, including mammals. Nanotechnology is still in a discovery phase in which novel materials are first synthesized in small scale in order to identify new properties and further applications.

Therefore, detail understanding of their sources, release interaction with environment, and possible risk assessment would provide a basis for safer use of ENPs with minimal or no hazardous impact on environment

In order to do that, future directions such as the inclusion of regulatory and knowledge gaps within the risk identification framework should be designed and applied.

Thus, the evaluation of potential health impact as well as an enhanced design of the production of higher performance nanomaterials is mandatory; as well as defining criteria that distinguish between technologies and products more or less likely to present a health risk to avoid inappropriate and possibly deleterious sweeping conclusions regarding potential impact.

It is required to study their release, uptake, and mode of toxicity in the organisms. Furthermore, to understand the long-term effect of ENPs on the ecosystem, substantial information is required regarding their persistence and bioaccumulation.

3.1. Safe polymer-metal nanocomposites

A massive industrial production of nanomaterials in the near future may result in the appearance of both NPs and the waste generated during their production in various environments, yielding the possibility that humans could be exposed to these NPs through inhalation, dermal contact or ingestion and absorption through the digestive tract. Nowadays, there is claim for more restrictive legislation that would allow a better protection for both human beings and the global environment.

In this sense, for a comprehensive knowledge of properties of these materials (both physical and chemical), it is important to find standards and control materials to work with as reference models (such those from the British Standards Institute and International Standards Organisation).[49]

An investigation into nanomaterials toxicity involves: a determination of the inherent toxicity of the material, their interaction with living cells and the effect of exposure time.[50] It should be noted that the doses or exposure concentrations used for in vitro and in vivo toxicological studies are most often extraordinarily high in comparison with possible accidental human exposure.[51, 52]

Consequently, more research is needed before generalized statements can be made regarding NPs ecotoxicology.

Unfortunately, only few initiatives in this direction have been started so far. For instance, the German Federal Ministry for Education and Research, together with industry, has established the research programme NanoCare. This programme has a budget of €7.6 million and aims to assess and communicate new scientific knowledge of the effects of NPs on health and the environment.[53]

Scientists and technologists in this area have to deal with NPs presence in the environment but very often they do not have the appropriate tools and analytical methods for NPs detection and quantification to guarantee a satisfactory detection.[4]

Thus, it is of vital relevance to dedicate those efforts towards this direction, as we have not yet invented a so-called "Geiger counter" for NPs.

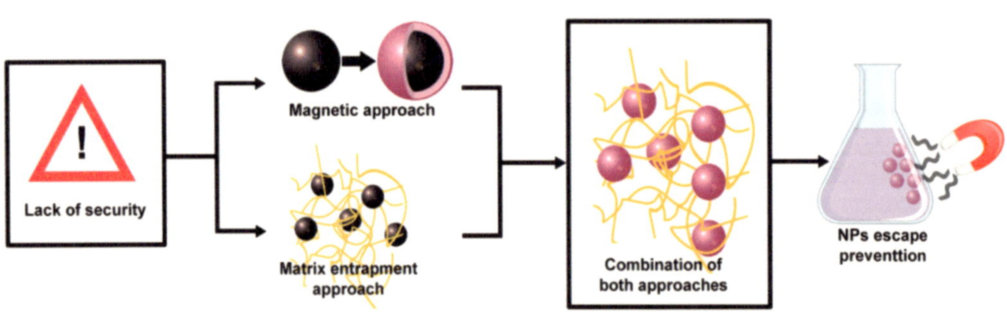

Figure 13. Schema describing the concepts involved in the design of safe polymer-metal nanocomposites.

As a consequence, the prevention of NPs escape into the environment is currently most likely the best approach that can be considered. In this regard, a possible solution appeared through the development of this project, which describes the results obtained by developing environmentally safe polymer–metal nanocomposite materials exhibiting magnetic properties. These materials prevent NPs escape by profiting of the embedding of NPs into organic matrices and the use of magnetism. [54] As a result, NPs reduce their mobility and, in case of leakage, NPs could be easily recovered by using simple magnetic traps.[13, 55-57]

4. Characterization of Polymer-Stabilized Nanoparticles (PSMNPs)

One of the main features for the development of polymer stabilized metal nanoparticles (PSMNPs) and nanocomposites is their detailed characterization. Specific techniques have to be applied in order to better understand the parameters affecting their synthesis, explain their properties and to adequate them to their final application.

In this sense, several techniques can be used, involving techniques for the chemical characterization and the typical techniques applied in Material Science to determine the MNPs composition, size and shape and their distribution into the matrix as well as the nanocomposite morphology and their special properties (such as magnetism, biocidal, electrocatalytic and catalytic activity among others). These parameters can be studied by using some of the techniques explained in this chapter (or by a combination of some) which include:

- Scanning Electron Microscopy (SEM)
- Transmission Electron Microscopy (TEM)
- Atomic Force Microscopy (AFM)
- Nuclear Magnetic Resonance (NMR)
- X-Rays Diffraction (XRD)
- Impedance Spectroscopy (IS)
- X-rays Photoelectron Spectroscopy (XPS)
- X-Rays Energy Dispersion Spectroscopy (EDS)
- Infrared-Attenuated Total Refraction (IR-ATR)
- Inductively Coupled Plasma Atomic Emission Spectrometry (ICP-AES)
- and many others.

4.1. Characterization of MNPs

When synthesizing PSMNPs the first parameter to determine is the metal content of the polymer-metal nanocomposite.

The composition of MNPs (even that of a single nanoparticle) can be determined by High Resolution TEM or SEM analysis coupled with EDS techniques. The microscopic techniques allow for the selection of the nanoobject(s) whereas the EDS provides the composition analysis. These methods usually gives qualitative or semi-quantitative results which can be useful to have an estimation of the composition differences in a sole sample. An example of this analysis is showed in in Figure 14, where a crossed-section resin bead sample is shown, prepared as described afterwards.

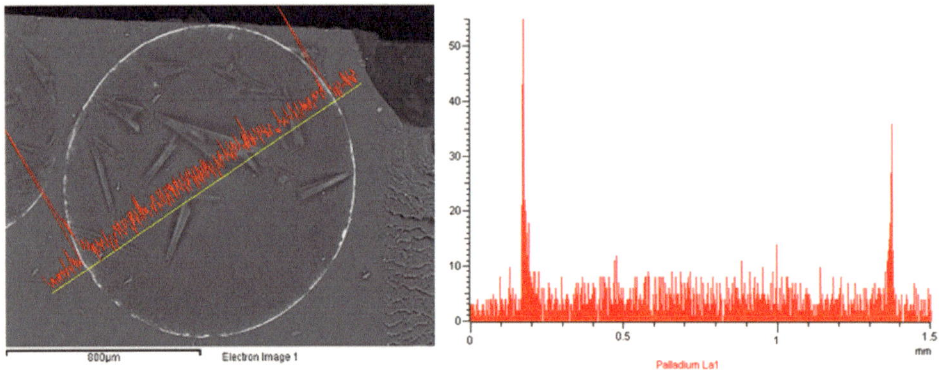

Figure 14. SEM image of an Ion Exchange resin with PdNPs and the corresponding EDS analysis.

If a quantitative analysis is required, the total metal content in the matrix can be determined by using and ICP coupled to an Atomic Emission Spectrometer (for metal concentrations between 0.1 to 500 ppb, depending on the metal) or coupled to a Mass Spectrometer (for metal concentrations between 0.1 to 10 ppb, depending on the metal)[58] to analyse the solution obtained after the treatment of a known amount of the nanocomposite with aqua regia to completely dissolve the MNPs and degrade the polymeric matrix.

The size and the shape of the MNPs obtained are important parameters allowing determining the nanocomposite characteristics. Further development of Nanotechnology needs a better understanding of nanomaterial properties and implies a better characterization of the above parameters, which are evaluated as a rule by using TEM technique.

In this sense, depending on the solubility properties of the polymeric matrices different sample preparation methodologies have to be considered. If the polymeric matrix is soluble in a volatile organic solvent (e.g. DMF, $CHCl_3$, THF) it is possible to prepare MNPs suspensions or "inks" (5% mass solutions) in adequate organic solvents which can be deposited onto a TEM grid to perform the microscopic characterization. On the contrary, if the MNPs are stabilized in a non-soluble matrix (e.g. Nafion), the preparation of the sample is more difficult (although allows us to determine the distribution of the MNPs in the matrix), since it requires to cut thin or ultrathin slices (thickness about 1μm or less) of the nanocomposite material and deposit them onto a TEM grid. An example of this procedure and the final image obtained is showed in Figure 15.

Unfortunately, in practice, in many instances the quality of TEM images appear to be quite low due to the high noise and low contrast, making their processing a challenging task to accomplish. Also, the quantitative treatment of TEM images is often carried out by manual measurements of high number of nanoparticles, a task that is highly subjective and time consuming. During the last years, several computer imaging particle analysis software tools have been conceived to achieve a more accurate assessment of the size and frequency (size distribution) of nanoparticles.

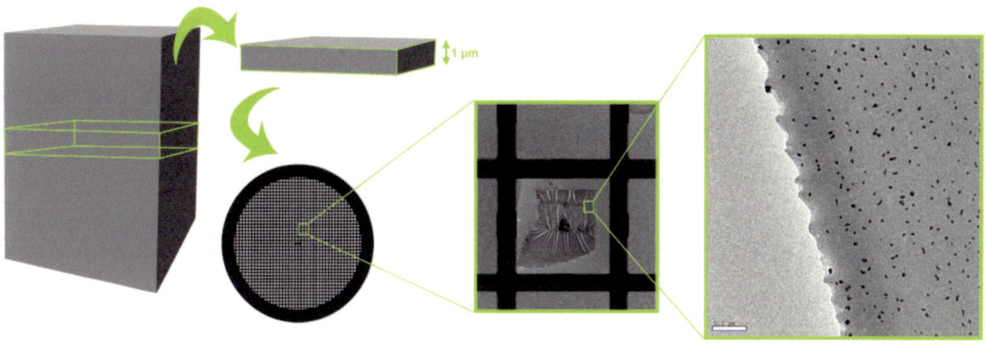

Figure 15. Sample preparation of cross-sectioned with microtome of a sulfonated polyetehersulphone (SPES-C) membrane with AgNPs and the corresponding TEM image obtained.

Through the image analysis of TEM micrographs (either manually or automatically) it is possible to make size distribution histograms from the sample data as the one shown in Figure 16. The obtained data can be used fitted to a 3-parameter Gaussian curve (14) where *a* is the height of Gaussian peak, d_m is the position of the center of the peak (corresponding to the most frequent diameter), and σ is the standard deviation.

$$y=a\cdot\exp\left[-0.5\cdot\left(\frac{d-d_m}{\sigma}\right)^2\right] \tag{14}$$

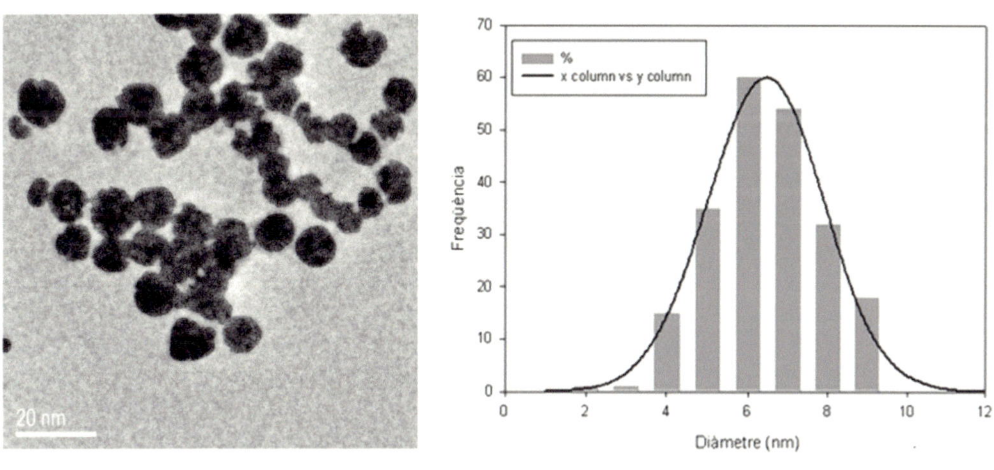

Figure 16. TEM image of an ink of PdNPs obtained in a SPES-C membrane with the corresponding histogram adjusted by a 3 parameter Gaussian curve.

4.2. Metal–polymer nanocomposite morphology

The loading of MNPs in the polymeric matrix may cause important changes in the polymer morphology. For this reason, a characterization of the matrix previous and after the stabilization of the MNPs on it is required.

In this sense and among other techniques, SEM has been established as a referent in the field of the surface characterization. Because polymeric matrices usually are not conductive, in some cases (in matrices without MNPs or with a low content of MNPs) it is imperative to prepare the sample for the study, by a sputter coating with gold, carbon or palladium layers of about 50Å of thickness.

For membranes and films, cross-section images can be obtained by cutting the samples under liquid N_2. For resin nanocomposites, it is necessary to embed the material in an epoxy resin to cut it transversally with a microtome, as shown in Figure 17.

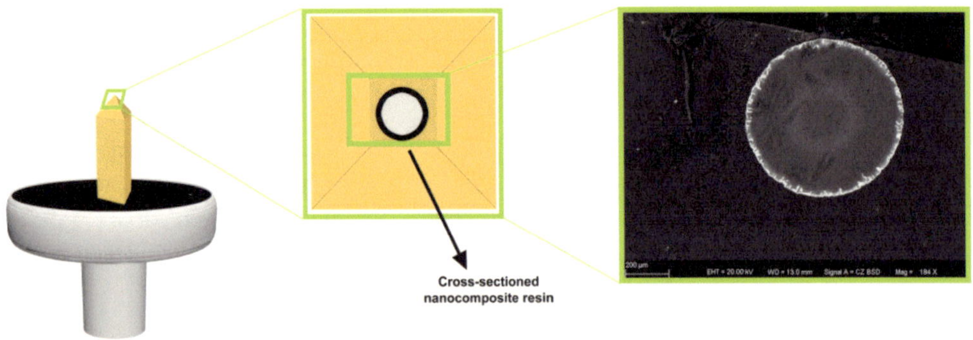

Figure 17. Sample preparation of a resin bead for SEM characterization.

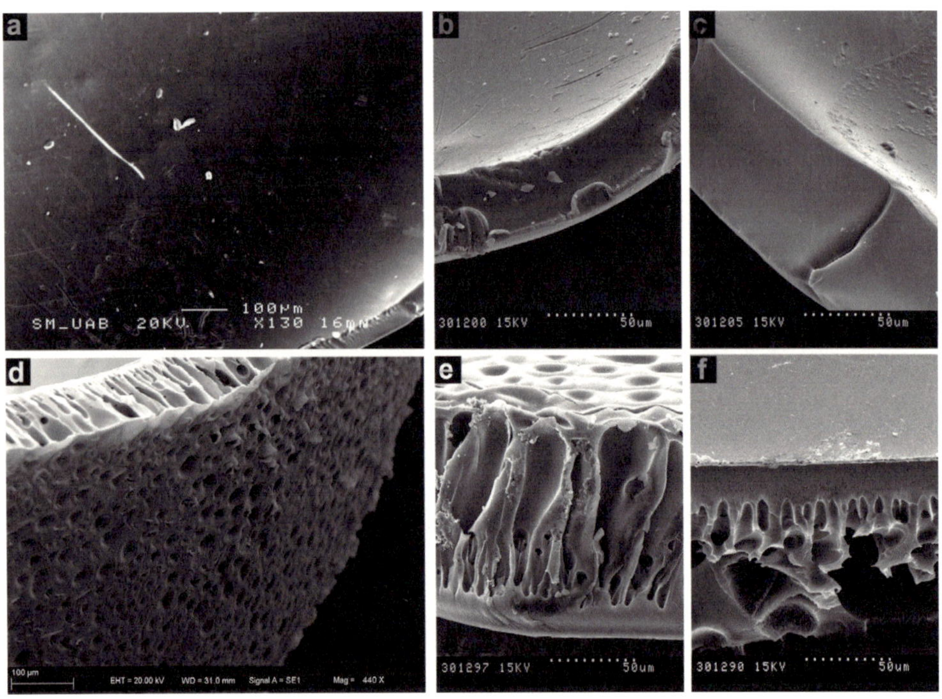

Figure 18. SEM images of (a) SPES-C membrane surface, (b) bare SPES-C membrane, (c) SPES-C membrane with PdNPs, (d) Blend membrane surface, (e) bare Blend membrane and (f) Blend membrane with PdNPs.[32]

Figure 18 shows SEM images of bare and coated SPES-C and Blend membranes (prepared with sulfonated and non sulfonated Polyethersulphone). Whereas in the case of SPES-C membranes there is an absence of porosity and surface defects in the case of Blend membranes it can be observed a finger structure porosity, and the existence of defects on the surface. In both cases, the load with PdNPs do not affect the final structure of the membrane.

The difference in porosity between the two types of membranes can be explained by an increase in the hidrofobicity of the final polymer: adding PES-C polymer increases the capacity of repulsion of H_2O molecules thus, when preparing the membranes by wet phase inversion method by immersion in a non-solvent such as water, pores are generated during the precipitation of the membrane.

In this way, changing the ratio of PES-C / SPES-C in the final polymer blend membranes with different morphology and, therefore, different final application can be obtained.

4.3. Magnetic properties

The metallic nanoparticles have larger magnetization compared to metal oxides, which is interesting for many applications. But metallic magnetic nanoparticles are not air stable, and are easily oxidized, resulting in changes or loss in their magnetization properties.

Thus, IMS of magnetic NPs open a new range of research. Lack of stability of this kind of nanoparticles finds a counterpart by their stabilization on a polymeric matrix.

Magnetic properties of metallic nanoparticles are dependent on the oxidative state of the NPs components. Therefore, the true knowledge of the degree of nanoparticle oxidation is necessary for the forecasting of magnetic characteristics of the obtained samples. This is not easy, but techniques such as XANES (X-Ray Absorption Near Edge Structure) may do it achievable by interaction of the atom core with the source of energy. By comparison with previously placed and analyzed patterns, information about chemical bonding and oxidative states is obtained. Figure 19 shows XANES analysis of nanocomposites containing either Ag or Ag@Co MNPs (with a superparamagnetic Co^0-core) on sulfonic resin. Standard elements spectra were linearly combined and fitted with the sample in order to determine the oxidative state of each element in the sample. The linear combination results are also included (normalized) inset. Ag@Co NPs in sulfonated matrices showed an average Co spectrum similar to that recorded by the Co^0 standard. In fact, linear combination fitting results confirmed that all the Co present in that sample was Co^0.

Furthermore, to characterize magnetic behavior SQUID (Superconducting Quantum Interference Device) magnetization curves are obtained. In order to do that, sample is placed in a changing magnetic field over at room temperature. Magnetization loops are registered through the overall process.

In general, ferromagnetic species have normally evident hysteresis curves. However, superparamagnetic materials (frequently, ferromagnetic materials at nanometer scale) shows a lack of hysteresis but high magnetic saturation. Figure 20 presents SQUID magnetization curves for Pd@Co MNPs supported on a sulfonic (C100E) and carboxylic(C104E) cation exchange resin.

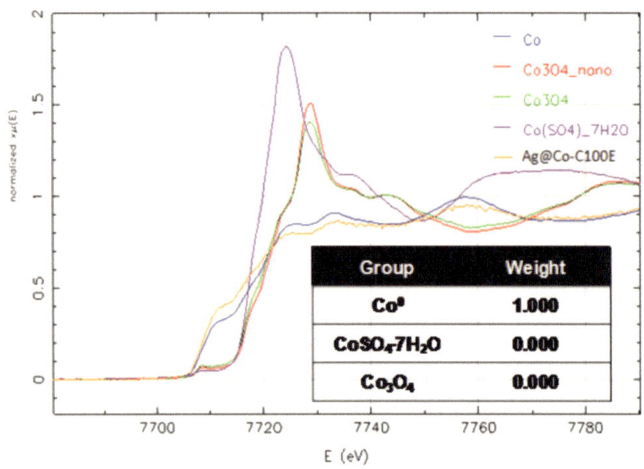

Figure 19. XANES spectra of Ag in comparison with Ag standards for Ag@Co-C100E sample (blue line in the graphic) and the linear combination fitting among all of the compounds analyzed is also shown in a fitting range from -20 to 30 eV. Ag^0 (red) and $AgNO_3$ (green) are standards.

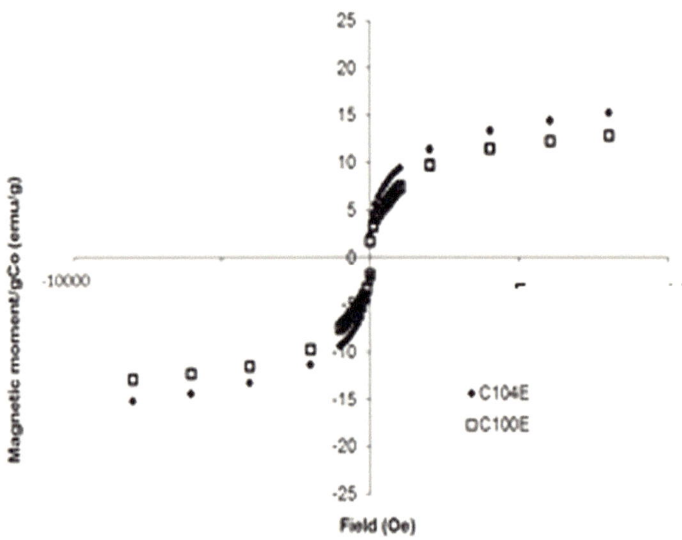

Figure 20. SQUID Magnetic curves obtained of Pd@Co-NPs stabilized in C100E and C104E supports

5. Applications

The use of nanomaterials to assemble architectures of defined size, composition and orientation allows researchers to utilize the particular electrical, optical, catalytic and magnetic properties of those materials for:

- the creation of functional materials, devices, and systems through control the matter on the nanometer scale and,
- the exploitation of novel phenomena and properties at that scale.

To achieve these goals, it is necessary to use multidisciplinary approaches; inputs from physicists, biologists, chemists, and engineers[59-61] are required to advance our understanding of nanomaterials.[62] In fact, NPs properties are already used for developing new products[63, 64] such as paints, where they serve to break down odour substances, on surgical instruments in order to keep them sterile, in highly effective sun creams, slow release pharmaceuticals and many others.[50, 65-67] Bench-marketing studies on main current industries[53, 68] revealed that market opportunities that are illustrated in Figure 21.

As it is clearly shown, most of these market opportunities are involved in the development of new materials. Moreover, among the applied NPs, metal nanoparticles and metal oxide nanoparticles are of the most importance. However, because of the aforementioned toxicity and stability concerns have driven the research to develop nanomaterials with higher levels of safety. In this sense, here two different examples of safe and stable nanocomposites applications are discussed: their use for organic catalysis or for water disinfection.

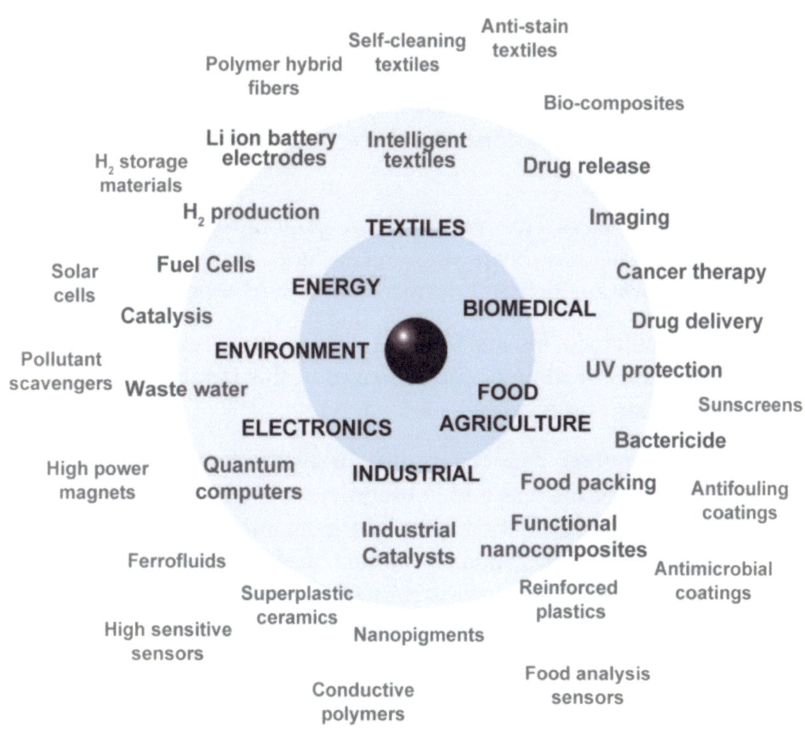

Figure 21. Possible marketable applications of nanocomposites in different fields.

5.1. Bactericidal activity of MNPs

Among the currently known nanomaterials, it is well-known that AgNPs have unique antimicrobial properties.[69] Textiles, keyboards, wound dressings, and biomedical devices now contain AgNPs that continuously release a low level of Ag ions to provide protection against bacteria. Even if Ag has been known to be a bactericidal element for at least 1200 years, considering the unusual properties of nanometric scale materials, largely different from those of their bulk counterparts[70], it is not surprising that AgNPs have been found significantly more efficient than Ag^{+} ions in mediating their antimicrobial activities.[71-75] All in all, the exact antibacterial action of AgNPs is still under debate.

Conversely, in many countries the microbial contamination of potable water sources poses a major threat to public health and the emergence of microorganisms resistant to multiple antimicrobial agents increases the demand for improved disinfection methods.[76] The importance of potable water for people in some countries dictates the need for the development of innovative technologies and materials for the production of safe potable water. This type of application can be a perfect niche for nanomaterials containing AgNPs. However, it is necessary to develop ecologically-safe nanomaterials that prevent the post-contamination of the used samples.[77] In this sense and as it has been already stated, functionalized polymers are currently acquiring a prominent role as NPs stabilizers for their excellent performance.[78, 79]

It is worthy to note here that ion-exchange materials are already widely used for various water treatment processes, mainly to eliminate undesired or toxic ionic impurities including hardness ions, iron, heavy metals, and others. The stabilization and immobilization of Ag-NPs in such matrices is very promising since using this approach, two complementary water treatment steps could be performed with a single material and the safety of the nanocomposites could be increased.

In our research group, the surface modification of ion-exchange materials used for traditional water treatment has been undertaken and promising results have been patented.[80] Such modification included the incorporation of either Ag or Ag@Co NPs.

As an example of the obtained results, Table 3 shows the synthetic conditions and the corresponding compositions of some nanocomposites of this family, probing the feasibility of the synthesis of both pure and core-shell nanoparticles.

To evaluate the efficiency of these nanocomposites for disinfection procedures, an increasing amount of nanocomposites beads was added to individual wells containing 10^{5} CFU/mL of *E. coli* suspension in LB medium. After overnight incubation, bacterial proliferation was evaluated by measuring the optical density of each well at 550 nm (this wavelength is indicative of bacterial proliferation). The bactericidal activity of the Ag, Co and Ag@Co nanocomposites (in all the polymeric granulated matrices studied) was determined and raw materials were used as control (results are shown in Figure 22).

As it was expected, Ag and Ag@Co NPs containing sulfonated granulted materials increased their activity due to the presence NPs, but the enhancement was slightly higher for modified Ag@Co NPs. Anyhow, this proof of concept demonstrates the rightness of the approach.

Matrix	IEC / meq·g^{-1}	NPs	[Ag^+]/M	[Co^{2+}]/M	$mmol_{Ag}$/meq	$mmol_{Co}$/meq
C100E	2.3	Ag	0.01	--	0.064	--
(-SO_3^-)		Ag@Co	0.01	0.01	0.069	0.061
C104E	6.0	Ag	0.01	--	0.010	--
(-COO^-)		Ag@Co	0.01	0.01	0.010	0.017

Table 3. Metal content in granulated nanocomposites containing Ag- or Ag@Co-NPs and, analyzed by ICP-MS.

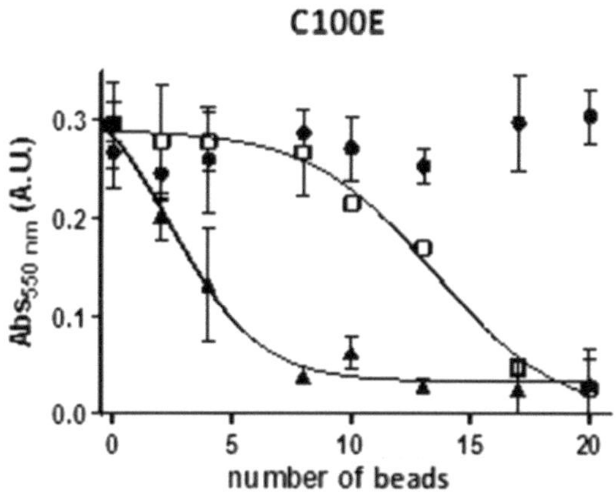

Figure 22. Variation of the absorbance at 550 nm with the number of polymer beads for (●) the raw material, (□) Ag- and, (▲) Ag@Co-nanocomposites (3 replicates).

5.2. Nanocatalysts for organic synthesis

Nanoparticles are increasingly used in catalysis, where the large surface area per unit volume of the catalyst may enhance reactions. This enhanced reactivity significantly reduces the quantity of catalytic materials required to carry out the reactions. Particular industries, including the oil and the automobile ones, are interested in this area for the use of NPs in catalytic converters.[81] As a prove of their industrial potentiality, many big companies, including BASF, Johnson Matthey and 3M, have interests in developing commercial applications for AuNPs catalysts.

In the last decade, heterogeneous catalysts have attracted much interest because of their general advantages that have been boosted thanks to the use of nanomaterials.[82, 83] One crucial property for catalysts is their recovery and, in this sense, magnetic nanocatalysts present some outstanding advantages because they can be conveniently recovered by using an external magnetic field.[84]

On the one hand, Platinum Group Metals (PGMs) are well-known as highly selective catalysts and are widely used in organic synthesis, chemical industry and other areas like dehalogenation, hydrodechlorination, carbonylation or oxidation.[85-87] Concerning the potential applications, Pd, Pt, Rh, and Au-NPs have proven to be very versatile as they are efficient and selective catalysts for several types of catalytic reactions, including olefin hydrogenation and C-C coupling such as Heck, Suzuki and Sonogashira reactions.[88-90] Among them, Pd-catalyzed cross-coupling has emerged as an effective synthetic methodology that is employed in both academic and industrial sectors.[91] Despite such progress, a number of challenges still remain unknown, including the dilucidation of highly efficient and selective catalysts able to react with multiple reactive C-H or N-H bonds.

On the other hand, several types of magnetic materials have been used, including magnetite, hematite, maghemite, wüstite.[92] Magnetic aggregation and their need for functionalization do still hinder the application of magnetic NPs in industry. Thus, searching for more suitable magnetic materials to overcome these restrictions is still a challenge for realizing practical catalytic applications. Yet, for catalytic purposes, magnetic NPs surface is often chemically functionalized with molecular catalytic complexes because of the poor catalytic properties of the bare Fe oxides or other catalytic materials (*e.g.*, Co).[93, 94]

Therefore, and taking into account, the demonstrated efficiency of PGMs and the advantages of magnetic nanoparticles, it has been possible to apply develop Pd@Co-based nanocomposites for a typical C-C coupling reaction: the Suzuki reaction.[95] The preparation of such catalyses has only been possible thanks to the characteristics of IMS procedure, which allows the combination of a magnetic nanocore (made of Co) with the catalytic activity of a shell (Pd). The resulting material can be separated by simple filtration methods and, moreover, NPs can be re-covered and re-used by their retention under a magnetic field (Figure 23).[96]

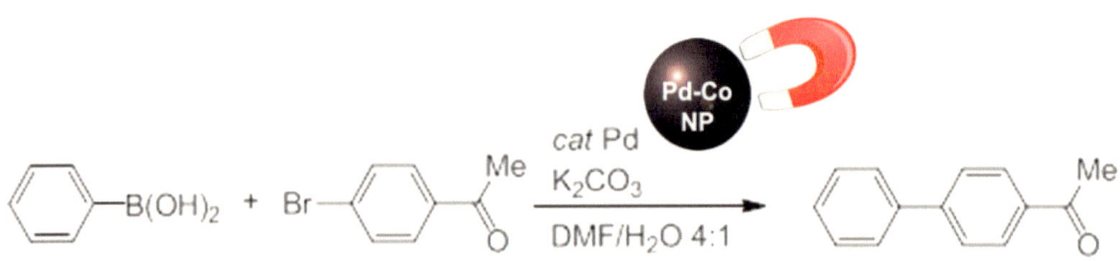

Figure 23. Suzuki reaction by using Pd or Pd@Co nanocomposites with optimized conditions.

Previous results with Pd@Co-NPs incorporated to fibrous materials showed the feasibility of this approach.[37] Differently, as it can be seen in Table 4, the catalytic efficiency of granulated polymers (containing either sulfonic or carboxylic groups) was very scarce, very likely due to the low metal immobilization achieved. However, two interesting results can be withdrawn: a substantial increase of the reaction yield was obtained when using the nanocomposite samples with higher Pd-content and the reaction yield could increase for consecutive runs.

Even if low conversions where achieved, this results are not discouraging since granulated polymeric matrices are still interesting for industrial applications because of their mechanical properties. Their high mechanical resistance leads to obtain higher reproducibility in synthesis as well as easier manipulation. Moreover, granulated polymer industry is big enough to pay attention on it. Further research is needed to successfully increase the amount of immobilised metals what, very probably will provide better catalytic nanocomposites.

Samples		Metal Content		Yield (%)		
Matrix	MNPS	Pd (mg/gNC)	Co (mg/gNC)	#0	#1	#2
C100E	Pd	57	-	44	-	-
	Pd	30	50	8.6	-	-
C104E	Pd@Co	47	48	9.8	8.4	-
	Pd@Co	30	7,6	4	**26**	28

Table 4. Suzuki reaction yields (in %) for the Pd- and Pd@Co-nanocomposites

6. Conclusions

The conclusions derived from the results presented in this chapter can be briefly formulated as follows:

1. The ion-exchange assisted Intermatrix Synthesis (IMS) technique represents one of the most promising techniques that allows for the production of a large variety of polymer–metal nanocomposites of practical importance for different fields of modern science and technology.

   The attractiveness of this technique is basically determined by its relative simplicity in comparison with other methods used for production of nanocomposite materials and also by its flexibility and the possibility of tuning the specific properties of the final nanocomposites to meet the requirements of their final applications. IMS technique gives a unique possibility of production of nanocomposites containing MNPs of various composition and structure (for example, monometallic, bimetallic or polymetallic MNPs with core-shell, core-sandwich and even more complex structures) for the applications of interest.

2. The spectrum of polymers applicable for IMS of PSMNPs is quite wide and includes various functionalized polymers, i.e. those bearing functional groups in the form of granules, fibers or membranes, which are capable to bind either metal or reducer ions prior to the metal reduction inside the polymer matrix (IMS of PSMNPs).

   The dissociated ionogenic functional groups of the polymer bearing positive or negative charges provide a possibility to couple IMS technique with Donnan exclusion effect. In case of polymers with negatively charged groups the IMS technique consists of the metal loading stage followed by metal reduction inside the polymer. When polymer bears positively charged groups the IMS procedure starts with the reducer loading followed by the simultaneous metal loading-reduction stage.

3. It has been shown that the use of, for example, anionic reducing agents (e.g., borohydride) for the synthesis of PSMNPs by using IMS coupled with Donnan effect in both cation exchange and anion exchange matrices results in their formation mainly near the surface of the polymer. This type of MNP distribution is favorable for many practical applications of polymer–metal nanocomposites such as reagent-free water disinfection, catalysis and some others.
4. The general properties of polymer-metal nanocomposite are not determined only by the properties of the MNPs. The formation of MNPs within the polymer matrices may strongly modify the polymer morphology, for example due to the appearance of nanoporosity in gel-type polymers, which enhances the rate of mass transfer inside the nanocomposites as well as some other structural parameters of great importance in their practical applications.
   The polymer matrix also serves as the MNPs stabilizing media preventing their aggregation and release to the medium under treatment. The functional properties of the nanocomposites (e.g., catalytic or bactericide) are mainly determined by the properties of MNPs immobilized inside the matrix.
5. The repetitive metal-loading-reduction (in case of cation exchangers) or reducer-loading-metal-reduction (in case of anion exchangers) permits to synthesize MNPs of core-shell or even more complex structure provided with additional functional properties.
   For example, the IMS of core shell PSMNPs consisting of superparamagnetic cores coated with functional shells having for example, catalytic or bactericide properties provides the polymer–metal nanocomposites with additional advantages. In the case of catalytic applications, the nanocomposite can be easily recovered from the reaction mixture and reused. In water treatment applications, the magnetic nature of MNPs permits for preventing their uncontrollable escape into the treated water by using simple magnetic traps.
6. Finally, the last and the most important in our opinion conclusion concern the general strategy in the development of novel nanocomposite materials. This strategy has to be focused not only on the desired properties of material, which is dictated by its further practical applications, but also on the material safety in both environmental and human health senses. The last point seems to be of particular importance for the further development of Nanoscience and Nanotechnology.

Author details

Berta Domènech, Julio Bastos-Arrieta, Amanda Alonso, Maria Muñoz and Dmitri N. Muraviev
Analytical Chemistry Division, Chemistry Department, Autonomous University of Barcelona, Barcelona, Spain

Jorge Macanás
Department of Chemical Engineering, Universitat Politècnica de Catalunya, Barcelona, Spain

7. References

[1] Ajayan, P.M., Schadler, L.S. and Braun, P.V., *Nanocomposite Science and Technology*. 2003: Wiley-Vch.

[2] *Metallic Nanoparticles*, in *Handbook of Metal Physics*, A.B. John, Editor. 2008, Elsevier. p. iii.

[3] Campelo, J.M., Luna, D., Luque, R., Marinas, J.M. and Romero, A.A., *Sustainable Preparation of Supported Metal Nanoparticles and Their Applications in Catalysis.* ChemSusChem, 2009. 2(1): p. 18-45.

[4] Giannazzo, F., Eyben, P., Baranowski, J., Camassel, J. and Lanyi, S., *Advanced materials nanocharacterization.* Nanoscale Research Letters, 2011. 6(1): p. 107.

[5] Hassan, M.H.A., *Small Things and Big Changes in the Developing World.* Science, 2005. 309(5731): p. 65-66.

[6] Wang, F.-C., Zhang, W.E., Yang, C.H., Yang, M.J., Bennett, B.R., Wilson, R.A. and Stone, D.R., *A tunneling field-effect transistor with 25 nm metallurgical channel length.* Applied Physics Letters, 1997. 70(22): p. 3005-3007.

[7] Davis Paul, H., Morrisey Cody, P., Tuley Sean, M.V. and Bingham Chris, I., *Synthesis and Stabilization of Colloidal Gold Nanoparticle Suspensions for SERS*, in *Nanoparticles: Synthesis, Stabilization, Passivation, and Functionalization*. 2008, American Chemical Society. p. 16-30.

[8] *Nanoparticles: Synthesis, Stabilization, Passivation, and Functionalization, Copyright, Foreword*, in *Nanoparticles: Synthesis, Stabilization, Passivation, and Functionalization*. 2008, American Chemical Society. p. i-v.

[9] Nagarajan, R., *Nanoparticles: Building Blocks for Nanotechnology*, in *Nanoparticles: Synthesis, Stabilization, Passivation, and Functionalization*. 2008, American Chemical Society. p. 2-14.

[10] Li, Q., El Khoury Jouliana, M., Zhou, X., Urbas, A., Qu, L. and Dai, L., *Synthesis of Thiol Surfactant with Tunable Length as a Stabilizer of Gold Nanoparticles*, in *Nanoparticles: Synthesis, Stabilization, Passivation, and Functionalization*. 2008, American Chemical Society. p. 41-54.

[11] Gaponik, N., Eychmüller, A., Schmid, G., Talapin, D.V. and Shevchenko, E.V., *Organization of Nanoparticles*, in *Nanoparticles*. 2010, Wiley-VCH Verlag GmbH & Co. KGaA. p. 311-370.

[12] Astruc, D., *Transition-metal Nanoparticles in Catalysis: From Historical Background to the State-of-the Art*, in *Nanoparticles and Catalysis*. 2008, Wiley-VCH Verlag GmbH & Co. KGaA. p. 1-48.

[13] Jorge, M., Patricia, R., Amanda, A., María, M. and Dmitri, M., *Ion Exchange-Assisted Synthesis of Polymer Stabilized Metal Nanoparticles*, in *Ion Exchange and Solvent Extraction*. 2011, CRC Press. p. 1-44.

[14] Gerion, D., Parak, W.J., Williams, S.C., Zanchet, D., Micheel, C.M. and Alivisatos, A.P., *Sorting Fluorescent Nanocrystals with DNA.* Journal of the American Chemical Society, 2002. 124(24): p. 7070-7074.

[15] Zhao, X., Yin, M., Ma, L., Liang, L., Liu, C., Liao, J., Lu, T. and Xing, W., *Recent advances in catalysts for direct methanol fuel cells.* Energy & Environmental Science, 2011. 4(8): p. 2736-2753.

[16] Astruc, D., Lu, F. and Aranzaes, J.R., *Nanoparticles as Recyclable Catalysts: The Frontier between Homogeneous and Heterogeneous Catalysis.* Angewandte Chemie International Edition, 2005. 44(48): p. 7852-7872.
[17] Spagnoli, D., Banfield, J.F. and Parker, S.C., *Free Energy Change of Aggregation of Nanoparticles.* The Journal of Physical Chemistry C, 2008. 112(38): p. 14731-14736.
[18] Rozenberg, B.A. and Tenne, R., *Polymer-assisted fabrication of nanoparticles and nanocomposites.* Progress in Polymer Science, 2008. 33(1): p. 40-112.
[19] Imre, Á., Beke, D.L., Gontier-Moya, E., Szabó, I.A. and Gillet, E., *Surface Ostwald ripening of Pd nanoparticles on the MgO (100) surface.* Applied Physics A: Materials Science & Processing, 2000. 71(1): p. 19-22.
[20] Carotenuto, G., Nicolais, L., Martorana, B. and Perlo, P., *Metal-Polymer Nanocomposite Synthesis: Novel ex situ and in situ Approaches*, in *Metal–Polymer Nanocomposites*. 2005, John Wiley & Sons, Inc. p. 155-181.
[21] Luckham, P.F., *Manipulating forces between surfaces: applications in colloid science and biophysics.* Advances in Colloid and Interface Science, 2004. 111(1–2): p. 29-47.
[22] Kamat, P.V., *Photophysical, Photochemical and Photocatalytic Aspects of Metal Nanoparticles.* The Journal of Physical Chemistry B, 2002. 106(32): p. 7729-7744.
[23] Jeon, S.-H., Xu, P., Zhang, B., Mack, N.H., Tsai, H., Chiang, L.Y. and Wang, H.-L., *Polymer-assisted preparation of metal nanoparticles with controlled size and morphology.* Journal of Materials Chemistry, 2011. 21(8): p. 2550-2554.
[24] Visser, J., *The concept of negative hamaker coefficients. 1. history and present status.* Advances in Colloid and Interface Science, 1981. 15(2): p. 157-169.
[25] Pomogailo, A. and Kestelman, V.N., *Metallopolymer nanocomposites*. Springer series in materials science, 81. 2005, New York.
[26] Henderson, J., *The analysis of ancient glasses part II: Luxury Roman and early medieval glasses.* JOM Journal of the Minerals, Metals and Materials Society, 1996. 48(2): p. 62-64.
[27] Walter, P., Welcomme, E., Hallégot, P., Zaluzec, N.J., Deeb, C., Castaing, J., Veyssière, P., Bréniaux, R., Lévêque, J.-L. and Tsoucaris, G., *Early Use of PbS Nanotechnology for an Ancient Hair Dyeing Formula.* Nano Letters, 2006. 6(10): p. 2215-2219.
[28] Meurdrac, M. and Jacques, J., *La chymie charitable et facile en faveur des dames : 1666*. 1999, Paris: CNRS.
[29] Diderot, D.D.A., M., *Encyclopédie ou Dictionnaire raisonné des sciences, des arts et des métiers*. 1751: Tome troisième: Paris.
[30] Mirguet, C., Fredrickx§, P., Sciau, P. and Colomban, P., *Origin of the self-organisation of Cu°/Ag° nanoparticles in ancient lustre pottery. A TEM study.* Phase Transitions, 2008. 81(2-3): p. 253-266.
[31] Mills, G.F., *Ammine Exchange Resins.* Ind. Eng. Chem, 1949. 41(12).
[32] Domènech, B., Muñoz, M., Muraviev, D.N. and Macanás, J., *Catalytic membranes with palladium nanoparticles: From tailored polymer to catalytic applications.* Catalysis Today, (0).
[33] Park, J.-I., Kim, M.G., Jun, Y.-w., Lee, J.S., Lee, W.-r. and Cheon, J., *Characterization of Superparamagnetic "Core–Shell" Nanoparticles and Monitoring Their Anisotropic Phase Transition to Ferromagnetic "Solid Solution" Nanoalloys.* Journal of the American Chemical Society, 2004. 126(29): p. 9072-9078.

[34] Bastos-Arrieta, J., Shafir, A., Alonso, A., Muñoz, M., Macanás, J. and Muraviev, D.N., *Donnan exclusion driven intermatrix synthesis of reusable polymer stabilized palladium nanocatalysts.* Catalysis Today, (0).

[35] Alonso, A., Macanas, J., Shafir, A., Munoz, M., Vallribera, A., Prodius, D., Melnic, S., Turta, C. and Muraviev, D.N., *Donnan-exclusion-driven distribution of catalytic ferromagnetic nanoparticles synthesized in polymeric fibers.* Dalton Transactions, 2010. 39(10): p. 2579-2586.

[36] Muraviev, D.N., Macanás, J., Farre, M., Muñoz, M. and Alegret, S., *Novel routes for intermatrix synthesis and characterization of polymer stabilized metal nanoparticles for molecular recognition devices.* Sensors and Actuators B: Chemical, 2006. 118(1–2): p. 408-417.

[37] Alonso, A., Shafir, A., Macanás, J., Vallribera, A., Muñoz, M. and Muraviev, D.N., *Recyclable polymer-stabilized nanocatalysts with enhanced accessibility for reactants.* Catalysis Today, (0).

[38] Barbaro, P. and Liguori, F., *Ion Exchange Resins: Catalyst Recovery and Recycle.* Chemical Reviews, 2008. 109(2): p. 515-529.

[39] Harland, C.E. and Grimshaw, R.W., *Ion exchange : theory and practice*. 1994, Cambridge: Royal Society of Chemistry.

[40] Myers, R.J., Eastes, J.W. and Myers, F.J., *Synthetic Resins as Exchange Adsorbents.* Industrial & Engineering Chemistry, 1941. 33(6): p. 697-706.

[41] Abrams, I.M. and Millar, J.R., *A history of the origin and development of macroporous ion-exchange resins.* Reactive and Functional Polymers, 1997. 35(1–2): p. 7-22.

[42] Xu, T., *Ion exchange membranes: State of their development and perspective.* Journal of Membrane Science, 2005. 263(1–2): p. 1-29.

[43] Sata, T., *Ion Exchange Membranes: Preparation, Characterization, Modification and Application*. 2004, Cambridge: The Royal Society of Chemistry 2004.

[44] Huang, R.Y.M., Shao, P., Burns, C.M. and Feng, X., *Sulfonation of poly(ether ether ketone)(PEEK): Kinetic study and characterization.* Journal of Applied Polymer Science, 2001. 82(11): p. 2651-2660.

[45] Xing, P., Robertson, G.P., Guiver, M.D., Mikhailenko, S.D., Wang, K. and Kaliaguine, S., *Synthesis and characterization of sulfonated poly(ether ether ketone) for proton exchange membranes.* Journal of Membrane Science, 2004. 229(1–2): p. 95-106.

[46] Domènech, B., Muñoz, M., Muraviev, D.N. and Macanás, J., *Polymer-stabilized palladium nanoparticles for catalytic membranes: ad hoc polymer fabrication.* Nanoscale Research Letters, 2011. 6(406).

[47] Blanco, J.F., Nguyen, Q.T. and Schaetzel, P., *Novel hydrophilic membrane materials: sulfonated polyethersulfone Cardo.* Journal of Membrane Science, 2001. 186(2): p. 267-279.

[48] Blanco, J.-F., Sublet, J., Nguyen, Q.T. and Schaetzel, P., *Formation and morphology studies of different polysulfones-based membranes made by wet phase inversion process.* Journal of Membrane Science, 2006. 283(1–2): p. 27-37.

[49] Maynard, R.L. and Howard, C.V., *Particulate Matter: Properties and Effects upon Health.* 1999, Oxford: Bios Scientific Publishers.

[50] Bottero, J.-Y., Rose, J. and Wiesner, M.R., *Nanotechnologies: Tools for Sustainability in a New Wave of Water Treatment Processes.* Integrated Environmental Assessment and Management, 2006. 2(4): p. 391-395.

[51] Borm, P.J.A. and Berube, D., *A tale of opportunities, uncertainties, and risks.* Nano Today, 2008. 3(1–2): p. 56-59.

[52] Abbott, L.C. and Maynard, A.D., *Exposure Assessment Approaches for Engineered Nanomaterials.* Risk Analysis, 2010. 30(11): p. 1634-1644.

[53] Schulenburg M., N.s.t., big effects Opportunities and risks., *Nanoparticles – small things, big effects Opportunities and risks.* . 2008, Bonn: Bundesministerium für Bildung und Forschung (BMBF) / Federal Ministry of

[54] n and Research.

[55] Vatta, L.L., Sanderson, R.D. and Koch, K.R., *Magnetic nanoparticles: Properties and*

[56] *applications.* Pure Appl. Chem., 2006. 78(9).

[57] Talapatra, S., Ganesan, P.G., Kim, T., Vajtai, R., Huang, M., Shima, M., Ramanath, G., Srivastava, D., Deevi, S.C. and Ajayan, P.M., *Irradiation-Induced Magnetism in Carbon Nanostructures.* Physical Review Letters, 2005. 95(9): p. 097201.

[58] Hyeon, T., *Chemical synthesis of magnetic nanoparticles.* Chemical Communications, 2003(8): p. 927-934.

[59] Ding, Y., Hu, Y., Zhang, L., Chen, Y. and Jiang, X., *Synthesis and Magnetic Properties of Biocompatible Hybrid Hollow Spheres.* Biomacromolecules, 2006. 7(6): p. 1766-1772.

[60] Bower, N.W., *Principles of Instrumental Analysis. 4th edition (Skoog, D. A.; Leary, J. J.).* Journal of Chemical Education, 1992. 69(8): p. A224.

[61] Schmid, G., *Conclusions and Perspectives*, in *Nanoparticles*. 2010, Wiley-VCH Verlag GmbH & Co. KGaA. p. 513-515.

[62] Fendler, J.H. and Tian, Y., *Nanoparticles and Nanostructured Films: Current Accomplishments and Future Prospects*, in *Nanoparticles and Nanostructured Films*. 2007, Wiley-VCH Verlag GmbH. p. 429-461.

[63] Plieth, W.J., *Electrochemical properties of small clusters of metal atoms and their role in the surface enhanced Raman scattering.* The Journal of Physical Chemistry, 1982. 86(16): p. 3166-3170.

[64] Weller, H., *Clusters and Colloids: From Theory to Applications. Edited by G. Schmid, VCH, Weinheim 1994, 555 pp., hardcover, DM 248.00, ISBN 3–527–29043–5.* Advanced Materials, 1995. 7(1): p. 94-94.

[65] Milev, A.S., Kannangara, G.S.K., Wilson, M.A., Sydney, U.o.W., College of Science, T., Environment, School of Science, F. and Horticulture, *Nanotechnology*.

[66] Lane, N., *Nanotechnologies meet market realities.* Chem Eng News 2002: p. 17.

[67] Hillie, T. and Hlophe, M., *Nanotechnology and the challenge of clean water.* Nat Nano, 2007. 2(11): p. 663-664.

[68] Ju-Nam, Y. and Lead, J.R., *Manufactured nanoparticles: An overview of their chemistry, interactions and potential environmental implications.* Science of The Total Environment, 2008. 400(1–3): p. 396-414.

[69] Narayan, R., *Use of nanomaterials in water purification.* Materials Today, 2010. 13(6): p. 44-46.

[70] Theron, J., Walker, J.A. and Cloete, T.E., *Nanotechnology and Water Treatment: Applications and Emerging Opportunities.* Critical Reviews in Microbiology, 2008. 34(1): p. 43-69.

[71] Li, W.R., Xie, X.B., Shi, Q.S., Zeng, H.Y., Ou-Yang, Y.S. and Chen, Y.B., *Antibacterial activity and mechanism of silver nanoparticles on Escherichia coli.* Applied microbiology and biotechnology, 2010. 85(4): p. 1115-1122.
[72] Kelly, K.L., Coronado, E., Zhao, L.L. and Schatz, G.C., *The Optical Properties of Metal Nanoparticles: The Influence of Size, Shape, and Dielectric Environment.* The Journal of Physical Chemistry B, 2002. 107(3): p. 668-677.
[73] Kumar, R., Howdle, S. and Münstedt, H., *Polyamide/silver antimicrobials: Effect of filler types on the silver ion release.* Journal of Biomedical Materials Research Part B: Applied Biomaterials, 2005. 75B(2): p. 311-319.
[74] De Gusseme, B., Sintubin, L., Baert, L., Thibo, E., Hennebel, T., Vermeulen, G., Uyttendaele, M., Verstraete, W. and Boon, N., *Biogenic Silver for Disinfection of Water Contaminated with Viruses.* Applied and Environmental Microbiology, 2010. 76(4): p. 1082-1087.
[75] Dibrov, P., Dzioba, J., Gosink, K.K. and Häse, C.C., *Chemiosmotic Mechanism of Antimicrobial Activity of Ag+ in Vibrio cholerae.* Antimicrobial Agents and Chemotherapy, 2002. 46(8): p. 2668-2670.
[76] Lok, C.-N., Ho, C.-M., Chen, R., He, Q.-Y., Yu, W.-Y., Sun, H., Tam, P.K.-H., Chiu, J.-F. and Che, C.-M., *Proteomic Analysis of the Mode of Antibacterial Action of Silver Nanoparticles.* Journal of Proteome Research, 2006. 5(4): p. 916-924.
[77] Silvestry-Rodriguez, N., Bright, K.R., Slack, D.C., Uhlmann, D.R. and Gerba, C.P., *Silver as a Residual Disinfectant To Prevent Biofilm Formation in Water Distribution Systems.* Applied and Environmental Microbiology, 2008. 74(5): p. 1639-1641.
[78] Ruparelia, J.P., Chatterjee, A.K., Duttagupta, S.P. and Mukherji, S., *Strain specificity in antimicrobial activity of silver and copper nanoparticles.* Acta Biomaterialia, 2008. 4(3): p. 707-716.
[79] Charnley, M., Textor, M. and Acikgoz, C., *Designed polymer structures with antifouling–antimicrobial properties.* Reactive and Functional Polymers, 2011. 71(3): p. 329-334.
[80] Murthy, P.S.K., Murali Mohan, Y., Varaprasad, K., Sreedhar, B. and Mohana Raju, K., *First successful design of semi-IPN hydrogel–silver nanocomposites: A facile approach for antibacterial application.* Journal of Colloid and Interface Science, 2008. 318(2): p. 217-224.
[81] Childs, W.R., Motala, M.J., Lee, K.J. and Nuzzo, R.G., *Masterless Soft Lithography: Patterning UV/Ozone–Induced Adhesion on Poly(dimethylsiloxane) Surfaces†.* Langmuir, 2005. 21(22): p. 10096-10105.
[82] D.N., M. 2010.
[83] Bonet, F., Grugeon, S., Herrera Urbina, R., Tekaia-Elhsissen, K. and Tarascon, J.M., *In situ deposition of silver and palladium nanoparticles prepared by the polyol process, and their performance as catalytic converters of automobile exhaust gases.* Solid State Sciences, 2002. 4(5): p. 665-670.
[84] Liu, G., Gu, H., Sun, Y., Long, J., Xu, Y. and Li, H., *Magnetically Recoverable Nanoparticles: Highly Efficient Catalysts for Asymmetric Transfer Hydrogenation of Aromatic Ketones in Aqueous Medium.* Advanced Synthesis & Catalysis, 2011. 353(8): p. 1317-1324.
[85] Kidambi, S., Dai, J., Li, J. and Bruening, M.L., *Selective Hydrogenation by Pd Nanoparticles Embedded in Polyelectrolyte Multilayers.* Journal of the American Chemical Society, 2004. 126(9): p. 2658-2659.

[86] Brock, S.L. and Senevirathne, K., *Recent developments in synthetic approaches to transition metal phosphide nanoparticles for magnetic and catalytic applications.* Journal of Solid State Chemistry, 2008. 181(7): p. 1552-1559.

[87] Wilson, O.M., Knecht, M.R., Garcia-Martinez, J.C. and Crooks, R.M., *Effect of Pd Nanoparticle Size on the Catalytic Hydrogenation of Allyl Alcohol.* Journal of the American Chemical Society, 2006. 128(14): p. 4510-4511.

[88] Umpierre, A.P., Machado, G., Fecher, G.H., Morais, J. and Dupont, J., *Selective Hydrogenation of 1,3-Butadiene to 1-Butene by Pd(0) Nanoparticles Embedded in Imidazolium Ionic Liquids.* Advanced Synthesis & Catalysis, 2005. 347(10): p. 1404-1412.

[89] Durand, J., Teuma, E. and Gómez, M., *An Overview of Palladium Nanocatalysts: Surface and Molecular Reactivity.* European Journal of Inorganic Chemistry, 2008. 2008(23): p. 3577-3586.

[90] Moreno-Mañas, M. and Pleixats, R., *Formation of Carbon–Carbon Bonds under Catalysis by Transition-Metal Nanoparticles.* Accounts of Chemical Research, 2003. 36(8): p. 638-643.

[91] Torborg, C. and Beller, M., *Recent Applications of Palladium-Catalyzed Coupling Reactions in the Pharmaceutical, Agrochemical, and Fine Chemical Industries.* Advanced Synthesis & Catalysis, 2009. 351(18): p. 3027-3043.

[92] Diéguez, M., Pàmies, O., Mata, Y., Teuma, E., Gómez, M., Ribaudo, F. and van Leeuwen, P.W.N.M., *Palladium Nanoparticles in Allylic Alkylations and Heck Reactions: The Molecular Nature of the Catalyst Studied in a Membrane Reactor.* Advanced Synthesis & Catalysis, 2008. 350(16): p. 2583-2598.

[93] Hoffmann, R., Imamura, A. and Hehre, W.J., *Benzynes, dehydroconjugated molecules, and the interaction of orbitals separated by a number of intervening sigma bonds.* Journal of the American Chemical Society, 1968. 90(6): p. 1499-1509.

[94] de Dios, A.S. and Díaz-García, M.E., *Multifunctional nanoparticles: Analytical prospects.* Analytica Chimica Acta, 2010. 666(1–2): p. 1-22.

[95] Hu, A., Yee, G.T. and Lin, W., *Magnetically Recoverable Chiral Catalysts Immobilized on Magnetite Nanoparticles for Asymmetric Hydrogenation of Aromatic Ketones.* Journal of the American Chemical Society, 2005. 127(36): p. 12486-12487.

[96] Gleeson, O., Tekoriute, R., Gun'ko, Y.K. and Connon, S.J., *The First Magnetic Nanoparticle-Supported Chiral DMAP Analogue: Highly Enantioselective Acylation and Excellent Recyclability.* Chemistry – A European Journal, 2009. 15(23): p. 5669-5673.

[97] M., P.C., S., K.M. and J., P., *Palladium Nanoparticles in Polymers: Catalyst for Alkene Hydrogenation, Carbon–Carbon Cross-Coupling Reactions, and Aerobic Alcohol Oxidation.* Synthesis, 2006.

[98] C., Z., D., X., W., J.-w. and X., H., *An efficient and recyclable ionic liquid-supported proline catalyzed Knoevenagel Condensation.* ISRN Organic Chemistry, 2011: p. 5.

Chapter 4

Preparation of Ionic Polysilsesquioxanes with Regular Structures and Their Ion-Exchange Behaviors

Yoshiro Kaneko

Additional information is available at the end of the chapter

http://dx.doi.org/10.5772/45732

1. Introduction

Silsesquioxanes (SQs: $RSiO_{1.5}$), one class of the materials containing Si–O framework, have attracted much attention in the research fields of materials chemistry for academic and application reasons (Baney et al., 1995; Loy et al., 2000) because they can contain the various functional groups as side chains (R) and are inorganic materials indicating the remarkable compatibility with organic materials such as polymers (Choi et al., 2001; Kim & Chujo, 2003; Yu et al., 2010; Wang et al., 2011), in addition to having superior thermal, mechanical, and chemical properties derived from Si–O framework with high bond energy compared with C–C bonds.

However, even though various kinds of polyhedral oligomeric SQs (POSS) are known as SQs with controlled molecular structures (Scheme 1b) (Feher & Wyndham, 1998; Laine et al., 1998; Zhang et al., 2007; Cordes et al., 2010; Tanaka et al., 2010; Tanaka & Chujo, 2012), the regularly structured polySQs (PSQs) have only been obtained in the limited cases. This is because PSQs are prepared by polycondensation of the trifunctional silane monomers such as organotrialkoxysilanes and organotrichlorosilanes. These multifunctional monomers generally result in the formation of insoluble polymers with irregular three-dimensional network structures of Si–O–Si bonds (Scheme 1a) (Loy et al., 2000). If the molecular, conformational, and higher-ordered structures of PSQs can be controlled, they are expected to be applicable for a wide range of materials research fields.

Ladder-structured PSQs are one of a few PSQs with controlled molecular structures (Scheme 1c) (Brown et al., 1960; Brown et al., 1964; Unno et al., 2005; Zhang et al., 2006; Seki et al., 2010). These structures are classified into two types: "perfect ladder structure" (Scheme 1c-i) and "ladder-like structure" (Scheme 1c-ii) (Abe & Gunji, 2004). The latter has slight defects

INTECH
open science | open minds

© 2012 Kaneko, licensee InTech. This is a paper distributed under the terms of the Creative Commons Attribution License (http://creativecommons.org/licenses/by/3.0), which permits unrestricted use, distribution, and reproduction in any medium, provided the original work is properly cited.

in Si–O–Si bond main-chain. Even though oligomeric SQ with "perfect ladder structure" could be synthesized by a step-by-step method and its characterization methods were established (Unno et al., 2002), it is difficult to prepare PSQs with such structures and to characterize them. In most cases, the ladder-structured PSQs probably do not have "perfect ladder structure" but have "ladder-like structure" as shown in Scheme 1c-ii.

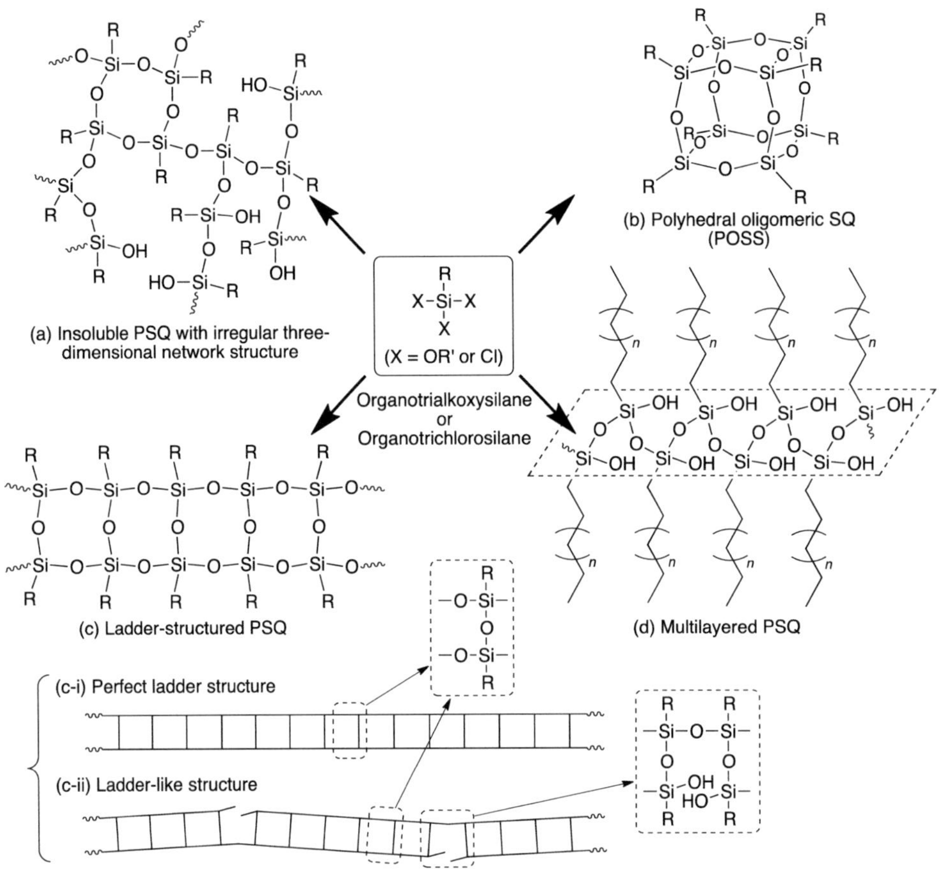

Scheme 1. Preparation of silsesquioxanes (SQs) with various structures.

The ladder-like PSQs exhibit rigidity and anisotropy in addition to the aforementioned superior physical properties of SQs because they have one-dimensional network structures of Si–O–Si bonds. Furthermore, ladder-like PSQs are colorless materials due to no absorptions of Si–O–Si bonds in the visible light region and can be used as transparent solutions owing to good solubility in any solvents. These properties of ladder-like PSQs would be useful for versatile applications, in particular, as inorganic fillers due to hybridization with organic functional materials.

Furthermore, control of the higher-ordered structures (nanostructures) of PSQs would also be significant to apply to various supramolecular organic–inorganic hybrid materials. For example, it has been reported that trifunctional silane monomers containing long alkyl chains were hydrolyzed to form amphiphilic molecules having silanol groups, resulting in the formation of multilayered PSQs by polycondensation (Scheme 1d) (Parikh et al., 1997; Shimojima et al., 1997). Another method for controlling higher-ordered structures of PSQs is sol–gel reaction (hydrolytic polycondensation) of 1,4-bis(trialkoxysilyl)benzene as a monomer in the presence of surfactants (Inagaki et al., 2002). The resulting material has a hexagonal array of mesopores and crystal-like frameworks. Self-organization of long alkyl chains by hydrophobic interactions is a driving force to form such regular higher-ordered structures.

From the aforementioned background, it is evident that the development of PSQs with regularly controlled molecular and higher-ordered structures is one of the important issues for research fields of SQs. In this chapter, therefore, the author first describes the preparation of cationic PSQs with controlled molecular and higher-ordered structures by simple sol–gel method and their detailed characterizations in section 2. The resulting PSQs indicated the anion-exchange properties due to existence of cationic functional groups as side-chains. Then, the author also describes the anion-exchange behaviors with various organic and inorganic compounds such as anionic surfactants (section 3), a polymer (section 4), layered clay minerals (section 5), and a dye molecule (section 6) to obtain the functional hybrid materials.

2. Preparation of ammonium group-containing ladder-like PSQs with hexagonally stacked structures

The author and co-workers have developed a preparation method for ammonium group-containing PSQs with controlled molecular and higher-ordered structures by hydrolytic polycondensation (sol–gel reaction) of organotrialkoxysilane monomers containing amino groups in aqueous inorganic acids (Kaneko et al., 2004a; Kaneko et al., 2005a; Kaneko & Iyi, 2007; Kaneko & Iyi, 2010).

2.1. Sol–gel reaction of 3-aminopropyltrimethoxysilane

Preparation of ladder-like PSQs with hexagonally stacked structures was achieved by sol–gel reaction of organotrimethoxysilanes containing amino groups. The first example of monomer to form these materials was 3-aminopropyltrimethoxysilane (APTMOS) (Kaneko et al., 2004a).

The sol–gel reaction of APTMOS was performed by stirring in an acid, *e.g.*, a hydrochloric acid, aqueous solution at room temperature for 2 h, followed by heating (*ca.* 50–60 ºC) in an open system until the solvent was completely evaporated (Scheme 2). The resulting product was dissolved in water and this aqueous solution was lyophilized to obtain ammonium chloride group-containing PSQ (PSQ-$NH_3^+Cl^-$). Here, a feed molar ratio of the acid to APTMOS is very important factor for the formation of the regular structures of PSQ. The higher-ordered structure of the product was characterized by the X-ray diffraction (XRD), transmission electron microscopy (TEM), scanning electron microscopy (SEM), and nitrogen adsorption–desorption isotherm measurements, and the molecular structure was characterized by ^{29}Si NMR, XRD, and static light scattering (SLS) measurements.

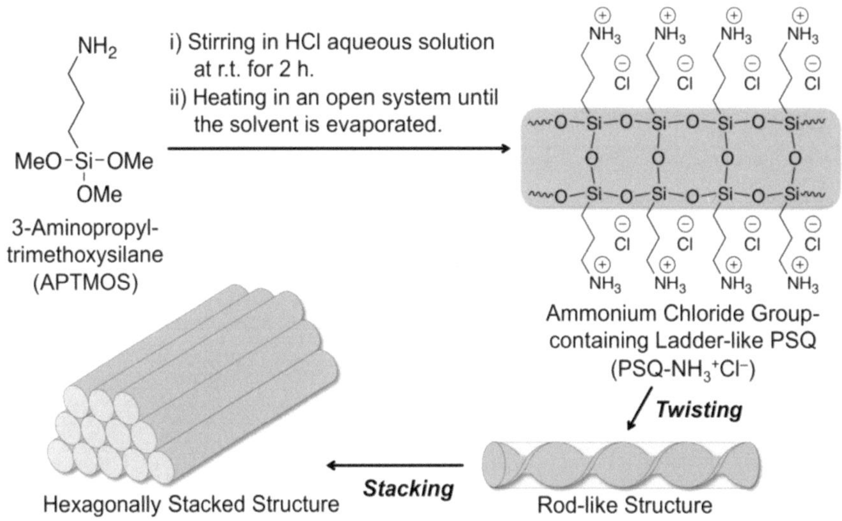

Scheme 2. Preparation of an ammonium chloride group-containing ladder-like PSQ (PSQ-$NH_3^+Cl^-$) with a hexagonally stacked structure by sol–gel method.

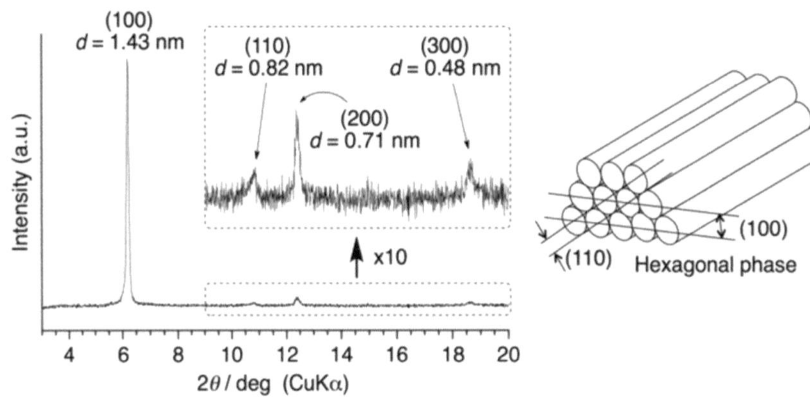

Figure 1. XRD pattern of PSQ-$NH_3^+Cl^-$. Relative humidity (RH) during XRD measurements was 50%.

RH (%)	*d*-Value		
	(100)	(110)	(200)
20	1.35	–	–
30	1.37	0.79	0.69
40	1.40	0.81	0.70
50	1.43	0.82	0.71
60	1.45	0.84	0.73
70	1.48	0.86	0.74
80	1.52	0.88	0.76
90	1.58	0.92	0.79

Table 1. *d*-Values of diffraction peaks in the XRD patterns of PSQ-$NH_3^+Cl^-$ under various RH conditions.

For the XRD measurements, the films of the products on the glasses were obtained by drying the aqueous product solutions spread on flat glass substrates. The XRD pattern of PSQ-$NH_3^+Cl^-$ film showed diffraction peaks with the *d*-value ratio of 1 : 1/√3 : 1/2 : 1/3, strongly indicating that PSQ-$NH_3^+Cl^-$ had a hexagonal phase (Fig. 1). These three peaks were assigned to the (100), (110), (200), and (300) peaks, respectively. However, based on only these data, it could not be determined whether this hexagonal phase originated from a porous-type structure or a stacking of a rod-like polymer. Therefore, the influence of relative humidity (RH) was investigated on the *d*-value in the XRD measurements of PSQ-$NH_3^+Cl^-$. As shown in Table 1, the *d*-values of the diffraction peaks changed by varying RH, *i.e.*, the *d*-value increased for a high RH and decreased for a low RH, although the *d*-value ratios of (110)/(100) and (200)/(100) did not change. Such a behavior cannot be expected for hexagonal-structured porous materials. Therefore, it was assumed that this hexagonal phase originated from the stacking of rod-like polymers.

The TEM image of the product showed a stripe pattern, indicating that the rod-like PSQs were stacked parallel. In addition, the SEM image showed the aggregate, which lined up in a regular direction. This indicates that nano-ordered regular structure of PSQ affected the structures in the micro-ordered aggregate. Furthermore, the real image of PSQ-$NH_3^+Cl^-$ obtained by lyophilization seemed to be reflected by the regular nano- and micro-structures. The Brunauer–Emmett–Teller (BET) surface area of the product evaluated by the nitrogen adsorption–desorption isotherm measurement at 77K was much too small (*ca.* 6 m^2/g), indicating that the rod-like PSQ stacked densely. These results supported the regular higher-ordered structure of PSQ-$NH_3^+Cl^-$ characterized by the aforementioned XRD measurements, *i.e.*, rod-like structure forming hexagonal phase.

So far, decisive analysis methods to prove the regular molecular structure such as ladder-like structure of PSQ have not been established. Therefore, necessary evidences for confirmation of the structures were collected by performing multiple analyses. The ladder-like PSQ simultaneously satisfies the following conditions: i) to be soluble in any solvents, ii) relatively high molecular weight, iii) relatively narrow molecular width, and iv) observation of large T^3 peak and small T^2 peak in ^{29}Si NMR spectrum.

The PSQ-$NH_3^+Cl^-$ had a rod-like structure with relatively small diameter (*ca.* 1.6–1.7 nm, estimated from *d*-value of (100) peak in XRD pattern of Fig. 1) in spite of forming highly dense Si–O–Si bond network structure, which was confirmed by the observation of large T^3 peak in the ^{29}Si NMR spectrum of Fig. 2. In addition, the PSQ-$NH_3^+Cl^-$ was soluble in water, despite its M_w was relatively high (*ca.* 12000, estimated by *Zimm* plot method using the SLS), which indicated no formation of three-dimensional network structure. These results satisfy the aforementioned conditions. All these things considered, it is reasonable to assume that the present PSQ has one-dimensional ladder-like molecular structure as shown in Scheme 1c-ii.

It was supposed that ladder-like PSQ-$NH_3^+Cl^-$ was twisted to form rigid rod-like structure in the solid state, resulting in the formation of the hexagonally stacked structure. The driving force for the formation of the twisted conformation is probably intramolecular charge repulsion between the ammonium cations in side-chain groups of PSQ-$NH_3^+Cl^-$. Therefore, to investigate correlation between intramolecular charge repulsion and regular structure of PSQ-$NH_3^+Cl^-$, the XRD measurements were performed with changing RH. Stability of the ammonium cations is affected with RH because of hydration with water molecules. The XRD patterns of PSQ-$NH_3^+Cl^-$ with RH higher than 30% indicated three diffraction peaks derived from hexagonal phase, meanwhile those with RH lower than 20% did not show such the diffraction peaks. Because these results indicate the presence of correlation between the formation of the hexagonally stacked structure and higher RH, it is assumed that the twisted structure caused by intramolecular charge repulsion between the ammonium cations is plausible conformation of PSQ-$NH_3^+Cl^-$.

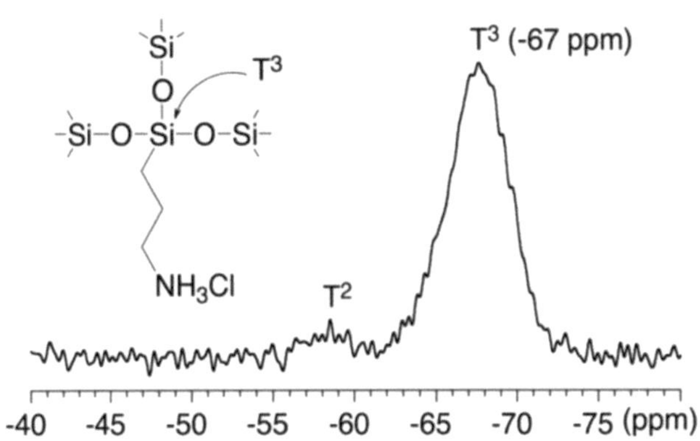

Figure 2. Solid-state ^{29}Si NMR spectrum of PSQ-$NH_3^+Cl^-$.

On the basis of the aforementioned results, it was considered that self-organization of the ion pair composed of ammonium cations in the side-chains of the PSQ and chloride anions was the driving force for the formation of regular molecular, conformational, and higher-ordered structures.

2.2. Sol–gel reaction of (3-(2-aminoethylamino)propyl)trimethoxysilane

As another monomer, organotrialkoxysilane containing two amino groups, *i.e.*, (3-(2-aminoethylamino)propyl)trimethoxysilane (AEAPTMOS), was employed for the preparation of rod-like PSQ with hexagonally stacked structure (Kaneko et al., 2005a). The procedure for sol–gel reaction of AEAPTMOS was similar to that of APTMOS.

The XRD pattern of the resulting product film showed three diffraction peaks with the *d*-value ratio of $1 : 1/\sqrt{3} : 1/2$, indicating that the product had a hexagonal phase. The *d*-values of the diffraction peaks were shifted by varying the RH, although the *d*-value ratios of (110)/(100) and (200)/(100) did not change (Table 2). The same behavior was observed with the aforementioned PSQ-$NH_3^+Cl^-$ (Table 1). Therefore, the PSQ containing double-ammonium groups in one repeating unit also had a hexagonal phase, which originated from the stacking of rod-like polymers. The *d*-value of (100) peak of this PSQ (1.85 nm for RH of *ca.* 50%) was larger than that of PSQ-$NH_3^+Cl^-$ (1.43 nm for RH of *ca.* 50%). This is because of difference in side-chain lengths between these PSQs.

RH (%)	*d*-Value		
	(100)	(110)	(200)
20	1.68	0.97	0.84
30	1.74	1.01	0.87
40	1.80	1.04	0.89
50	1.85	1.07	0.93
60	1.93	1.12	0.97
70	2.00	1.16	0.99
80	2.06	1.20	1.03
90	2.15	–	–

Table 2. *d*-Values of diffraction peaks in the XRD patterns of the PSQ obtained from AEAPTMOS under various RH conditions.

3. Ion-exchange behaviors of ladder-like PSQ with fatty acids

Because PSQ-$NH_3^+Cl^-$ has ammonium groups as side-chains and chloride anions (Cl^-) as counterions, the anion-exchange property can be expected. Therefore, the ion-exchange reaction of the PSQs was performed with anionic organic compounds and the higher-ordered structures of the resulting products were characterized. First, a fatty acid salt such as *n*-octanoic acid sodium salt was employed as an anionic compound (Scheme 3a) (Kaneko et al., 2004a). By pouring PSQ-$NH_3^+Cl^-$ aqueous solution into aqueous solution of *n*-octanoic acid sodium salt, precipitation immediately occurred. The XRD pattern of the resulting water-insoluble product showed peaks for a typical hexagonal phase and the *d*-value of (100) peak increase more than those of the original PSQs (PSQ-$NH_3^+Cl^-$), indicating that the diameter of the rod-like PSQ increased when the Cl^- as the counterion was exchanged with the bulky *n*-octanoate. Thus, the hexagonal phase of the rod-like PSQ was maintained, in spite of the increase in the *d*-value by ion-exchange reaction with *n*-octanoic acid sodium salt.

On the other hand, when the ion-exchange reactions were performed using the fatty acid salts containing longer alkyl chains (*n*-decanoic acid sodium salt, *n*-dodecanoic acid sodium salt, and *n*-tetradecanoic acid sodium salt), the peaks due to the typical hexagonal phase were not obtained. This is because the hydrophobic interaction between the guest fatty acid salts containing longer alkyl chains was too strong to maintain the hexagonally stacked structure of rod-like PSQ.

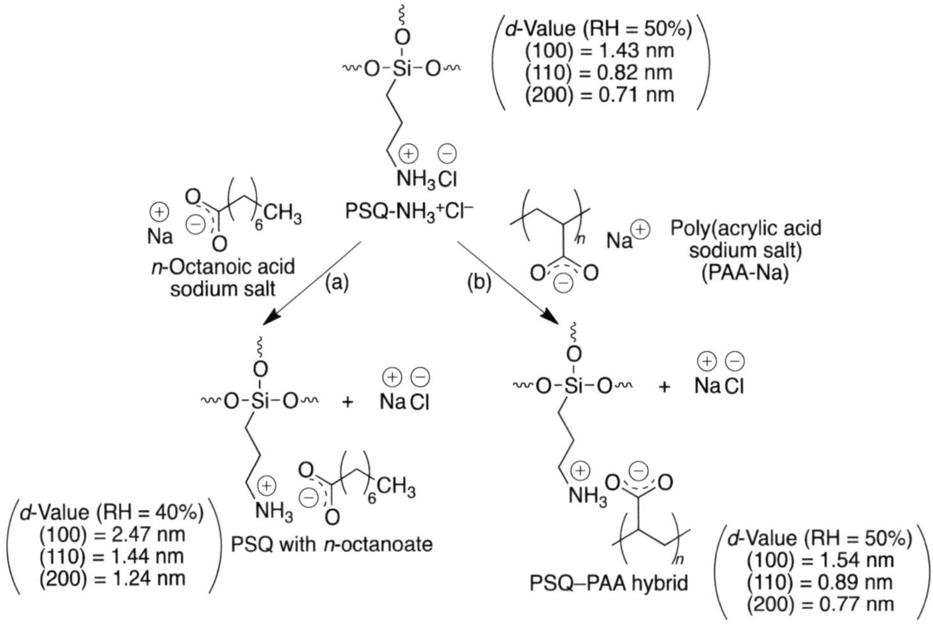

Scheme 3. Ion-exchange reaction of PSQ-$NH_3^+Cl^-$ with (a) *n*-octanoic acid sodium salt and (b) poly(acrylic acid sodium salt).

4. Preparation of organic–inorganic polymer hybrids by ion-exchange reaction of ladder-like PSQ with organic polymer

Hybrids composed of organic and inorganic materials usually exhibit improved performance properties compared with conventional composites, mixtures on a micrometer scale (μm), due to their unique phase morphology and improved interfacial properties. For these reasons, nanostructured organic–inorganic hybrids have attracted considerable attention from both fundamental research and applications points of view (Usuki et al., 1993; Chujo, 1996). In particular, the organic–inorganic hybrids obtained from the synthetic polymers as organic species are important industrial materials.

In this section, the author describes the preparation of an organic–inorganic polymer hybrid (PSQ–PAA hybrid) with regular higher-ordered structure composed of the aforementioned PSQ-$NH_3^+Cl^-$ as the inorganic species and poly(acrylic acid sodium salt) (PAA-Na) as the organic species by the ion-exchange reaction (Kaneko et al., 2005b). To obtain PSQ–PAA hybrid, the ion-exchange reaction was performed by pouring PSQ-$NH_3^+Cl^-$ aqueous solution into PAA-Na aqueous solution (Scheme 3b).

The IR spectrum of PSQ–PAA hybrid displayed absorptions at 1044 and 1144 cm^{-1} assigned to the Si–O bond of the PSQ, and absorptions at 1407 and 1558 cm^{-1} due to the COO^- of the PAA, indicating that the product consisted of both organic and inorganic polymers. The CHN elemental analysis data showed that the C/N molar ratio for the product was 5.92. From this value, the ratio of the functional groups of the two polymers, *i.e.*, NH_3^+ and COO^-, was calculated to be *ca.* 1 : 1.

The XRD pattern of the resulting insoluble product showed formation of a hexagonal phase and the *d*-values of the diffraction peaks were different from those of PSQ-$NH_3^+Cl^-$(Scheme 3b), indicating formation of not a macroscopic mixture but the molecular-scale mixture of two polymers, *i.e.*, organic–inorganic hybrid.

5. Preparation of nanoporous materials composed of ladder-like PSQs and layered clay minerals by ion-exchange reaction

There have been intense research activities on layered silicates pillared with inorganic or organic clusters, which are called pillared interlayer clays (PILCs) (Otsuka, 1997; Kloprogge, 1998; Ding et al., 2001). PILCs have much higher surface areas and pore volumes than the original clays. Such properties make them useful catalysts, ion-exchangers, and adsorbents. These materials are normally prepared by the intercalative ion-exchange of layered clay minerals with a variety of nano-sized pillars, such as organic ions (Ogawa et al., 1994), inorganic ions (Pichowicz & Mokaya, 2004), and sol particles (Yamanaka et al., 1992).

Even though the preparation of various polymers/clay hybrids (Pinnavaia & Beall, 2000) has already been reported, there have been a few studies on the preparation of PILCs using polymers. Because the polymers usually have a flexible structure, expansion of the interlayer space of clays by polymer incorporation is not enough for providing more free space.

The aforementioned PSQ-$NH_3^+Cl^-$ has motivated the author and co-workers to develop new PILCs, because it has rigidity and bulkiness to expand the interlayer of clays and ability to intercalate molecules into the interlayer of anionic clays by ion-exchange reaction due to the presence of ammonium cations in side-chains of PSQ-$NH_3^+Cl^-$. In this section, therefore, the author describes the preparation of a clay pillared with rod-like cationic PSQ-$NH_3^+Cl^-$ (Scheme 4) (Kaneko et al., 2004b).

The synthesis was performed by pouring the PSQ-$NH_3^+Cl^-$ aqueous solution into an aqueous suspension of Na-saturated saponite (Na-SAP). After the mixture was vigorously stirred for 2 h at room temperature, the product was collected by filtration, washed with water, and then dried under reduced pressure at room temperature to obtain the clay pillared with PSQ-$NH_3^+Cl^-$ (PSQ–SAP). The resulting product was characterized by IR, CHN elemental analysis, XRD, and nitrogen adsorption–desorption isotherm measurements.

The IR spectrum of PSQ–SAP showed an absorption band at *ca.* 1515 cm^{-1} assigned to the ammonium ion of the PSQ component, indicating that the PSQ was inserted into the interlayer of SAP. On the basis of the CHN analysis data, the exchange amount of a repeating unit of the PSQ component in PSQ–SAP was calculated to be 126 meq/100 g SAP. This value is higher than that of the cation exchange capacity (CEC) of Na-SAP (92 meq/100 g SAP) (Bujdák et al., 2002). The distance between the charges of PSQ would be shorter than that of SAP. Therefore, excess ammonium groups and counter anions (Cl^-) of PSQ-$NH_3^+Cl^-$ were inserted into the interlayer of SAP, which was confirmed by a Cl elemental analysis. The XRD pattern of PSQ–SAP was completely different from that of Na-SAP and PSQ-$NH_3^+Cl^-$. Accordingly, PSQ–SAP was not a mixture, but an intercalated nano-order material, *i.e.*, a hybrid.

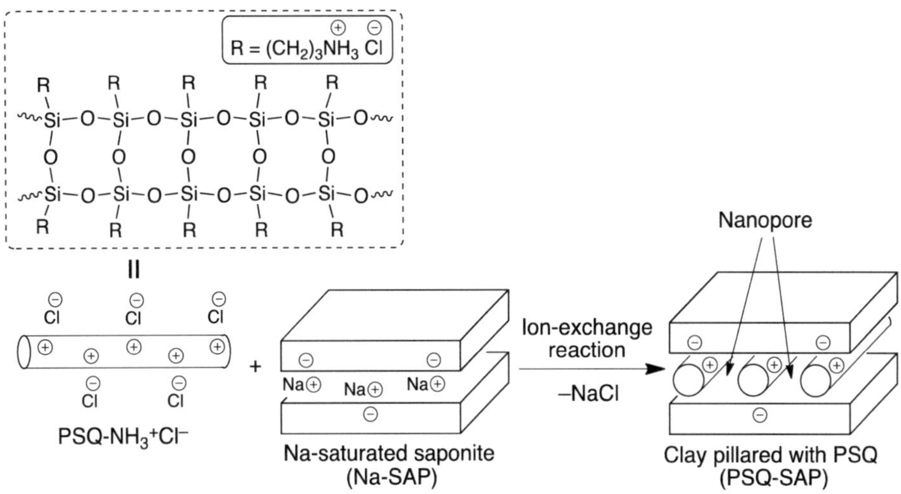

Scheme 4. Preparation of a clay pillared with PSQ by ion-exchange reaction of PSQ-$NH_3^+Cl^-$ with Na-saturated saponite.

From the nitrogen adsorption–desorption isotherms at 77K, the surface area and pore volume of PSQ–SAP derived from the *t*-plot were estimated to be 370m^2/g and 0.15 cm^3/g, respectively. This indicates that a porous material was prepared from the starting materials with dense structures (BET surface areas of Na-SAP and PSQ-$NH_3^+Cl^-$ were *ca.* 26 and 5 m^2/g, respectively).

When a clay mineral with high CEC such as Li-saturated taeniolite was employed, such a porous material was not obtained by combination with PSQ-$NH_3^+Cl^-$ (BET surface area of the resulting product was *ca.* 53 m^2/g), although a sufficient interlayer spacing existed as confirmed by the XRD measurement (*d*-value of the product was *ca.* 1.83 nm). Because the distance between the PSQs in the interlayer of taeniolite is short due to the higher CEC of the Li-saturated taeniolite (exchange amount of a repeating unit of PSQ was calculated to be 140 meq/100 g taeniolite), sufficient space was not provided. Furthermore, when polyallylamine hydrochloride (PAA-Cl) —a common cationic polymer— was used for pillaring in the SAP interlayer, a porous structure was not obtained (BET surface area of the product was *ca.* 52 m^2/g). It was difficult for PAA-Cl to pillar the interlayer of SAP due to the lack of rigidity and bulkiness. From these results, it was considered that the rigidity and bulkiness of the guest polymers and a sufficient distance between charges in the host layered clay minerals are necessary for preparing clays pillared with polymers.

6. Preparation of chiral ladder-like PSQs and hybridization with dye compounds by ion-exchange reaction

Self-assembled hybrids formed by noncovalent interactions between photofunctional compounds and chiral molecules have attracted much attention because of their potential applications in circularly polarized luminescent (CPL) materials. To achieve the preparation of these hybrids by chiral induction from chiral molecules to photofunctional compounds, several combinations have been investigated with respect to supramolecular organization, *e.g.*, anionic dye/cationic chiral surfactants (Franke et al., 2006), laser dye/cholesteric liquid crystal (Uchimura et al., 2010), pyrene derivatives/cyclodextrins (Kano et al., 1988), porphyrins/helical polyacetylene (Onouchi et al., 2006), and π-conjugated polymers/polysaccharides such as amylose (Ikeda et al., 2006), and schizophyllan (Li et al., 2005). On the other hand, there have been no reports regarding hybridization using inorganic compounds such as siloxane (SiO)-based materials as chiral inductors. The chiral inductors derived from SiO-based materials may enable the development of durable and thermostable hybrids with photofunctional compounds because the SiO-based materials exhibit superior thermal, mechanical, and chemical stabilities. In this section, therefore, the author describes the preparation of chiral ladder-like PSQs as the SiO-based chiral inductors and their chiral induction behaviors into dye compounds.

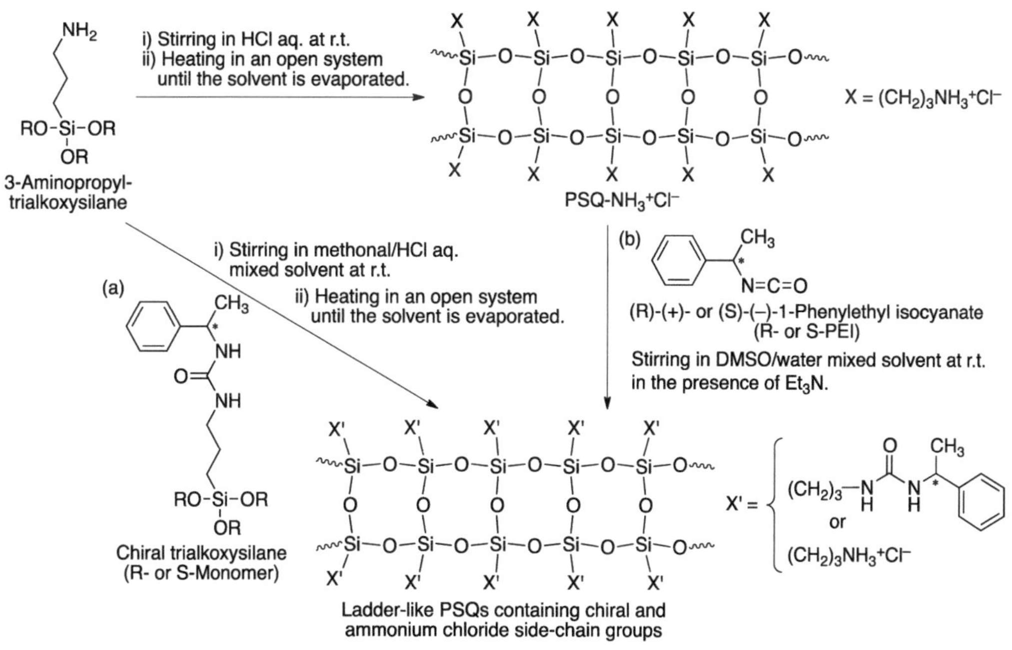

Scheme 5. Preparation of ladder-like PSQs containing chiral and ammonium chloride side-chain groups by (a) copolycondensation method and (b) polymer reaction method.

6.1. Copolycondensation of 3-aminopropyltrialkoxysilane with chiral group-containing trialkoxysilane

To prepare chiral ladder-like PSQs, the introduction of a chiral moiety into the PSQs was first investigated by acid-catalyzed copolycondensation of two organotrialkoxysilanes containing chiral and amino groups, respectively (Kaneko & Iyi, 2009).

The chiral trialkoxysilanes (R- and S-Monomers) were synthesized by reaction of 3-(triethoxysilyl)propyl isocyanate with (R)-(+)- or (S)-(–)-1-phenylethylamine in dichloromethane at room temperature for 15 min, respectively, followed by evaporation of the dichloromethane. The sol–gel copolycondensation of the resulting chiral trialkoxysilanes with 3-aminopropyltriethoxysilane (APTEOS) (feed molar ratio is 1 : 9) was performed in a mixed solvent of aqueous hydrochloric acid and methanol by heating in an open system until the solvent was completely evaporated (Scheme 5a). The products were isolated as the fractions insoluble in acetone, washed with acetone and chloroform, and then dried under reduced pressure at room temperature to yield the white powdered PSQs.

The ^{1}H NMR spectra in D_2O at 60 ºC of the products showed the signals indicating the presence of both the components of the chiral trialkoxysilanes and APTEOS; this indicates that the products were copolymers composed of both trialkoxysilanes. The unit ratio of the chiral groups to the ammonium chloride groups in the products was calculated to be 6 : 94. Therefore, the resulting PSQs are denoted as R_6- and S_6-PSQs, according to the stereostructure and functionality of the chiral groups. The IR spectra of R_6- and S_6-PSQs showed absorptions at 1620 cm^{-1} attributed to the C=O bond of the urea groups, indicating the existence of the units of chiral groups in the products.

In the IR spectra of R_6- and S_6-PSQs, large absorption bands at 1135 and 1040 cm^{-1} assigned to the Si–O–Si bonds were observed, indicating the formation of PSQs. In addition, the ^{29}Si NMR spectra in DMSO-d_6 at 40 ºC of these PSQs exhibit the large peaks in the regions of T^3. These results indicate that the complete progress of sol–gel copolycondensations of the chiral trialkoxysilanes and APTEOS, and the formation of Si–O–Si bonds. The M_ws of R_6- and S_6-PSQs estimated by the *Zimm* plot method using SLS apparatus were assessed to be 10700 and 9800, respectively, indicating that the products were not oligomeric compounds but polymers.

The XRD patterns of the films of R_6- and S_6-PSQs showed three diffraction peaks with the *d*-value ratio of $1 : 1/\sqrt{3} : 1/2$, indicating the formation of the hexagonal phases. Because these PSQs were soluble in water and DMSO, it was supposed that these hexagonal phases originated not from porous-type structures but from the stacking of rod-like polymers. The diameters of the rod-like PSQs calculated from *d*-values of (100) peaks (1.47–1.48 nm) were assessed to be *ca.* 1.7 nm.

As aforementioned, R_6- and S_6-PSQs had rod-like structures with relatively small diameter (*ca.* 1.7 nm) in spite of the presence of large T^3 peak in the ^{29}Si NMR spectrum. In addition, these PSQs were soluble in water and DMSO, despite the M_ws of the PSQs were relatively high (*ca.* 9800–10700). These results satisfy the aforementioned conditions for ladder-like structure of PSQ. Therefore, the present chiral PSQs would also have one-dimensional ladder-like structure.

The vibrational circular dichroism (VCD) spectroscopy, which is the extension of the electronic circular dichroism (ECD) into the IR region, is powerful technique to obtain conformational information of chiral molecules (Tang et al., 2007). The VCD spectra of R_6- and S_6-PSQs showed the reversed absorptions at *ca.* 1140–1165 cm^{-1}, respectively (Fig. 3), corresponding to the absorptions assigned to Si–O–Si bond of polymer main-chains. These results indicate that R_6- and S_6-PSQs had chiral conformations of main-chains.

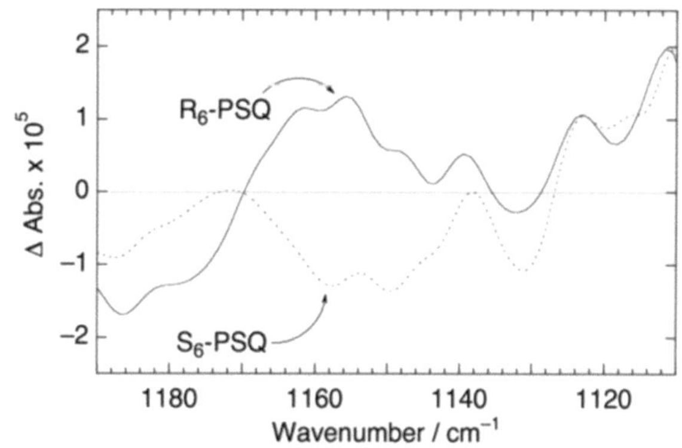

Figure 3. Vibrational circular dichroism (VCD) spectra in DMSO of R_6- and S_6-PSQs.

6.2. Polymer reaction of chiral group-containing compound with PSQ-$NH_3^+Cl^-$

To prepare ladder-like PSQs containing higher compositional ratios of chiral side-chain groups, the aforementioned acid-catalyzed copolycondensation was investigated with a higher feed molar ratio of chiral trialkoxysilanes (>20%). However, the resulting PSQs were insoluble in all solvents owing to the formation of irregular three-dimensional cross-linked network structures. This is because the number of ion pairs formed by the amino groups of APTEOS and the acid-catalysts decreased.

Therefore, to prepare soluble ladder-like PSQs containing higher ratio of chiral groups, *i.e.*, a lower ratio of ammonium chloride groups, the introduction reaction (polymer reaction) of chiral isocyanate compounds into the aforementioned PSQ-$NH_3^+Cl^-$ was investigated (Kaneko et al., 2011).

Synthesis was performed by reaction of (R)-(+)- or (S)-(−)-1-phenylethyl isocyanate (R- or S-PEI) with PSQ-$NH_3^+Cl^-$ in the presence of triethylamine in DMSO/water (9 : 1 (v : v)) mixed solvent at room temperature for 10 min (Scheme 5b). The products were isolated by precipitation in acetone. The compositional ratios of chiral groups to ammonium chloride groups in the resulting products were estimated from their ^{1}H NMR spectra and were found to depend on the feed molar ratio of PEI to ammonium chloride group of PSQ-$NH_3^+Cl^-$. Here, soluble PSQs with the compositional ratio of chiral groups to ammonium chloride groups = *ca.* 80 : 20 were prepared. These PSQs are denoted as R_{80}- and S_{80}-PSQs, respectively. The M_ws of R_{80}- and S_{80}-PSQs were estimated by the *Zimm* plot method using SLS apparatus and were assessed to be *ca.* 54000 and 46000, respectively.

The diffraction peaks in the XRD patterns of the PSQs were broadened compared with those of PSQ-$NH_3^+Cl^-$. This is due to the decrease in the number of ion pairs, *i.e.*, ammonium chloride groups. The ion pair has an important role in the construction of a regular higher-ordered structure. However, because the XRD pattern of the product film showed diffraction peaks with *d*-value of 1.80 nm, indicating a relatively regular stacking structure, the rigid structures of PSQs would be maintained.

The chiral conformations of many kinds of helical polymers are stabilized by intramolecular interaction, *e.g.*, hydrogen bonding (Zhao et al., 2004). Therefore, specific rotations of these polymers are generally changed by varying the solvents because their intramolecular interactions are affected by the nature of the solvent. The specific rotations $[\alpha]_D^{22}$ of R_{80}- and S_{80}-PSQs in methanol were +17.4º and –18.9º, respectively, while those in DMF were +8.6º and –8.5º, respectively. Because these PSQs have urea groups as side chains, which are involved in intramolecular hydrogen bonding, their $[\alpha]_D^{22}$ values were probably affected by solvent effects. Such a solvent effect on specific rotations indicates the presence of chiral conformations of these PSQs.

6.3. Chiral induction behavior of chiral PSQ to dye compound

These chiral ladder-like PSQs were applied to chiral inductors for hybridization with a dye compound such as an anionic achiral porphyrin by the ion-exchange reaction (Kaneko & Iyi, 2009; Kaneko et al., 2011). In this section, the author describes the chiral induction behavior from the aforementioned chiral ladder-like PSQs into an anionic achiral porphyrin such as tetraphenylporphine tetrasulfonic acid (TPPS).

In the UV–Vis spectra of TPPS/R_6- and S_6-PSQs aqueous mixtures (4 μmol/L and 100 μmol unit/L, respectively), absorptions due to the Soret band of TPPS in these mixtures were blue-shifted (to 400 nm) compared with that of TPPS alone indicated monomeric state with protonated (at 434 nm) and deprotonated (at 414 nm) species (Fig. 4a). These results indicate that the negatively charged TPPS formed H-aggregates along the positively charged ammonium groups as side-chains of the PSQs.

The ECD spectra of these TPPS/PSQs aqueous mixtures showed the reversed absorptions due to the Soret bands of TPPS-aggregates, corresponding to R_6- and S_6-PSQs as templates, respectively (Fig. 4b), indicating that TPPS-aggregates have chiralities induced from the chiral PSQs.

However, these mixtures did not show fluorescence emissions due to self-quenching of the excited state of the TPPS-aggregate. Therefore, to inhibit the formation of H-aggregates of TPPS by extension of the distance between the ammonium chloride side chains of PSQs, same analyses were performed using R_{80}- and S_{80}-PSQs as chiral inductors.

ECD spectra of TPPS/R_{80}- and S_{80}-PSQs mixtures in methanol (4 μmol/L and 100 μmol unit/L, respectively) exhibited the reverse absorptions at 418 nm (Fig. 4d), corresponding to the absorptions assigned to the Soret bands of TPPS in the UV–Vis spectrum (Fig. 4c). The absorption wavelength of this mixture was almost same as that of dilute methanol solution of sole TPPS (concentration = 4 μmol/L), indicating that TPPS maintained the monomeric state in the mixture. These results indicate that the chiral induction from PSQs to TPPS was achieved without formation of H-aggregate of TPPS. In the present mixture, because of no formation of H-aggregate, the fluorescence spectrum excited at 420 nm showed an emission peak at 654 nm.

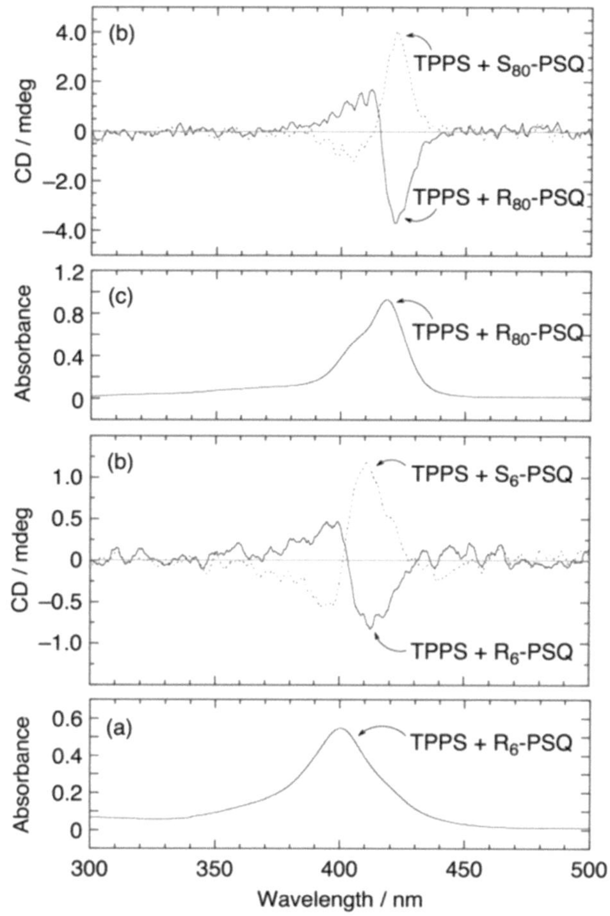

Figure 4. (a) UV–Vis spectrum of aqueous mixture of TPPS/R_6-PSQ, (b) ECD spectra of aqueous mixtures of TPPS/R_6- and S_6-PSQs, (c) UV–Vis spectrum of mixture in methanol of TPPS/R_{80}-PSQ, and (d) ECD spectra of mixtures in methanol of TPPS/R_{80}- and S_{80}-PSQs.

7. Conclusion

In this chapter, the author described the preparation of cationic ladder-like PSQs with hexagonally stacked structures by the sol–gel reaction of amino group-containing trialkoxysilane monomers using acid-catalysts. Then, their detailed characterizations were performed. The resulting PSQs indicated the anion-exchange properties due to existence of cationic functional groups as side-chains. Therefore, it is also described that the anion-exchange behaviors with various organic and inorganic compounds such as anionic surfactants, a polymer, layered clay minerals, and a dye molecule to obtain the functional hybrid materials.

Author details

Yoshiro Kaneko
Graduate School of Science and Engineering, Kagoshima University, Japan

Acknowledgement

The author gratefully acknowledges Dr. Nobuo Iyi of National Institute for Materials Science (NIMS) for causing my research interests in the sol–gel chemistry, silsesquioxanes, and organic–inorganic hybrid materials. The author is indebted to the co-workers, whose names are found in references from this chapter, for their enthusiasistic collaborations.

8. References

Abe, Y. & Gunji, T. (2004). Oligo- and polysiloxanes, *Prog. Polym. Sci.* 29: 149-182.

Baney, R. H., Itoh, M., Sakakibara, A., & Suzuki, T. (1995). Silsesquioxanes, *Chem. Rev.* 95: 1409-1430.

Brown, J. F. Jr., Vogt, L. H. Jr., Katchman, A., Eustance, J. W., Kiser, K. M., & Krantz, K. W. (1960). Double chain polymers of phenylsilsesquioxane, *J. Am. Chem. Soc.* 82: 6194-6195.

Brown, J. F. Jr., Vogt, L. H. Jr., & Prescott, P. I. (1964). Preparation and characterization of the lower equilibrated phenylsilsesquioxanes, *J. Am. Chem. Soc.* 86: 1120-1125.

Bujdák, J., Iyi, N., Hrobáriková, J., & Fujita, T. (2002). Aggregation and decomposition of a pseudoisocyanine dye in dispersions of layered silicates, *J. Colloid Interface Sci.* 247: 494-503.

Choi, J., Harcup, J., Yee, A. F., Zhu, Q., & Laine, R. M. (2001). Organic/inorganic hybrid composites from cubic silsesquioxanes, *J. Am. Chem. Soc.* 123: 11420-11430.

Chujo, Y. (1996). Organic–inorganic hybrid materials, *Curr. Opin. Solid State Mater. Sci.* 1: 806–811.

Cordes, D. B., Lickiss, P. D., & Rataboul, F. (2010). Recent developments in the chemistry of cubic polyhedral oligosilsesquioxanes, *Chem. Rev.* 110: 2081-2173.

Ding, Z., Kloprogge, J. T., Frost, R. L., Lu, G. Q., & Zhu, H. Y. (2001). Porous clays and pillared clays-based catalysts. Part 2: A review of the catalytic and molecular sieve applications, *J. Porous Mater.* 8: 273-293.

Feher, F. J. & Wyndham, K. D. (1998). Amine and ester-substituted silsesquioxanes: synthesis, characterization and use as a core for starburst dendrimers, *Chem. Commun.* 323-324.

Franke, D., Vos, M., Antonietti, M., Sommerdijk, N. A. J. M., & Faul, C. F. J. (2006). Induced supramolecular chirality in nanostructured materials: Ionic self-assembly of perylene-chiral surfactant complexes, *Chem. Mater.* 18: 1839-1847.

Ikeda, M., Furusho, Y., Okoshi, K., Tanahara, S., Maeda, K., Nishino, S., Mori T., & Yashima, E. (2006). A luminescent poly(phenylenevinylene)–amylose composite with supramolecular liquid crystallinity, *Angew. Chem. Int. Ed.* 45: 6491-6495.

Inagaki, S., Guan, S., Ohsuna, T., & Terasaki, O. (2002). An ordered mesoporous organosilica hybrid material with a crystal-like wall structure, *Nature* 416: 304-307.

Kaneko, Y., Iyi, N., Kurashima, K., Matsumoto, T., Fujita, T., & Kitamura, K. (2004a). Hexagonal-structured polysiloxane material prepared by sol–gel reaction of aminoalkyltrialkoxysilane without using surfactants, *Chem. Mater.* 16: 3417-3423.

Kaneko, Y., Iyi, N., Matsumoto, T., & Kitamura, K. (2004b). Preparation of a clay pillared with rodlike cationic polysiloxane, *Chem. Lett.* 33: 1486-1487.

Kaneko, Y., Iyi, N., Matsumoto, T., & Kitamura, K. (2005a). Synthesis of rodlike polysiloxane with hexagonal phase by sol–gel reaction of organotrialkoxysilane monomer containing two amino groups, *Polymer* 46: 1828-1833.

Kaneko, Y., Iyi, N., Matsumoto, T., & Kitamura, K. (2005b). Preparation of higher-ordered inorganic–organic nanocomposite composed of rodlike cationic polysiloxane and polyacrylate, *J. Mater. Chem.* 15: 1572-1575.

Kaneko, Y. & Iyi, N. (2007). Sol–gel synthesis of rodlike polysilsesquioxanes forming regular higher-ordered nanostructure, *Z. Kristallogr.* 222: 656-662.

Kaneko, Y. & Iyi, N. (2009). Sol–gel synthesis of ladder polysilsesquioxanes forming chiral conformations and hexagonal stacking structures, *J. Mater. Chem.* 19: 7106-7111.

Kaneko, Y. & Iyi, N. (2010). Sol–gel synthesis of water-soluble polysilsesquioxanes with regular structures, *Kobunshi Ronbunshu* 67: 280-287.

Kaneko. Y., Toyodome, H., & Sato, H. (2011). Preparation of chiral ladder-like polysilsesquioxanes and their chiral induction to anionic dye compound, *J. Mater. Chem.* 21: 16638-16641.

Kano, K., Matsumoto, H., Yoshimura, Y., & Hashimoto, S. (1988). Binding sites of pyrene and related-compounds and chiral excimer formation in the cavities of cyclodextrins and branched cyclodextrins, *J. Am. Chem. Soc.* 110: 204-209.

Kim, K. M. & Chujo, Y. (2003). Organic-inorganic hybrid gels having functionalized silsesquioxanes, *J. Mater. Chem.* 13: 1384-1391.

Kloprogge, J. T. (1998). Synthesis of smectites and porous pillared clay catalysts: A review, *J. Porous Mater.* 5: 5-41.

Laine, R. M., Zhang, C., Sellinger, A., & Viculis, L. (1998). Polyfunctional cubic silsesquioxanes as building blocks for organic/inorganic hybrids, *Appl. Organometal. Chem.* 12: 715-723.

Li, C., Numata, M., Bae, A. H., Sakurai, K., & Shinkai, S. (2005). Self-assembly of supramolecular chiral insulated molecular wire, *J. Am. Chem. Soc.* 127: 4548-4549.

Loy, D. A., Baugher, B. M., Baugher, C. R., Schneider, D. A., & Rahimian, K. (2000). Substituent effects on the sol–gel chemistry of organotrialkoxysilanes, *Chem. Mater.* 12: 3624-3632.

Ogawa, M., Takahashi, M., Kato, C., & Kuroda, K. (1994). Oriented microporous film of tetramethylammonium pillared saponite, *J. Mater. Chem.* 4: 519-523.

Ohtsuka, K. (1997). Preparation and properties of two-dimensional microporous pillared interlayered solids, *Chem. Mater.* 9: 2039-2050.

Onouchi, H., Miyagawa, T., Morino, K., & Yashima, E. (2006). Assisted formation of chiral porphyrin homoaggregates by an induced helical poly(phenylacetylene) template and their chiral memory, *Angew. Chem. Int. Ed.* 45: 2381-2384.

Parikh, A. N., Schivley, M. A., Koo, E., Seshadri, K., Aurentz, D., Mueller, K., & Allara, D. L. (1997). n-Alkylsiloxanes: From single monolayers to layered crystals. The formation of crystalline polymers from the hydrolysis of n-octadecyltrichlorosilane, *J. Am. Chem. Soc.* 119: 3135-3143.

Pichowicz, M. & Mokaya, R. (2004). Stability of pillared clays: Effect of compaction on the physicochemical properties of Al-pillared clays, *Chem. Mater.* 16: 263-269.

Pinnavaia, T. J. & Beall, G. W. (2000). *Polymer–Clay Nanocomposites,* John Wiley & Sons, Chichester, UK

Seki, H., Kajiwara, T., Abe, Y., & Gunji, T. (2010). Synthesis and structure of ladder polymethylsilsesquioxanes from sila-functionalized cyclotetrasiloxanes, *J. Organomet. Chem.* 695: 1363-1369.

Shimojima, A., Sugahara, Y., & Kuroda, K. (1997). Inorganic–organic layered materials derived via the hydrolysis and polycondensation of trialkoxy(alkyl)silanes, *Bull. Chem. Soc. Jpn.* 70: 2847-2853.

Tanaka, K., Ishiguro, F., & Chujo, Y. (2010). POSS ionic liquid, *J. Am. Chem. Soc.* 132: 17649-17651.

Tanaka, K. & Chujo, Y. (2012). Advanced functional materials based on polyhedral oligomeric silsesquioxane (POSS), *J. Mater. Chem.* 22: 1733-1746.

Tang, H. Z., Garland, E. R., Novak, B. M., He, J. T., Polavarapu, P. L., Sun, F. C., & Sheiko, S. S. (2007). Helical polyguanidines prepared by helix-sense-selective polymerizations of achiral carbodiimides using enantiopure binaphthol-based titanium catalysts, *Macromolecules* 40: 3575-3580.

Uchimura, M., Watanabe, Y., Araoka, F., Watanabe, J., Takezoe, H., & Konishi, G. (2010). Development of laser dyes to realize low threshold in dye-doped cholesteric liquid crystal lasers, *Adv. Mater.* 22: 4473-4478.

Unno, M., Suto, A., & Matsumoto, H. (2002). Pentacyclic laddersiloxane, *J. Am. Chem. Soc.* 124: 1574-1575.

Unno, M., Chang, S., & Matsumoto, H. (2005). *cis-trans-cis-*Tetrabromotetramethylcyclotetrasiloxane: A versatile precursor of ladder silsesquioxanes, *Bull. Chem. Soc. Jpn.* 78: 1105-1109.

Usuki, A., Kojima, Y., Kawasumi, M., Okada, A., Fukushima, Y., Kurauchi, T., & Kamigaito, O. (1993). Synthesis of nylon 6-clay hybrid, *J. Mater. Res.* 8: 1179-1184

Wang, F., Lu, X., & He, C. (2011). Some recent developments of polyhedral oligomeric silsesquioxane (POSS)-based polymeric materials, *J. Mater. Chem.* 21: 2775-2782.

Yamanaka, S., Inoue, Y., Hattori, M., Okumura, F., & Yoshikawa, M. (1992). Preparation and properties of clays pillared with SiO_2–TiO_2 sol particles, *Bull. Chem. Soc. Jpn.* 65: 2494-2500.

Yu, X. F., Zhong, S., Li, X. P., Tu, Y. F., Yang, S. G., Van Horn, R. M., Ni, C. Y., Pochan, D. J., Quirk, R. P., Wesdemiotis, C., Zhang, W. B., & Cheng, S. Z. D. (2010). A giant surfactant of polystyrene-(carboxylic acid-functionalized polyhedral oligomeric silsesquioxane) amphiphile with highly stretched polystyrene tails in micellar assemblies, *J. Am. Chem. Soc.* 132: 16741-16744.

Zhang, X., Xie, P., Shen, Z., Jiang, J., Zhu, C., Li, H., Zhang, T., Han, C. C., Wan, L., Yan, S., & Zhang, R. (2006). Confined synthesis of a *cis*-isotactic ladder polysilsesquioxane by using a π-stacking and H-bonding superstructure, *Angew. Chem. Int. Ed.* 45: 3112-3116.

Zhang, Z., Liang, G., & Lu, T. (2007). Synthesis and characterization of cage octa(aminopropylsilsesquioxane), *J. Appl. Polym. Sci.* 103: 2608-2614.

Zhao, H., Sanda, F., & Masuda, T. (2004). Transformation of helical sense of poly(N-propargylamides) controlled by competition between structurally different enantiomeric amino acids, *Macromolecules* 37: 8888-8892.

Chapter 5

Carbon Nanomaterials – A New Form of Ion Exchangers

Kriveshini Pillay

Additional information is available at the end of the chapter

http://dx.doi.org/10.5772/54114

1. Introduction

Carbon nanomaterials are a new form of carbon based materials which have been receiving much attention for their interesting properties over the past three decades. The materials were initially discovered in 1952 [1] but were only scientifically recognized in 1991 when Iijima submitted a report [2]. An interesting feature of these allotropes of carbon has been the different forms in which they can exist. These include single-walled carbon nanotubes (SWCNTs) and multi-walled carbon nanotubes (MWCNTs). These structures are depicted in **Figure 1**. The fundamental difference between these unique forms of carbon lies in their structural properties. SWCNTs and MWCNTs differ with respect to number of graphene carbon sheets.

Nevertheless, these carbon based materials have demonstrated a vast array of physical and chemical properties. Such properties include outstanding mechanical properties, electrical properties, chemical and thermal stability and large specific surface areas [4]. Consequently, much emphasis has been placed on synthesizing these materials and using them in a number of applications which include nanodevices [5], field emissions [6], plasma apheresis [7], catalyst supports [8], biosensors [9] and chemosensors [10,11]. However, one application of much interest has been their use in environmental remediation. Such an application has arisen from their well-defined porosity and functionality [12]. This property has enabled them to demonstrate superior adsorption capabilities to that of conventional adsorbents like activated carbon [12]. Among, the numerous adsorption applications examined, metal ion uptake by these materials has featured quite prominently. However, the actual mechanism of metal ion adsorption by these materials has been poorly understood.

This chapter therefore serves to examine the ion-exchange properties of carbon nanomaterials. The ability of wide range of nanomaterials namely carbon nanotubes to act as anion and cation exchangers are discussed. Such a discussion includes potential

INTECH
open science | open minds

© 2012 Pillay, licensee InTech. This is a paper distributed under the terms of the Creative Commons Attribution License (http://creativecommons.org/licenses/by/3.0), which permits unrestricted use, distribution, and reproduction in any medium, provided the original work is properly cited.

modifications to these materials improve the ion exchange capacities. Such modifications include heteroatom doping and functionalisation of the surfaces of these materials.

With regards to functionalisation acid-base treatment is discussed. The focus in this section of the chapter is mainly on the how the introduction of functional groups alters the surface of the nanomaterials thereby improving the ion exchange capacities. The influence of various functional groups on the point of zero charge is examined. Such effects are then extended to the effect of heteroatom doping such as sulphur-doping and acid-base treatment.

Lastly, the ability of carbon nanomaterials to act as selective ion-exchangers is reviewed. This includes a discussion on how these materials can function as ion-exchangers in complex matrices. The use of these materials as ion-exchangers in real industrial effluents is also discussed. The way forward with respect to using carbon nanomaterials in analytical ion-exchange applications is discussed with some conclusions and recommendations for future research.

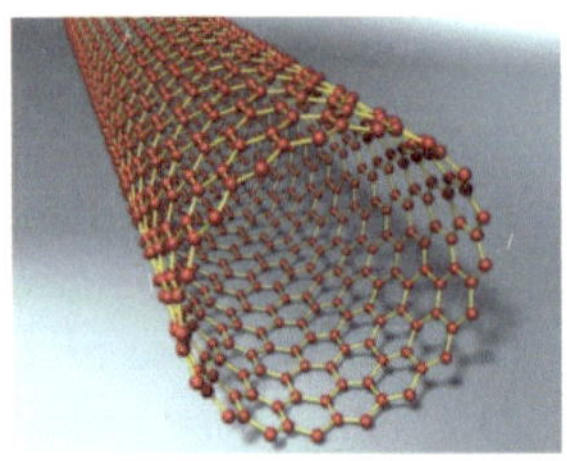

A)

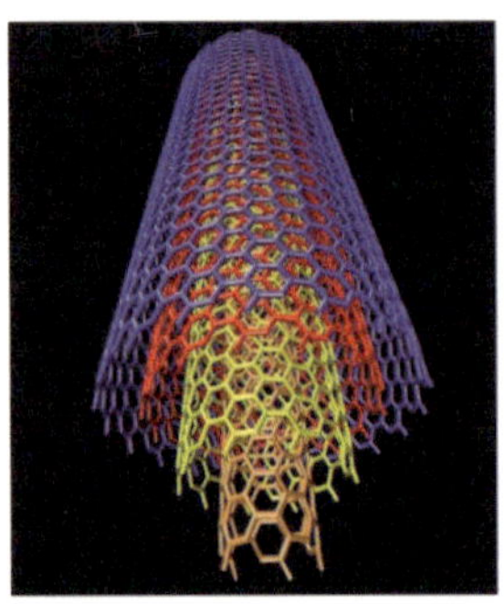

B)

Figure 1. The two main types of carbon nanotubes a) Single –walled carbon nanotubes (b) Multi-walled carbon nanotubes (from: http://www.cnano-rhone-alpes.org/spip.php?article57&lang=en [3]

2. Problem statement

Both SWCNTs and MWCNTs have been extensively used in a number of environmental applications. These include the removal of both organic and inorganic contaminants by adsorption. However, as far as the uptake of ions (especially metal ions) is concerned, ion-exchange has been identified as one of the key mechanisms governing this process. The key question to be addressed here is how can these carbon nanotubes act as ion-exchangers? What properties of these materials enable them to possess ion-exchange characteristics? A number of researchers have attempted addressing these questions and the findings of the research conducted to date are reviewed below.

3. Carbon nanomaterials as anion and cation exchangers

In some recent studies conducted it has been shown that surface modifications to carbon nanotubes have played a key role influencing their ion exchange properties. It has been said that oxidized CNT's show a better potential for cation uptake than unoxidised CNTs [4]. Conversely, it has also been shown that unoxidised MWCNTs are more effective for uptake of anions such as dichromate than oxidized MWCNTs [13]. In such observations it has been argued that oxidation has an impact on the point of zero charge (pH_{PZC}) of these materials which ultimately governs the overall surface charge [4,13,14].

Typically, oxidation lowers the pH_{PZC} of these materials thereby resulting in a predominantly negatively charged surface which is more favorable for cation uptake. Likewise, unmodified or unoxidised CNTs have a higher pH_{PZC} which results in a predominantly positively charged surface which is more favorable for anion uptake. This is supported conclusively by experimental evidence where Rao et al. [4] have reported that oxidized SWCNTs showed a better uptake for Ni(II) and Zn(II) ions than unmodified MWCNTs. Further, Pillay et al. [13] have also shown conclusively that unmodified MWCNTs were more effective for removal of Cr(VI) than oxidized MWCNTs due to a higher (pH_{PZC}).

The question of how these surface modifications can be achieved so as to control the ion-exchange properties of these materials has thus arisen. This can be achieved in two ways. One way includes the addition of surface functional groups and the other is via heteroatom doping. These methods are discussed below.

4. Functionalisation of carbon nanotubes

A number of functional groups have been added to the surface of CNTs. These have included mostly oxygen-containing functional groups which take on the form of both –OH and carboxyl groups. This type of functionalisation has been primarily via oxidation of the CNTs in acid which results in the introduction of carboxyl groups. Consequently, these types of CNTs have been extensively used in the uptake of a number cations. These studies are summarized in **Table 1**.

Most of the above results conclusively support the fact that interaction between the functional groups and the cations are the primary means by which the cations are retained. The fact that an ion exchange mechanism predominates is also supported by the fact the uptake of cations has been pH dependent in some cases. In general it has been observed that cation uptake is low at low pH and high at high pH. This suggests that electrostatic repulsions between the cations and excess H^+ ions at low pH hinder the uptake of cations whereas electrostatic attractions between the cations and excess OH^- ion at high pH enhances the uptake of the cations.

TYPE OF NANOTUBE	POLLUTANT	MAIN OBSERVATIONS	COMMENTS	REFERENCE
SWCNTs and MWCNTs purified with a sodium hypochlorite solution	Zn(II).	1. This purification method rendered the CNTs more favourable for the uptake of Zn(II) by making the CNTs more hydrophilic. 2. The adsorption was also pH dependent. 3. Both types of CNTs were better adsorbents than powdered activated carbon (PAC)	1. The concentration range studied was 10-80 mg L^{-1}.	Lu and Chiu[12]
Oxidised MWCNTs	Ni(II)	Up to 93% Ni(II) could be desorbed from the oxidised MWCNTs at pH< 2 using distilled water as the desorbing agent. 2. Lack of dependence on ionic strength. 3. Equilibrium is attained within 40 min of contact time		Chen and Wang[15]
As grown and oxidised MWCNTs	Ni(II)	1. Adsorption capacity of oxidised MWCNTs was superior. 2. A short time was required to attain equilibrium and noted that the adsorption capability	These workers also neglected to study the effect of competing cations and did not conduct any desorption experiments.	Kandeh and Meunier[16]

TYPE OF NANOTUBE	POLLUTANT	MAIN OBSERVATIONS	COMMENTS	REFERENCE
		of this material was also superior to that of peanut shell		
NaOCl oxidised SWCNTs and MWCNTs	Ni(II)	1. More efficient at adsorbing and desorbing over 10 cycles than granular activated carbon.	None	Lu et al. [17]
Oxidised MWCNTs	Cd(II)	1. MWCNTs oxidised with KMnO4 gave the best results. 2. Oxidation also lowered the pHpzc of these materials and increased the quantity of surface functional groups.	Did not examine the effects of competing anions or determine the regenerability of these materials.	Li et. al[18]
Acid-treated MWCNTs	Pb(II).	1. The adsorption capacity reported was 85 mg g-1 which is higher than that reported for unmodified MWCNTs and activated carbon. 2. The equilibrium adsorption time was shortened to 20 minutes. 3. Provided experimental evidence that chemical interactions between Pb(II) and oxygen containing functional groups accounts for 75,3% of the Pb(II).	The adsorbed Pb(II) aggregates occur mainly on the ends and defect sites of the MWCNTs.	Wang et. al [19]
Oxidised MWCNTs	Ni(II)	1. The presence of polyacrylic acid	The strength of the complexes of Ni(II)	Yang et. al.[20]

TYPE OF NANOTUBE	POLLUTANT	MAIN OBSERVATIONS	COMMENTS	REFERENCE
		(PAA) enhanced the uptake at low pH. 2. That the adsorption was independent of the addition sequence of PAA and Ni(II). 3. The presence of foreign ions had no effect at low pH.	with PAA adsorbed onto the MWCNTs was higher than the strength between Ni(II) and oxygen containing functional groups.	
MWCNTs	Cu(II)	Adsorption was positively affected by the presence of humic and fulvic acids at pH>7.5	Adversely affected at pH < 7.5.	Sheng et. al [21]
Activated carbon, carbon nanotubes and carbon encapsulated nanoparticles	Co(II) and Cu(II).	Both types of carbon nanomaterials had a superior adsorption capability to that of activated carbon.	The carbon nanomaterials only lost 5-11% of their adsorption efficiency in saline environments where competing ions exist while activated carbon lost 30-50% of its adsorption efficiency.	Pyrzynska and Bystrzejewski[22]

Table 1. Some Cation Uptake Studies by Oxidised CNTs (from Pillay [14])

Another interesting feature of the above results has been the general observation that surface modified CNTs show superior adsorption capabilities than that of conventional adsorbents like activated carbon and other low cost adsorbents. Furthermore, the methods of functional group introduction have varied from acid-treatment to purification with sodium hypochlorite. The primary atom that seems to be governing the surface charge hence the pH_{PZC} seems to be the oxygen atom. Such an electron rich atom would impart a negative surface charge to these materials. However, it is by no means the only atom which can achieve this.

Recently Cech et al. [23] showed that sidewall thiolation of MWCNTs can be achieved by treatment with P_2S_5. and Pillay et. al [24] modified this method to produce sulphur containing MWCNTs which showed improved and selective uptake of Hg(II). Thus, both functional group and heteroatom introduction can be controlled by well-defined chemical treatment of CNTs. Such a method has been identified as acid and base treatment of the CNTs.

5. Acid and base treatment of carbon nanotubes

As discussed above acid treatment of CNTs leads primarily to the introduction of oxygen containing functional groups such as hydroxyl and carboxyl groups. The introduction of these functional groups subsequently leads to a negative surface charge which favours cation uptake. However, very little knowledge on the effect of base treatment is available. Pillay [14] attempted studying the effect of strong and weak acid mixtures on the uptake of both cations and anions. Here it was observed that treatment in both strong acid and base mixtures resulted in the introduction of oxygen-containing functional groups which favoured cation uptake by lowering the pH_{PZC} of these materials. Conversely, treatment in a weak base like NH_3 resulted in the introduction of nitrogen-containing functional groups which favoured anion uptake by increasing the pH_{PZC}. Thus treatment in acid and base depends on the strength of the acid or base and the types of heterotoms which can be introduced. The ion exchange properties of the CNTs have therefore become dependent ultimately on heteroatom doping. The methods of heteroatom include functional group addition but the effects of individual heteroatoms depends on the individual atoms added the method of introduction.

6. The effect of heteroatom doping on the ion exhange properties of carbon nanotubes

Thus far it has been noted that the addition of oxygen-containing functional groups and oxygen as a heteroatom favours cation uptake by lowering the pH_{PZC}. Likewise sulphur-containing functional groups and sulphur as a heteroatom have been shown to have a similar effect for the uptake of Hg(II) [24]. However, the effect of other heterotoms such as the nitrogen atom have been poorly understood. Nitrogen –doped carbon nanotubes (N-CNTs) have been receiving much attention recently. Although the properties such as the semi-conducting properties of these materials are well understood [25] little is known about their function in metal ion exchange. Although Perez-Aguilar et al. [26] showed that oxidized nitrogen-doped carbon nanotubes were more effective for the uptake of Cd(II) and Pb(II) than undoped unoxidised MWCNTs the actual role played by the introduction of nitrogen as a heteroatom in this study was not clear. Pillay [14] investigated this further and found that the role played by the nitrogen atom depends largely on the form in which nitrogen occurs on the carbon nanotubes, For instance if quartenary nitrogen is present this can result in positively charged nitrogen centres which increase the pH_{PZC} and favour anion uptake. On the other hand if nitrogen is occurring in conjunction with oxygen this can lower the pH_{PZC} thereby favouring cation uptake.

Some emphasis has therefore been placed on the introduction of heteroatoms such as oxygen, sulphur and nitrogen atoms. The question of whether the introduction of other heteroatoms particularly less electronegative atoms can also influence the ion-exchange properties of CNTs has arisen. Again this is an area which has not been extensively explored. Although Aguiar et al [27] showed the SWCNTs doped with iron were effective for the removal of benzonitrile, the question of how this would affect metal ion exchange still needs to be investigated. Furthermore the introduction of other electronegative atoms such as the halogens needs to be explored. Tan et al [28] also showed the introduction of surfactants to these materials also influence the ion-exchange properties where such a system provided a counterion system for the uptake of Ni(II). The issue of whether the introduction of such atoms or groups of atoms can also impact on the selectivity of CNTs for specific target pollutants also needs to be addressed.

7. Carbon nanotubes as selective ion exchangers

Carbon nanotubes are therefore a new form of ion-exchangers which have demonstrated potential to extract a number of cations and anions based on the functional groups and heteroatoms which are present. However, the selectivity of these materials for specific pollutants is still questionable and limited studies in this area have been conducted. Li et al. [29] attempted addressing this issue by showing that MWCNTs are able to adsorb Pb(II) more efficiently in the presence of other competing cations such as Cd(II) and Cu(II). However, complete selectivity was not achieved. Pillay et al. [13] also showed that unfunctionalised MWCNTs were not selective to the uptake of Cr(VI) in the presence of competing anions such as chloride and sulphate ions. This also applied to industrial effluents where sulphites are present [14]. In fact, the only study to date which been the study in which sulphur-doped MWCNTs showed selectivity for Hg(II) in the presence of competing ions [24]. Thus the selectivity of these ion-exchangers needs improvement.

8. Conclusion

From the information reviewed in this chapter it is evident that carbon nanomaterials are a new form of ion-exchangers based on the diverse number of metal ion uptake studies which have been conducted using these materials. An interesting feature of these materials has been the fact that they can be modified to function as both cation and anion exchangers. This is carefully controlled by both the introduction of functional groups and heteroatoms which ultimately impact on the surface charge. However, the selectivity of these materials for specific ions still needs improvement. Furthermore, the heteroatoms introduced have been restricted to oxygen, sulphur and nitrogen. The effect of other heteroatoms also need to be explored. Even acid and base treatments contribute mainly to the introduction of oxygen and nitrogen atoms. Such topics now lend themselves to future research perspectives.

Nomenclature:

CNTs	Carbon Nanotubes
SWCNTs	Single-walled Carbon Nanotubes

MWCNTs	Multi-walled Carbon Nanotubes
S-MWCNTs	Sulphur-containing Multi-walled Carbon Nanotubes
N-MWCNTs	Nitrogen-doped Multi-walled Carbon Nanotubes

Author details

Kriveshini Pillay
Department of Applied Chemistry, University of Johannesburg, Johannesburg, South Africa

9. References

[1] Monthioux M., Kunetsov V. Who should be given the credit for the discovery of carbon nanotubes? Carbon 2006; 44 1621.
[2] Iijima S. Helical microtubes of graphitic carbon. Nature 1991; 354 56-58.
[3] http://www.cnano-rhone-alpes.org/spip.php?article57&lang=en (accessed 10 December 2009)
[4] Rao, G.P., Lu, C. Su, F. Sorption of divalent metal ions from aqueous solution: *A review.* Separation and Purification Technology 2007; 58 224-231.
[5] Collins, P.G., Zettl, A., Bando, H., Thess, A., Smalley, R.E. Nanotube nanodevice. Science 1997; 278 100-103.
[6] Wang, Q.H., Setlur, A.A., Lauerhaas, J.M., Dai, J.Y., Seelig, A.W., Chang, R.P.H. A nanotube-based field emission flat panel display. Applied Physical Letters 1998; 72 2912-2913.
[7] Ando, K., Shinke, K., Yamada, S., Koyama, T., Takai, T., Nakaji, S., Ogino, T. Fabrication of carbon nanotube sheets and their bilirubin adsorption capacity. Colloids and Surfaces B: Biointerfaces, 2009;71 255-259.
[8] Planeix, J.M., Coustel, N., Coq, B., Brotons, V., Khumbar, P.S., Futartre, R., Geneste, P., Bernier, P., Ajayan, P.M. Application of carbon nanotubes as supports in heterogeneous catalysis. Journal of American Chemical Society. 1994; 116 7935-7936.
[9] Dai, L., He, P., Li, S. Functionalised surfaces based on polymers and carbon nanotubes for some biomedical and optoelectronic applications. Nanotechnology 2003; 14 1081-1097.
[10] Kombarrakaran, J. Clewett, C.F.M. Pietra, T. Ammonia adsorption on multi-walled carbon nanotubes. Chemical Physical Letters 2007; 441 282-285.
[11] Hu, C., Yang, C., Hu, S. Hydrophobic adsorption of surfactants on water-soluble carbon nanotubes; A simple approach to improve sensitivity and antifouling capacity of carbon nanotube-based electrochemical sensors. Electrochemistry Communications 2007; 9 128-134.
[12] Lu, C., Chiu, H. Adsorption of zinc(II) from water with purified carbon nanotubes. Chemical Engineering Science, 2006; 61 1138-1145.
[13] Pillay, K., Cukrowska E.M., Coville, N.J. Multi-walled carbon nanotubes as adsorbents for the removal of parts per billion levels of hexavalent chromium from aqueous solution. Journal of Hazardous Materials 2009; 169, 1067-1075.

[14] Pillay K. Nanomaterials for the Removal and Recovery of Heavy Metal Ions from Industrial Effluents. PhD. thesis. University of the Witwatersrand; 2011.
[15] Chen, C., Wang, X. Adsorption of Ni(II)f from aqueous solution using oxidized multi-walled carbon nanotubes. Ind. Eng. Chem. Res. 2006; 45 9144-9149.
[16] Kandah, M.I., Meunier, J.L. Removal of nickel ions from water by carbon nanotubes. Journal of Hazardous Materials 2007;146 283-288.
[17] Lu, C., Liu, C., Rao, G.P. Comparison of sorbent cost for the removal $Ni^{2+}$from aqueous solutions by carbon nanotubes and granular activated carbon. Journal of Hazardous Materials 2008; 151 239-246.
[18] Li, Y.H., Wang, S., Luan, Z., Ding, J., Xu, C., Wu, D. Adsorption of Cd(II) from aqueous solution by surface oxidized carbon nanotubes. Carbon 2003; 41, 1057-1062.
[19] Wang, H.J., Zhou, A.L., Peng, F., Yu, H. Chen, L.F. Adsorption characteristic of acidified carbon nanotubes for heavy metal Pb(II) in aqueous solution. Materials Science and Engineering A 2007; 466 201-206.
[20] Yang, S., Li, J., Shao, D., Hu, J. Wang, X. Adsorption of Ni(II) on oxidized multi-walled carbon nanotubes: Effect of contact time, pH , foreign ions and PAA Journal of Hazardous Materials 2009; 166 109-116.
[21] Sheng, G., Li. J., Shao, D., Hu, J., Chen, C., Chen, Y. Wang, X. Adsorption of copper(II) on multi-walled carbon nanotubes in the absence and presence of humic or fulvic acids. Journal of Hazardous Materials 2010; 178 333-340.
[22] Pyrzynska, K., Bystrzejewski, M. Comparative study of heavy metal ions sorption onto activated carbon, carbon nanotubes and carbon-encapsulated magnetic nanoparticles. Colloids and Surfaces A: Physiochemical and Engineering Aspects 2010; 362 pp. 102-109.
[23] Cech, J., Curran, S.A., Zhang, D., Dewald, J.L., Avandhandula, A., Kandadai, M., Roth S. Functionalisation of Multi-walled carbon nanotubes: Direct proof of sidewall thiolation. Phys. Stat. Sol. 2006; 243 (13) 3221-3225.
[24] Pillay, K., Cukrowska E.M., Coville, N.J. Improved uptake of mercury by sulphur-containing carbon nanotubes. Submitted to Microchemical Journal.
[25] Nxumalo, E.N., Coville, N.J. Nitrogen Doped Carbon Nanotubes from Organometallic Compounds: A Review. Materials 2010; 3 2141-2171.
[26] Perez-Aguilar, M.V., Munoz-Sandoval, E., Diaz-Florez, P.E., Rangel-Mendez, J.R. Adsorption of Cadmium and Lead onto oxidized nitrogen-doped mutiwalled carbon nanotubes: Equilibrium, and Kinetics. Journal of Nanoparticle Research, 2010; 12 467-480.
[27] Aguiar, A.L., Fagan, S.B., da Silva, L.B., Mendez Filho, J., Souza Filho, A.G. Benzonitrile adsorption on Fe-doped carbon nanostructures. J. Phys. Chem. C. 2010; 114 10790-10795.
[28] Tan, X., Fang, M., Chen. C., Yu, S., Wang, X. Counterion effects of nickel and sodium dodecylbenzene sulfonate adsorption to multiwalled carbon nanotubes in aqueous solution. Carbon 2006; 46 1741-1750.
[29] Li, Y.H., Ding, J., Luan, Z., Di, Z., Zhu, Y. Xu, C., Wu, D., Wei, B. Competitive adsorption of Pb^{2+}, $Cu^{2+}$and Cd^{2+} from aqueous solution by multi-walled carbon nanotubes. Carbon 2003; 41 2782-2792.

Chapter 6

Investigation of Sorption and Separation of Lanthanides on the Ion Exchangers of Various Types

Dorota Kołodyńska and Zbigniew Hubicki

Additional information is available at the end of the chapter

http://dx.doi.org/10.5772/50857

1. Introduction

Rare earth elements of high purity play a significant role in many areas of contemporary techniques. They also have many scientific applications. For example, their compounds are used as catalysts in the production of petroleum and synthetic products, lanthanides are used in lamps, lasers, magnets, phosphors, motion picture projectors, and X-ray intensifying screens. The addition of the pyrophoric mixed rare-earth alloy called Mischmetal or lanthanide silicates improves the strength and workability of low alloy steels. Therefore, the preparation of high purity rare earth elements is very important for such technologies.

Mainly cation exchangers and elution by complexing agents are used for separation of rare earth elements using ion exchange methods. In this process the order of elution of individual rare earth(III) elements depends on the values of stability constants of formed complexes. They generally increase from light lanthanides(III) to heavy lanthanides(III). Ion exchange of rare earth elements in the presence of chelating ligands on anion exchangers is still a poorly studied field. However, the papers published during the last few years show particular applicability of anion exchangers to this end. The anion exchangers have been used so far for separation of thorium(IV) and uranium(IV,VI) from lanthanides(III) with mineral acid solutions and for the studies of chromatographic separation of rare earth(III) elements. As for isotopes of these elements separation processes were mainly of analytical or physicochemical character. For separation of rare earth elements the impregnating resins are also used.

In separation of rare earth(III) element complexes with chelating organic ligands strongly basic, gel polystyrene anion exchangers of quaternary ammonium groups proved to be the most effective. The research carried out dealt mainly with application of such complexing

INTECH
open science | open minds

© 2012 Kołodyńska and Hubicki, licensee InTech. This is a paper distributed under the terms of the Creative Commons Attribution License (http://creativecommons.org/licenses/by/3.0), which permits unrestricted use, distribution, and reproduction in any medium, provided the original work is properly cited.

agents as: ethylenediaminetetraacetic acid (EDTA), iminodiacetic acid (IDA), N′-(2-hydroxoethyl)ethylenediamine-N,N,N′-triacetic acid (HEDTA) and *trans*-1,2-cyklohexanediaminetetraacetic acid (CDTA) in lanthanides separation.

Particular attention is also paid to separation and removal of rare earth(III) elements nitrate complexes by means of frontal analysis from the polar organic solvent-H_2O-HNO_3 on anion exchangers of various types. The affinity series of rare earth(III) elements nitrate complexes depends on the kind of functional groups, kind of the skeleton, porosity of skeleton, cross linking degree of anion exchanger skeleton as well as kind and concentration of polar organic solvent, concentration of nitric acid, addition of another organic solvent and concentration of rare earth(III) elements.

In the paper the research on the applicability of different types of anion exchangers for the separation of rare earth elements in the presence of the complexing agents IDA, HEDTA and CDTA will be presented. The effect of the addition of a polar organic solvent (methanol, ethanol, acetone, 1-propanol, 2-propanol) on separation of rare earth(III) elements in such system will be also discussed. The examples of the removal of rare earth(III) elements nitrate complexes from the polar organic solvent-H_2O-HNO_3 will also be presented in detail.

2. Rare earth elements occurrence

The rare earth elements (REE) are an unusual group of metallic elements with unique properties: chemical, catalytic, magnetic, metallurgical and phosphorescent which consists of seventeen elements belonging to lanthanides. The lanthanide group includes rare earth elements with the atomic number (Z) from 57 to 71 which are: lanthanum (La), cerium (Ce), praseodymium (Pr), neodymium (Nd), promethium (Pm), samarium (Sm), europium (Eu), gadolinium (Gd), terbium (Tb), dysprosium (Dy), holmium (Ho), erbium (Er), thulium (Tm), ytterbium (Yb) and lutetium (Lu). Yttrium (Y) and scandium (Sc) belonging to the scandium subgroup are grouped with rare earth elements because of their similar physicochemical properties (Spedding & Daane, 1961; Powell, 1964; Gschneidner, 1981; Stasicka, 1990)

Generally, lanthanide elements with low atomic numbers are more abundant in the earth crust than those with high atomic numbers. Those with even atomic numbers are two to seven times more abundant than the adjacent lanthanides with odd atomic numbers. The lanthanide elements traditionally have been divided into two groups: the light rare earth elements group (LREEs) which contains elements from lanthanum to europium (Z from 57 to 63) and the heavy rare earth elements group (HREEs) which contains elements from gadolinium to lutetium (Z from 64 to 71). Although yttrium is the lightest rare earth element, it is usually grouped with the HREEs to which it is chemically and physically similar (Kumar, 1994; Robards et al. 1998; Moustafa & Abdelfattah, 2010).

The geochemical studies have revealed that rare earth elements are actually not rare at all. Rare earth elements as the lithophilous ones occur mainly as phosphates and silicates. Due to their chemical similarity, lanthanides occur side by side in the scattered form in about 200 own minerals or as admixtures in the minerals of other elements. They occur in the crust of the earth in higher concentrations than Bi, I and Ag. For example, cerium, which is the most abundant and its average amount is equal to $6.0 \cdot 10^{-3}$% occurs in higher concentrations than Sn, Pb, Co and Mo. Thulium the least abundant of them ($4.8 \cdot 10^{-5}$%) is still more abundant than the platinum group metals (PGM). As for spread in the earth crust scandium occupies position 31 ($2.2 \cdot 10^{-3}$%) and occurs in higher concentrations than Pb, Cu and Ag, whereas the average content of yttrium is $3.3 \cdot 10^{-3}$%.

There are known abort 250 rare earth element minerals of which 10-20 are found to be useful and only 5 practically applicable. Over 90% of the world's economically recoverable rare earth elements are found in primary mineral deposits i.e. in bastnaesite ores which are located in China and at Mountain Pass in California (USA) (Fig.1). Monazite deposits in Australia, South Africa, China, Brazil, Malaysia, India and Russia are the second largest concentrations of rare earth elements. Concerns over radioactive hazards associated with monazites because of thorium presence and high costs associated with its disposal have nearly eliminated it as a rare earth element source in the USA. Additional rare earth elements reserves and resources are found in Colorado, Idaho, Montana, Missouri and Utah. HREEs dominate in the Quebec-Labrador (Strange Lake) and Northwest Territories (Thor Lake) areas of Canada. There are high-grade deposits in Bayan Obo, Inner Mongolia, China and lower-grade deposits in South China provinces providing a major source of the HREEs. The areas considered to be attractive for rare earth elements development include also Karonga, Burundi and Wigu Hill in Southern Tanzania.

1 Mountain Pass, USA; 2 Pajarito Mountain, USA; 3 Gallinas Mountains, USA; 4 Iron Hill, USA; 5 Bald Mountain, USA; 6 Bear Lodge, USA; 7 Pea Ridge, USA; 8 Green Cove Springs, USA; 9 Carolina placers, USA; 10 Lemhi Pass, USA; 11 Snowbird, USA; 12 Rock Canyon Creek, Canada; 13 Hoidas Lake, Canada; 14 Thor Lake, Canada; 15 Elliot Lake, Canada; 16 Strange Lake, Canada; 17 Ilimaussaq complex, Greenland; 18 Araxa, Brazil; 19. Barro do Itapirapua, Brazil; 20 Lovozero and Khibina complexes, Russia; 21 Tamazeght complex, Morocco; 22 Bou Naga, Mauritania; 23 Nile Delta and Rosetta, Egypt; 24 Etaner, Nambia; 25 Okorous, Namibia; 26 Steenkampskraal , South Africa; 27 Zandkopsdrif, South Africa; 28 Plinesberg Complex, South Africa; 29 Naboomspruit, South Africa; 30 Palabora, South Africa; 31 Richards Bay, South Africa; 32 Kangankunde, Malawi; 33 Congolone, Mozambique; 34 Karonge, Burundi; 35 Chavara, India; 36 Amba Dongar, India; 37 Orrisa, India; 38 Perak, Malaysia; 39 Maoniuping/Dalucao, China; 40 Bayan Obo, China; 41 Weishan, China; 42 Xunwu/Longnan, China; 43 Dong Pao, Vietnam; 44 Eneabba, Australia; 45 Jangardup, Australia; 46 Mount Weld, Australia; 47 Brockman, Australia; 48 Nolans Bore, Australia; 49 Mary Kathleen, Australia; 50 Olympic Dam, Australia; 51 WIM 150, Australia; 52 Dubbo Zirconia, Australia; 53 Fraser Island, Australia.

Figure 1. The rare earth deposits (http://www.bgs.ac.uk/research/highlights/2010 /rare _earth_elements.html).

As for their distribution in the environment and in living organisms it should be stressed that they are found, as mentioned above, in the earth crust in a relatively wide range (Hedrick, 1993; Hedrick, 1995). Moreover, they are found in the North Atlantic Ocean waters in very low concentrations. The predominant species are carbonates, such as $La_2(CO_3)_3$ with the concentration 0.002-0.005 ppb in the case of La(III), Ce(III) and Nd(III) and from 4 to 20 times less in the case of other lanthanides. The studies of the pathways of La(III), Ce(III), Th(IV) and Sm(III) from the soil to plants and farm animals show that sorption and soil abundance decrease in the following order: Ce(III) > La(III) > Th(IV) > Sm(III) (Linsalata, et al. 1986). The levels of lanthanides in healthy human tissues have been reported as follows: liver 0.005 μg/g of ash, kidney 0.002 μg/g, lung 0.004 μg/g, bones 0.2-1.0 μg/g (Goering, et al. 1991).

In the paper by Du & Graedel (2011) the first quantitative life cycles (for the year 2007) for ten rare earth elements: La(III), Ce(III), Pr(III), Nd(III), Sm(III), Eu(III), Gd(III), Tb(III), Dy(III) and Y(III) were presented. In the charts it was shown that after extraction from ores using different separation processes the mixed rare earth element concentrates are produced. Thereafter, they are separated from each other into individual rare earth element compounds (i.e., oxides, chlorides, fluorides). The compounds are converted into pure metals or alloys and further transformed into intermediate products. Purification of metals is by electrolysis or vacuum reduction. Production of alloys is either by direct co-reduction of the rare earth element compounds or by melting and casting of metals. The intermediate products are manufactured into final goods. When the products containing rare earth elements are discarded at the end-of-life (EOL), the quantity of rare earth element material in use is lost unless recycling occurs. The idea of generic scheme for the REE cycle is presented in Fig.2. Losses of rare earth elements occur at five points in the cycles: mining, separation, fabrication, manufacturing, and waste management. The authors also emphasize that recycling of rare earth elements is challenging and it appears possible for metallurgical applications, automobile catalysts, magnets in wind turbines and automobiles, in which REEs are used in fairly large quantities.

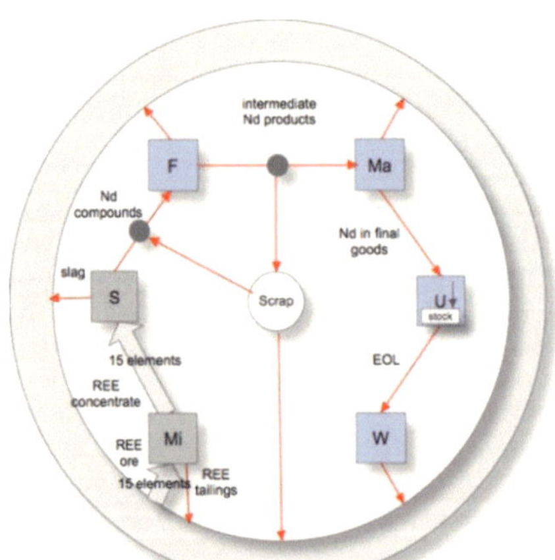

Figure 2. The generic scheme of REE cycle (Du & Graedel, 2011)

3. Rare earth elements – physicochemical properties

Lanthanides are characterized by great similarity with respect to chemical properties. This similarity is considered by approximate electron structure of exterior coating and ionic radii. Generally, the electron configuration of scandium, yttrium and lanthanum can be written as $(n\text{-}1)d^1ns^2$. The elements occurring after lanthanum do not develop the subcoating *5d* (except gadolinium and lutetium, which possess the electrons $5d^1$) but the subcoating *4f*. Considering lanthanides as *f*-electron elements, the configuration $4f^{1\text{-}14}5d^{0\text{-}1}6s^2$ can be ascribed to them. According to some authors also lanthanum can be considered as *f*-electron element (Charewicz, 1990).

The electrons reaching the subcoating *4f* do not affect significantly chemical properties of the elements with the increasing atomic number. All lanthanides have the same oxidation number +3. Passing into the three positive ions, the lanthanide atom loses *6s* electrons and one *5d* electron (if it possesses) or *4f* electron (in the case *5d* electron does not occur). Cerium, praseodymium and terbium as well as neodymium and dysprosium can also have the oxidation number +4 but samarium, europium and ytterbium the oxidation state +2. Also cerium, neodymium and thulium form low stability compounds with the oxidation number +2.

Passing from scandium to yttrium the atomic radius increases from 164.1 pm to 180.1 pm. Also the radius of the ions of the above mentioned elements increases with the oxidation number +3 (from 88.5 pm to 104 pm). In the case of lanthanides the situation is different. The number of electron coatings does not change with the increasing atomic number.

The effect of electrons reaching the *4f* subcoating on the atom size is small. An important factor is the increasing atomic number as well as increased attraction of valency electrons by the nuclei with its charge increase. This leads to the decrease of atomic and ionic radii. This phenomenon is called *lanthanide contraction.* The atomic radius decreases from 187.7 pm (for La) to 173.4 pm (for Lu) in the lanthanide series. Europium whose atomic radius is 204.2 pm and ytterbium of the radius 194.0 pm are the exceptions (Cotton, 2006).

The contraction phenomenon is characteristic of lanthanide ions with different oxidation numbers and it is the most evident for the ions Ln^{3+}. The ionic radius decreases more quickly from lanthanum to gadolinium reaching the value from 115 pm to 107.8 pm and more slowly from gadolinium to lutetium reaching the values: 106.3 pm for terbium, 105.2 pm for dysprosium, 104.1 pm for holmium, 103 pm for erbium, 102 pm for thulium, 100.8 pm for ytterbium and 100.1 pm for lutetium. In the case of gadolinium there occurs the so called *gadolinium break.*

4. Rare earth occurrence and the market

Consumption of rare earth elements in individual countries all over the world is the measure of their technological level and modernity. This is evidenced by concurrence of intensive development of production of many new materials and possession of rare earth elements separation and purification technology. This is also reflected in the total number of papers registered in Chemical Abstracts of American Chemical Society (CA) which gradually increases (Fig.3). In 2007 the total number of papers in CA was around 1 million and the percentage of papers related with rare earths was about 3%.

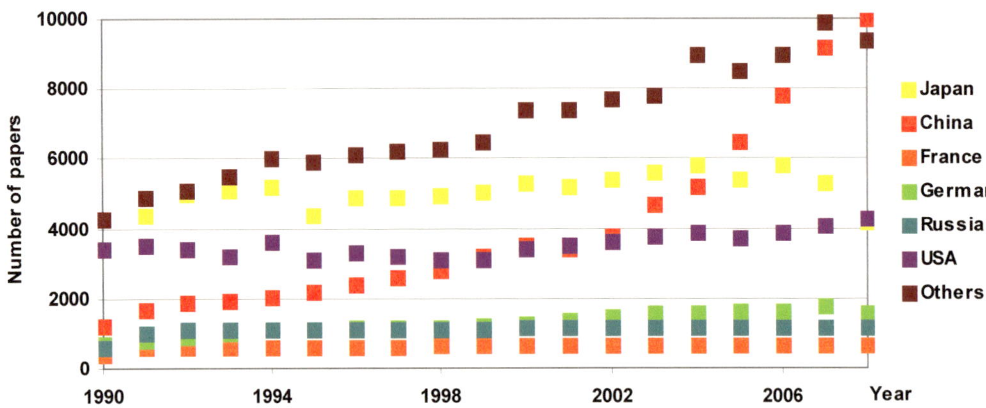

Figure 3. The number of papers related to rare earths in Chemical Abstracts of American Chemical Society (Adachi, et al. 2010).

As follows from Fig.3 there has been the high leap of increase of the number of Chinese papers and a sharp decline in the Japanese one since 2001 probably due to greater research funding and reforming of research organization systems in China such as the foundation of the State Key Laboratory of Rare Earth Materials and Application and the State Key Laboratory of Rare Earth Resource Utilization. According to Adachi et al. (2010) in 2008 in such fields of research as: separation, complexes, electrolysis, oxides, spectroscopy, RKKY interaction, Kondo effect, organometallics, magnet China, magnetism, crystal fields, superconductors, battery, hydrides, phosphors, capacitors, polishing agents, environment, recycling, catalysis and catalysts China is the leader in 16 of them (even in such sections as separation of rare earths, complexes, electrolysis, oxides and spectroscopy) and the growth is faster than in other countries (Fig.4a-d).

In recent few years China has contributed about 97% of the supply on the world market of rare earth elements, whereas according to the United States Geological Survey in 2009 the total resources were about 99 million tons of rare earth. In that China had about 37%, CIS (Commonwealth of Independence States-the former Soviet Union) had about 19% and USA about 13% (Fig.5).

This is reflected in a new supply risk index for chemical elements or element groups which are of economic value published by British Geological Survey (2011) (Table 1). The risk list highlights a group of elements where global production is concentrated in a few countries and which are at risk of supply disruption. On it rare earths (risk index 8.0, fifth position), antimony, platinum group metals, mercury, tungsten and niobium are included. The list also shows the current importance of China in production of many metals and minerals.

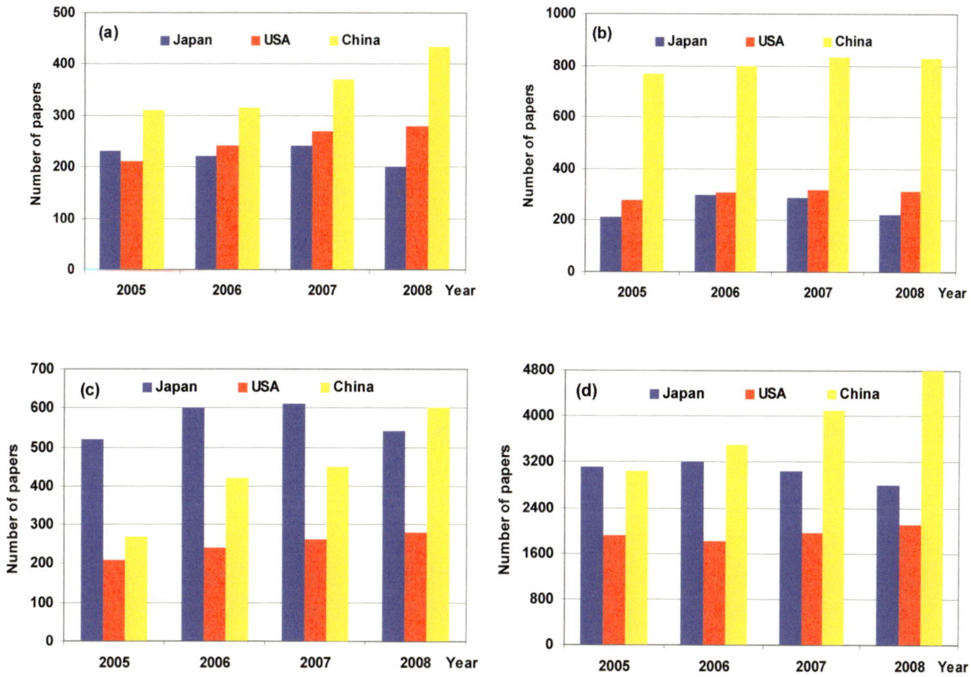

Figure 4. The number of papers related to (a) separation, (b) complexes, (c) electrolysis and (d) oxides of REEs published by Japan, USA and China (Adachi, et al. 2010).

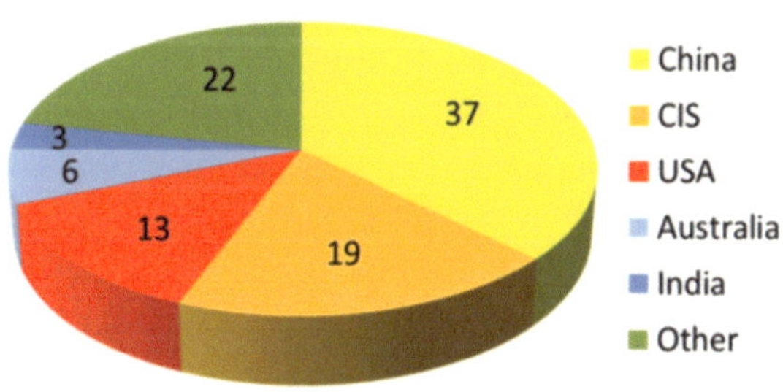

Figure 5. The global reserves of REEs (2009) (US Geological Survey, USGS).

Element or element group	**Symbol**	**Risk index**	**Leading producer**
antimony	Sb	8.5	China
platinum group elements	PGE	8.5	South Africa
mercury	Hg	8.5	China
tungsten	W	8.5	China
rare earth elements	REEs	8.0	China
niobium	Nb	8.0	Brazil

Table 1. The risk index for chemical element group which are of economic value.

It should be mentioned that the rare earth elements resources were discovered by the Chinese scientists in Bayan Obo (Inner Mongolia) in 1927. The rare earth elements production started in 1957. At present, among 21 of Chinese Provinces and Autonomous Regions possessing rare earth elements resources (Fujian, Gansu, Guangdong, Guangxi, Guizhou, Hainan, Henan, Hubei, Hunan, Jiangxi, Jilin, Liaoning, Nei Mongol, Qinghai, Shaanxi, Shandong, Shanxi, Sichuan, Xinjiang, Yunnan, and Zhejiang) the most important are *Fujian, Guangdong, Jiangxi, Sichuan and Nei Mongol Autonomous Region.*

Among North American companies with combined resources of $ 52.7 billion the following should be mentioned: *Molycorp Inc., Avalon Rare Metals Ltd., Quest Rare Minerals Ltd.* and *Rare Element Resources Ltd.*

From the mid-1960s to the 1980s, Molycorp's Mountain Pass mine was the world's dominant source of rare earth oxides. In 2002 they ceased production and in 2008 resumed it based on rare earth oxides from stockpiled concentrates derived from the rare earth ore that was previously mined at Mountain Pass. Since that they have been testing the innovative new processes on a commercial scale. Three facilities e.g. Molycorp Mountain Pass (Califorina), Molycorp Sillamae (Estonia) and Molycorp Tolleson (Arizona) produced 5,000 tons in 2011 and currently have been predicted to reach 8,000-10,000 tons in 2012. *Avalon Rare Metals Ltd.* from Canada it is a mineral developmentcompany with a primary focus on the rare metals and minerals at (Thor Lake, Ontario) which has one of the highest concentrations of heavy rare earth oxides (HREOs) in the world with 26.1% of their 4.298 million tons of total rare earth oxides (TREO) composed of more expensive HREOs. Moreover, *Quest Rare Minerals Ltd.,* Canadian based the exploration company focused on the identification and discovery of new world class rare earth deposit opportunities. Their 'Strange Lake' project in Northern Quebec is believed to hold at least 2.1 million tons of TREOs in their important new rare earth elements mineralized zone named the B Zone, of which 39% is estimated to be HREOs. They expect to start production in mid-2015 to 2016, with the output initially expected to be 12,000 tons of REO/yr. On the other hand, *Rare Element Resources Ltd.* the core project 'Bear Lodge' is believed to be one of the richest LREO deposits in the USA. And with the company expecting to begin production in 2015, with an anticipated output of 11,400 tons of REO/yr, it could be well-positioned to meet growth in demand for LREOs.

The advanced stage projects of rare earth elements production in 2012 and 2013 are also proposed by Lynas Corporation Ltd. (Mt. Weld, West Australia) based on the richest known deposit of rare earth elements in the world, and a state-of-the-art Rare Earths processing plant, the Lynas Advanced Materials Plant (LAMP), currently under construction near Kuantan (Pahang, Malaysia). At the spearhead there are also Silmet (Estonian Republic) which is one of the biggest rare metal and rare earth metal producers in Europe.

In Poland rare earth elements resources do not occur (Charewicz 1990; Paulo, 1993; Paulo 1999). Since 1987 the apatite concentrates from the deposits of Chibiński Massif on Kola Peninsula have been imported to Poland as phosphorous raw material. Apatite deposits contain about 1% of REO. Therefore, the secondary resources of rare earth elements are in the form of phosphogypsum dump in the area of the former Chemical Plant 'Wizów' near Bolesławiec. In 1948 the firm was set up as a producer of sulphric acid. In 1969 -1979 it started to produce phosphoric acid and then phosphoric salts. Up to the 80s the plant was only producer of tripolyphosphate. In its area there is localized a dump of phosphogypsum including mainly calcium sulphate(VI) originating from the extraction of phosphoric acid (obtained from the apatite raw material) and on the average 0.5% of rare earth element oxides. At present over 5 million tons of the waste are found in the dump.

Despite the fact the Chinese resources are estimated to be 37% (Fig.5), due to intensive promotion of their exploitation and large expenditure of money on investigations, at present China is a monopolist imposing the prices of these elements. The latest policy of China – decrease of export by 40% and temporary ban of export to Japan is reflected in an immediate rise in raw materials prices (Fig.6).

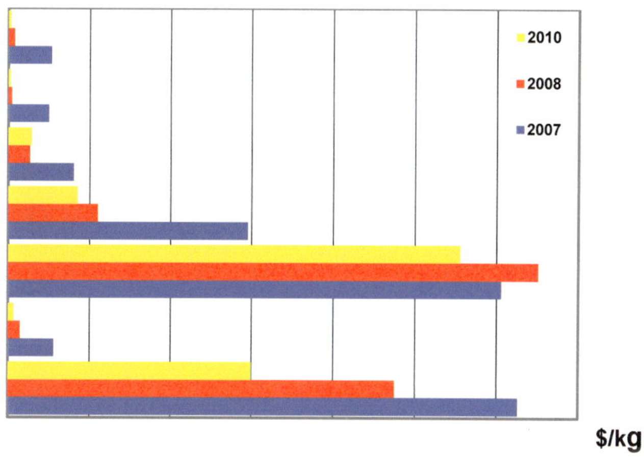

Figure 6. REO prices from 2007 to 2010 (Congressional Research Service report).

High prices mostly affected HREO. The price of terbium oxide used in production of hybrid cars or solar systems increased from $600 in the end of 2010 to $ 3200 for a kilogram at the beginning of 2011. As follows from the data of the firm Shanghai Metals Market, despite over 20% decrease in the prices in the end of 2011, temporary stoppage of production by the largest Chinese firms (among others, Inner Mongolia Baotou Steel Rare-Earth, Hi-Tech Co.) is to stabilize the market and prevent from further drop in prices. According to the China authorities the high cost of mining is connected with huge environmental cost of rare earth processing. On one hand, the mining of rare earths causes the erosion of land and water, on the other, the emission after the mining further damages the environment. Additionally, the China government will not approve of any new rare earth separation projects before 2015. To protect rare earth resources, rare earth producers will be required to have a minimum mine output capacity of 300,000 tons/yr of ore for light rare earths and 3,000 tons/yr (REO) for ion adsorption rare earths. The Chinese government will ban monazite mining if the monazite contains radioactive elements. For rare-earth separation, producers will be required to have a separation output capacity of 8,000 tons/yr (REO) of mixed rare earths, 5,000 tons/yr (REO) of bastnaesite, and 3,000 tons/yr (REO) of ion-adsorption rare earths.

Metal smelting producers must have an output capacity of 1,500 tons/yr. Rare earth producers will be required to meet the environmental emission standards; otherwise, they will be shut down (China Ministry of Environmental Protection, 2011; Tse, 2011). Therefore, the countries, which are potential miners of rare earth elements start to search for their deposits and exploitation. For example, based on the released data from USGS (2010) there is an interesting situation concerning rare earth elements sources from Brazil.

5. Application of rare earth elements

The rare earth elements export reduction introduced by China in 2009 caused anxiety among many world economies including the largest ones as these in the USA or Japan. The USA from being self-reliant in domestically produced rare earth elements over past 15 years became 100% reliant on imports, primary from China. Japan-based firms and the Japanese government made a number of joint venture agreements and potential partnerships around the world to secure supplies of rare earth elements (Sumitomo Corp. and Kazakhstan National Mining Co.; Toyota Tsusho and Sojitz with Vietnam's Dong Pao to produce LREEs, Japan's JOGMEC had also decided to seek investments in Australia's Lynas Corporation).

As follows from the market analysis for several years the exploitation of rare earth elements has been on the same level but the demand for rare earth element compounds has been growing. The total demand for rare earth elements is expected to grow from 128,000 tons in 2011 to 170,000 tons by 2015 and to 255,000 tons in 2020 and a growth rate of about 7-10%/yr is estimated according to the data presented in Table 2. According to the Chinese Rare Earth Industry Association the global demand for rare earth elements may even reach 210,000 tons/yr. However, the Industrial Minerals Company of Australia (IMCOA) estimates that the demand will be 185,000 tons in 2015. China's output may reach 140,000 tons/yr in 2015 as China's annual demand is estimated to rise from 73,000 tons. Such great demand for rare earth elements results mainly from their potential application in many fields of human life (Bünzli et al., 2007).

Application	Growth rate [% p.a.]	Demand in 2015 [tones]
Catalysts	0	25,500
Glass additive	0	10,000
Polishing powder	5-10	23-30,000
Metal alloys	4-8	36-40,000
Permanent magnets	10-15	40-45,000
Phosphors and pigments	4-8	13-15,000
Ceramics	5-8	9-10,000
Other	8-12	12-14,000
Total	**7-10**	**170-190,000**

Table 2. Probable total demand for REE in 2015.

Rare earth elements find application in many advanced materials such as high performance magnets ($Nd_2Fe_{14}B$ and $SmCo_5$, Sm_2Co_{17}) which revolutionized technology by miniaturization of hard disc drives, in fluorescent materials, chemical sensors, high temperature superconductors, magnetoptical discs and rechargeable nickel-metal hybride batteries (NiMH). The alloy used for the battery's positive pole consists of rare earth metals and this makes up 26%of the weight of the battery. Taking into account consumer products they are used in TVs, computer hard drives, plasma and LCD screens, laptop computers, cell phones, DVD players, cameras, electric motors and generators of hybrid cars (Nd-Fe-B magnets and La batteries). For example, in Toyota Prius, Chevy Volt, Nissan Leaf etc. motors and generators of hybrid auto use approximately 25-27 kilograms of rare earth elements and only Toyota plans to sell 1 million of these cars till the end of 2012. In cars, rare earth elements are also used in fuel injectors, airbags and seat belt sensors, anti-lock brakes, power steering and seat adjustment motors and even in the fuel and car catalytic converters (Kucharczyk & Zabrzeski, 2001). The next example, are new wind turbines that are 70% more efficient than the standard ones. However, each requires about two tons of rare earth magnets (Nd-Fe-B magnets). The wind power industry is experiencing explosive growth. The Chinese have spent $150 billion and they plan to install 330 GW of the additional wind powers in the next 10 years. This would require about three times the world's total annual production and as follows from the World Wind Energy Report projects, the global wind power capacity will grow by almost four times in the next 10 years. Rare earth elements are also used in catalysts that can increase the capacity of refinery equipment up to 30% and gasoline production by 5% (in the USA that is additionally 18 million gallons of production per day), commercial air conditioners, medical magnetic resonance imaging equipment, for peptide preparation as well as microfertilizers that can increase certain crop yields by 15% etc (Chegwidden & Kingsnorth, 2002; Xu et al. 2002; O'Driscoll, 2003; Fricker, 2006; Li et al. 2010).

The main groups of rare earth elements application are presented in Fig. 7.

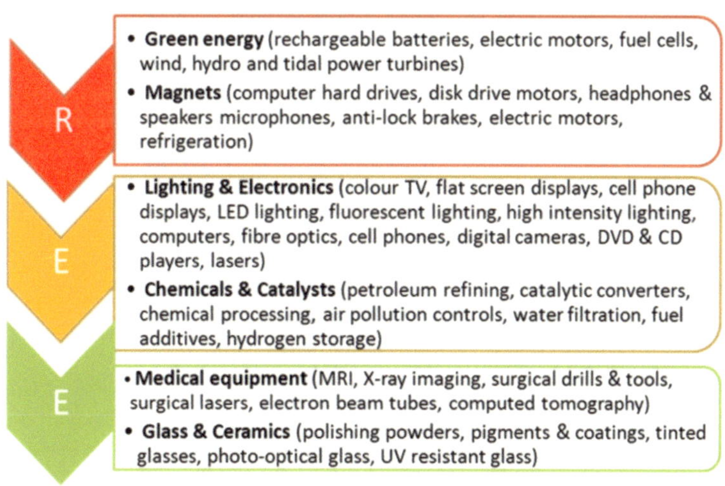

Figure 7. The uses of REEs.

6. Separation of rare earth elements

Lanthanides separation and preconcentration of high purity compounds is one of the most difficult problems in inorganic chemistry as it makes use of the subtle differences between the physicochemical properties of these elements and their compounds like solubility, basicity, volatility and possibility of occurrence with different oxidation numbers (Kowalczyk & Mazanek, 1990; Jimenez-Reyes, 1993; Uda et al. 2000).

Most frequently used methods in the separation of rare earth elements are - fractional crystallization, selective precipitation, oxidation-reduction methods, ion exchange and liquid-liquid extraction. As the rare earth ions can be substituted for readily in crystal lattices and the most precipitates consist of crystals of almost the same rare earth mixture, the fractional precipitation is used for the nitrate(V), suphate(VI) and bromate solutions. However, fractional separation for adjacent heavy rare earths is extremely slow and tedious. If the lanthanides are differentiated in terms of their atomic number, their separation is simplified. It takes only a few partial precipitations, for example, to obtain a lanthanum–cerium–praseodymium fraction completely free of erbium, thulium, ytterbium, and lutetium. The separation of La(III) and Ce(IV) is even easier. In this case only a few fractions should be enough for their separation. Consequently, fractional precipitation is used in operations of rare earth concentrate pre-treatment and pure lanthanum and cerium compounds have been commercially available for many years (Nash, 1994).

The method of crystallization of phosphorus containing lanthanides at a temperature of 150-200 ^{o}C depends on the acid concentration. This type of technology is used in the processing of apatites. The phosphate type mineral, apatite used in the production of phosphoric acid has rare earth oxide content between 0.4 and 0.9%. There are two main types of processing of apatites to phosphoric acid and fertilizers based on the decomposition of raw material with nitric(V) or sulphuric(VI) acids. In terms of recovery of rare earth elements favourable conditions occur during apatite dissolution in nitric(V) acid because this method provides a quantitative transition of lanthanides to liquid phase containing phosphoric(V) and nitric(V) acids as well as calcium nitrate(V). The concentration of lanthanide in this method can be up to 0.5% Ln_2O_3 in solution. However, with the continuous precipitation method to separate Ln(III) ions high concentrations of Ca(II) ions will result in their co-precipitation.

It should be also mentioned that in the case of application of sulphuric(VI) acid, depending on temperature, acid concentration, the ratio of the liquid phase to the solid phase as well as the phase contact time $CaSO_4{\cdot}2H_2O$, $CaSO_4{\cdot}0{,}5H_2O$ or $CaSO_4$ are obtained as byproducts. The concentration of lanthanides in phosphoric(V) acid is equal to 0.1%, while in the phosphogypsum exceeds 0.3%. One method of further processing consists in combining the process of hydration of $CaSO_4{\cdot}0{,}5H_2O$ to $CaSO_4{\cdot}2H_2O$ with the simultaneous process of lanthanides extraction using e.g. D2EHPA. Also application of resin in the leaching (RIL) process instead of solvent extraction eliminates the need for a costly solid/liquid separation unit operation (Padayachee et al. 1996). The authors found that hydrocycloning gypsum to concentrate the leachable rare earth elements into a smaller and finer particle size mass fraction resulted in the increasing rare earth elements concentration from about 2500 g/kg to about 9000 g/kg. Additionally, the application of ion exchange resins such as aminophosphonic Duolite ES-467 and sulphonic Duolite C 20MB to extract rare earth elements from the cyclone gypsum shifted the equilibrium reaction and allowed to increase their leaching efficiency up to five times. The technology developed by the Institute of Chemistry and Inorganic Technology of the University of Technology in Cracow (Poland) eliminates the storage of phosphogypsum, by processing prostproduction wastes and stocks into commercial products: anhydrous calcium sulphate (anhydrite) and rare earth metal oxide concentrates of the content up to 99% recalculated into La_2O_3 (Kowalczyk & Mazanek, 1987; Kowalczyk & Mazanek, 1990). The resulting product, which is a building material, maintains parameters of the cement anhydrite produced from natural materials, and other parameters (such as strength or white colour). The resulting anhydrite, among others, can be applied for producing self-leveling floor screeds, fully meeting all requirements of this type of materials. This technology was tested on the experimental scale processing with 1ton/h phosphogypsum. The technology allows for the elimination of phosphogypsum storage and elimination of the existing waste dumps of phosphogypsum (Kowalski et al. 2006).

7. Cation exchange

Ion exchange separation of rare earth elements was initiated by Spedding and Powell to separate fission products obtained from nuclear reactors (Spedding et al. 1956; Powell, 1961, 1964). For several years the cation exchange was the primary method used to obtain individual lanthanides(III). In previous years and at present the development of extraction methods for separation of rare earth elements(III) proceeded simultaneously to the ion exchange method which is the most successful to obtain these elements with a high degree of purity, as the final product of the concentrates obtained in the extraction process [Preston 1996; Preston et all, 1996].

In the process of cation exchange separation of rare earth elements(III) the polystyrene-sulphonic cation exchangers are most often used and rare earth cations are exchanged with H^+, ammonium ion or other cations derived from the ion exchange phase. The charge, size and degree of hydration of the exchanged ions are the most important factors affecting their affinity for the cation exchanger. In the case of ions with the same charge, the affinity depends on their size and degree of the hydration.

In the lanthanide(III) group with the increasing atomic number decrease of the ionic radius is observed. However, due to similar values of ionic radii of individual lanthanide(III) ions there are not significant differences in their affinity for the polystyrene cation exchangers. Therefore, attempts to obtain individual rare earth elements from the solutions of mineral acids (HCl, HBr, HNO_3 and H_2SO_4) did not yield positive results (Nelson et al. 1964; Korkish, 1967; Korkish & Ahluwalia, 1967). HCl and HNO_3 can be used for the separation of lanthanide(III) from other metal ions occurring in the lower oxidation states (Strelow & Bothma, 1964). This relationship is also used for the separation of cerium(IV) from other rare earth elements using nitric acid(V). Some improvement of the separation of rare earth elements(III) can be obtained by using mineral acid solutions containing organic solvents. In this case rare earth elements(III) are much harder sorbed on cation exchangers than using aqueous solutions of these acids (Starý, 1966).

Therefore, both on a laboratory scale and in industrial separations the elution technique is usually applied. The complexing agents used as eluents form complexes with rare earth elements with different values of stability constants. Separation of rare earth elements, without the introduction of a complexing agent is not possible due to small differences in the values of the separation coefficients. In the case of the complexation of cations by the anionic ligands cation exchanger prefers a cation which forms a complex anion with the lowest average number of ligands and in a series of analogous complexes the one which forms the weakest complex.

The effectiveness of rare earth elements(III) separation on cation exchangers using complexing agents as eluents depends on both the affinity of a given element for the cation exchanger as well as on the kind of complexing agent. As the affinity of rare earths(III) elements is similar to the cation exchanger, the order of elution depends on the stability constants of complexes of individual elements. Thus the separation rate depends on the ratio of stability constants of these complexes. In this group: aminopolycarboxylic acids (EDTA, NTA, HEDTA, DTPA, CDTA), carboxylic acids (acetic acid, malonic acid, maleic acid, phtalic acid), hyroxylicacids (α-hydroxyisobutyric acid, citric acid, lactic acid), ketoacids (pyruvic acid), aldehydeacids (glyoxilic acid), tioacids (tiodiglicolic acid), phosphonic (1-hydroxyethane-1,1-diphosphonic, HEDP) and aminophosphonic acids should be mentioned. The eluent selection and elution conditions are largely dependent on the composition of the mixture of separated rare earth elements.

Among eluents used in the process of cation exchange separation of rare earth elements EDTA and NTA were of the greatest industrial application. Stability constants of the formed complexes generally increase from light to heavy lanthanides(III) because lanthanides(III) are eluted in the order of decreasing atomic numbers. Y(III) location in the elution sequence changes with the change of stability constants of its complexes, therefore, is dependent on the type of eluent, for example, yttrium(III) elutes between Dy(III)-Tb(III) with 1% EDTA solution at pH 3.5, near Nd(III) with DTPA, near Pr(III) with HEDTA, near Eu(III) with citrate 10-20° C and Dy(III)-Ho(III) with citrate at 87-100° C, near Ho(III)-Dy(III) with lactate, near Dy(III)-Ho(III) with thiocyanate and between Sm(III)-Nd(III) with acetate (Powell, 1964).

The advantages of EDTA, in comparison with other complexing agents, are its high efficiency separation of adjacent pairs of rare earth elements(III) with the exception of the pair Eu(III)-Gd(III) ($K_{Eu(III)}=2.24 \cdot 10^{17}$; $K_{Gd(III)}=2.34 \cdot 10^{17}$). The separation factors (β) of rare earth elements using EDTA are as follows: La(III)-Ce(III) 3.3; Ce(III)-Pr(III) 2.4; Pr(III)-Nd(III) 2.0; Nd(III)-Pm(III) 1.9; Pm(III)-Sm(III) 1.8; Sm(III)-Eu(III) 1.5; Eu(III)-Gd(III) 1.1; Gd(III)-Tb(III) 3.5; Tb(III)-Dy(III) 2.7; Dy(III)-Ho(III) 2.0; Ho(III)-Er(III) 2.0; Er(III)-Tm(III) 2.0; Tm(III)-Tb(III) 1.8 and Yb(III)-Lu(III) 1.6. EDTA is readily available, inexpensive and easy to regenerate. However, due to its low solubility elution can not be conducted in an acidic environment (pH <3) and on the cation exchanger in the hydrogen form. Using a solution of EDTA at higher pH values the separation process is carried out with a cupric-ion retaining bed. Of the ions proposed by Spedding, Krumholz and Powell the most relevant retaining ions are Cu(II) and Zn(II) (Spedding et al. 1956; Powell, 1961).

The elution can be carried out even at elevated temperatures, which creates the possibility of recovery of EDTA and increases the separation factor of the Gd(III)-Eu(III) and Eu(III)-Sm(III) pairs (Powell & Burkholder, 1967). The partial complexation method proved to be advantageous for obtaining concentrated heavy lanthanides(III) as well as for separation of lanthanum(III) from other rare earth elements(III). NTA as EDTA is available, although it is more expensive (Fitch et al. 1951, Courtney et al. 1958; Mąkowska, 1970).

Of the group of salts of aminopolycarboxylic acids used as eluents in the process of cation exchange separation of rare earth elements(III), HEDTA proved to be effective for the separation of light and heavy lanthanides(III) (exhibits high separation efficiency of a mixture containing Er(III), Tm(III), Yb(III) and Lu(III)) and for the separation of La(III) from other rare earth elements(III). However, it is completely useless for the separation of medium lanthanides (from Sm(III) to Ho(III)) (Powell, 1961). In the case of separation of rare earth elements(III) with the buffered solution of HEDTA on the cation exchanger Dowex 50 there was reported over 7-fold reduction of the number of theoretical plates with the reduction in the degree of cross linking of the cation exchanger (from 12% to 2% DVB DVB). In this system, the decrease of the number of theoretical plates as a result of the addition of neutral salt solution such as LiCl, NaCl or KCl to the eluent and with the increasing concentration of these salts (Merciny & Duyckaerts, 1966). However, so far, there has not been theoretical explanation of this phenomenon.

The other complexing agent - DTPA proved to be particularly useful for the separation of Y(III) from heavy lanthanides(III), since in a elution series with DTPA Y(III)occupies a position near the Nd(III) (Hale & Hammer, 1972). Of the group of carboxylic acids (acetic, malonic, maleic, phthalic acid) ammonium acetate is the cheapest and easily regenerated complexing agent. Elution using ammonium acetate gives good results in the separation of light lanthanides(III), yttrium(III) from the light and medium lanthanides(III). The best effects of separation were achieved using a solution of ammonium acetate at pH 6.8-6.9 and the gradient concentration 0.45-1.0 M. Yttrium(III) in the elution series occupies a position between Nd(III) and Sm(III).

In the group of hydroxyacids (α-hydroxyisobutiric (α-HIBA), citric and lactic acids) used as eluents, of significant importance are α-hydroxyisobutiric and α-hydroxy-2-methtlbutiric acids (Faris, 1967). α-HIBA is one of the most favourable eluents in this group. In comparison to citric or lactic acids, using α-HIBA high rates of separation of neighbouring pairs of rare earth elements(III) were achieved (Smith & Hoffman, 1956; Choppin & Chopoorian, 1961). This also gives good results of separation of Gd(III)-Eu(III) pair (Hubicka & Hubicki, 1982). The eluent can be applied at room temperature using the ion exchangers of low cross linking, such as Dowex 50x4 and Dowex 50x8 (Smith & Hoffman, 1956). It should be mentioned that α-hydroxyisobutiric acid is not used to on a commercial scale. The most favourable ion exchange separation of lanthanides(III), compared to the α-HIBA can be achieved using 2-hydroxy-2-methylbutyric acid.

As eluents of rare earth elements also other complexing agents such as pyruvic, glyoxylic and thiodiglycolic and 1-hydroxyethane-1,1-diphosphonic as well as aminophosphonic acids were used (Jegorov & Makarova, 1971; Hubicka & Hubicki, 1983a; Hubicka & Hubicki, 1983b).

Of them, special attention should be paid to pyruvic acid (Hubicka & Hubicki, 1983a; Hubicka & Hubicki, 1983b). The pyruvic acid solutions at the concentration 0.15-0.4 M at pH 3.5 and 5.0 proved to be useful for separating such pairs of elements as Y(III)-Nd(III), Sm(III)-Nd(III) as well as for separation of lanthanum(III) from other light lanthanides(III), yttrium(III) from heavy lanthanides using the cation exchanger Wofatit KPS with 4 and 8% DVB. Using pyruvic acid the elution of rare earth elements proceeds in the order of decreasing atomic numbers. Yttrium becomes similar to the medium lanthanides(III).

In the case of thiodiglycolic acid application for the separation of rare earth elements, the unusual position of Y(III) in the elution series can be seen. It can be as follows: Sm(III), Eu(III), Gd(III), Nd(III), Pr(III), Dy(III), Ho(III), Er(III), Yb(III), Lu(III), Y(III), La(III). Yttrium(III) elutes after heavy lanthanides, which enables its separation from Dy(III). The thiodiglycolic acid solution at the concentration 0.15 M and pH 5.5 can be applied for the separation of Y(III) from Nd(III); Sm(III) from light lanthanides(III) and Y(III) as well as Y(III) from Sm(III), Eu(III) and Gd(III).

Using α-hydroxoethylideno-1,1-diphosphonic acid proved that Y(III) behaves as a medium lanthanide(III) and can be separated from heavy lanthanides(III) (especially from Lu(III), Yb(III) and Tm(III)) as well as from Nd(III) (Hubicka & Hubicki, 1980). Availability and low price of this acid also provides an opportunity to use it as an eluent in the purification process of lanthanum(III).

The disadvantage of ion exchange separation of mixtures of rare earth elements on the polystyrene-sulphonic cation exchangers is the lack of universal eluent, which would allow for selective separation of light, medium and heavy lanthanides(III) as well as to achieve high concentrations in the eluate.

8. Anion exchange

In the separation and preparation of rare earth elements with a high degree of purity and separation of macro quantities from micro quantities practical application of anion exchangers to this end took place much later than that of cation exchangers because the mechanism of the processes involved in anion exchangers was more complex and for a long time was not fully explained.

Rare earth elements(III) show little tendency to form anionic complexes with simple inorganic ligands and are poorly sorbed on the anion exchangers from aqueous solutions of hydrochloric and nitric(V) acids. They are also weakly sorbed from sulphuric(VI), phosphoric(V) and mixture of hydrochloric and hydrofluoric acid solutions (Jegorov & Makarova, 1971). Much better results of rare earth elements(III) sorption were obtained from the solutions of such salts as chlorides, nitrates(III), nitrates(V), sulphates(IV), sulphates(VI), thiocyanates, thiosulphates and carbonates (Marcus & Nelson, 1959), which for the chloride system was interpreted by HCl_2^- formation which as a stronger anion than

HCl acid has greater affinity for the anion exchanger than Cl^- ion (Minczewski et al. 1982). It was also shown that using the gradient elution of 6-3 M LiCl solutions at 351 K the separated elements are eluted in the order: Cs(I), Ba(II), Yb(III), Eu(III), Sm(III), Nd(III), Pr(III), Ce(III), La(III). For 3 M solution of $Mg(NO_3)_2$ there was obtained the analogous elution series: Gd(III), Eu(III), Sm(III), Nd(III), Pr(III), Ce(III), La(III) and the heavy lanthanides are poorly separated.

In practice, the anion exchangers were therefore used only for the separation of thorium(IV) and uranium(IV, VI) from rare earth elements(III) from the solutions of mineral acids (Buddery et al. 1959; Marhol, 1982). Thorium(IV) in the nitric(V) acid and uranium(VI) in hydrochloric acid medium form stable, anionic complexes and therefore their separation from rare earth elements(III) forming less stable, cationic, neutral or anionic complexes is possible. Using 7 M HNO_3 solution selective separation of thorium(IV) from rare earth elements(III) was achieved (Danon, 1960).

Significant improvement in the sorption and separation processes of nitrate(V) complexes of rare earth elements(III), due to higher rates of separation, may be obtained by the addition of methanol to the aqueous solution of nitric(V) acid (Stewart & Faris, 1956, Faris & Wharton, 1962). A similar role is also played by higher order alcohols, derivatives of ethylene glycol, dioxane, acetone and tetrahydrofurane.

In the case of such eluents as bufforic solutions of citric and α-hydroxyisobutiric acids rare earth elements from strongly basic anion exchangers are eluted in the reverse order than in the case of their elution from the cation exchangers: La(III), Ce(III), Pr(III), Nd(III), Sm(III), Eu(III), Gd(III), Tb(III), Dy(III), Ho(III), Er(III), Tm(III), Yb(III), Lu(III). Yttrium(III) occupies the position near dysprosium(III).

The studies begun by Dybczyński (Minczewski & Dybczyński, 1962a; Minczewski & Dybczyński, 1962b) on application of complexing agents from the group of aminopolycarboxylic acids (EDTA, CDTA) for the micro quantities of rare earth elements separation reveal that their separation mechanism depends on the factors affecting differentiated anion exchange affinity in the studied system.

Determined by Dybczyński for radioactive indicators the order of elution of rare earth elements(III) using EDTA from the strongly basic anion exchanger Dowex 1x4 in the EDTA form is as follows: Lu(III), Yb(III), Tm(III), Sc(III), Er(III), Y(III), Ho(III), La(III), Dy(III), Ce(III), Tb(III), Pr(III), Nd(III), Gd(III), Pm(III), Eu(III), Sm(III). While the elution order for radioactive indicators of rare earth elements(III) with CDTA from the same anion exchanger in the CDTA form is as follows: Lu(III), Yb(III), Tm(III), Er(III), Y(III), Ho(III), Dy(III), Tb(III), Sc(III), La(III), Gd(III), Eu(III), Ce(III), Pr(III), Sm(III), Nd(III), Pm(III).

Selectivity coefficients of complexes of rare earth elements(III) with EDTA and CDTA in both systems initially grow with the increasing atomic number, pass through a maximum and then decrease (Dybczyński, 1964; Wódkiewicz & Dybczyński, 1968; Wódkiewicz & Dybczyński, 1972). The change of the maximum values of selectivity coefficients from the position of Sm(III) (when EDTA was used as the eluent) to the position of Pm(III) (when

CDTA was used as the eluent) can be conditioned by the necessity of the presence of metal ion with a larger ionic radius, in order to ensure optimal packing of ligands around the central ion. Selectivity coefficients of complexes of rare earth elements(III) with CDTA, in most cases, are lower than in the case of rare earth elements(III) complexes with EDTA. The higher values of the theoretical plates designated for the system with CDTA also demonstrated a less favorable kinetic reaction of ion exchange than in the case of EDTA.

It was also shown that not only the type of eluent used affects the quality of the separation of the above mentioned complexes of rare earth elements(III). In the selected chromatographic system, the values of the separation coefficients can also be modified by changes in temperature and the degree of cross linking of the anion exchanger (Dybczyński, 1964).

Increasing the temperature generally results in improvement of the kinetics of ion exchange (an increase of diffusion coefficients in the ion exchange phase and solution), which significantly reduces the height of the theoretical plates. The temperature rise also affects the selectivity coefficients and separation factors, for example for CDTA the selectivity coefficients increase with the increasing atomic number from La(III) to Pm(III) and then decrease with further increase in the atomic number. At higher temperatures they reach the maximum value for Nd(III). The order of elution of rare earth elements(III) at 365 K is thus as follows: Lu(III), Yb(III), Tm(III), Er(III), Y(III), Ho(III), Dy(III), Tb(III), Gd(III), La(III), Eu(III), Sm(III), Ce(III), Pm(III), Pr(III), Nd(III).

The temperature rise does not always lead to improvement of the separation process. Distribution coefficients and the values of the theoretical plates can either increase, decrease, or remain constant with the increasing temperature, depending on the system. For elution of rare earth elements(III) with EDTA on the anion exchanger Dowex 1x4 in the EDTA form, for some pairs of elements as, for example, Pm(III)-Eu(III) and La(III)-Tb(III) reverse of selectivity takes place (Minczewski et al. 1982).

Dybczyński (1970) examined the impact of the degree of cross linking of the anion exchanger Dowex 1 on the effectiveness of separation of rare earth elements(III). He stated that the separation coefficients generally increase on a regular basis with the increasing degree of cross linking. However, for the complexes $[Ln(edta)]^-$ the change in the degree of cross linking from 4 to 16% of DVB causes increase of the value of the theoretical plate up to two orders of magnitude. This has an important impact on the separation. Resolving power, which is good for anion exchange resins with the optimal degree of cross linking (4% DVB for Dowex 1 - which corresponds to the smallest value of the theoretical plate) is less than unity for the anion exchange resins with a high degree of cross linking. This is related to the exclusion of large ions from the anion exchange phase by the 'sieve effect'.

On the basis of the thermodynamic studies of the anion exchange process with application of EDTA and CDTA it was stated that the unusual and non-monotonic affinity of the rare earth elements(III) complexes with the above mentioned complexing agents is associated with differences in the structure of these complexes and therefore with their different hydration (Surls & Choppin, 1957).

It is not bound, as in the case of cation exchange, with different values of their stability constants. Sizes of complex ions such as [Ln(edta)]⁻ and [Ln(cdta)]⁻ do not change on a regular basis with the change of the central ion radius. On a regular basis, the degree of hydration of these ions does also not change as evidenced by the designated value of the standard thermodynamic potential (ΔG^o). The hydration degree decreases (from La(III) to Nd(III)) and then increases (from Pm(III) to Ho(III)), then it increases but much more slowly, from Ho(III) to Lu(III) .

The papers by Dybczyński (Minczewski & Dybczyński, 1962a) confirmed the hypothesis made by Speeding and Mackcy according to which, together with the decreasing ionic radius of lanthanide(III) in the yttrium group a decrease of dentate of EDTA is observed. $Edta^{4-}$ and $cdta^{4-}$ anions act as pentadentate ligands for light and heavy(III) lanthanide(III) while their hexadentate character is enhanced in the group of medium lanthanides (Fig.8).

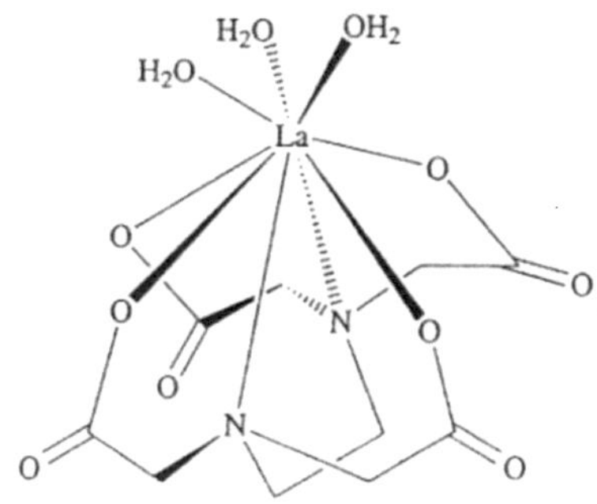

Figure 8. The structure of $[La(edta)(H_2O)_3]^-$ complex.

The proposed model is also suggested by comparing the solubility of the complex salts of [MLn(edta)] and [MLn(cdta)] types (where M = Na^+ or K^+) to the ion exchange affinity of the complexes $[Ln(edta)]^-$ and $[Ln(cdta)]^-$ (Minczewski & Dybczyński, 1962a). Ion affinities of these complexes are arranged in the opposite order to the solubility of the corresponding salts. The selectivity coefficients increase from lanthanum(III) to europium(III), and then decrease to lutetium(III), while the solubility of salt decreases from lanthanum(III) to samarium(III), europium(III) and then rises to lutetium(III).

Non-monotonic and different affinity of the complexes of rare earth elements(III) with EDTA and CDTA for strongly basic anion exchangers was used by Hubicka (Hubicka & Hubicki, 1986; Hubicka, 1989a) to separate the pairs of complexes of rare earth elements(III) in the macro-micro component system using the frontal analysis technique. It was shown that the efficiency of separation of rare earth elements(III) complexes with EDTA of $[Ln(edta)]^-$ type is affected by not only the type of anion functional groups, but also their form and degree of cross linking and porosity of the used anion exchanger. Good results were obtained by purification of yttrium(III) from neodymium(III), samarium(III) and terbium(III) as well as lanthanum(III) from neodymium(III) and erbium(III) from dysprosium(III). For the pair Y(III)-Nd(III), the influence of the type of functional groups, degree of cross linking and anion exchanger form as well as structure and porosity of the skeleton was determined. Among the tested anion exchangers Dowex 2x8 and Dowex 1x8 in the form of EDTA proved to be most advantageous. Good results were also obtained for the macroporous, weakly basic anion exchanger Lewatit MP-7080 in the EDTA form.

Studies on the separation of anionic complexes of rare earth elements(III) with CDTA $[Ln(cdta)]^-$ type, which is an analogue of EDTA showed that in the macro-micro system Y(III) from Sm(III), Eu(III) and Nd(III) can be separated (Hubicka, 1989a). It was found that the process of separation of these complexes affects the degree of cross linking and the anion exchanger form. As follows the best results of separation of Y(III) from Nd(III) and Sm(III) were obtained on Dowex 1x4 in the acetate form.

Application of anion exchangers for separation of rare earth elements(III) complexes with EDTA and CDTA by the frontal analysis technique allows, in comparison with cation exchangers and elution process, to reduce the consumption of these chelating agents, obtaining higher concentrations of rare earth elements(III) in the eluate and shortening the process time which is important from the economical point of view. Additionally, alkali and alkaline earth metal ions forming unstable complexes with EDTA, and sometimes accompanying rare earth elements(III) have no effect on the separation result.

Kutun and Akseli (1999, 2000) for the separation of milligram quantities (5 mg) of rare earth elements(III) in the anion exchange process used the solution of sodium trimethaphosphate as the eluent. The elution was carried out with a gradient of 0.007-0.01 M concentration on the strongly basic polystyrene anion exchangers of types 1 and 2.

The advantage of anion exchangers over cation exchangers in the separation of rare earth elements(III) using aminopolycarboxylic acids as complexing agents is associated with their lower consumption in comparison to other eluents used, much faster process time, achieving higher concentrations in the eluate and the lack of negative impact of alkali(I), Ca(II) and Mg(II) ions on separation.

A particular attention has been paid to separation and removal of rare earth(III) elements nitrate complexes by means of frontal analysis from the polar organic solvent-H_2O-HNO_3 on anion exchangers of various types. The addition of organic solvent to an aqueous solution of rare earth complexes generally improves their ability of separation. Selection of an organic component and its concentration in the mixture is to a large extent arbitrary. The separation process is frequently carried out with the HNO_3, H_2SO_4, NH_4SCN and CH_3COOH solutions. The examples of such systems are presented in (Marcus, 1983).

9. Separation using chelating ion exchangers

Contrary to the cation and anion exchangers chelating ion exchangers have varying degrees of affinity and selectivity with respect to rare earth elements(III). Their properties depend mainly on the nature of the functional group and less on the beads size and other physicochemical properties. The sorption selectivity particularly affects the relative position of functional groups, their spatial configuration, etc. while the less important are the properties of the matrix. Ion exchange capacity of ion exchange resins depends on the content of these groups and the pH of the solution. A negative feature is their low rate of sorption (Kunin & Gustafson, 1969).

For the separation of rare earth elements(III) and their purification from uranium(IV, VI), thorium(IV), iron(III), chromium(III), copper(II), nickel(II), cobalt(II), manganese(II) the chelating ion exchangers with the following functional groups: phosphinic -PO(OH), phosphonic-$PO(OH)_2$, phosphate -$OPO(OH)_2$, iminodiacetate -$CH_2N(CH_2COOH)_2$, iminodiacetate and aminoacetate > N-CH_2COOH, -aminophosphonic -$CH_2NHCH_2PO(OH)_2$ carboxylic -COOH, etc are used.

It is worth mentioning that the ion exchangers with the phosphonic functional groups in the medium acidic system effectively absorb the rare earth elements(III), uranium(VI) and iron(III). According to Hubicki (1989), uranium(VI) and thorium(IV) can be selectively separated from rare earth elements(III) on the phosphonic ion exchanger Duolite ES-63 by both the frontal analysis technique and elution with mineral acids. It was found that on this ion exchanger in the macro-micro component system Y(III) from Lu(III), Yb(III), Tm(III), Er(III), Dy(III) and Ho (III); La(III) from Pr(III) and Nd(III); Ce(III) from Pr(III), Nd(III), Sm(III), Eu(III), Y(III), Dy(III), Ho(III), Er(III), Tm(III), Yb(III) and Lu(III) can be separated (Hubicka & Hubicki, 1978a,b). For Duolite ES-63 the affinity series of rare earth elements(III) can be as follows: Sc(III) > Lu(III) > Yb(III) > Tm(III) > Er(III) > Ho (III) > Dy(III) > Tb(III) > Gd(III) > Eu(III) > Sm(III) > Nd(III) > Pr(III) > Ce(III) ≈ La (III).

Very interesting are the results of the studies on selective separation of micro quantities of scandium(III) on Duolite ES-63 from other rare earth elements(III) at the maximum concentration of 500 g Ln_2O_3/dm^3 (Hubicki, 1990). This ion exchanger is particularly useful for separation of micro quantity of scandium(III) and the preparation of scandium(III) with a high degree of purity. In the technology of processing of rare earth elements(III) concentrates also phosphonic ion exchanger KFP of Russia production was used for the selective removal of micro quantities of Th(IV) (Sozański, 1981).

The selectivity of phosphonic ion exchangers can be increased by connecting different substituents to the phosphorus atom, which affects its electro-donor strength. In the case of the alkylphosphonic ion exchangers affinity for rare earth elements(III) increases with the increasing atomic number of the element, which can be used for the separation of individual rare earth elements(III) by the elution using mineral acid solutions. Separation coefficients of the heavy lanthanides (Lu(III), Yb(III), Tm(III), Er(III)) for alkylphosphonic ion exchangers are from 1.3 to 1.4. The ion exchangers with the phosphine groups are characterized by stronger affinity for lanthanide(III) than the phosphonic ion exchangers. The percentage of the complexation of europium(III) and ytterbium(III) in 0.5 M HCl using the phosphonic ion exchanger is 70% and 82%, while for the ion exchanger with the phosphonic groups is equal to 25% and 50%, respectively.

Marhol (1982) which synthesized a number of ion exchange resins with the functional groups containing phosphorus found that the phosphinic ion exchangers are characterized by strong affinity for scandium(III) in agreement with the affinity series: Sc (III)> Fe (III)> In (III)> Ga (III)> Al(III) > La(III).

Szczepaniak and Siepak (1973) as a result of the cross linking reaction of (5% DVB) vinylbenzylamine with formaldehyde and phosphoric acid(III) in the hydrochloric acid medium obtained the ion exchanger containing aminomethylphosphonic groups. The obtained K-AMF ion exchanger can be defined as a phosphonic analogue of the iminodiacetic ion exchanger Dowex A-1 (Chelex 100). It can be recommended for separation of cations on the +2 and +3 oxidation states. Based on the volume distribution coefficients the affinity series of selected rare earth elements(III) for this aminophosphonic ion exchanger is as follows: La (III) > Gd (III) > Sm (III) > Nd (III) > Dy(III) > Pr(III) > Y(III). This series is not consistent with that obtained for the commercial ion exchange resins such as Duolite ES-467 and Lewatit OC-1060. These ion exchangers are used for the purification of micro quantities of lanthanum(III) from praseodymium(III) and neodymium(III) by means of both the frontal analysis technique and elution with mineral acids.

The aminophosphonic ion exchangers were used for selective purification of rare earth elements(III) from U(VI), Th(IV), Fe(III), Zn(II), Co(II) and Ni(II); purification of yttrium (III) from micro quantities of ytterbium(III), erbium(III), dysprosium(III) and holmium(III) as well as cerium(III) from rare earth elements(III). Of special interest are the values of separation coefficients determined by Van der Walt and Coetzee (1996) for the commercial ion exchanger Purolite S-950, which shows that rare earth elements(III) can be selectively separated from a number of other elements.

As for the phosphate ion exchangers cellulose phosphate is commercially produced. For cellulose phosphate, the mass distribution coefficients were determined for Cs(I), Li(I), Ba(II), Ca(II), Sr(II), Co(II), Ni(II), La(III), Pr(III), Yb(III), Cr(III) and Al(III) ions depending on the concentration of hydrochloric acid. There was made quantitative separation of rare earth elements(III) and earth metals(II); rare earth elements(III) and earth metals(II) and aluminum(III) as well as rare earth elements(III) and alkali and alkaline earth metals. High selectivity can be found in the case of cellulose phosphate ion exchangers. Passing the solution obtained from decomposition of monazite through such ion exchanger a relatively selective sorption of thorium(IV) is achieved with the capacity of the ion exchanger of about 300 g/kg with respect to Th(IV). In the four-column battery (bed height 340 cm, flow rate 15 cm^3/min) of which two columns worked in the sorption cycle and two were eluted, the recovery of thorium(IV) exceeds 99% (Brown and Coleman, 1956). The recovery of pure mixed rare earth elements as well as Th(IV) and U(VI) from the monazite concentrate is based on the reaction with sulphuric acid. Then, after adding hydrazine sulphate, Ce(IV) is reduced to Ce(III). This treatment is also based on the fact that Th(IV) ion with sulphate(VI) ions forms anionic $[(ThSO_4)_4]^{4-2n}$, that could pass freely through a cation exchange resin while the rare earth elements are completely adsorbed (Sherief & Almasy, 1968). Also the uranyl ionUO_2^{2+} forms anionic complexes of the type $[UO_2(SO4)_n]^{2-2n}$ where n=1, 2 or 3 (Preuss & Kunin, 1958).

The polyfunctional ion exchanger Diphonix Resin® is commonly used for separation of rare earth elements(III) (Fig. 9). It contains phosphonic, carboxylic and sulphonic groups. Another ion exchanger is Diphonix A with the phosphonic and ammonium or pyridine functional groups (Horwitz et al., 1993).

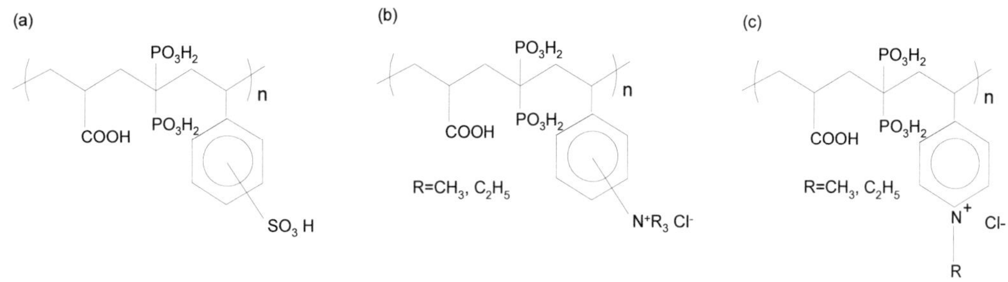

Figure 9. The structure of Diphonix®Resin, Dipxonix A (type 1) and Diphonix A (type 2).

These ion exchangers are characterized by high affinity for uranium(VI), plutonium(IV), neptunium(IV), thorium(IV), americium(III), europium(III) and a number of outside transient elements. The removal factor of europium(III) from 1 M HNO_3 solution at the phase contact time 30 min. for Diphonix®Resin is 98.3%.

An important group of chelating ion exchangers used for lanthanide separations are those with the iminodiacetate functional groups. As follows from the literature data in the pH range 2.5-3.5 with the Ln(III) ions the cationic complexes are formed and in the pH range 4.0-4.5 the anionic ones.

Christell et al. (1961) showed that the standard iminodiacetic ion exchanger Chelex 100 has stronger affinity for La(III) than Lu(III), which is opposed to the values of stability constants of iminodiacetate complexes of these elements. Affinity for Chelex 100 in a lanthanide series increases with the increasing atomic number of La(III) to Eu(III), and then decreases to Lu(III). By contrast, based on the distribution coefficient determined by Schrobilgen and Lang (1968) for the iminodiacetic ion exchanger Dowex A-1 the affinity series is as follows: La(III) $<$ Pr(III) $<$ Nd(III) $>$ Sm(III) $>$ Gd(III) $>$ Tb(III) $\approx$ Dy(III) $<$ Er(III) $>$ Yb(III).

The gel and macroporous ion exchangers, iminodiacetate ion exchangers Amberlite IRC 718, Diaion CR-10, Duolite ES 466, Lewatit TP 207 and Lewatit 208 as well as Wofatit MC 50 were tested for separation of rare earth elements(III) by both the frontal analysis technique and elution. The results of these studies revealed their great applicability in the process of rare earth elements separation as well as for obtaining individual lanthanides with a high degree of purity. Of the group of the above mentioned ion exchangers Wofatit MC-50 proved to be the most useful. Particularly favourable results were obtained for purification of the concentrate of yttrium(III) from ytterbium(III) (Hubicka & Hubicki, 1978; Hubicka, 1989b), which is one of the most common contaminants of yttrium(III). It was also very useful for the ion exchange purification of samarium(III) from ytterbium(III); samarium(III) from europium(III); yttrium(III) from neodymium(III); lanthanum(III) from praseodymium(III) and neodymium(III); cerium (III) from praseodymium(III), neodymium(III), samarium(III), europium(III), yttrium(III), dysprosium(III), holmium(III), erbium(III), thulium(III), ytterbium(III) and lutetium(III) as well as scandium(III) from other rare earth elements. The determined affinity series for heavy lanthanides(III), neodymium(III) and yttrium(III) can be as follows: Yb(III) $>$ Er(III) $>$ Dy(III) $>$ Ho(III) $>$ Nd(III) $>$ Y(III) and this is consistent with the values of stability constants of their complexes with iminodiacetic acid (Inczedy, 1972).

Noteworthy is the unusual position of yttrium(III). It was also found that Yb(III) has higher affinity for this ion exchanger than Er(III), Dy(III), Ho(III) and Tb(III), which is not consistent with the data obtained for Dowex A-l (Schrobilgen & Lang, 1968).

The selectivity of the carboxylic acid ion exchange resins in relation to the rare earth elements (III) is highly variable. Arnold and Son Hing (1967) set the separation coefficients and investigated the mechanism of sorption of lanthanides on the carboxylic ion exchangers Amberlite IRC-50 and Amberlite XE-89. They showed that with the decreasing ionic radii of rare earth elements(III) their affinity for Amberlite IRC 50 increases reverse to that in the case of polystyrene-sulphonic cation exchangers. The determined separation coefficients for the selected pairs of elements are equal to: Ce(III)-La(III) 1.86, Pr(III)-La(III) 2.40; Nd(III)-La(III) 2.60; Nd(III)-Pr(III) 1.1; Pm(III)-La(III) 3.50; Tb(III)-La(III) 5.60; Tb(III)-Ce(III) 3.00, respectively.

Both small ion exchange rate (compared to the polystyrene-sulphonic cation exchangers) and high affinity for the ion H^+ excludes practical application of the carboxylic ion exchange resins for the separation of rare earth elements(III) in the acidic media. However, carboxylic ion exchangers, especially phenol-carboxylic ion exchangers can be used for separation of uranium(VI) from rare earth elements(III).

A relatively small number of selective ion exchange resins is produced commercially. Therefore, the authors propose various modifications of ion exchangers by sulphonated aromatic chelating agents. The development of new functional resins which have chelating properties, prepared by simple immobilization of complexing organic reagents by ion exchange or adsorption onto conventional anion exchange resins or non ionic adsorbents has acquired great importance.

These modified resins can react with RRE ions by complex formation and can be used to preconcentrate their traces. For example, the research carried out by Hubicki (1989a) connected with the selective separation of micro quantities of scandium(III) from macro quantities of yttrium(III) and lanthanum(III) (50g/dm^3) on strongly basic anion exchangers with gel and macroporous skeleton modified with sulphonated organic reagents as prototypes of the new chelating ion exchange resins should be mentioned. To this end the anion exchangers were modified by alizarin S, arsenazo I, arsenazo III, beryllonite II, thymol blue, phenol red, cresol red, pyrogallol red, chrome azurol S, 8-hydroxyquinoline-5-sulphonic acid, sulphosalicilic acid, nitroso-R-soli, R-salt, SPANDS, tirone, torone as well as orange xylene. The best results of separation of Sc(III) from Y(III) and La(III) were obtained on the anion exchanger Merck MP-5080 in the chrome azurol S form. In addition, modified different types of anion exchangers were used for the purification of macro quantities of lanthanum chloride from micro quantities of Eu(III), Tb(III), Yb(III) and Lu(III). Of the anion exchangers tested for this purpose the most preferred proved to be the anion exchanger modified by 8-hydroxyquinoline-5-sulphonic acid (Hubicki, 1989a; Hubicki, 1989b).

Amberlite XAD-4 functionalized with o-vanillinsemicarbazone has been applied for the preconcentration of lanthanum, cerium, thorium and uranium ions (Jain et al., 2001). Complexing properties of the XAD-4 resin functionalized with the bicine ligand (N,N-bis(2-

hydroxyethyl) glycine) were investigated for La(III), Nd(III), Tb(III), Th(IV) and uranium(VI). Polydithiocarbamate chelating resin has been applied for the preconcentration of rare earth elements with Fe(III), Fe(II), Cr(VI), Cr(III), V(V), V(IV), Ti(IV), Mo(VI), W(VI), Th(IV) (Miyazaki & Barnes 1981). In the paper by De Vito et al. (1999) it was found that Amberlite XAD-7 resin with (o-[3,6-disulpho-2-hihydroxy-1-naphthylazo]-benzenearsonic acid) was successfully used for the separation and preconcentration of Sm(III), Eu(III) and Gd(III). The resin containing the fluorinated-β-diketone chelating group immobilized on solid support styrene divinyl benzene was applied for the simultaneous preconcentration of La(III), Ce(IV), Nd(III), Sm(III), Gd(III), Eu(III), Dy(III), Er(III), Yb(III) and Lu(III) (Rao & Kala, R. 2004; Waqar, et al. 2009). Preconcentration of La(III), Eu(III) and Yb(III) was also achieved by sorption of their 1-phenyl-3-methyl-4-benzoylpyrazol-5-one (PMBP) complexes on a silica gel column (Liang & Fa, 2005).

In the paper by Vigneau et al. (2001) the molecular imprinting resins containing DTPA and EDTA were also studied in lanthanides(III) separation. It was found that DTPA derivative monomers exhibited much higher Ga(III)-La(III) selectivity than EDTA ones. Also ionic imprinting resins can be used to this end. In the molecular imprinting process, the selectivity of a polymeric material is based on the size and shape of the template as well as the affinity between the host matrix and the guest molecule. More particularly in the case of ionic imprinting, the affinity partly depends on the number and the orientation of interaction points (ligand denticity) as well as on the counter ion. As described in the paper by (Krishna et al. (2005) ion imprinting allows significant enhancement in selectivity coefficients of neodymium(III) with respect to La(III), Ce(III), Pr(III), Sm(III) and Eu(III). The obtained selectivity coefficients are several times higher than those for the best extractant such as D2EHPA, for example selectivity coefficient for Nd(III) over Ce(III) and Pr(III) increases twentyfold, threefold over La(III) and Sm(III).

10. Separation of rare earth(III) by means of the extraction method

Ion exchange was of most significant importance in uranium production. However, success of ion exchange methods in uranium technology does not apply to that of other elements. Compared to the extraction method, rate of ion exchange and concentration of purified rare earth elements(III) salts are, as a rule, much smaller. Therefore large size of industry installations, generating large investment costs, is necessary. During extraction the elements are divided, according to the Nernst division law, into two immiscible with each other phases: organic and aqueous. Tri-n-butyl phosphate (TBP) and D2EHPA are commonly applied extracting agents in separation of rare earth elements(III) (Preston & Du Preez, 1990). Their application results mainly from their good extraction properties with a relatively law price. The advantages of this method are large concentration of elements in the organic and aqueous phases, thus large yield of multi-stage extraction. Others have discussed the use of bis(2-ethylhexyl)phosphonic acid (HDEHP) in rare earth elements separations (Jensen et al. 2001; Fontana & Pietrelli, 2009; Yin et al. 2010). HDEHP, a liquid cation exchanger that is also a chelating agent, most typically forms a tris complex with trivalent lanthanides in the organic phase, simultaneously releasing three H^+ for each trivalent metal ion transferred into the organic phase. The distribution ratios for the extraction of the lanthanides from mineral acid solutions vary by nearly 10^5 from La(III) to Lu(III) (Nilsson & Nash, 2007, Mel'nik et al. 1999).

As follows from the literature data (Yan et al. 2006) the novel solvent extraction process and its application in industry for separating HREEs (thulium(III), ytterbium(III) and lutetium(III)), yttrium(III) and scandium(III) has been developed recently. The most popular solvent extraction method is using PC-88A (2-ethylhexyl phosphonic acid mono-2-ethylhexyl ester) or Cyanex 272 (bis(2,4,4-trimethylpentyl)phosphinic acid) as the extractant with lower equilibrium acidity. The additives of modified PC-88A were also adopted to separate Lu(III) in practice. Naphthenic acid or related carboxylic acid derivative were also used to separate the heavy rare earth elements in a very long cascade. Although the efficiency was not satisfactory, the chemical consumption was economical enough. The electrochemical method is also piloted to obtain Yb(III). This makes the separation of Tm(III), Yb(III) and Lu(III) more effective by solvent extraction.

As for the purification of Y(III) its chloride salt is separated from heavy rare earth elements using PC-88A - HCl process. The contaminants of light rare earth elements and bivalent ions such as Ca(II) were removed in PC-88A-HCl cascade and then yttrium (≥99.999%) is co-precipitated with Eu(III), Tb(III) and Zr(IV) by purified oxalic acid (Yan, et al. 2006).

Solvent extraction is also the most commonly used and effective technique in Ce(IV) separation. Cerium(III) is most likely to be oxidized to a tetravalent state either by bubbling oxygen during rare earth hydroxide precipitation or by drying the rare earth hydroxide in the presence of air. In acidic solutions, the oxidation of Ce(III) to Ce(IV) may occur by chemical oxidation with strong oxidants, such as peroxosulphates(VI), permanganates, lead oxide or silver oxide or by electrochemical oxidation or photochemical oxidation. Separation of insoluble cerium(IV) from rare earth elements can be carried out by selective dissolution of the trivalent rare earth hydroxides or through its selective precipitation from acid solution (Ura et al. 2005; Zhang et al. 2008; Luna et al. 2011). Many studies have been conducted on Ce(IV) separation using various extractants such as Cyanex 923, TBP, primary amine N1923, Aliquat 336, synergistic extractant and ionic liquids. The recovery of Ce(IV) and Th(IV) from rare earths(III) with Cyanex 923 has been applied in a industrial process. However, there are some disadvantages of solvent extraction of Ce(IV), for example such as a third phase formation.

The extraction of rare earth elements by amine and quaternary ammonium salts has been investigated in detail by many authors. As follows from the paper by Kovalancik and Galova (1992) low values of separation factors in the rare earth elements extraction with amines may be enhanced by the addition of a complexing agent. The greatest differences in the stability constant values are found in the case of CDTA and HEDTA.

In the extraction process the separation of Eu(III) from Gd(III) is exception. Similar to Zr(IV)-Hf(IV) this pair is the most difficult one to separate. Among rare earth elements, which can be reduced to divalent ions, Eu(III) has the highest standard redox potential, which makes its selective reduction and recovery from a mixture containing the other trivalent rare earth ions possible (Morais & Ciminelli, 1998). It can be accomplished by several techniques, such as chemical reduction by Zn or Zn-Hg, photochemical reduction and electrochemical reduction (Atanasyants & Seryogin, 1995; Preston et al. 1996). The recovery is finally accomplished by precipitation with sulphate, based on the fact that the chemical properties of Eu(III) are similar to those of the alkaline earth ions. Among different sulphate sources $(NH_4)_2SO_4$, K_2SO_4, Na_2SO_4, $NaHSO_4$ and H_2SO_4 can be used as precipitating agents. As follows from the Morais and Ciminelli paper (2001) the precipitation with sulphuric(VI) acid led to higher-grade europium oxide by keeping pH in a range that does not favour gadolinium co-precipitation. The continuous addition of sulphuric(VI) acid is mainly responsible for the improvement of europium recovery. Maximum recovery was achieved within 2 h or more. Based on the experiments the product assaying 99.99% Eu_2O_3 can be obtained from the feed containing 5.0 Eu_2O_3 g/dm^3 and 138.2 Gd_2O_3 g/dm^3 in two stages of reduction–precipitation. The overall recovery is about 94%.

To overcome inconveniences of the extraction of rare earth elements there should be used various impregnating substances. Solvent impregnated resins (SIRs) were developed by Warshawsky (1981). It is now well known that the extraction of metal ions with macroporous polymeric supports impregnated by extractants is an attractive method for the separation and preconcentration of metal ions (Schmidt, 1987; Cortina & Warshawsky, 1997; Horwitz and Schulz, 1999). Solvent impregnated resins for lanthanides separation are produced by adsorbing extractants such as TBP or D2EHPA on a macroporous resin without any functional groups. Different types of highly selective resins can be obtained by adsorbing various extractants. However, it should be mentioned that SIRs are characterized by smaller adsorption capacity and shorter life compared with the common ion exchange resins (Shibata & Matsumoto, 1998). The disadvantage of impregnating substances is slow kinetics due to the limited size of interfacial area and wet ability with the aqueous solution. An essential drawback is the irreversible loss of active substances. However, using Cyanex 302 in industrial processes of scandium(III) purification should be highlighted.

In the paper by Turanov (2010) bis(diphenylphosphoryl-methylcarbamoyl)alkanes were synthesized and studied as extractants for La(III), Ce(III), Pr(III), Nd(III), Sm(III), Eu(III), Gd(III), Tb(III), Dy(III), Ho(III), Er(III), Tm(III), Yb(III), Lu(III) and Y(III) from perchloric acid solutions. The influence of both $HClO_4$ concentration in the aqueous phase and that of the extractant in the organic phase on the extraction of metal ions was considered. The stoichiometry of the extracted complexes has also been determined. Bis(diphenylphosphoryl-methylcarbamoyl)alkanes possess a higher extraction efficiency towards Ln(III) than their monoanalogue $Ph_2P(O)CH_2C(O)NHC_9H_{19}$.

In the paper the research on the applicability of different types of anion exchangers for the separation of rare earth elements in the presence of the complexing agents IDA, HEDTA and CDTA is presented. The effect of the addition of a polar organic solvent (methanol, ethanol, acetone, 1-propanol, 2-propanol) on separation of rare earth(III) elements in such system is discussed. The examples of the removal of rare earth elements nitrate complexes from the polar organic solvent-H_2O-HNO_3 are discussed in detail.

11. Experimental

The strongly basic anion exchangers Dowex 1x1, Dowex 1x2, Dowex 1x4, Dowex 1x8 (type 1), Dowex 1x16, Dowex 2x8 (type 2) and macroporous Dowex MSA-1 produced by the Dow Chemical Company (USA) as well as Lewatit MonoPlus M 500, Lewatit MonoPlus M 600, Lewatit MonoPlus MP 500, Lewatit MonoPlus MP 64 and Lewatit MP 62 produced by Lanxess (Germany) were used in the investigations. From the group of polyacrylate anion exchangers Amberlite IRA 458, Amberlite IRA 68 and Amberlite IRA 958 (Rohm and Haas, France) were selected. Additionally, in the case of separations of rare earth elements in the polar organic solvent systems there were used the following: Zerolite FF IP 2-3%, Zerolite FF IP 3-5%, Zerolite FF IP 7-9%, Lewatit MP 5080, Purolite A 850, Permutit SK, Wofatit SBWx2%, Wofatit SBWx4%, Wofatit SBWx6%, Wofatit SBWx8%, Wofatit SBWx12%, Wofatit SBWx16%, Wofatit SBKx7%. The bead size of these anion exchangers was 0.15-0.3 mm. They were used in the appropriate form.

In the case of rare earth elements separation in the presence of complexing agents (L=IDA, HEDTA, CDTA or EDTA) 1.5 or 2.5 g of the oxides of rare earth elements with 2% excess of stoichiometric quantities of complexing agent solution (in the Ln(III):L=1:1 or Ln(III):L=1:2 systems) were mixed while heated. Moreover, in the case of the systems in the polar organic solvent the rare earth element solution of 2.5-100 g Ln_2O_3/dm^3 for separation of Y(III) from Nd(III) and 1.5-5.3 g Ln_2O_3/dm^3 for separation of Sm(III) from Y(III) were prepared by dissolving the appropriate rare earth element oxides in HNO_3 and then the organic solvent was added. In order to measure affinity series the breakthrough curves were determined using the solutions of rare earth elements at a concentration 0.01 M or 0.004 M. These solutions were passed continuously through 2 cm i.d. glass columns packed with the suitable ion exchanger keeping the flow rate 0.2 $cm^3/cm^2 \cdot min$. The breakthrough curves were normally obtained using 40-80 cm^3 of the anion exchanger in the appropriate form. The effluent was collected as fractions of 15-10 cm^3 from which oxalates were precipitated and converted to oxides.

The percentage of micro component in macro component was determined by the spectrophotometric analysis using a SPECORD M 42 spectrophotometer (Zeiss, Germany). The determination was made by a direct method using the neodymium adsorption maximum λ=793.4 nm and the praseodymium adsorption λ=444 nm. Calculations were made by using the background adsorption elimination method. In some cases the determination was made using the XRF spectrometer (Canberra Packard).

There were determined the affinity of rare earth element complexes and the effects of kind of functional groups (basicity), kind of skeleton, porosity of skeleton (microporous and macroporous), cross linking degree of anion exchanger skeleton as well as kind and concentration of complexing agent and polar organic solvent (methanol, ethanol, 1-propanol, 2-propanol, acetone, dimethylformamide, dimethylsulphoxide). The effect of the nitric acid concentration, the addition of another organic solvent and the concentration of rare earth elements(III) (up to 100 g Ln_2O_3) on effectiveness of their sorption and separation were also studied.

12. Results

As follows from the literature data little attention was paid to the purification of rare earth elements using complexing agents and separation of formed anionic complexes on strongly and weakly basic anion exchangers (Hubicka & Hubicki, 1992a; Hubicka & Hubicki, 1992b). For this aim such complexing agents as iminodiacetic acid (IDA), N′-(2-hydroxoethyl)ethylenediamine-N,N,N′-triacetic acid (HEDTA), *trans*-1,2-cyklohexane-diaminetetraacetic acid (CDTA) and ethylenediaminetetraacetic acid (EDTA) can be proposed.

It was worth mentioning that iminodiacetic acid (IDA) as the eluent is not applied in the separation of rare earth elements(III) on cation exchangers. In the aqueous solution and the solid state the following complexes of $[Ln(ida)]^{+}$, $[Ln(ida)_2]^{-}$, [Ln(ida)OH], $[Ln(ida)_2OH]^{2-}$ and $[Ln(ida)_3]^{3-}$ types with IDA are formed. Based on the breakthrough curves found the affinity series in the Ln(III):IDA=1:2 system and at pH 5.07-6.0 for most rare earth element complexes was found for the anion exchanger Dowex 1 in the acetate and IMDA forms to be as follows: **Ho(III) ≈Dy(III) > Gd(III) > Eu(III) > Er(III) > Y(III) > Yb(III) > Sm(III) > Tm(III) > Nd(III) > Pr(III) >>La(III)** (Hubicka & Drobek, 1998).

The affinity series for the anion exchanger in the acetate and IDA forms is the same, but for the acetate form greater differentiation in affinity of individual lanthanides was found. Additionally, it can be stated that the affinity of rare earth elements for the anion exchangers does not depend on the stability constants of Ln(III)-IDA complexes and the obtained series is different from that of rare earth element complexes with EDTA, CDTA and HEDTA. Taking into account the position of individual lanthanides in the above mentioned series the possibility of purification of La(III) from Nd(III) or Pr(III); Sm(III) from Ho(III); Nd(III) from Y(III); Y(III) from Dy(III), Ho(III) as well as Yb(III) from Ho(III) and Er(III) is possible.

As follows from the obtained data the most effective separations of La(III) from Nd(III) or Pr(III) were achieved on Dowex 2 (type 2) with 8% DVB. Using Dowex 2x8 in the acetate form with an anion exchanger bed of 1 dm^3, over 1285 g La_2O_3 was obtained, in which the Nd_2O_3 content (0.52%) was reduced to below 0.005% and 1100 g La_2O_3 in which the Pr_6O_{11} content (0.50%) was reduced to below 0.01% (Fig.10). It is also worth noting that the yield of obtained La_2O_3 is high about 96%. Comparing the earlier results of La(III) purification from Nd(III) on the anion exchanger of the same type with those obtained for EDTA, it follows that using IDA increases the amount of purified La_2O_3 by over 20 times. The analogous results were obtained on Dowex 1x8 (Hubicka & Drobek, 2000b).

Satisfactory results were also obtained on Dowex MSA-1. On 1 dm^3 of this anion exchanger in the acetate form 1100 g of La_2O_3 was purified from Nd_2O_3 and 600 g La_2O_3 from Pr_6O_{11}. Despite slightly worse performance separation of those pairs on Dowex MSA-1, in comparison with the analogous results obtained for Dowex 1x8 and Dowex 2x8, due to their favourable physicochemical properties, such as large resistance to chemical agents, well-developed surface area and availability of all ion exchange groups, it can be recommended for the use on a macro scale (Hubicka & Kołodyńska, 2000).

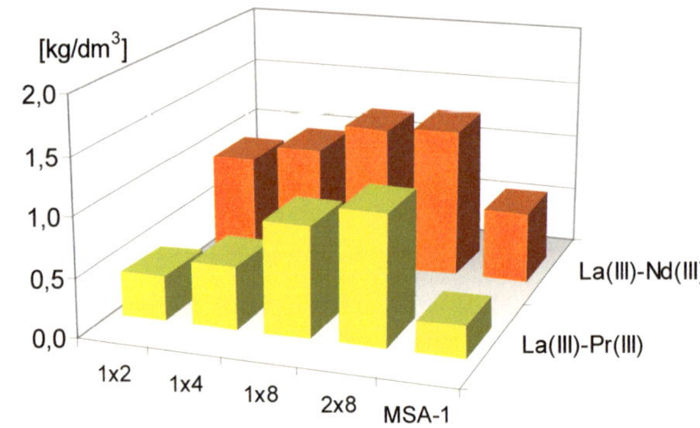

Figure 10. Separation of La(III) from Nd(III) (0.52%) and La(III) from Pr(III) (0.50%) on Dowex 1x2, Dowex 1x4, Dowex 1x8 and Dowex 2x8 as well as Dowex MSA-1 in the acetate form (1 dm^3 of anion exchanger in the Cl^- form, Ln(III):IDA=1:2, pH 6.0, 2.5g Ln_2O_3/dm^3).

In next step the optimal conditions for sorption and separation processes of Sm(III)-Ho(III) on the polystyrene anion exchangers were studied. It was found that in the Sm(III) from Ho(III) separation the most effective was Dowex 2x8 (type 2). On 1dm^3 of Dowex 2x8 about 221 g of Sm_2O_3 was purified from Ho_2O_3 (Fig.11).

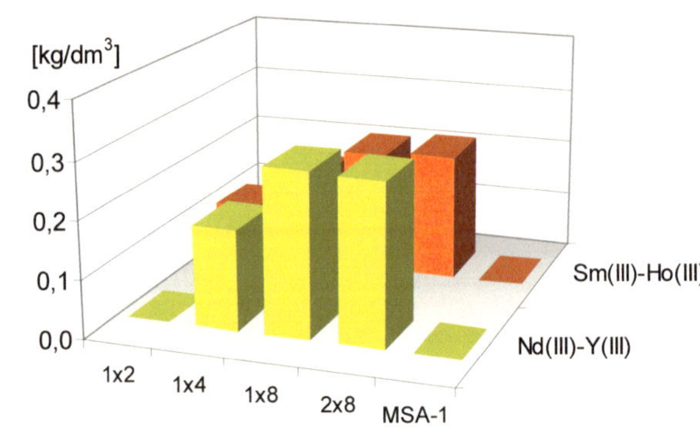

Figure 11. Separation of Nd(III) from Y(III) (0.35%) and Sm(III) from Ho(III) (0.33%) on Dowex 1x2, Dowex 1x4, Dowex 1x8 and Dowex 2x8 as well as Dowex MSA-1 in the Ac form (1 dm^3 of anion exchanger in the Cl^- form, Ln(III):IDA=1:2, pH 5.0, 1.5g Ln_2O_3/dm^3).

It was found that for Dowex 1 the effectiveness of Sm(III)-Ho(III) pair separation (as pairs of LaIII)-Nd(III) and La (III)-Pr (III)) increases with the increasing degree of cross linking from 2 to 8% DVB. The Ho(III) complexes with IDA exhibit the greatest affinity for the anion exchanger Dowex 1 in the acetate form of cross linking 8 % DVB and the smallest for Dowex 1 in the acetate form of cross linking 2 % DVB.

As for the Nd(III) separation from Y(III) it should be emphasized that in the ion exchange processes of rare earth elements(III) separation on cation exchangers using such eluents as ammonium acetate or DTPA, the yttrium(III) elutes close to neodymium(III), and partially overlaps the neodymium(III) (Fig.11). The above method obtaining high purity neodymium(III) on Dowex 1x8 or Dowex 2x8 can be complementary to obtaining this element from the neodymium concentrates in the process of lanthanide(III) elution on the polystyrene-sulphone cation exchangers (Hubicka et al., 1998).

Due to the increasing demand for yttrium(III) in many branches of industry and technology, purification of yttrium(III) from heavy lanthanides(III) can be of significant applicative character. Its advantage is a simple method and low consumption of iminodiacetic acid. Based on the results of purification of the studied pairs of rare earth elemet(III) complexes on anion exchangers of polystyrene skeleton, it was proved that the anion exchangers Dowex 1 and Dowex 2 of the cross linking 8% DVB in the acetate form are the most effective. On 1 dm^3 of strongly basic anion exchangers Dowex 1x8 and Dowex 2x8 it can be obtained more than:

- 113 g and 116 g Y_2O_3 in which the Ho_2O_3 content can be reduced from 0.34% to 0.05%;
- 115 g and 108 g Y_2O_3 in which the Dy_2O_3 content can be reduced from 0.365% to 0.035%;
- 103 g and 91 g Y_2O_3 in which the Er_2O_3 content can be reduced from 0.35% to 0.05%.

The best results were obtained during purification of Y(III) from Ho(III) and Y(III) from Dy(III). However, the amount of Y(III) purified from Er(III) is smaller which is in accordance with the erbium position in the affinity series of rare earth element complexes with IDA compared to the strongly basic gel anion exchanger Dowex 1x4.

In the process of ion exchange separation of rare earth elements on cation exchangers, most eluents washe out ytterbium(III) in the first fractions. However, if the content of ytterbium(III) is small, it is eluted along with heavy lanthanides(III) and yttrium(III). Whereas separation of ytterbium(III) to ytterbium(II) using reduction with metallic limestone at elevated temperature in the argon atmosphere or using the reduction electrolytic method is a multi-stage expensive process (Hubicka & Drobek, 1999; Hubicka & Drobek, 2000a; Hubicka & Drobek, 2000b).

The affinity series of anionic complexes of rare earth elements(III) with IDA for the strongly basic gel anion exchanger Dowex 1 in both acetate and iminodiacetate forms indicates atypical position of ytterbium(III) in it, which creates possibility of Yb(III) purification from accompanying heavy lanthanides(III) such as e.g. Ho(III) or Er(III). The process of Yb(III) separation from Ho(III) and Yb(III) from Er(III) in the acetate form was studied using the anion exchangers Dowex 1x4 and Dowex 1x8 by means of the frontal analysis technique (Fig.12).

Assuming that 1:2 complexes are formed the sorption process of Ln(III)-IDA complexes can be written as:

$$RCH_3COO + [Ln(ida)_2]^- \rightleftarrows R[Ln(ida)\,CH_3COO] + Hida^-$$

and

$$RCH_3COO + [Ln(ida)_2]^- \rightleftarrows R[Ln(ida)_2] + CH_3COO^-$$

as well as for the IDA form

$$RHida + [Ln(ida)_2]^- \rightleftarrows R[Ln(ida)_2] + Hida^-$$

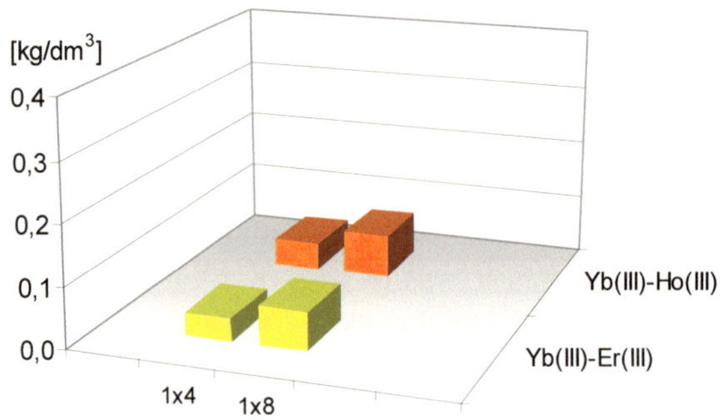

Figure 12. Separation of Yb(III) from Er(III) (0.36%) and Yb(III) from Ho(III) (0.35%) on Dowex 1x4 and Dowex 1x8 in the acetate form (1 dm^3 of anion exchanger in the Cl^- form, Ln(III):IDA=1:2, pH 5.0, 1.5g Ln_2O_3/dm^3).

The acetate form of the studied anion exchangers shows greater differentiation in the affinity of individual complexes than in the iminodiacetate form. Structure and porosity of the skeleton of the anion exchanger and cross linking degree play a significant role in the sorption process. Taking into consideration the kind of functional groups, it was shown that in the separation process the anion exchanger Dowex 2x8 proved to be a little more effective for the pair La(III)-Pr(III) whereas the results are similar for all other pairs.

HEDTA is more readily soluble in water than EDTA. It forms with rare earth element complex compounds in the ratio 1:1. For example, the following complexes are known: $[Ln(hedta(OH)]^-$, $[LnH(hedta)_2]^{2-}$ and $[Ln(hedta)_2]^{3-}$. The studies of sorption of the Ln(III) complexes with HEDTA on anion exchangers were begun by (Brücher et al. 1975). The affinity of the rare earth element complexes with HEDTA at pH 7.5 (Ln(III):HEDTA=1:1) for the anion exchanger Dowex 1x2 in the HEDTA form is: **Dy(III) > Ho(III) > Er(III) > Gd(III) > Y(III) > Tm(III) > Tb(III) > Eu(III) > Sm(III) > Yb(III) > Nd(III) > Pr(III) ≈ La(III)** (Hubicka & Drobek, 1997a) and differs from that of rare earth element complexes with EDTA and CDTA.

Based on the determined affinity the possibility of separation of Sm(III) from Ho(III) and Sm(III) from Y(III); Y(III) from Ho(III); Y(III) from Er(III); Y(III) from Dy(III); Nd(III) from Y(III) as well as Yb(III) from Ho(III) and Yb(III) from Er(III) complexes with HEDTA in the macro-micro component system by the frontal analysis technique is possible (Hubicka & Drobek, 1997b).

Separation of Sm(III) from Ho(III) (Ln(III):HEDTA=1:1, pH 4.0 and 7.5) in the HEDTA form was conducted on the strongly basic polystyrene anion exchangers Dowex 1 with cross linking 2, 4 and 8% DVB as well as Dowex MSA-1.

As follows from the comparison of the obtained data, the process of purification of Sm(III) from Ho(III) and Sm(III) from Y(III) is the most effective on the anion exchanger Dowex 1x4 in the HEDTA form at solution pH 7.5 (Fig. 13). On 1 dm^3 of this anion exchanger it is possible to obtain about 48 g Sm_2O_3 in which Ho_2O_3 content can be reduced from 0.33% to the value below 0.05% and about 15 g Sm_2O_3 in which Y_2O_3 content can be reduced from 0.37% to the value below 0.03%. Using the macroporous Dowex MSA-1 gave less advantageous results of Sm(III) purification from Ho(III).

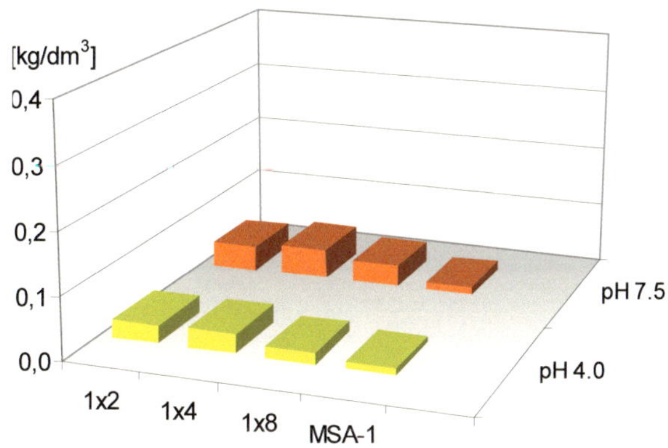

Figure 13. Separation of Sm(III) from Ho(III) (0.33%) on Dowex 1x2, Dowex 1x4, Dowex 1x8 and Dowex MSA-1 in the HEDTA form (1 dm^3 of anion exchanger in the Cl^- form, Ln(III):HEDTA=1:1, pH 4.0 and 7.5, 1.5g Ln_2O_3/dm^3).

A smaller amount of Sm(III) purified from Y(III) compared with that of Sm(III) purified from Ho(III)confirms that the Ho(III) complexes exhibit greater affinity than that of Y(III) complexes which is in agreement with the determined affinity series of the rare earth elements(III) complexes with HEDTA.

As follows from the series of rare earth elements(III) complexes with HEDTA for the anion exchanger Dowex 1x2 in the HEDTA form, also the Y(III) complexes are characterized by greater affinity for this anion exchanger than corresponding Nd(III) complexes which indicates possibility of purification of Nd(III) from Y(III) in the macro-micro component system by the frontal analysis technique. In the process of Nd(III) purification from Y(III) the best separation results were obtained using the anion exchanger Dowex 1 of the 4% DVB cross linking. The amount of neodymium(III) purified from yttrium(III) (under the same conditions) obtained using the anion exchangers Dowex 1x2 and Dowex 1x8 is comparable and lower than using the anion exchanger Dowex 1x4. However, the anion exchanger Dowex 2x8proved to be useless in the process of Nd(III) purification from Y(III) (Fig. 14).

The affinity of the complexes of the $[LnH(hedta)_2]^{2-}$ type is greater for heavy lanthanides(III) such as Dy(III), Ho(III) and Er(III) than for Y(III) which suggests possibility of purification of Y(III) macro quantities from Dy(III), Ho(III) and Er(III) micro quantities. As follows from the studies, similar to the pair Nd(III)-Y(III) the effectiveness of the pairs Y(III)-Ho(III) and Y(III)-Er(III) separation on strongly basic gel polystyrene anion exchangers is greater at solutions pH 7.5 than pH 4.0. Of all studied anion exchangers Dowex 1x4 gave the best results in purification of the above mentioned rare earth elements(III) complexes with HEDTA. The anion exchanger Dowex 1x4 was also used for separation of the pair Y(III)-

Dy(III). The obtained results show that on 1 dm^3 of this anion exchanger in the HEDTA form it is possible to obtain 16 g of yttrium(III) (counted over Y_2O_3) in which the dysprosium(III) content can be reduced over 10 times i.e. from 0.365% Dy_2O_3 to the value below 0.035%.

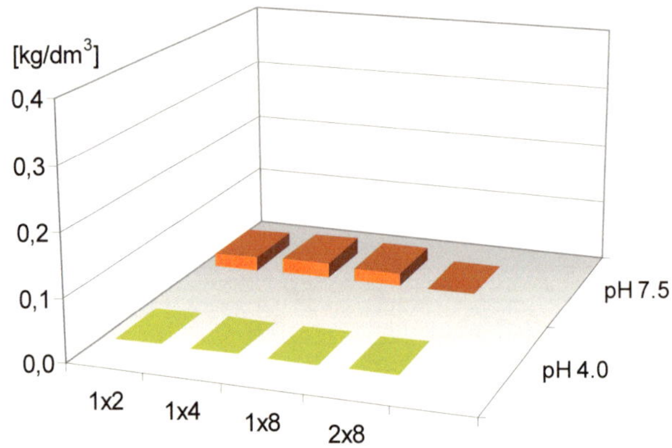

Figure 14. Separation of Nd(III) from Y(III) (0.35%) on Dowex 1x2, Dowex 1x4, Dowex 1x8 and Dowex 2x8 in the HEDTA form (1 dm^3 of anion exchanger in the Cl^- form, Ln(III):HEDTA=1:1, pH 4.0 and 7.5, 1.5g Ln_2O_3/dm^3).

In the determined affinity series of rare earth elements(III) with HEDTA for strong basic anion exchangers of the polystyrene skeleton one can see atypical position of Yb(III). Using the strongly basic anion exchanger Dowex 1 with different DVB cross linking, it was proved that for Dowex 1x2 in the HEDTA form, in the affinity series position of Yb(III) is near light rare earth: **Yb(III) > Nd(III) > Pr(III) > La(III).** However, its position is different for Dowex 1x4 in the HEDTA form: **Ho(III) > Y(III) > Sm(III) > Nd(III) > Yb(III)** and Dowex 1x8 in the HEDTA form **Ho(III) > Y(III) > Yb(III) > Sm(III) > Nd(III)**, which indicates that the purification process of Yb(III) from Ho(III) should be more effective on Dowex 1x4.

Based on the results of Yb(III) separation from Ho(III), Yb(III) purification from Er(III) as a micro component was carried out at pH 7.5 using Dowex 1x2 and Dowex 1x4 in the HEDTA form. In this case the anion exchanger Dowex 1x4 also proved to be more effective than Dowex 1x2 (Fig.15).

The obtained results of separation of the pairs Yb(III)-Ho(III) and Yb(III)-Er(III) are in agreement with the breakthrough curves and calculated values of distribution coefficients of the Yb(III), Ho(III) and Er(III) with HEDTA on the anion exchanger Dowex 1x2 in the HEDTA form (Hubicka & Drobek, 1997c). The agreement of the results in Yb(III) purification from Ho(III) and Er(III) with the affinity series of rare earth complexes with HEDTA leads to the suggestion that it is also possible to purify Yb(III) macro quantities from Dy(III), Tm(III) and Y(III) (Hubicka & Drobek, 1997c).

Non-monotonic and atypical the affinity series of anion lanthanide complexes with CDTA was also obtained for isotopes of these elements on the strongly basic anion exchanger Dowex 1x4 in the H_2cdta^{2-} form by Wódkiewicz and Dybczyński (1968): **Pm(III) > Nd(III) > Sm(III) > Pr(III) > Ce(III) > Eu(III) > Gd(III) > La(III) > Sc(III) > Tb(III) > Dy(III) > Ho(III) >**

Y(III) > Er(III) > Tm(III) > Yb(III) > Lu(III). Higher affinity of the Nd(cdta)- and Sm(cdta)-complexes for the anion exchangers than that of [Y(cdta)]- complexes was used for yttrium purification in the macro-micro component system by the frontal analysis technique on the polystyrene anion exchangers Dowex 1x2 and Dowex 1x4 (Hubicka, 1989a). Based on the obtained results it was pointed out that the affinity depends not only on the type of complexes, their structure but also on physicochemical properties of anion exchangers such as their form. The most effective form in the separation of $[Y(cdta)]^-$ from $[Nd(cdta)]^-$, $[Sm(cdta)]^-$ and $[Eu(cdta)]^-$ complexes proved to be the acetate from rather than the H_2cdta^{2-} one but the chloride form of the anion exchangers was completely useless.

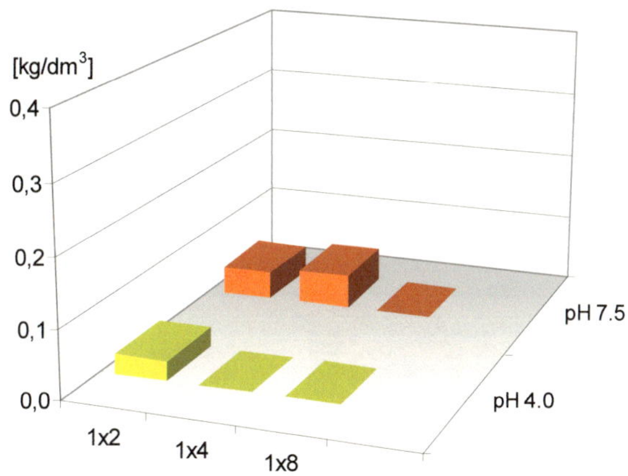

Figure 15. Separation of Yb(III) from Ho(III) (0.34%) on Dowex 1x2, Dowex 1x4 and Dowex 1x8 in the HEDTA form (1 dm^3 of anion exchanger in the Cl^- form, Ln(III):HEDTA=1:1, pH 4.0 and 7.5, 1.5g Ln_2O_3/dm^3).

Taking the above into consideration it was interesting to study applicability of the anion exchanger of the polyacrylate skeleton for separation of rare earth elements(III) complexes with CDTA (Hubicka & Kołodyńska, 2003). As follows from the breakthrough curves the Nd(III) and Sm(III) complexes with CDTA exhibit higher affinity for both strongly basic and weakly basic polyacrylate anion exchangers Amberlite IRA 458, Amberlite IRA 958 for the corresponding Y(III) complexes and their affinity is arranged in the same order as for strongly basic polystyrene anion exchangers.

Assuming that the complexes 1:1 are formed, the anion exchange reaction can be written as:

$$RCH_3COO + Ln(cdta)^- \rightleftarrows RLn(cdta) + CH_3COO^-$$

The obtained data indicate that the strongly basic, polyacrylate, gel anion exchanger Amberlite IRA 458 is more effective in purification of Y(III) from Nd(III) in comparison with the strongly basic, macroporous anion exchanger Amberlite IRA 958.

It was shown that the weakly basic polyacrylate gel anion exchanger Amberlite IRA 68 is more effective than the strongly basic, gel anion exchangers of polyacrylate and polystyrene types (Fig. 16). On 1 dm^3 of this anion exchanger it is possible to obtain about 90 g Y_2O_3 in which the Nd_2O_3 content was reduced from 0.35% to below 0.005% and about 81 g Y_2O_3 in which the Sm_2O_3 content was reduced from 0.36% to below 0.015%. Therefore, polyacrylate anion exchangers with respect to their applicability in purification of Y(III) from Nd(III) and Sm(III) complexes with CDTA can be arranged as follows: weakly basic, gel > strongly basic, gel > strongly basic, macroporous (Hubicka & Kołodyńska, 2003; Hubicka & Kołodyńska, 2004).

The analogous results were obtained using the monodisperse anion exchangers Lewatit MonoPlus M 500, Lewatit MonoPlus M 600, Lewatit MonoPlus MP 500, Lewatit MonoPlus MP 64 and the heterodisperse anion exchanger Lewatit MP 62 (Hubicka & Kołodyńska, 2008). The data indicate that Lewatit MonoPlus M 500 (type 1) in the acetate form is more effective in purification of Y(III) from Nd(III) and Y(III) from Sm(III) in comparison with Lewatit MonoPlus M 600 (type 2).

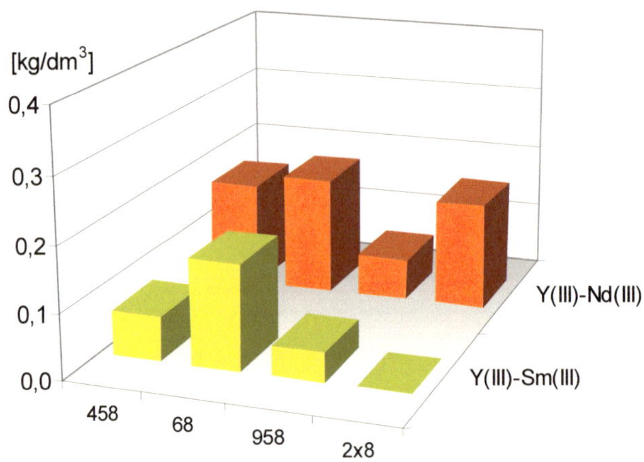

Figure 16. Separation of Y(III) from Sm(III) (0.36%) and Y(III) from Nd(III) (0.35%) on Amberlite IRA 458, Amberlite IRA 68, Amberlite IRA 958 and Dowex 2x8 in the CDTA form (1 dm^3 of anion exchanger in the Cl^- form, Ln(III):CDTA=1:1, pH 4.8, 1.5g Ln_2O_3/dm^3).

On 1 dm^3 of this anion exchanger, it is possible to obtain approximately 79 g Y_2O_3 in which the Nd_2O_3 content was reduced from 0.35% to below 0.005% and approximately 70 g Y_2O_3 in which the Sm_2O_3 content was reduced from 0.35% to below 0.015% (data not presented). It was also shown that purification of Y(III) from Nd(III) in the system with CDTA on the polystyrene anion exchangers is less effective than in the system with EDTA (Hubicka & Kołodyńska, 2007). Smaller differentiation in affinity of the $[Ln(cdta)]^-$ complexes than the $[Ln(edta)]^-$ complexes on anion exchangers can be justified by larger 'flexibility' of EDTA compared with the 'rigidity' structure of CDTA.

Ion exchange reactions proceed not only in the aqueous system but also in the non-aqueous solvents such as alcohols, ketones, glycols, etc. as well as in the mixed systems, e.g. water-alcohol, water-ketone, etc. Application of non-aqueous and mixed solvents in the anion exchange of metal complexes has increased considerably by a number of possible separations. In most cases the distribution coefficients of metal complexes in mixed media like water-alcohol, water-acetone etc. are larger than those in pure water solutions (Moody & Thomas, 1968; Marcus, 1983). Therefore, using an unusual order of affinity series of anion lanthanide complexes with EDTA of $Ln(edta)^-$ type for the strongly basic anion exchanger in the H_2edta^{2-} form (Dybczyński, 1964; Dybczyński, 1970): **Sm(III) > Eu(III) > Gd(III) > Nd(III) > Pr(III) > Tb(III) > Ce(III) > Dy(III) > La(III) > Ho(III) > Y(III) > Er(III) > Sc(III) > Tm(III) > Yb(III) > Lu(III)** and higher affinity of the $[Nd(edta)]^-$ complexes than that of $[Y(edta)]^-$ complexes the yttrium(III) purification as a macro component from neodymium(III) by the frontal analysis technique in the solvent organic system were also carried out. The results of separation of Y(III) from Nd(III) (Nd_2O_3 0.35%) in the presence of EDTA without and with 10%(v/v) and 20%(v/v) addition of methanol, 1-propanol, 2-propanol and acetone on Lewatit MonoPlus M 500, Lewatit MonoPlus M 600, Lewatit MonoPlus MP 500, Lewatit MonoPlus MP 64 and Lewatit MP 62 are presented in Fig. 17.

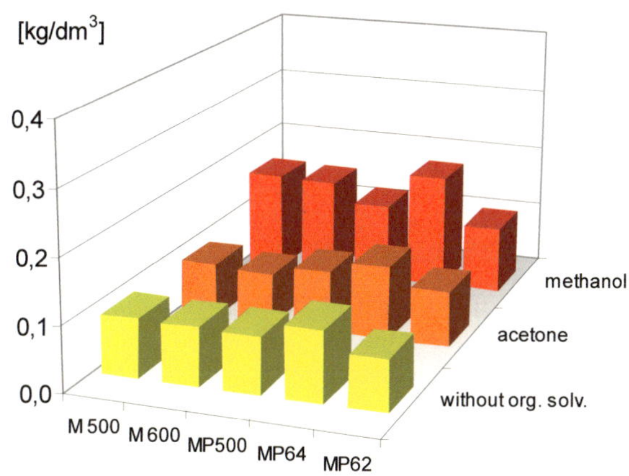

Figure 17. Separation of Y(III) from Nd(III) (0.35%) on Lewatit MonoPlus M 500, Lewatit MonoPlus M 600, Lewatit MonoPlus MP 500, Lewatit MonoPlus MP 64 and Lewatit MP 62 in the H_2O-20 % (v/v) polar organic solvent systems (1 dm^3 of anion exchanger in the Cl^- form, Ln(III):EDTA=1:1, pH 4.8, 1.5g Ln_2O_3/dm^3).

Taking into consideration the effect of the addition of polar organic solvent on the effectiveness of separation of rare earth element(III) complexes with EDTA on the above mentioned anion exchangers the best results of Y(III) from Nd(III) purification are obtained in 20% (v/v) methanol system on Lewatit MonoPlus M 500 and Lewatit MonoPlus MP 64. Taking 1 dm^3 bed of Lewatit MonoPlus M 500 it is possible to obtain above 158 g Y_2O_3 in which the Nd_2O_3 content can be reduced from 0.35% to below 0.005%, whereas 182 g Y_2O_3 for Lewatit MonoPlus MP 64. The yield obtained in these processes is above 70-100% higher than that obtained in the aqueous solutions. However, the increase of methanol addition up to 20% (v/v) increases accordingly the yield of purified Y_2O_3 only by about 15%, but the increase of methanol addition up to 50% (v/v) already affects insignificantly the yield increase in this process (Hubicka & Kołodyńska, 2005; Hubicka & Kołodyńska, 2007).

Analogous studies on the separation of Y(III)-Nd(III) with EDTA were also carried out in the presence of ethanol, 1-propanol, 2-propanol and acetone. However, only 20% (v/v) addition of acetone to the system increases effectiveness of separation process compared to the aqueous solutions.

Moreover the analogous studies were carried out for the IDA complexing agent. The data obtained indicate that Lewatit MonoPlus M 500 and Lewatit MonoPlus M 600 in the acetate form in aqueous solutions are the most effective in separation of the $[Sm(ida)_2]^{-}$-$[Ho(ida)_2]^{-}$ pair (Fig.18).

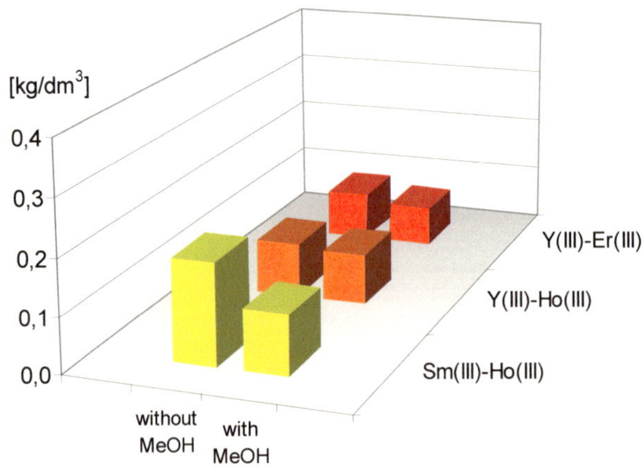

Figure 18. Fig. 18. Separation of Sm(III) from Ho(III) (0.35%), Y(III) from Ho(III) (0.36%) and Y(III) from Er(III) (0.35%) on Lewatit MonoPlus M 500 in the H_2O-20 % (v/v) CH_3OH (1 dm^3 of anion exchanger in the Cl^- form, Ln(III):IDA=1:1, pH 5.0, 1.5g Ln_2O_3/dm^3).

Taking into account the effect of types of functional groups $-N^+(CH_3)_3$ (Lewatit MonoPlus M 500) and $-N^+(CH_3)_2C_2H_4OH$ (Lewatit MonoPlus M 600) on the results of separation of rare earth element(III) complexes with IDA both for the Sm(III)-Ho(III) and Y(III)-Ho(III) pairs, the anion exchanger Lewatit MonoPlus M 500 (type 1) of slightly larger basicity of the functional group compared with the anion exchanger Lewatit MonoPlus M 600 (type 2). However, the addition of 10% (v/v) or 20% (v/v) methanol proved to be disadvantageous both for Lewatit MonoPlus M 500 and Lewatit MonoPlus M 600.

Due to the fact that the addition of polar solvent improves the efficiency of the separation of rare earth elements(III) only for the selected complexing agents, the possibility of selective separation of nitrate complexes of rare earth elements(III) in a polar organic solvent-H_2O-HNO_3 by the frontal analysis was also examined in the case of the absence of complexing agents. In the studies the following parameters were taken into account: type of the functional groups, type of the anion exchanger skeleton, the degree of cross linking, porosity and grain size as well as the effect of type and concentration of polar organic solvent, the presence of another organic solvent and the concentration of nitric(V) acid as well as the concentration of rare earth elements(III) in solution (Hubicki & Olszak, 1994; Hubicki et al. 1995; Hubicki & Olszak, 1996a; Hubicki & Olszak, 1996b; Hubicki et al. 1996).

Taking into account the obtained affinity series for the anion exchangers Wofatit SBWx4%, Wofatit SBKx7% and Wofatit SBWx6%in the systems:

90% v/v CH_3OH-10% v/v 7M HNO_3 – Wofatit SBWx4%

Nd(III) > Pr(III) > **Sm(III)** > Ce(III) > La(III) > Eu(III) > Gd(III) > Tb(III) > **Y(III)** > Dy(III) > Tm(III) ≥ Ho(III) ≥ Er(III) > Yb(III),

90% v/v CH_3OH-10% v/v 7M HNO_3 – Wofatit SBKx7%

Pr(III) ≥ **Nd(III)** > La(III) > **Sm(III)** > Eu(III) > Gd(III) > Tb(III) > Dy(III) ≥ Ho(III) = Er(IIII) = Tm(III) = Yb(III) > **Y(III),**

90% v/v C_2H_5OH-10% v/v 7M HNO_3 – Wofatit SBWx4%

Nd(III) > La(III) ≥ Pr(III) > **Sm(III)** > Eu(III) > Gd(III) > Tb(III) > Dy(III) > Ho(III) ≥ Er(III) > Yb(III) ≥ **Y(III)** ≥ Tm(III),

90% v/v C_2H_5OH-10% v/v 7M HNO_3 – Wofatit SBKx7%

Yb(III) ≥ La(III) ≥ **Y(III)** = Pr(III) = **Nd(III)** = Tb(III) = Er(III) ≥ Eu(III) = Dy(III) = Ho(III) ≥ **Sm(III)** = Gd(III) = Tm(III),

90% v/v CH_3COCH_3-10% v/v 7M HNO_3 – Wofatit SBWx4%

Nd(III) > La(III) ≥ Pr(III) > **Sm(III)** > Eu(III) > Gd(III) > Tb(III) > Er(III) > Dy(III) > Ho(III) = **Y(III)** ≥ Tm(III) > Yb(III),

90% v/v CH_3COCH_3-10% v/v 7M HNO_3 – Wofatit SBKWx7%

Nd(III) > La(III) ≥ Pr(III) > **Sm(III)** > Eu(III) > Gd(III) > Tb(III) > Dy(III) > Ho(III) ≥ Er(III) > Yb(III) ≥ **Y(III)** ≥ Tm(III),

it was found that the distribution coefficients have the largest values in the CH_3OH system.

The differences in affinity of rare earth(III) elements nitrate complexes for anion exchangers in individual systems, for example Y(III)-Nd(III), Sm(III)-Nd(III), are probably due to a different structure of complexes (degree of their solvation) or different stability or kinetics of their formation in the anion exchanger phase. Probably in the resin phase the $[Nd(NO_3)_5]^{2-}$ complexes are formed, whereas Y(III) forms the complexes of $[Y(NO_3)_4]^-$ (Korkish, 1968). It was also found that macroporous anion exchangers give better separation results whereas using strongly basic anion exchangers with the pyridine functional group Permutit SK the obtained results were worse. The analogous situation was in the case of Amberlite IRA 938 and Wofatit SBWx2% (Hubicki et al. 1996; Hubicki and Olszak, 1998a; Hubicki and Olszak, 1998b; Hubicki and Olszak, 1998c; Hubicki and Olszak, 1998d; Hubicki and Olszak, 1998e).

Additionally, it should be emphasized that these non-typical affinity series give the opportunity to obtain ion exchange separation of rare earth elements. Very interesting is the position of Y(III) in the obtained affinity series on Wofatit SBKx7%in the 90% v/v CH_3OH-10% v/v 7M HNO_3 system (Hubicki and Olszak, 1994; Hubicki et al. 1995). As follows from the theory of yttrium(III) migration in the lanthanide series, it behaves like pseudolanthanide(III) when the central ion-ligand bonding in the complex is covalently shortened and behaves like heavy lanthanides(III) when this bonding is of ionic character. In the studied system yttrium(III) behaves like heavy lanthanide(III).

Noteworthy are also the results of treatment of macro quantities of yttrium(III) from neodymium(III) in the 90% v/v CH_3OH–10% v/v 7 M HNO_3 system on the strongly basic anion exchanger Wofatit SBW with the degree of cross linking 4 and 6% DVB. On this anion exchanger, the effect of nitric(V) acid concentration in the 90% v/v CH_3OH–10% v/v 0.1-9 M HNO_3 system as well as the effect of the addition of ammonium nitrate(V) in the 90% v/v CH_3OH–10% v/v 0,1 M HNO_3–1-5 M NH_4NO_3 system were also studied (Fig.19).

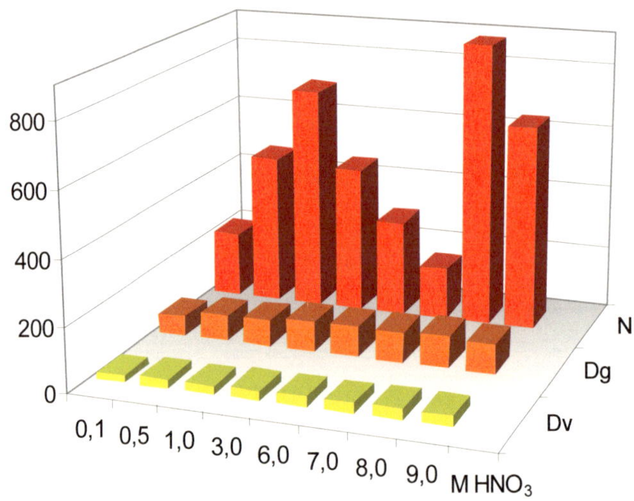

Figure 19. Mass (D_g) and volume (D_v) distribution coefficients as well as the number of the theoretical plates (N) of Wofatit SBWx6% in the dependence of HNO_3 concentration (0.1-9.0 M).

It was shown that with the increasing concentration of nitric(V) acid, the efficiency of sorption and separation of nitrate complexes of rare earth elements(III) initially increases in the systems 7-8 M HNO_3–CH_3OH, 3 M HNO_3–C_2H_5OH and 7 M HNO_3–CH_3COCH_3 and then decreases. The results of the effect of HNO_3 concentration on the efficiency of the purification process of Y(III) from Nd (III) (<0.001%Nd_2O_3) on the anion exchanger Wofatit SBWx4 with a varyingdegree of the cross linking is shown in Fig.20a. The analogous results obtained for the Sm(III)-Nd(III) pair are shown in Fig. 20b.

The type of skeleton of the used ion exchanger is also an important parameter. For the polyacrylate anion exchangers Amberlite IRA 458, Amberlite IRA 958 and Purolite A 850 the efficiency of the sorption process is low. Mass of the purified yttrium oxide(III) (<0.001%Nd_2O_3) was equal to 0.18 kg/dm^3 for Amberlite IRA 458; 0.26 kg/dm^3 for Amberlite IRA 958 and 0.09 kg/dm3 for Purolite A 850. The anion exchanger Amberlite IRA 68 proved to be completely useless. However, the best results were obtained on the strongly basic anion exchangers of type 1. For example, using Lewatit MP 5080 with the macroporous 25 skeleton structure there was achieved 1.9 kg Y_2O_3/dm^3 (<0.001%Nd_2O_3) in the methanol system while in the acetone system 2.3 kg Y_2O_3/dm^3 (<0.001%Nd_2O_3) (Fig.21).

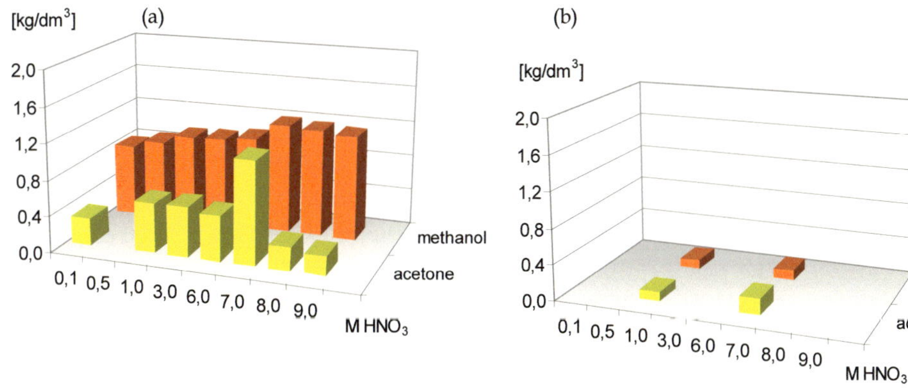

Figure 20. a-b. Effect of the HNO_3 concentration on the effectiveness of Y(III) from Nd(III) separation (<0.001%Nd_2O_3) on Wofatit SBWx4 (a) and Sm(III) from Nd(III) (<0.001%Nd_2O_3) on Wofatit SBWx6 (b) at different HNO_3 concentrations in the methanol and acetone systems.

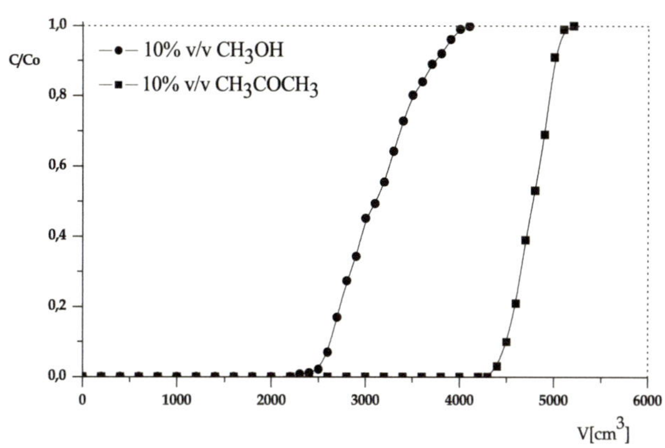

Figure 21. Separation of Y(III) fromNd(III) (<0.001%Nd_2O_3) on Lewatit MP 5080 in the 90% v/v CH_3OH–10% v/v 7 M HNO_3 and 90% v/v CH_3COCH_3–10% v/v 7 M HNO_3 systems.

Good results were also obtained on Zerolit FF IP 2-3%, Zerolit 3-5% FF IP and Wofatit SBWx6%. In in the case of Zerolit FF IP 7-9% more than two fold decrease of the efficiency of purification process of Y(III) from Nd(III) was obtained (Hubicki et a. 1994).

The influence of cross linking degree on the effectiveness of separation of Y(III)-Nd (III) pair was also studied. For the ion exchanger Dowex it increases from 2% to 6% DVB and then decreases. For the systems with methanol this effect is observed at 4% DVB. For Zerolit FF IP the highest sorption capacities were obtained at the degree of cross linking equal to 2-3%, which may be associated with the sieve effect (Hubicki & Olszak, 1996; Hubicki, et al. 1996). In comparison with the separation of yttrium(III) from neodymium(III) the effects of the purification process of samarium(III) from neodymium(III) are much lower. This is due to a smaller difference between the distribution coefficients of Sm(III)-Nd(III) in comparison with the Y(III)-Nd(III) pair. Taking into account the effectiveness of this sorption process gel and macroporous anion exchangers can be arranged as follows: in the $Ln(NO_3)_3$–90% v/v

CH_3OH–10% 7 M HNO_3 system: Lewatit MP 5080 (0.26 kg/dm^3) > Wofatit SBWx6% (0.16 kg/dm^3) > Dowex 1x4 (0.11 kg/dm^3) > Wofatit SBKx7% (0.07 kg/dm^3) > Wofatit SBWx7% (0.04 kg/dm^3) > Dowex 1x8 (0.031 kg/dm^3) > Dowex 1x2 (0.03 kg/dm^3) > Wofatit SBWx4% (0.02 kg/dm^3) > Dowex 1x10 (0.01 kg/dm^3) > Dowex 1x1 (0.009 kg/dm^3) (Hubicki & Olszak, 2000a; Hubicki & Olszak, 2001). The strongly basic anion exchanger Lewatit MP 5080 proved to be the most effective (Hubicki & Olszak, 2001). Depending on the separation system the obtained results are in the range from 0.26 to 0.24 kg of Sm_2O_3 (0.001 % Nd_2O_3) on 1 dm^3 of the ion exchanger. The smaller results were obtained on Wofat SBWx6 - 0.18 kg Sm_2O_3 (0.001% Nd_2O_3)/dm^3). Much smaller yield of Sm(III) from Nd(III) purification compared with that of Y(III) from Nd(III) is caused by a smaller difference between the distribution coefficients of Sm(III)-Nd(III) compared with Y(III)-Nd(III). Different affinities of neodymium(III) and samarium(III) nitrate complexes in the systems $Ln(NO_3)_3$–90% v/v CH_3OH–10% 1 M or 7 M HNO_3 as well as in $Ln(NO_3)_3$–90% v/v CH_3COCH_3–10% 1 M or 7 M HNO_3 are probably caused by the formation of neodymium nitrate complexes with a higher negative charge, e.g. $[Nd(NO_3)_5]^{2-}$ than that of samarium(III) complexes , e.g. $[Sm(NO_3)_4]^-$) or by their different structures (possibly by the differences in the degree of solvation) and the kinetics of formation of neodymium(III) and samarium(III) anionic nitrate complexes (Hubicki, et al. 1996; Hubicki and Olszak, 2000a; Hubicki and Olszak, 2000b; Hubicki and Olszak, 2001; Hubicki and Olszak, 2002).

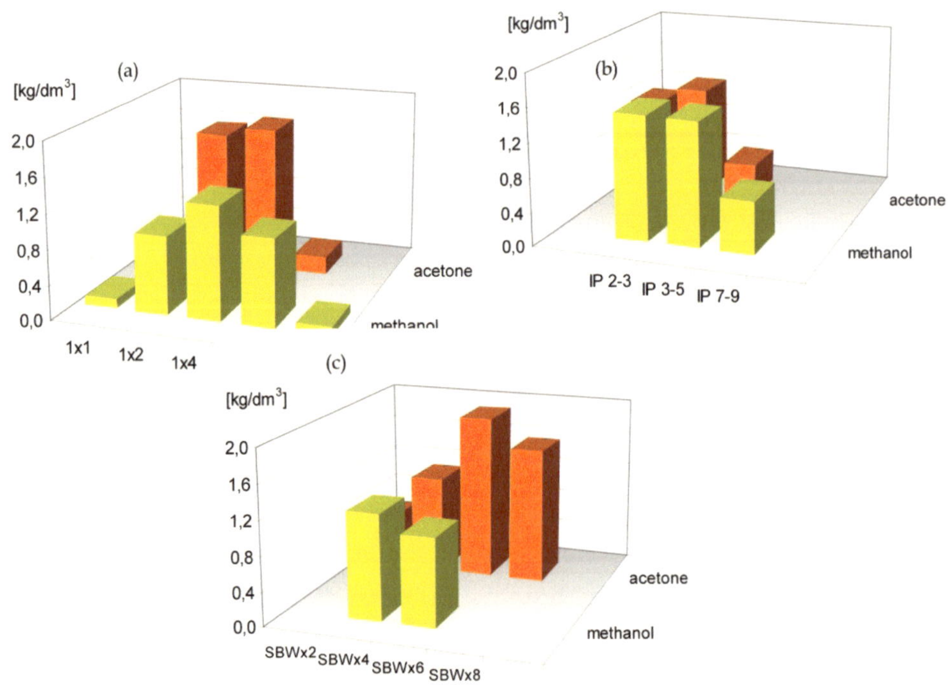

Figure 22. a-c. Separation of Y(III) from Nd(III) (<0.001%Nd_2O_3) on Dowex 1 (a), Zerolit FF IP (b) and Wofatit SBW (c) with different cross linking in the methanol and acetone systems.

Determination which of the used organic solvents possesses the best sorption properties is not possible.

However, it should be noted that using 1-propanol and 2-propanol the efficiency of the separation process is lower than for methanol and acetone (data not presented). It was also shown that for dimethylosulphoxide (DMF) in a 90% v/v DMF-10% v/v 7 M HNO_3 system separation does not occur.

In industrial processes of ion exchange separation of rare earth elements(III) on cation exchangers using aminopolycarboxylic acids or hydroxyacids as eluents, the concentration of lanthanides(III) in the eluate rarely exceeds a few grams per liter. In the present study, the effect of concentration of rare earth elements(III) on the efficiency of Y(III)-Nd(III) pair separation in the macro-micro component was investigated. It was found that in the methanol and ethanol systems the increase in the concentration of purified yttrium(III) to 50 g Ln_2O_3/dm^3 practically does not affect the efficiency of the process. However, further increase of Ln_2O_3 concentration decreases the efficiency of the ion exchange column. Very good results were obtained at a concentration of 50 g Ln_2O_3/dm^3 in the acetone system.

The effect of the percentage concentration of Nd(III) on the effectiveness of the separation process was also examined. Both, in the case of yttrium(III) and samarium(III) separation at 0.1% of neodymium(III) content the effectiveness of the process is much better.

The sorption of nitrate(V) complexes of rare earth elements(III) is also dependent not only on the type of solvent used but also on its concentration in the mixture. This fact was confirmed by the results of the separation of yttrium(III) from neodymium(III). The addition of an organic solvent such as methanol or acetone (with a lower polarity and dielectric constant than water) generally reduces the degree of dissociation of rare earth elements(III) and increases their tendency to form ion complexes. At the same time the distribution coefficients of lanthanides(III) on anion exchangers are also changed.

It should be mentioned that the anion exchangers can be regenerated with water in the amount of 2–4 bed volumes after the purification process of rare earth elements. The relatively high yield of rare earth elements purification, the low costs and the simple and cheap regeneration of the anion exchanger bed when the purification process is over, as well as the possibility of methanol recovery from the eluate by using the distillation method suggest the possibility of using this process in technologies for high purity rare earth production.

13. Conclusions

Rare earth elements(III) are extremely important for the development of economy and technology. Among others, they are used for the production of electronic equipment, in automotive, aerospace, missile, military industries and even in medical diagnostics.

Atypical affinity series create new possibilities of ion exchange separation of rare earth elements(III) which is very significant from a practical point of view. The results of rare earth elements nitrate complexes separation in the micro-macro component system as well as in the presence of aminopolycarboxylic acids and organic solvents can be successfully applied in production of rare earth elements(III), particularly yttrium(III), ytterbium(III), samarium(III) and lanthanum(III) of a large purity degree. It is worth mentioning that regeneration of anion exchangers in these systems is very economical using distilled water as a regenerating factor.

Author details

Dorota Kołodyńska and Zbigniew Hubicki
Maria Curie-Skłodowska University, Faculty of Chemistry, Department of Inorganic Chemistry, Lublin, Poland

14. References

Adachi G.Y., Imanaka N., Tamura S. (2010) Research trends in rare earths: A preliminary analysis, *Journal of Rare Earths*, Vol. 28, pp. 843-846

Arnold R., Son Hing, L.B. (1967) Selectivity of carboxylic ion-exchange resin for lanthanide ions, *Journal of the Chemical Society A: Inorganic, Physical, Theoretical*, pp. 306-308.

Atanasyants, A.G., Seryogin, A.N. (1995) The reaction of the electrochemical reduction Eu (III)+e^{-}→Eu(II) in hydrochloric solution, *Hydrometallurgy*, Vol. 37, pp. 367-374.

British Geological Survey (2011) http://www.bgs.ac.uk/research/highlights/2010/rare_earth_elements.html.

Brücher, E., Király, R., Ngypál, I. (1975) Equilibrium relations of rare earth ethylenediaminetetraacetate complexes in the presence of a ligand excess, *Journal of Inorganic and Nuclear Chemistry*, Vol. 37, pp. 1009-1012.

Buddery, J.H., Jamrack, W.D. ,Wells, R.A. (1959) The extraction of thorium chemistry and industry, *Chemical Industry*, Vol. 8, pp. 235-244.

Bünzli, J.C.G., Comby, S., Chauvin, A.S., Vandevyver, C.D.B (2007) New opportunities for lanthanide luminescence, *Journal of Rare Earths*, Vol. 25, pp. 257-274.

Charewicz, W. (1990) *Rare Earth Elements*, WNT, Warsaw (in Polish).

Cortina, J.L.; Warshawsky, A. (1997) *Developments in solid-liquid extraction by solvent impregnated resins.* In Ion exchange and solvent extraction, Marinsky, J.A.; Marcus, Y., (Eds.) Marcel Dekker, Inc., New York, Vol. 13, pp. 195-293.

Cotton, S. (2006) *Lanthanide and actinide chemistry*, Wiley-VCH, Chichester, England.

Courtney, R.C., Gustafson, R.L., Chaberek, S., Martell A.E. (1958) Hydrolytic tendencies of metal chelate compounds. II. Effect of metal ion , *Journal of American Chemical Society*, Vol. 80, pp. 2121-2128.

Chegwidden, J., Kingsnorth, D.J. (2002) Rare earth supply and demand. A European market focus, *Industrial Minerals*, Vol. 4, pp. 52-61.

China Ministry of Environmental Protection, 2011, Emission standards of pollutants from rare earths industry GB 26451–2011: Beijing, China, China Ministry of Environmental Protection Standard GB 26451–2011, January 24.

Choppin, G.R., Chopoorian J.A. (1961) Complexes of the lanthanide elements with α-hydroxy carboxylate ligands, *Journal of Inorganic and Nuclear Chemistry*, Vol. 22, pp. 97-113.

Christell, R., Forberg, S., Westermark, T. (1961) Some experiments on the use of the chelating ion exchanger Dowex A-1 in nuclear chemistry, *Journal of Inorganic and Nuclear Chemistry*, Vol.17 pp. 187-189

Danon, J. (1960) Determination of the stability constants of thorium nitrate complexes with anion-exchange resins, *Journal of Inorganic and Nuclear Chemistry*, Vol. 13, pp. 112-118.

De Vito, I.E., Masi, A.N., Osina, R.A. (1999) Determination of trace rare earth elements by X-ray fluorescence spectrometry after preconcentration on a new chelating resin loaded with thorin, *Talanta*, Vol. 49, pp. 929-935.

Dev, K., Pathak, R., Rao, G.N. (1999) Sorption behaviour of lanthanum(III), neodymium(III), terbium(III), thorium(IV) and uranium(VI) on Amberlite XAD-4 resin functionalized with bicine ligands, *Talanta*, Vol. 48, pp. 579-584.

Du, X. Graedel, T.E. (2011) Uncovering the global life cycles of the rare earth elements, Scientific Reports, doi: 10.1038/srep00145.

Dybczyński, R. (1964) Separation of rare earths on anion exchange resins. IV. Influence of temperature on anion exchange behaviour of the rare earth ethylenediaminetetracetates, *Journal of Chromatography*, Vol. 14, pp. 79-96.

Dybczyński, R. (1970) Effect of resin cross-linking on the anion exchange separation of rare earth–EDTA complexes, *Journal of Chromatography*, Vol. 50, pp. 487-503.

Faris, J.P. (1967) Separation of the rare earths by anion exchange in water-solvent mixtures in the presence of α-hydroxyisobutyric acid, *Journal of Chromatography*, Vol. 26, pp. 232-238.

Faris, J.P., Wharton, W.J. (1962) Anion exchange resin separation of the rare earths, yttrium, and scandium in nitric acid-methanol mixtures, *Analytical Chemistry*, Vol. 34, pp. 1077-1080.

Fitch, F.T., Russel, D.S., Mitchell, J. (1951) Determination of lanthanum in rare earth mixtures, *Analytical Chemistry*, Vol. 23, pp. 1469-1473.

Fricker, S.P. (2006) The therapeutic application of lanthanides, *Chemical Society Reviews*, Vol. 35, pp. 524-533

Fontana D., Pietrelli, L. (2009) Separation of middle rare earths by solvent extraction using 2-ethylphosphonic acid mono-2-ethylhexyl ester as an extractant, *Journal of Rare Earths*, Vo. 27, pp. 830-833.

Gschneidner, K.A. Jr. (1981) *Industrial application of rare earth*, American Chemical Society, Washington.

Goering, P.L., Fisher, B.R., Fowler, B.A., (1991) *The lanthanides*, In: Metals and their compounds in the environment, Merian, E. (ed.) VCH, Weinhein, New York, Basel, Cambridge, pp. 959-970.

Hale, W.H., Jr., Hammer, C.A. (1972) Cation exchange elution sequence with DTPA, *Ion Exchange Membranes*, Vol. 1, 380-381.

Hedrick, J.B. (1993) *Rare earths. The lanthanides, yttrium and scandium*, U.S. Department of the Interior, Bureau of Mines, Washington.

Hedrick, J.B. (1995) The global rare earth cycle, *Journal of Alloys and Compounds*, Vol. 225, pp. 609-618.

Hedrick, J.B. (2009) Rare earths: U.S. Geological Survey Minerals Yearbook, 2007, Vol. I, pp. 60.1-60.19. (http://minerals.er.usgs.gov/minerals/pubs/commodity/rare_earths/myb1-2007-raree.pdf.)

Horwitz, E.P., Chiarizia, R., Diamond, H., Gatrone, R.C., Alexandratos, S.D., Trochimczuk, A.W., Crick, D.W. (1993) Uptake of metal ions by a new chelating ion exchange resin. Part. 1: Acid dependences of actinide ions, *Solvent Extraction and Ion Exchange*, Vol. 11, pp. 943-966.

Horwitz, E.P., Schulz, W.W. (1999) *Solvent extraction in the treatment of acidic high-level liquid waster: where do we stand?* In: *Metal ion separation and preconcentration: progress and opportunities*; Bond, A.H., Dietz, M.L., Rogers, R.D. (eds.) ACS Symposium Series 712, American Chemical Society, Washington, D.C., pp. 20–50.

Hubicka H. (1989a) Studies on separation of rare earth element complexes with trans-1,2-diaminocyclohexane-N,N,N′,N′-tetraacetic acid on strongly basic anion exchangers, *Hungarian Journal of Industrial Chemistry*, Vol. 17, pp. 257-265.

Hubicka, H. (1989b) Studies on purification of yttrium from heavy lanthanides using chelating ion exchangers on amino acid and phosphonic types, *Hungarian Journal of Industrial Chemistry*, Vol. 17, pp. 355-363.

Hubicka, H., Drobek, D. (1997a) Studies of ion-exchange purification process of Sm(III) complexes with HEDTA from Ho(III) on anion-exchangers, *Hungarian Journal of Industrial Chemistry*, Vol. 25, pp. 99-102.

Hubicka, H., Drobek, D. (1997b) Studies of the effect of strongly basic anion-exchanger cross-linking on separation of Y(III) complexes with HEDTA from Ho(III) and Er(III), *Hungarian Journal of Industrial Chemistry*, Vol. 25, 215-218.

Hubicka, H., Drobek D. (1997c) Anion-exchange method for separation of ytterbium from holmium and erbium, *Hydrometallurgy*, Vol. 47, pp. 127-136.

Hubicka, H., Drobek D. (1998) Studies of separation of iminodiacetate complexes of lanthanum(III) from neodymium(III) and praseodymium(III) on anion-exchangers, *Hydrometallurgy*, Vol. 50, pp. 51-60

Hubicka, H., Drobek D., Kogut, E. (1998) Studies on purification of neodymium(III) from yttrium(III) complexes with IMDA and HEDTA on anion-exchangers, *Chemical and Environmental Research*, Vol. 7, pp. 229-241

Hubicka, H., Drobek D. (1999) Separation of Y(III) complexes from Dy(III), Ho(III) and Er(III) complexes with iminodiacetic acid on the anion-exchangers type 1 and type 2, *Hydrometallurgy*, Vol. 53, pp. 89-100.

Hubicka, H., Drobek D. (2000a) Studies on separation of intermediate and heavy lanthanide complexes with iminodiacetic acid on anion-exchangers, *Chemical and Environmental Research*, Vol. 9, pp. 245-257

Hubicka, H., Drobek D. (2000b) Study on separation of lanthanum from praseodymium complexes with IMDA by gel and macroporous anion-exchangers, *Journal of Rare Earths*, Vol. 18, pp. 90-96.

Hubicka, H., Hubicki, Z. (1978a) Selective separation of ytterbium from rare earth elements on chelate ion exchangers, *Materials Science*, Vol. 4, pp. 39-41.
Hubicka, H., Hubicki, Z. (1978b) , Separation of Yb^{3+},Tb^{3+}, Dy^{3+}, $Ho3^{+}$, Er^{3+} from Y^{3+} on various types of chelating ion exchangers, *Science Fascide, Silesia Technical University, Chemistry*, Vol. 86, pp. 157-168 (in Polish).
Hubicka, H., Hubicki, Z. (1980) Ion exchange separation of rare earth elements by α-hydroxyethyl-deno-1,1-diphoshonic acid, *Scientific Fascide, Silesia Technical University, Chemistry*, Vol. 100, pp. 61-68 (in Polish).
Hubicka, H., Hubicki, Z. (1982) Preparation of gadolinium of > 99.9% purity by ion exchange method, *Metal Ores*, Vol. 1, pp. 32-34 (in Polsh).
Hubicka, H., Hubicki, Z. (1983a) Ion exchange separation of rare earth elements by pyruvic acid, Part I., *Metal Ores*, Vol. 28, pp. 160-163 (in Polsh).
Hubicka, H., Hubicki, Z. (1983b) Ion exchange separation of heavy lanthanides from yttrium using pyruvic acid, Part. II., *Metal Ores*, Vol. 28, pp. 220-222 (in Polsh).
Hubicka, H., Hubicki, Z. (1986) Separation of rare earth - polyaminocarboxylic acids complexes on various types of anion-exchangers, *Solvent Extraction and Ion Exchange*, Vol. 4, pp. 383-399.
Hubicka, H., Hubicki, Z. (1992a) Studies on separation of yttrium(III) complexes from Tb(III) with nitryltriacetic acid on strongly basic anion exchangers, *Hungarian Journal of Industrial Chemistry*, Vol. 20, pp. 113-116.
Hubicka, H., Hubicki, Z. (1992b) Studies on separation of pair Y(III)-Nd(III) on chelating ion-exchangers of aminoacid type using aminopolyacetic acids as eluents, *Hungarian Journal of Industrial Chemistry*, Vol. 20, pp. 249-254.
Hubicka, H., Hubicki, Z. (1994) Separation of pair La(III)-Nd(III) on chelating ion exchanger of amino acid and phosphonic types, *Hungarian Journal of Industrial Chemistry*, Vol. 22, pp. 171-176.
Hubicka, H., Kołodyńska D. (2000) Investigation into the use of macroporous anion-exchangers for the sorption and separation of iminodiacetate complexes of lanthanum(III) and neodymium(III), *Adsorption Science and Technology*, Vol. 18, pp. 719-726.
Hubicka, H., Kołodyńska D. (2003) Separation of $Y(dcta)^{-}$ complexes from $Nd(dcta)^{-}$ and $Sm(dcta)^{-}$ complexes on polyacrylate anion-exchangers, *Journal of the Serbian Chemical Society*, Vol. 68, pp. 183-190.
Hubicka, H., Kołodyńska D. (2004) Separation of rare earth element complexes with trans-1,2-diaminocyclohexane-N,N,N',N'-tetraacetic acid on the polyacrylate anion-exchangers, *Hydrometallurgy*, Vol. 71, pp. 343-350.
Hubicka, H., Kołodyńska D. (2004) Studies of applicability of strongly and weakly basic polystyrene and polyacrylate anion exchangers for separation of Y(edta)- from Sm(edta)- complexes, *Chemical and Environmental Research*, Vol. 13, pp. 73-85.
Hubicka, H., Kołodyńska D. (2005) Studies of separation of Nd(III), Sm(III) and Y(III) complexes on polystyrene anion exchangers with EDTA in the H_2O – polar organic solvent, *Chemical and Environmental Research*, Vol. 14, pp. 23-36.

Hubicka, H., Kołodyńska D. (2007) Application of strongly basic monodisperse anion exchangers in sorption and separation of Y(edta)$^-$ from Nd(edta)$^-$ complexes, *Chemical and Environmental Research*, Vol. 16, pp. 79-93.

Hubicka, H., Kołodyńska D. (2008) Application of monodisperse anion exchangers in sorption and separation of Y^{3+} from Nd^{3+} complexes with DCTA, *Journal of Rare Earths*, Vol. 26, pp. 619-625.

Hubicki, Z. (1989a) Anion-exchange resin modified with sulfonic derivatives of organic complexing reagents as a new type of functional resin for the selective separation of scandium(III) from yttrium(III) and lanthanum(III), *Hungarian Journal of Industrial Chemistry*, Vol. 17, pp. 51-60.

Hubicki, Z. (1989b) Studies of selective removal of U(VI), Th(IV) and Fe(III) from rare earth elements and their separation on chelating ion exchangers, *Applied Chemistry*, Vol. 33, pp. 205-218 (in Polish).

Hubicki, Z. (1990) Studies on selective separation of Sc(III) from rare earth elements on selective ion-exchangers, *Hydrometallurgy*, Vol. 23, pp. 319-331.

Hubicki, Z., Hubicka, H., Olszak, M. (1994) Investigations on separation of nitrate complexes of yttrium(III) from neodymium(III) on anion-exchangers of different cross-linking in the system CH_3OH-H_2O-HNO_3, *Hydrometallurgy*, Vol. 34, pp. 307-318.

Hubicki, Z. Hubicka, H., Olszak, M. (1994) Studies on separation of nitrate complexes of yttrium(III) from neodymium(III) on various types of anion-exchangers in the system CH_3OH-H_2O-HNO_3, *Hungarian Journal of Industrial Chemistry*, Vol. 22, pp. 177-182.

Hubicki, Z. Hubicka, H., Olszak, M. (1995) Investigations of the effect of nitric acid concentration on separation of yttrium(III) from neodymium(III) on the strongly basic anion-exchanger Wofatit SBWx4 in the CH_3OH-H_2O-HNO_3 system, *Hungarian Journal of Industrial Chemistry,* Vol. 23, pp. 5-10.

Hubicki, Z., Hubicka, H., Olszak, M. (1996a) Studies on separation of nitrate complexes of yttrium(III) from neodymium(III) on various type anion-exchangers in CH_3COCH_3-H_2O-HNO_3 system, *Hydrometallurgy,* Vol. 40, pp. 181-188.

Hubicki, Z., Hubicka, H., Olszak, M. (1996b) Studies of separation of Y(III) and Nd(III) nitrate complexes on the strongly basic anion-exchanger Wofatit SBW of the cross-linking 2-16% DVB in the CH_3OH-H_2O-HNO_3 System, *Hungarian Journal of Industrial Chemistry,* Vol. 24, pp. 51-56.

Hubicki, Z., Olszak, M. (1996) Studies of the effect of strongly basic anion exchanger Wofatit SBW cross-linking on the effectiveness of Y(III)-Nd(III) separation in the system C_2H_5OH-HNO_3, *Hungarian Journal of Industrial Chemistry,* Vol. 24, pp. 255-258.

Hubicki, Z., Olszak, M. (1998a) Studies on separation of rare earth element nitrate complexes on the C_2H_5OH-HNO_3 system on the strongly basic anion-exchanger Wofatit SBW, *Adsorption Science and Technology,* Vol. 16, pp. 487-492.

Hubicki, Z., Olszak, M. (1998b) Studies of lanthanum(III) nitrate complexes sorption on the strongly basic anion-exchanger Wofatit SBW of various cross-linking, *Adsorption Science and Technology,* Vol. 16, pp. 521-528.

Hubicki, Z., Olszak, M. (1998d) Studies of influence of percentage content of polar organic solvent on the sorption and separation of rare earth element nitrate complexes on anion-exchangers, *Adsorption Science and Technology,* Vol. 16, pp. 817-836.

Hubicki, Z., Olszak, M. (1998c) Studies on sorption and separation processes of rare earth element complexes on the anion-exchanger Wofatit in the CH_3OH-HNO_3 system, *Hydrometallurgy,* Vol. 50, pp. 261-268.

Hubicki, Z., Olszak, M. (1998e) Studies on separation of rare earth elements(III) from the organic-aqueous nitric acid on strongly basic anion-exchanger Wofatit SBK x 7, *Chemical and Environmental Research,* Vol. 7, pp. 333-340.

Hubicki, Z., Olszak, M. (2000a) Studies on separation of nitrate complexes of samarium(III) from neodymium(III) on the strongly basic anion-exchanger Wofatit SBW in the C_2H_5OH – 7 M HNO_3 system, *Adsorption Science and Technology,* Vol. 18, pp. 701-707.

Hubicki, Z., Olszak, M. (2000b) Studies on separation of samarium(III) nitrate complexes from neodymium(III) on strongly basic anion-exchangers, *Separation Science and Technology*, Vol. 35, 535-546.

Hubicki, Z., Olszak, M. (2001) Studies of the effect of crosslinking in the strongly basic anion exchanger Dowex1 and the HNO_3 concentration employed on the separation of the Sm(III)-Nd(III) pair in the polar organic solvent – H_2O – HNO_3 system, *Adsorption Science and Technology*, Vol. 19, pp. 219-228.

Hubicki, Z., Olszak, M. (2002) Studies on separation of rare earth elements on various types of anion exchangers in the C_3H_7OH-7 M HNO_3 systems, *Journal of Chromatography,* Vol. 955, pp. 257-262.

Inczedy, J. (1972) *Complexation equilibria in analytical chemistry*, PWN, Warsaw (in Polish).

Jain, V.K., Handa, A., Sait, S.S., Shrivastav, P., Agrawal, Y.K. (2001) Pre-concentration, separation and trace determination of lanthanum(III), cerium(III), thorium(IV) and uranium(VI) on polymer supported o-vanillinsemicarbazone, *Analytical Chimica Acta,* Vol. 429, pp. 237-246.

Jegorov, E., Makarova, A. (1971) Jonnyj obmien w radiochimii, Atomizdat, Moskov (in Russia).

Jensen, M.P., Chiarizia, R., Urban, V., Nash, K.L. (2001) Aggregation of the neodymium complexes of HDEHP, Cyanex 272, Cyanex 302, and Cyanex 301 in toluene, Proceedings of the International Symposium NUCEF 2001, Scientific Bases for Criticality Safety, Separation Process and Waste Disposal, Japan Atomic Energy Research Institute (JAERI), Tokai-mura, Ibaraki, Japan, Oct. 31–Nov. 2, JAERI-Conf, 2002–004, March 2001, pp. 281-288.

Jimenez-Reyes, M. (1993) PIXE analysis of rare earth elements, *International Journal of PIXE,* Vol. 3, pp.129-143.

Kołodyńska D., Hubicka, H. (2007) Application of macroporous anion exchangers in sorption and separation of Y(III) and Nd(III) complexes with EDTA, *Trends in Inorganic Chemistry,* Vol. 9, pp. 33-43

Korkish J. (1968) *Modern methods for the separation of rare metal ions,* Pergamon Press, Oxford 1968.

Korkish, J., Ahluwalia S.S. (1967) Cation-exchange behaviour of several elements in hydrochloric acid-organic solvent media, *Talanta*, Vol. 14, pp. 150-170.

Kowalczyk, J., Mazanek, Cz., (1987) Rare earths – problems to satisfy national economy needs, *Physicochemical Problems of Mineralurgy,* Vol. 35, pp. 144-147 (in Polish).

Kowalczyk, J., Mazanek, Cz., (1990) Methods of ores and rare earth concentrates processing, *Ores and Metals,* Vol. 35, pp. 144-147 (in Polish).

Kowalczyk, J., Mazanek, Cz., (1990) Processes of rare earth concentrates separation, *Ores and Metals,* Vol. 35, pp. 294-297 (in Polish).

Kowalski, Z., Kulczycka, J., Wirth, H. (2006) The concept of using the phosphogypsum dumps 'Wizów' as a source of secondary raw material, Czasopismo Techniczne, Vol .128, pp. 39-45 (in Polish).

Kovalancik, J., Galova, M. (1992) Extraction separation of rare earth elements by amines in the presence of complexing agents, *Journal of Radioanalytical and Nuclear Chemistry,* Vol. 162, pp. 35-46.

Kumar M. (1994) Recent trends in chromatographic procedures for separation and determination of rare earth elements, *Analyst*, Vol. 119, pp. 2013-2024.

Kunin, R., Gustafson, R.L. (1969), Ion Exchange, *Industrial and Engineering Chemistry Research,* Vol. 61, pp. 38-42.

Krishna, P.G., Gladis, J.M., Rao, T.P. (2005) selective recognition of neodymium(III) using ion imprinted polymer particles, *Journal of Molecular Recognition,* Vol. 18, pp. 109-116.

Kucharczyk, B., Zabrzeski, J. (2001) Influence of lanthanum oxide(III) on activity of monolithic catalysts in process of hydrocarbons combustion, *Przemysł Chemiczny,* Vol. 80, pp. 109-112 (in Polish).

Kutun, S. Akseli, A. (1999) New elution agent, sodium trimetaphosphate, for the separation and determination of rare earths by anion-exchange chromatography, *Journal of Chromatography,* Vol. 847, pp. 261-269.

Kutun, S., Akseli, A., (2000) Separation and determination of rare earth elements by Dowex 2-X8 resin using sodium trimetaphosphate as elution agent, *Journal of Chromatography,* Vol. 874, pp. 311-317.

Li, H., Wang, D., Wang, X., Xu, Y., Zhang, L., Liu, B. (2010) Study on preparation and properties of tannins immobilized chitosan-Ce^{4+} resins, *Journal of Rare Earths,* 28, 144-148.

Liang, P., Fa, W. (2005) Determination of La, Eu and Yb in water samples by inductively coupled plasma atomic emission spectrometry after solid phase extraction of their 1-phenyl-3-methyl-4-benzoylpyrazol-5-one complexes on silica gel column, *Microchimica Acta,* 150, pp. 15-19.

Linsalata, P., Eisenbud, M., Franca, E.P. (1986) Ingestion estimates of Th and the light rare earth elements based on measurements of human feces, *Health Physics,* Vol. 50, pp. 163-167.

Luna, J.S., Flores, A., Muniz, R., Fuentes, A.F., Torres, J., Rodriguez, R., Ortiz, J.C., Orozco, P. (2011) Cerium extraction by metallothermic reduction using cerium oxide powder injection, *Journal of Rare Earths*, Vol. 29, pp. 74.

Mąkowska, B. (1970) *Separation of lanthanum from rare earth mixture by preparation of complexes with NTA,* Doctoral Dissertation, UAM, Poznań 1970.

Marcus, J. (1983) *Ion exchange and solvent extraction,* in: Gmelins's Handbook of Inorganic Chemistry, Vol. D6.

Marcus, Y., Nelson, F. (1959) Anion-exchange studies. The rare earths in nitrate solutions, *Journal of Physical Chemistry,* Vol. 63, pp. 77-79.

Marhol, M. (1982) *Ion exchangers in analytical chemistry. Their properties and use in inorganic chemistry,* Academia, Prague 1982.

Mel'nik, M.I., Filimonov, V.T., Karelin, E.A. (1999) Mechanism of extraction of weighable amounts of lanthanides(III) with HDEHP from nitric and acetic acid solutions, *Radiochimica,* Vol. 41, pp. 67-70.

Merciny, E., Duyckaerts, G. (1966) Separation de lanthanides et des actinides par l'acide hydroxyéthylénediaminotriacétique : I. Étude de l'effet de sur l'efficacité de la collonne, *Journal of Chromatography,* Vol. 22, pp. 164-176.

Minczewski, J., Chwastowska, J., Dybczyński, R. (1982) *Separation and preconcentration methods in inorganic trace analysis,* Wiley-VCH, New York-Toronto.

Minczewski, J., Dybczyński, R. (1962a) Separation of rare earths on anion exchange resins II. Anion exchange behaviour of the rare earth complexes with ethylenediaminetetraacetic acid, *Journal of Chromatography,* Vol. 7, pp. 98-111.

Minczewski, J., Dybczyński, R. (1962b) Separation of rare earths on anion-exchange resins III. The position of scandium in the separation scheme of rare earth complexes with ethylenediaminetetraacetic acid, *Journal of Chromatography,* Vol. 7, pp. 564-567.

Miyazaki, A., Barnes, R.M (1981) Complexation of some transition metals, rare earth elements, and thorium with a poly(dithiocarbamate) chelating resin, *Analytical Chemisty,* Vol. 53, pp.299-304

Moody, G.J. Thomas, J.D.R. (1968) Inorganic ion exchange in organic and aqueous-organic solvents, *The Analyst,* Vol. 93, pp. 557-588.

Morais, C.A, Ciminelli, V.S.T. (1998) Recovery of europium from rare earth chloride solution, *Hydrometallurgy,* Vol. 49, pp. 167-177.

Morais, C.A, Ciminelli, V.S.T. (2001) Recovery of europium by chemical reduction of a commercial solution of europium and gadolinium chlorides, *Hydrometallurgy,* Vol. 60, pp. 247-253.

Moustafa, M.I., Abdelfattah, N.A. (2010) Physical and chemical beneficiation of the Egyptian beach monazite, *Resource Geology,* Vol. 60, pp. 288-299.

Nash, K.L. (1994) *Separation chemistry for lanthanides and trivalent actinides,* In: Handbook on the physics and chemistry of rare earths, K.A. Gschneidner Jr. et al., (eds.) Vol.18, pp.197-238.

Nelson, F., Murase, T., Kraus, K.A. (1964) Ion exchange procedures. I. Cation exchange in concentration HCl and $HClO_4$ solutions, *Journal of Chromatography,* Vol. 13, p. 503-535.

Nilsson, M., Nash, K.L. (2007) A review of the development and operational characteristics of the TALSPEAK process, *Solvent Extraction and Ion Exchange,* Vol. 25, pp. 665-701.

O'Driscoll, M. (2003) REgenerationn in autocatalyst, *IM,* Vol.1, pp. 24-33

Padayachee, A.M.A., Johns, M.W., Green, B.R. (1996) *The use of ion exchange resins to recover rare earths from apatite gypsum residue.* In. Ion Exchange Developments and Applications, J.A. Greig (ed.) The Royal Society of Chemistry, SCI, pp. 380-387.

Paulo, A. (1993) Why niobium and rare earth elements deposits should not be searched for in Poland, *Special Papers of PTM,* Vol. 3, pp. 55-77 (in Polish).

Paulo, A. (1999) Rare earth elements at the end of XX century, *Geological Review,* Vol. 47, pp. 34-41 (in Polish).

Powell, J.E. (1964) *The separation of rare earths by ion exchange.* In: Progress in the science and technology of the rare earths, Eyring L. (ed.), Pergamon Press, Oxford.

Powell, J.E. (1961) Symposium on rare earths, Rare earth (ed. E.V. Kleber), Macmillan, New York.

Powell, J.E. Burkholder H.R. (1967) Augmenting the separation of gadolinium and europim and europium and samarium mixtures in ion exchange elutions with EDTA, *Journal of Chromatography,* Vol. 29, pp. 210-217.

Preston, J.S. (1996) The recovery of rare earth oxides from a phosphoric acid byproduct. Part 4. The preparation of magnet-grade neodymium oxide from the light rare earth fraction, *Hydrometallurgy,* Vol. 42, pp. 151-167.

Preston J.S., Du Preez A.C. (1996) Synergistic effects in solvent-extraction systems based on alkylsalicylic acids. Part 2. Extraction of nickel, cobalt, cadmium and zinc in the presence of some neutral N-, O- and S-donor compounds, *Solvent Extraction and Ion Exchange,* Vol. 14, pp. 179-201.

Preston, J.S., Du Preez, A.C., Cole, P.M., Fox, M.H. (1996) The recovery of rare earth oxides from a phosphoric acid by-product. Part 3. The separation of the middle and light rare earth fractions and the preparation of pure europium oxide, *Hydrometallurgy,* Vol. 42, pp. 131-149.

Preuss, A.F., Kunin, R. (1958) *Uranium recovery by ion exchange.* In: Clegg, J. W. and Foley, D. D. (eds.) Uranium ore processing. Addison-Wesley, Reading, 191-236.

Rao, T.P., Kala, R. (2004) On-line and off-line preconcentration of trace and ultratrace amount of lanthanides. *Talanta,* Vol. 63, pp. 949-959.

Robards, K., Clarke, S., Patsalides, E. (1998) Advances in the analytical chromatography of the lanthanides. A rewiew, *Analyst,* Vol. 113, pp. 1757-1779.

Schmidt, V.S. (1987) Some problems of development of physicochemical fundamentals of modern extraction technology, *Uspehi Khimii,* Vol. 56, pp. 1387-1415 (in Russia).

Schrobilgen, G.J., Lang, C.E. (1968) Lanthanide distribution coefficients on Dowex chelating resin A-1, *Journal of Inorganic and Nuclear Chemistry,* Vol. 30, pp. 3127-3130.

Sherief, M.K.S., Almasy, A. (1968) Separation process of thorium from rare earth metals. US Patent No. 3383182.

Shibata, J. Matsumoto, S. (1998) Separation of Dy, Y, Tm and Y from heavier rare earth residue with solvent imprednated resins, The International Rare Earths Conference, 25-30 October, West Australia.

Smith, H.L., Hoffman, D.C. (1956) Ion-exchange separations of the lanthanides and actinides by elution with ammonium alpha-hydroxy-isobutyrate, *Journal of Inorganic and Nuclear Chemistry,* Vol. 3, pp. 243-247.

Sozański, A. (1981) Industrial methods for separation of thorium from rare earth elements, *Scientific Works, Institute of Inorganic Chemistry and Rare Earth Elements, Wrocław Technical University*, Wrocław 1981.

Spedding, F.H., Powell, J.E., Wheelwright, E.J. (1956) The stability of the rare earth complexes with N-hydroxyethylethylenediaminetriacetic acid, *Journal of American Chemical Society*, Vol. 78, pp.34-37.

Spedding, F.H., Daane, A.H. (1998) *The rare earth*, Wiley-VCH, New York.

Surls, J.P., Choppin, G.R. (1957) Equilibrium sorption of lanthanides. Americium, and cerium on Dowex-50 resin, *Journal of American Chemical Society*, Vol. 79, 855-859.

Starý J. (1966) Separation of transplutonium elements, *Talanta*, Vol. 13, pp. 421-437.

Stasicka, Z. (ed.) (1990) *Inorganic Chemistry nomenclature. Recommendations 1990.* Polish Chemical Society, Wrocław University Publishers, Wrocław (in Polish).

Stewart, D.C., Faris, J.P. (1956) Preparation of rare-earth concentrates, *Journal of Inorganic and Nuclear Chemistry*, Vol. 3, pp. 64-66.

Strelow, F.W.E., Bothma, C.J.C. (1964) Separation of scandium from yttrium, lanthanum, and the rare earths by cation exchange chromatography, *Analitical Chemistry*, Vol. 36, pp. 1217-1220.

Szczepaniak, W., Siepak, J. (1973) Chelating ion exchanger containing aminomethane-phosphonic acid groups, *Chemia Analityczna*, Vol. 18, pp. 1019-1025.

Tse, P.K. (2011) China's Rare-Earth Industry, U.S. Department of the Interior U.S. Geological Survey, Open File Report 2011–1042, pp. 1-11.

Turanov, A.N., Karandashev, V.K., Sharova, E.V., Artyushin, O.I., Odinets, I.L., Extraction of lanthanides(III) from $HClO_4$ solutions with bis(diphenylphosphorylme thylcarbamoyl) alkanes, *Solvent Extraction and Ion Exchange*, Vol. 28, pp. 579–595.

Uda, T., Jacob, K.T., Hirasawa, M. (2000) Technique for enhance rare earth separation, *Science*, Vol. 289, pp. 2326-232.

Ura, P., Prakorn, R., Weerawat, P. (2005) Purely extraction and separation of mixture of cerium(IV) and lanthanum(III) via hollow fiber supported liquid membrane, *Journal of Industrial Engineering and Chemistry*, Vol. 11, pp. 926-931.

Warshawsky, A. (1981) *Extraction with solvent impregnated resins.* In: Ion exchange and solvent extraction, Marinsky, J.A.; Marcus, Y., (eds.) Marcel Dekker, Vol. 8, pp.229-310.

Waqar F., Jan, S., Mohammad, B., Hakim, M., Alam, S., Yawara, W. (2009) Preconcentration of rare earth elements in seawater with chelating resin having fluorinated β-diketone immobilized on styrene divinyl benzene for their determination by ICP-OES, *Journal of the Chinese Chemical Society*, Vol. 56, pp. 335-340.

Wódkiewicz L., Dybczyński R. (1968) Anion exchange behaviour of the rare earth complexes with trans-1,2-diaminocyclohexane-N,N'-tetraacetic acid, *Journal of Chromatography*, Vol. 32, pp. 394-402.

Wódkiewicz, L., Dybczyński, R. (1972) Effect of resin cross-linking on the anion-exchange separation of rare earth complexes with DCTA, *Journal of Chromatography*, Vol. 68, pp. 131-141.

Van der Walt T.N., Coetzee, P.P. (1996) Distribution coefficients and ion exchange behaviour of some elements with Purolite S-950 in hydrochloric acid, *Fresenius' Journal of Analytical Chemistry,* Vol. 356, pp. 420-424.

Vigneau, O., Pinel, C., Lemaire, M. (2001) Ionic imprinted resins based on EDTA and DTPA derivatives for lanthanides(III) separation, Analytica Chimica Acta, Vol. 435, pp. 75-82.

Vito, I.E.D., Masi, A.N., Olsina, R.A. (1999) Determination of trace rare earth elements by X-ray fluorescence spectrometry after preconcentration on a new chelating resin loaded with thorin, *Talanta,* Vol. 49, pp. 929-935.

Xu, X., Zhu, W., Wang, Z., Witamp, G.J. (2002) Distributions of rare earths and heavy metals in field-grown maize after application of rare earth-containing fertilizers, *The Science of Total Environmental,* 293, 97-105.

Yan, Ch., Jia, J., Liao, Ch., Wu, S., Xu, G. (2006) Rare earth separation in China, *Tsinghua Science and Technology,* Vol. 11, pp. 241-247.

Yin, S., Wu, W., Zhang, B., Hang, F., Luo, Y., Li, S., Bian, X. (2010) Study on separation technology of Pr and Nd in D2EHPA-HCl-LA coordination extraction system, *Journal of Rare Earths,* Vol. 28, pp. 111-115.

Zhang, Z., Li, H., Guo, F., Meng, S., Li, D (2008) Synergistic extraction and recovery of cerium(IV) and fluorine from sulfuric solutions with Cyanex 923 and di-2-ethylhexyl phosphoric acid, *Separation and Purification Technology,* Vol. 63, pp. 348-352

Chapter 7

Ion Exchange in Glass – The Changes of Glass Refraction

Roman Rogoziński

Additional information is available at the end of the chapter

http://dx.doi.org/10.5772/51427

1. Introduction

The phenomenon of ion exchange in glasses in the practical application has been known since the Middle Ages when it was used for coloring glass. However, the application of this phenomenon for the production of changes in the glass refraction associates with the waveguide technology. The development of this technology has started in the second half of the 20th century and was dictated by the huge potential of the optical transmission of information in comparison to its classical form that uses the transmission of electrical signals through wired links. The optical transmission, in turn, uses dielectric fibers, known as waveguides. The materials that since the beginning have been used for producing the fiber waveguides are oxide glasses. However, their attenuation in this study period were of about *1000 dB/km*. In 1966 K.C. Kao and G.A. Hockman in their work [1] indicated the possibility of using for the near-infrared transmission a specially treated glass, devoid of impurities in the form of ions of iron, cobalt and copper, which are the main cause of the absorption of the energy in the propagating wave.

In 1972, Corning Glass company has developed a technology of the production of preforms for extracting the optical fibers, whose attenuation was approximately *4 dB/km*, using the technology of production of synthetic silica in the high temperature hydrolysis of silicon chloride, occurring in the presence of oxygen with admixtures of chlorides of boron, phosphorus and germanium (CVD technology - *Chemical Vapor phase Deposition*). In addition to the development of material technologies designed for the production of low-loss fiber waveguides, the research has begun on the production of passive waveguide structures on flat substrates made of oxide glasses of small size (area of a few to several cm^2), which, by definition, were not designed for the long-distance transmission of optical signals. Therefore, the requirements for the purity of ingredients (from which the glasses used as substrates were produced) have become less important. The method, by which such

INTECH
open science | open minds

© 2012 Rogoziński, licensee InTech. This is a paper distributed under the terms of the Creative Commons Attribution License (http://creativecommons.org/licenses/by/3.0), which permits unrestricted use, distribution, and reproduction in any medium, provided the original work is properly cited.

waveguide structures has started to be produced in glass, is called (from the physical phenomenon, which is its foundation) the *ion exchange method* [2,3]. This phenomenon is based on the mechanism of glass ionic conductivity [4], allowing in a sufficiently high temperature the substitution of natural components (called modifiers), by inserting, in the due process, the so-called admixture ions. In the result of their different properties, such as electric polarizability and ionic radius, the admixture ions introduced into the glass locally alter its optical properties (refractive index). Due to the diffusive nature of the ion exchange phenomenon and the fact that the admixture is introduced into glass by its surface, thus resulting waveguide structures are localized in the superficial area of glass substrate (planar structures), and the change in the glass refractive index has the gradient character.

The first waveguide structures were prepared by the ion exchange method in the early seventies of the last century [5-7]. The advantages of this method include first of all: the ease of implementation of the basic technological processes (a highly specialized technological apparatus is not required here), low production costs associated with the possibility of using commercially available glass as substrates and relatively cheap materials as a source of admixture, very good repeatability of obtained elements, a very good time stability, low attenuation ($<0.1\ dB/cm$), as well as good compatibility - in terms of producing changes in the refractive index - with the fiber waveguides used in telecommunications.

2. Ion exchange in glass phenomenon

The ion exchange in glass phenomenon is initiated in the phase boundary: glass - liquid admixture source. The system with a liquid source of admixture is the most common in the literature and its attractiveness is due to the main advantage: the simplicity of implementing such processes. In this case, subject to adequate conditions of the process realization, an unlimited source efficiency can be assumed in its theoretical description. For these reasons, the description of the ion exchange phenomena has been limited only to liquid phases of admixture sources.

The assumption of the binarity of the ion exchange process' character made in its description is rather a simplification. This postulate is commonly found in the literature [2,8-11] and as the only one is acceptable for the examination of the effects of this phenomenon by optical methods only. In the next part, for the description of the ion exchange phenomenon, the following designations has been adopted: (A) for the admixture ions and (B) for the modifier ions.

When discussing the ion exchange phenomenon in glass, which is realized through the contact of glass with liquid admixture at sufficiently high temperatures, the main aspects should be specified (Fig.1). These are:

- determination of the balance of ion exchange process in the phase boundary: glass-admixture source. The consequence of this balance is the relationship between the concentration of the admixture in the c_{As} source phase and the admixture concentration in the superficial area of glass c_{max},

- time dependence $c_{max}(t)$ in achieving the equilibrium value,
- phenomenon of transport of ions in the glass phase resulting from its structure, whose effect is the migration of admixture into the glass - a model description of this process,
- relationship between the concentration of admixture introduced into the glass $c_A(x)$ and the local change of the refractive index of glass $\delta n(x)$.

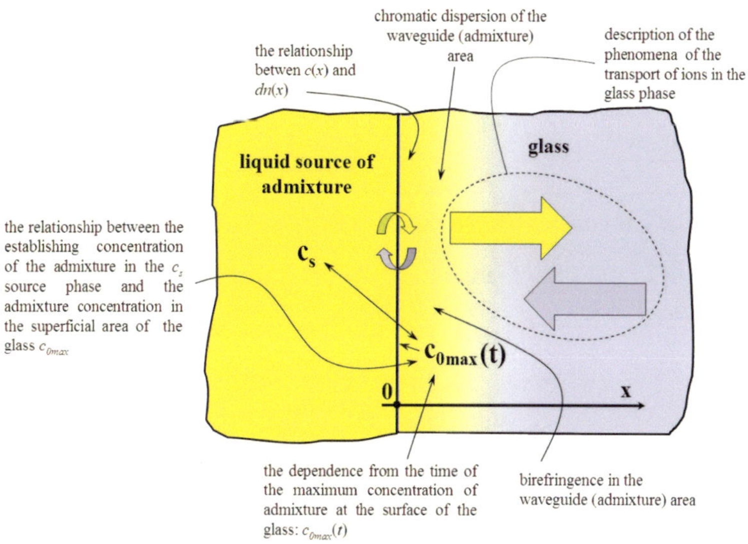

Figure 1. The main aspects of the ion exchange phenomenon in glass

The ion exchange phenomenon in the superficial area of glass, which is in contact with a source of admixture, can be described as a reaction in which the substrates are: in the source of ions *A* in the admixture source and ions *B* in the glass, and products are, respectively: ions *A* in the glass and ions *B* in the phase of the source. Thermodynamic equilibrium state of the exchange process at the phase boundary: glass - liquid source of admixture is then described by constant *K* [11,12]:

$$K = \frac{\bar{a}_A \cdot a_B}{a_A \cdot \bar{a}_B} \tag{1}$$

In the equation above $\bar{a}_A$ and a_A mean the thermodynamic activities of ions *A*, respectively: the glass phase and the source phase, likewise: $\bar{a}_B$ and a_B are the thermodynamic activities of ions *B* in the glass phase and the source phase.

Knowledge of the quantitative determination of the equilibrium in the ion exchange process at the phase boundary: liquid admixture source - glass, is essential for the theoretical description of the formation of spatial distribution of admixture introduced into the glass in the exchange process. The determination of constant equilibrium (at a given temperature of exchange process) for a particular glass-admixture system therefore requires the experimental determination of the relationship between the molar fraction of admixture ions in the liquid phase of source and its molar fraction providing at the glass surface.

In equation (1) the ratio of activities of exchanged ions in the glass phase can be expressed with the Rothmund - Kornfeld relationship [12,13]:

$$\frac{\bar{a}_A}{\bar{a}_B} = \left(\frac{\bar{N}_A}{\bar{N}_B}\right)^{\eta}, \tag{2}$$

where: $\bar{N}_i$ – means the molar fraction of ions in the glass phase ($i = A, B$) and η is expressed by the interaction energy of ions in the glass $W_{A\text{-}B}$:

$$\eta = 1 - \frac{W_{A-B}}{RT}, \tag{3}$$

where: R – universal gas constant, T – temperature.

In turn, the activities ratio of exchanged ions in the liquid phase of the source is given by their mole fractions N_i and activity coefficients γ_i:

$$\frac{a_B}{a_A} = \frac{N_B \gamma_B}{N_A \gamma_A} \tag{4}$$

After taking into account (2) and (4) in (1), we obtain:

$$\log\left(\frac{N_A}{N_B}\right) - \frac{Q_{A-B}}{RT}\left(1 - 2N_B\right) = \eta \cdot \log\left(\frac{\bar{N}_A}{\bar{N}_B}\right) - \log K \tag{5}$$

The equation above takes into account the relationship:

$$\log\left(\frac{\gamma_B}{\gamma_A}\right) = \frac{Q_{A-B}}{RT}\left(1 - 2N_B\right), \tag{6}$$

in which $Q_{A\text{-}B}$ is the energy of interaction of ions in the source phase.

The knowledge of $Q_{A\text{-}B}$ for the systems of liquid mixtures used as the admixture source allows on the basis of (5) to calculate the parameters η and K. The process of establishing the maximum concentration of admixture c_{Amax} introduced into the superficial area of glass will be characterized by a time constant $-\tau_d$. This should be considered in cases of carrying short-term exchange processes.

Glasses, which are used as the substrate of gradient waveguide structures, are, in the vast majority, the oxide glasses. The possibility of the realization of ion exchange phenomenon in glass is a consequence of its structure. There are many theories about the concept of this structure, among which one of the basic is Zachariasen theory [14] experimentally verified in studies of Warren [15]. It allows a pretty good description of the additive properties of glass, due to its composition. Based on this theory the mechanism of ion transport in glass can be explained, which is the basis of the ion exchange phenomenon [4]. This theory highlights the notion of a structural skeleton - called the glass network – which has features of low-range ordering. The elementary unit of this structure (in the case of silicate glasses) are the coordination polyhedra in the shape of tetrahedra that are connected by corners. In the vertices of this tetrahedron there are the anions O^{2-} surrouding the cation Si^{4+} which is located in the middle.

The rules under which a merger of elementary cells occurs are determined by the conditions of Zachariasen vitrification [14]:

- each oxygen anion O^{2-} is directly related to only two cations,
- the cation coordination number should be 3 or 4 at most,
- the coordination polyhedra connect to the neighbors only with corners, not edges or walls,
- at least three corners of each coordination polyhedron may be included in the other polyhedra

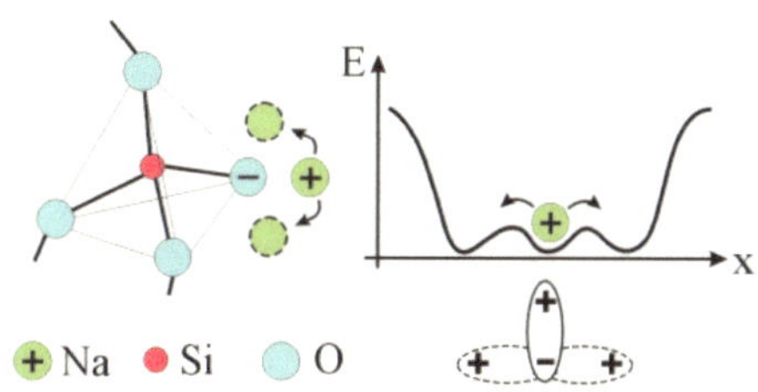

Figure 2. Possible energy positions of ion Na^+ in glass structure as well as changes of the orientational polarization of the structure due to its hopping (by [4])

In the multicomponent oxide glasses in addition to the glass network there are also so-called modifiers, which are alkali metal ions, whose presence in the glass is required by giving it the appropriate physicochemical properties. When the sodium oxide (Na_2O), for example, is introduced into the structure of glass in its melting process, the network connections in the emerging structure are reduced due to excess oxygen. So then the non-bridging oxygen arise. The most likely places for sodium in the structure of the glass are around non-bridging oxygen, which have uncompensated negative charge (Figure 2). For alkali ions, it is possible to address more than one energetically equivalent position near the non-bridging oxygen. However, a suitable location for sodium ions is determined not only by the same oxygen bridges, but also throughout their environment - giving in the final effect a system with the lowest possible energy [4]. These fragments of glass network, surrounded by the trapping energy ions of modifiers that can fluctuate within a structural unit as a result of thermal induction, are polar elements of the glass network. In the thermally induced structure of glass, in addition to hopping to the local structural unit, hoppings of long range outside the polar unit are also more likely to appear. This is accompanied by dissociation of the polar unit, and the modifier ion migrates inside the glass network. Non-bridging oxygen ion remains fixed in the glass network, making a vacancy in it. The directions of migration of mobile modifier depends on the local environment of polar units. Each of them is able to intercept mobile cations after the dissociation. Since the structural units in glass forming its glass network don't create an arrangement of long range, so a random network of the most probable directions of hopping arises. Those hoppings create (on a microscopic scale) the system of paths shown in Fig 3.

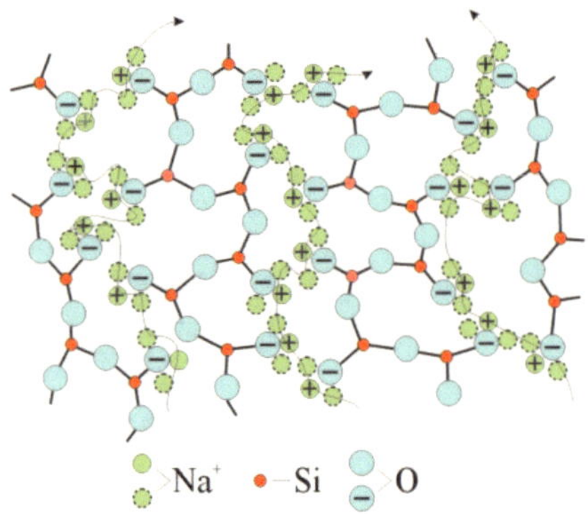

Figure 3. 2-D fragment of glass structure with marked preferential pathways of modifier ions (by [4])

The described ability of the modifier ions to move within the glass structure is the basis of the mechanism of glass ionic conductivity. In turn, this mechanism enables the ion exchange, which takes place in case of introducing other types of modifier ions (admixture) into the glass.

The appearance of admixture ions in the superficial area of glass creates the opposite directed concentration gradients of both: admixture ions and modifiers. Consequently, in the glass there are two divergent streams of ions: a stream of admixture ions directed into the glass and a stream of modifier ions directed to the glass surface.

Association of the diffusion coefficient D_σ with the electrical conductivity σ or mobility μ of exchanged ions in the crystal structures is described by the Nernst-Einstein equation:

$$D_\sigma = \frac{kT\sigma}{cq^2} = \frac{kT\mu}{q}, \tag{7}$$

where: c and q – are respectively: concentration and charge of diffusing ions, k is the Boltzmann constant, T - is the absolute temperature.

The relationship between the diffusion coefficient D_σ and the diffusion coefficient D^*, which is set by the tracer diffusion method, is determined by the correlation coefficient f, which, in the case of vacancy diffusion mechanism of ions in the three-dimensional glass network, is equal to the Haven ratio H [11.16]:

$$f = H = \frac{D^*}{D_\sigma} < 1 \tag{8}$$

So for the description of the two-component ion exchange process in glass the relationship between self-diffusion coefficients and electrical mobilities of both types of exchanged ions take the form:

$$D_i = \frac{H_i k T \mu_i}{q} \qquad (i=A,B) \tag{9}$$

The final effect of the ion exchange process in glass, which has practical application in the manufacture of optical structures, is the change of the refractive index within its admixture area. Local changes in the refractive index of glass, which are the effect of the ion exchange, are the result of the differences in their electric polarizability and ionic radii.

According to the principle of additivity [9,17,18] refraction index of oxide glass consisting of a metal oxide M_mO_n can be represented as:

$$n_b = 1 + \frac{R_0}{V_0} = 1 + \frac{\sum_i a_i N_i}{C + \sum_i b_i N_i}, \tag{10}$$

where: R_0, V_0 – are respectively: the refraction and the molar volume of oxide atoms constituting the glass, N_i – molar fraction of i-th component in the composition of the glass, a_i, b_i, C – empirical coefficients.

The ion exchange process starts with the glass surface, which contacts with the admixture source phase. Defining the fraction of exchanged ions in glass $u(x)$ as a function of depth measured from the surface of the glass, it can be written:

$$u(x) = \frac{N_A(x)}{N_0} = \frac{c_A(x)}{c_0}, \tag{11}$$

where: $N_A(x)$ – is the number of admixture ions (A), which replaced the glass ions at a depth x, N_0 – is the number of exchangeable ions in the whole glass volume, $c_A(x)$ and c_0 – are the concentrations corresponding to mentioned numbers of ions.

Therefore, the refractive index after the ion exchange, according to (10), reaches a value:

$$n(x) = 1 + \frac{R_0 + u(x)\Delta R}{V_0 + u(x)\Delta V}, \tag{12}$$

and its increase $\delta n(x)$, assuming $\Delta V << V_0$, can be expressed as:

$$\delta n(x) = n(x) - n_b = \frac{R_0 + u(x)\Delta R}{V_0 + u(x)\Delta V} - \frac{R_0}{V_0} = \ldots = \frac{u(x)}{V_0}\left(\Delta R - R_0 \frac{\Delta V}{V_0}\right) \tag{13}$$

Assuming that in the contact of the glass surface with clean (undiluted) source phase all of the mobile ions at the glass surface ($x = 0$) are exchanged for admixture ions, then $u(0)=1$ and the relation occurs:

$$\delta n(0) = \Delta n_s = \frac{1}{V_0}\left(\Delta R - R_0 \frac{\Delta V}{V_0}\right) \tag{14}$$

The linearity of the change of refractive index of the normalized concentration of admixture introduced into the glass $u(x)$, expressed by equation (13), is taken as the fundamental relation describing the ion exchange processes [9,10].

In the cases of diluted admixture sources, according to (11), the normalized concentration of ions introduced at the glass surface takes values $u(0)<1$.

3. Technological processes of producing ion exchange in glass

As mentioned before, the ion exchange phenomenon in glass substrates occurs as a result of introduction of foreign admixture ions into the glass. The used admixtures are alkaline metal ions: lithium [19], potassium [20-29], rubidium [30], cesium [31], and ions of heavier elements: copper [32,33], silver [6,34-43] and thallium [7.44]. The most common sources of these admixtures are molten nitrates. They are characterized by relatively low temperatures of the extent of their liquid state (Table 2.1). This is important for the ion exchange processes in oxide glasses, for which the limitation is the transition temperature T_g [45]. Among the admixture ions listed in Table 2.1 the largest changes of the refractive index Δn_s are achieved when using thallium ions. However, due to the high toxicity of its compounds, it is not widely used. The second, in turn, in terms of the size of obtained changes in the refractive index, is silver ion (Ag^+). It is often used because of the possibility of controlling the refractive index change by using matched solutions of silver nitrate $AgNO_3$ in sodium nitrate $NaNO_3$ as its sources.

ion	source	T_{min} (℃)	T_{max} (℃)	Δn_s
Li^+	$LiNO_3$	261	600	0.012
Na^+	$NaNO_3$	307	380	-
K^+	KNO_3	334	410	0.01
Rb^+	$RbNO_3$	310	370	0.015-0.02
Cs^+	$CsNO_3$	414	510	0.04
Ag^+	$AgNO_3$	212	444	0.1
Tl^+	$TlNO_3$	206	430	0.1-0.2

Table 1. Sources of admixture in the form of nitrates used in the ion exchange

In [42] the applicability of such solutions with mole fractions of silver nitrate contained in the interval: $2{,}48 \cdot 10^{-4} \div 5 \cdot 10^{-2}$ was presented. For such obtained admixture sources the refractive index changes at the surface of the waveguide in the terms of: $0.01 \leq \Delta n_s \leq 0.09$. The temperatures of ion exchange processes, with the use of these solutions, ranged from 315 ℃ to 375 ℃. Other mixtures of nitrates, enabling a lower temperature processes, are also used. The [46] presents results of research into the possibilities of low-temperature admixture sources based on the nitrates of potassium and silver. The mixtures of KNO_3:$AgNO_3$ (50:50% *mol*) in the temperature range of $160 \div 270$ ℃ were used. For those nitrates the eutectic mixture (62% *mol* $AgNO_3$) has a melting point of 131 ℃. Also in this work there are examples of the use of mixtures of $KNO_3:NaNO_3$ *(50:50% mol)*, with a melting point of *220* °C. For the latter mixtures the additives *0.001÷0.05% mol* $AgNO_3$ were also used changing the operating temperature of the source of the *230÷375* °C. It also shows the possibility of using the eutectic mixture of $AgNO_3:TlNO_3$ *(48% mol* $TlNO_3$*)*, whose melting point is *82.5* °C.

3.1. Processes of the preliminary thermal diffusion

The construction of the stand used for the thermal diffusion processes is shown in Fig.4a. It is an electric furnace. The molten salt is in the melting pot (crucible) with a glazed ceramic surface. The glass substrate is placed into a glass holder, at the same time fulfilling the role of the mixer (Fig.4b). This handle makes it easy to enter the plate into the crucible. Before each immersion of the substrate into the molten salt it is held over its surface for some time in order to equalize their temperatures. The connection of the tube holder with the pump generating pulsating pressure changes causes oscillating movements of molten salt in the holder, causing, as a result, mixing of the entire melted contents of the crucible. This way of implementation of the process ensures thermal and molar concentrations homogeneity of liquid admixture source. This is important during the long process of diffusion, because in no-mixed salt at the glass surface a layer with the poor admixture ions can be formed. A waveguide produced in such conditions would have a lowered refraction index at the glass surface. The element monitoring the temperature is a thermocouple placed in a silica glass tube, which is immersed directly into the crucible. During the implementation process the dependence of thermocouple voltage from time $U_T(t)$ is recorded. On its basis the average temperature throughout the duration of the process can be determined.

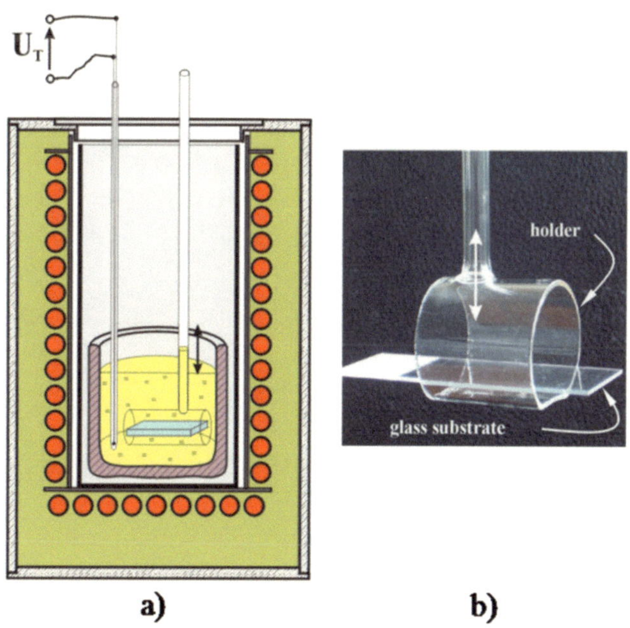

Figure 4. a) Cross section of the stands for the diffusion processes b) glass holder

3.2. Electrodiffusion processes

These are the most technologically advanced processes of glass doping in the case of the liquid sources of admixture usage. However, they give a lot of opportunities to influence the final form of the function describing the distribution of the refractive index profile in the produced waveguide structure. Scheme of the stand [47] is presented in Fig.5a. Molten salt, which is a source of admixture introduced into the glass, is located in the two vessels placed inside the furnace. Salt temperature is controlled with a thermocouple. An essential part of a stand is a system of two symmetric, aluminum crucibles (Fig.5b) between which a glass substrate is clenched. The proposed construction of two independent crucibles enables the electrodiffusion processes, where it is possible to change the polarization direction of the electric field. The seal between the glass and the two crucibles is achieved by using silicone paste resistant to temperatures up to *350* ℃. Both crucibles are mounted on a special holder which allows them a precise introduction into the furnace. Crucibles are attached to the tubes of silica glass. They are used for suction of the molten salt into the crucibles, using negative pressure generated by a vacuum pump. Each crucible is also connected to the lead supplying voltage from the power supply. Also the temperature of the crucible is monitored by thermocouples. Installation of the substrates between the crucibles is made outside the furnace. Special shackles enable quick disconnection of crucibles and removal of the substrate after the process. After entering the crucibles into the furnace chamber the glass substrate does not come into a contact with the molten salt until they reach the same temperature (which has a molten salt). The molten salt is then sucked with the vacuum pump into both crucibles. In this way on both sides of the glass substrate there is molten salt, which is a liquid source of admixture.

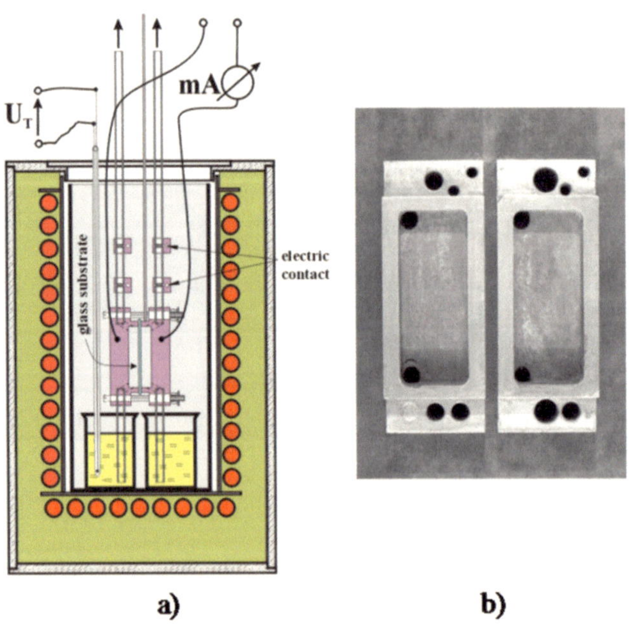

Figure 5. a) Cross section of the stand for the electrodiffusion processes b) the crucibles

The inclusion of appropriate polarization of crucibles enables conducting of the electrodiffusion process, during which both the direction of polarization and the value of voltage applied to the crucibles can be freely changed. During the process a current value flowing through the glass substrate is being recorded. To prevent a situation in which the molten salt could be sucked by the pump outside the chamber of the furnace, on the glass tubes between the two crucibles with a vacuum pump the special contacts are placed, which are capable of inspecting the level of molten salt in both crucibles. After completion of the process the vacuum pump is switched off and the molten salt returns to the vessels inside the furnace, and both crucibles are immediately lifted and discharged outside of the furnace. They are then separated and the glass substrate is released. With the applied solutions the release time of the plate does not exceed *1* minute.

3.3. Heating processes

Spatial distributions of admixture introduced into the glass, in the preliminary diffusion or the electrodiffusion, may be a subject of further modification. The process which is relatively easiest to use is called here the heating process. This is achieved as a secondary process, in relation to pre-doped glass. During the heating the glass is located in the furnace without any contact with the source of admixture.

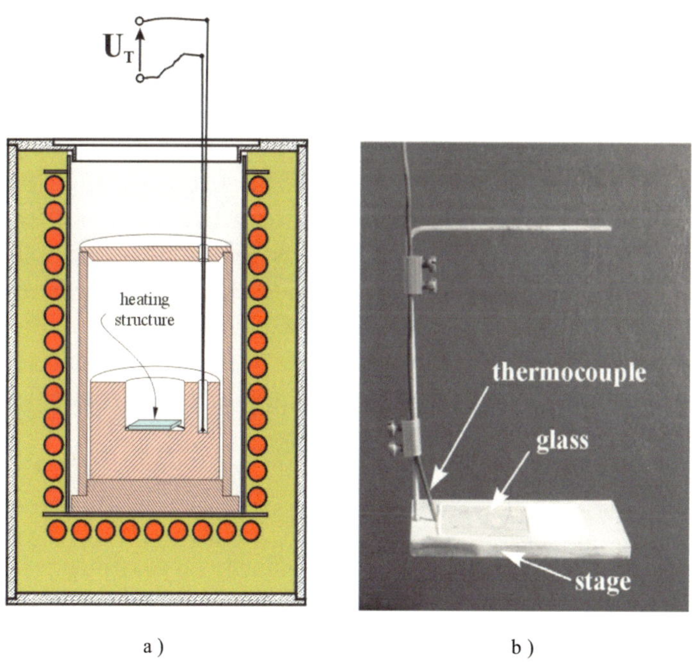

Figure 6. a) The stand cross section for heating processes b) the stage with a thermocouple

Then is a redistribution of the existing concentration of admixture in the glass occurs. The maximum value of the concentration at the glass surface is reduced while expanding its reach into the glass. During the heating of glass in which the waveguide is produced, it is desirable to maintain the process temperature at a given level. Then the final result of the heating will depend only on its duration. However, during the implementation of the heating process, when placing a heated structure into the furnace, a disturbance of the temperature is inevitable. The scale of this phenomenon obviously depends on the thermal inertia of the furnace. During the heating the glass substrate eventually reaches the maximum temperature corresponding to the conditions of the process. At this time, the diffusion processes are already taking place. In contrast to the initial diffusion, there is no possibility of early matching the temperature of the heated sample to the temperature of the furnace. The Fig.6a shows the stand used for the heating process. Its main part is the capsule made of aluminum alloy, which is placed into the furnace. It has a cover that closes it over the top. Inside there is a massive block (aluminum alloy), aimed at increasing the thermal capacity of the entire system. The thermocouple controlling the temperature is attached into this block. Into the cavity of the block the sample subject to the heating is inserted. This sample is located on a special stage (Fig.6b) to allow a quick entry and removing from the furnace. In order to obtain as precise control of the actual temperature of the heated sample as possible, there is an additional thermocouple mounted to the table.

4. Measurements of ion exchange in glass' effects using optical methods

Determination of refractive index profile is a key issue in the metrology of waveguide structures. For the gradient waveguides produced in the glass substrates by ion exchange, many optical measuring methods using different phenomena directly related to the wave propagation in the waveguide structure have been developed. Among these methods, the most preferred one is the waveguide method, based on the measurements of effective refractive indexes of the modes [48-55]. This is a non-destructive method that provides at the same time the high-precision of determination of refractive index profiles. The basis of this method is the measurement of the synchronous angles φ_m of the modes. Goniometric measurements of these angles, using a prism coupler, can be performed with very small uncertainties in the order of tens of seconds of arc. In terms of the effective refractive indexes of the modes N_m it gives the uncertainty of their determination $\Delta N_m \sim 10^{-4}$ [53,54]. Refractive index profiles of the tested waveguides are then reconstructed by means of procedures [48,49,51], based on the modal equation.

4.1. Prism coupler

Using a prism as an element enabling the introduction of an electromagnetic wave to a thin-film structures is described in the work of Tien [56]. In later works of Ulrich [57] and Kernsten [58] there were presented the possibilities of using a prism coupler for the measurement of the properties of thin films. The analysis of the prism coupler work in the measurements of propagation characteristics of planar waveguide structures can be found in the papers [57-59]. In the measuring practice it is a prism having a refractive index greater than the refractive index at the surface of the tested waveguide. To allow it to fulfill the role of the measuring element, the calibration is required.

It involves the determination, with the greatest possible precision, of the breaking angle δ (Fig.7) and the determination of the refractive index n_p for the wavelength used in the measurements. In the case of spectral measurements it is also necessary to know the dispersion curve $n_p(\lambda)$ of the glass of the prism. When measuring with the use of the prism coupler the beam falling on the front wall at an angle φ_m is refracted reaching its base which is bordering the waveguide at an angle θ_m (Fig.7). This angle is bigger than the critical angle on the boundary: prism - the environment (air).

In the presence of a third medium (waveguide), a phenomenon of tunneling the energy of the wave reflected in the prism to the area of the waveguide may occur. A necessary condition for the occurrence of this phenomenon is to provide a so-called optical contact between the prism base and the surface of the waveguide. This contact is produced right on the edge of the prism base with the use of a mechanical clamp (Fig.8a). Optical contact area is visible as a dark spot while watching to the inside of the prism in the scattered light. The pressure size of the prism should not be too big in order not to introduce significant stress in both: the prism and in the waveguide. In practice, the minimum value of the pressure is used, at which the realization of the measurements is still possible [57,59,60].

The thickness of the interspace s produced in the optical contact (Fig.7) is of the order of a fraction of the wavelength used in the measurements. Thus, the part of the evanescent field of the wave reflected from the prism base spans the area of the waveguide. It is presented in Fig.8b. The waves stimulated in the waveguide along the axis z through the beam illuminating the prism base interfere with each other. For certain angles of incidence θ_m the interference is constructive in character, resulting in efficient energy transfer into the waveguide - stimulation

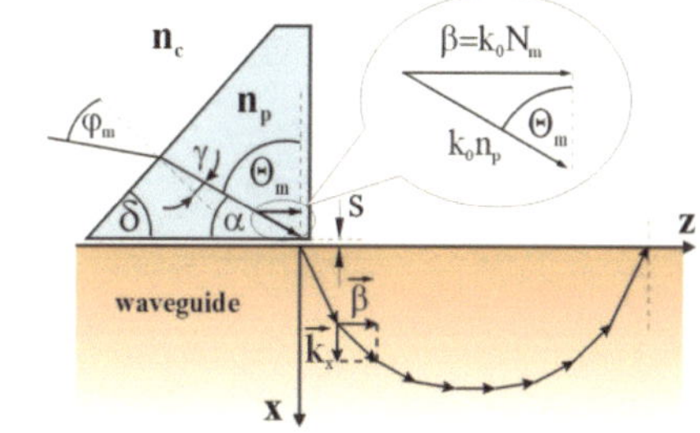

Figure 7. The geometry of the condition of the mode stimulation using the prism coupler

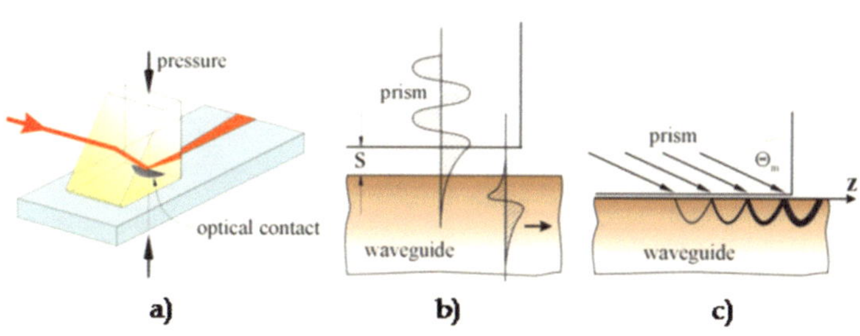

Figure 8. Coupling of a prism with a waveguide a), the penetration of the wave evanescent field in a prism to the waveguide area b), the resonant nature of the stimulation of the mode in the waveguide c)

of the mode (Fig.8c). This phenomenon is of a resonant character. The efficiency of energy tunneling into the waveguide modes, using a prism coupler, can reach a value of to tens of percent [57]. The coupler can also serve as an element, carrying a wave off of the waveguide.

The angles θ_m, for which there is a coupling, are called *the synchronous angles*. In Fig.9 a measuring system for recording these angles is shown. Still, the concept of synchronous angles θ_m will be identified with the values of the angles of incidence φ_m of the beam on the prism, because between these angles there is an unambiguous relationship expressed by the refraction index of the prism n_p and its breaking angle δ. Using the symbols of Fig.7 it can be easily shown that the equality holds:

$$n_p \cdot \sin\theta_m = \sin\delta \cdot \sqrt{n_p^2 - n_c^2 \cdot \sin^2\varphi_m} + n_c \cdot \cos\delta \cdot \sin\varphi_m \tag{15}$$

In the case of tunneling of the wave energy from the prism to the waveguide, the fulfillment of boundary conditions for the wave field on the bottom surface of the prism and the waveguide requires the equality of the wave vector components in both of them. This means that the *z* component of the wave vector in the prism is equal to the corresponding component of the wave vector in the waveguide. On the basis of the equation (15) the expression for the dependence of the effective refractive index N_m of the *m*-th mode from the synchronous angle φ_m is achieved:

$$N_m(\varphi_m) = \sin\delta \cdot \sqrt{n_p^2 - n_c^2 \cdot \sin^2\varphi_m} + n_c \cdot \cos\delta \cdot \sin\varphi_m, \tag{16}$$

where: δ – breaking angle of the prism, φ_m – synchronous angle calculated from the normal, n_p – the refraction index of the prism, n_c – the refraction index of the environment (n_c=*1* the air).

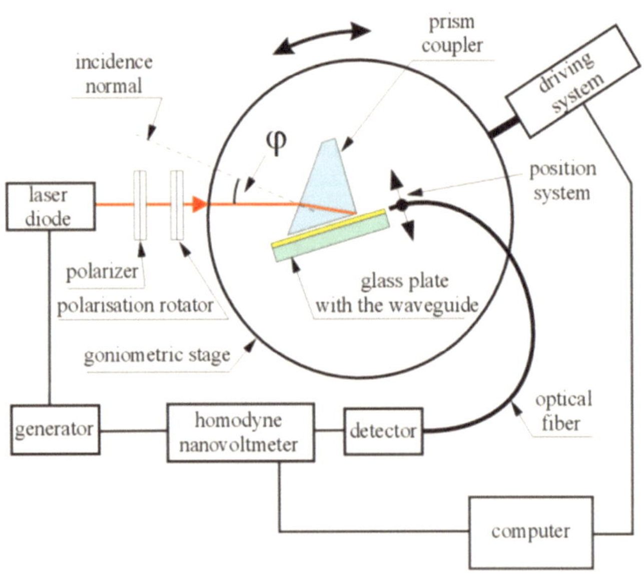

Figure 9. Scheme of a registration stand for synchronous angles of the waveguide modes

4.2. Procedure of refractive profile reconstruction based on the modal equation

The measurement of the synchronous angles by goniometric method allows to determine a set of effective refractive indices of the own modes of the gradient waveguide or the step-index type. Due to the ease of the measurements made and their high accuracy, this method is used as a routine for determining the refractive index profiles of the examined structures. For the waveguides characterized by a monotonic character of the refractive index, a set of designated effective indices $\{N_i\}$ will be able to determine the refractive index profile of the waveguide if it will have assigned a set of turning points $\{x_i\}$ of the corresponding modes. These are the distances x_i calculated from the waveguide surface, for which the wave vector of the propagating wave (of i-th mode) has the only component in the direction of z ($k_x=0$) (Fig.10).

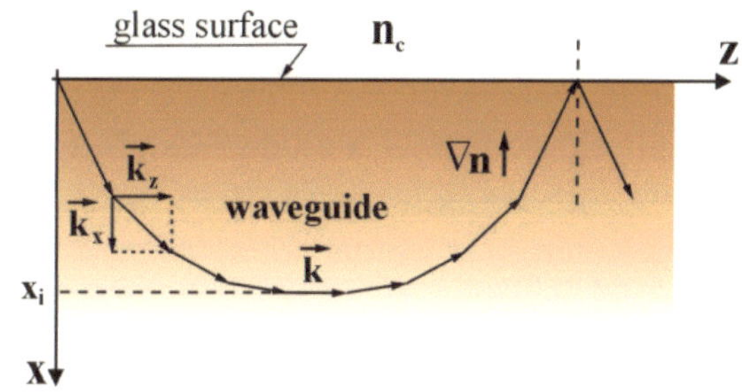

Figure 10. Determination of the turning point of the waveguide mode in the gradient waveguide of a monotonic profile

Procedure for calculating of these turning points for all modes of the waveguide, using the modal equation, was proposed in 1976 by White and Heidrich [48]. The reconstruction of the refractive index profile obtained in this way is its approximation, made in the number of points equal to the number of own modes. Thus, the precision of reconstruction of the refractive index profile, on the basis of the modal equation, depends on the number of modes that may propagate in the waveguide. For this reason, the reconstructing procedures of the refractive index profile of the waveguide based on the modal equation can not be used for the single-mode structures. Also, in the case of the multi-mode waveguides, their application requires a caution in inferring on the resulting refractive index. The refractive index profiles of the planar waveguides, characterized by a monotonic function of changes in the refractive index, are created in the processes of preliminary diffusion (electrodiffusion). The maximum value of refractive index, reflecting the distribution of admixture introduced into the glass, is right on its surface. Assuming the denominations of the refractive indices: n_b in the volume of the glass, n_s at the surface of the waveguide and n_c for the coating of the waveguide (Fig.11) and assuming the distribution of the refractive index in the glass given by a monotonic function $n(x)$, it is possible to write a modal equation for the gradient waveguide, indicating the condition of propagation of the mode of m-th order [48]:

$$k_0 \int_0^{x_m} \sqrt{n^2(x) - N_m^2}\, dx = \pi\left(m + \frac{1}{4}\right) + \operatorname{arctg}\left(r\sqrt{\frac{N_m^2 - n_c^2}{n_s^2 - N_m^2}}\right), \tag{17}$$

where: k_0 – wave number of the electromagnetic wave in free space, x_m –turning point of the mode of m-th order, $N_m=\beta_m/k_0$ – effective refractive index of the mode of m-th order, β_m – propagation constant of the mode of m-th order, $r=1$ (TE polarization), $r=(n_s/n_c)^2$ (TM polarization).

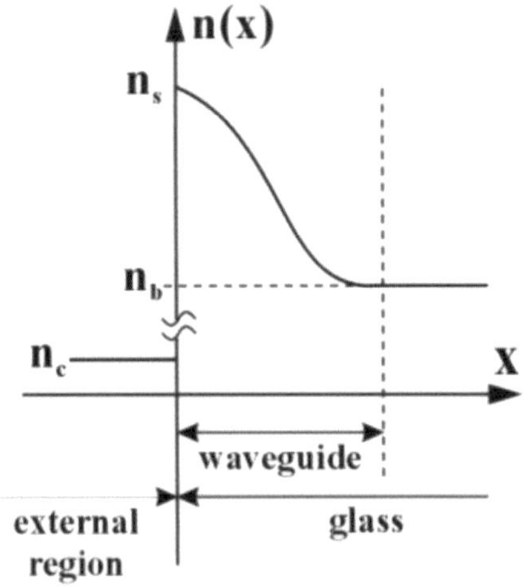

Figure 11. A gradient waveguide with a monotonic change of the refractive index

The White-Heidrich procedure is the simplest method that allows to specify monotonic profile of the planar waveguide $n(x)$. An approximation by a function linear in segments can be made. From the assumption itself: the greater the number of modes that can propagate in waveguide is, the more accurate such approximation will be. This limits the application of this procedure for the multimode waveguides only. Let the desired function $n(x)$ describing the distribution of the refractive index has the form (Fig.12):

$$n(x) = N_k + \frac{N_{k-1} - N_k}{x_k - x_{k-1}}(x_k - x) \tag{18}$$

for: $x_{k-1}<x<x_k$, where $k = 0,1,2 \ldots M-1$; M – is the number of the modes.

Substituting expression (18) into the modal equation (17) one receives the phrase, in which the variable of integration occurs as a function of the square root. Then the left side of the equation (17) can be represented as:

$$\int_0^{x_m} \sqrt{n^2(x) - N_m^2}\, dx = \sum_{k=0}^{m} \int_{x_{k-1}}^{x_k} \sqrt{\left[N_k + \frac{N_{k-1} - N_k}{x_k - x_{k-1}}(x_k - x)\right]^2 - N_m^2}\, dx \tag{19}$$

After calculating the integrals, the m-th component can be distinguished now from the sum. Then a recursive expression is obtained allowing to calculate the turning points of individual modes:

$$x_m = x_{m-1} + \frac{2\left(N_{m-1}-N_m\right)}{N_{m-1}\sqrt{N_{m-1}^2-N_m^2} - N_m^2 \times \ln\left(\frac{N_m}{N_{m-1}-\sqrt{N_{m-1}^2-N_m^2}}\right)} \times$$
$$\times\left\{\frac{\lambda}{2\pi}\left[\pi\left(m+0{,}25\right)+\mathrm{arctg}\left(r\sqrt{\frac{N_m^2-n_c^2}{n_s^2-N_m^2}}\right)\right] - \sum_{k=0}^{m-1}\frac{x_k-x_{k-1}}{2\left(N_{k-1}-N_k\right)}\times\right. \tag{20}$$
$$\left.\left[N_{k-1}\sqrt{N_{k-1}^2-N_m^2} - N_k\sqrt{N_k^2-N_m^2} - N_m^2\times\ln\left|\frac{\sqrt{N_k^2-N_m^2}-N_k}{\sqrt{N_{k-1}^2-N_m^2}-N_{k-1}}\right|\right]\right\}$$
$$\left(m=0,1...M-1\right)$$

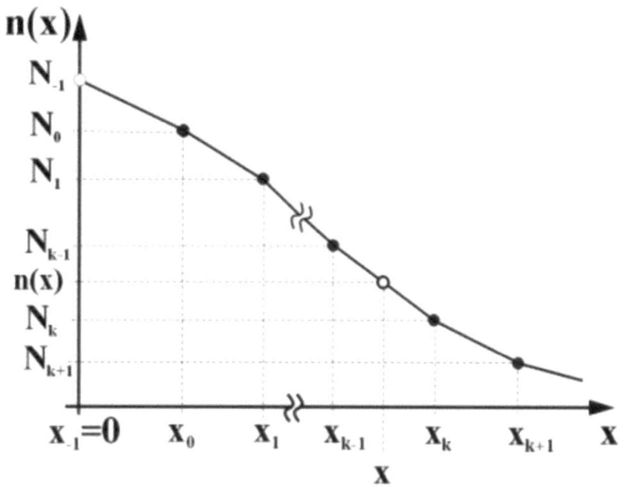

Figure 12. Rys.3.6. Approximation of a refractive index profile by a function linear in segments.

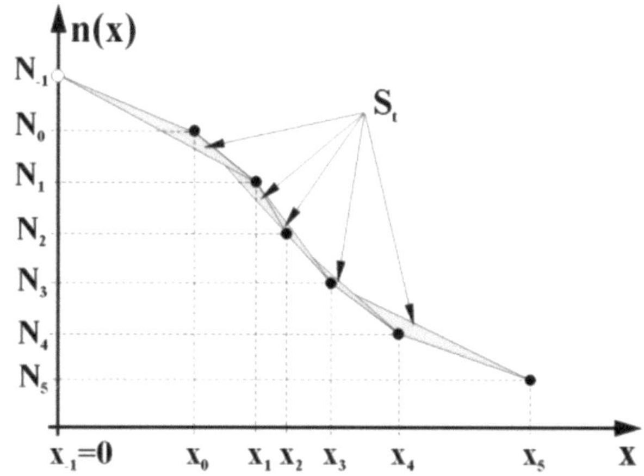

Figure 13. The criterion of smoothness of the reconstructed refractive index profile

for $m=0 \Rightarrow x_{-1}=0$ and $N_{-1}=n_s$:

$$x_0 = \frac{2(n_s - N_0)}{n_s\sqrt{n_s^2 - N_0^2} - N_0^2 \times \ln\left(\frac{N_0}{n_s - \sqrt{n_s^2 - N_0^2}}\right)} \times \frac{\lambda}{2\pi}\left[0{,}25\pi + \mathrm{arctg}\left(r\sqrt{\frac{N_0^2 - n_c^2}{n_s^2 - N_0^2}}\right)\right] \tag{21}$$

From equation (17) it follows that to determine the turning point x_0 of the zero-order mode it is needed to know the value of the refractive index at the surface of the waveguide ($N_{-1}=n_s$). In this method, it is not directly measured. This value is chosen here so that the designated points of the profile $\{x_k, N_k : k=-1..M-1\}$ meet certain criteria. As a physically reasonable criterion the smoothness of the resulting profile is assumed, as a quality of the diffusion processes. As shown in Fig.13, the minimization of the sum of the areas of all triangles S_t, spread on the three adjacent vertices of the reconstructed profile, was adopted as the criterion.

The Fig.14a illustrates a sample modes spectrum recorded in the measurement system of Fig.9. While the Fig.14b presents the refractive index profile of the waveguide calculated from the procedure mentioned above for the effective refractive indices resulting from the modes spectrum shown nearby.

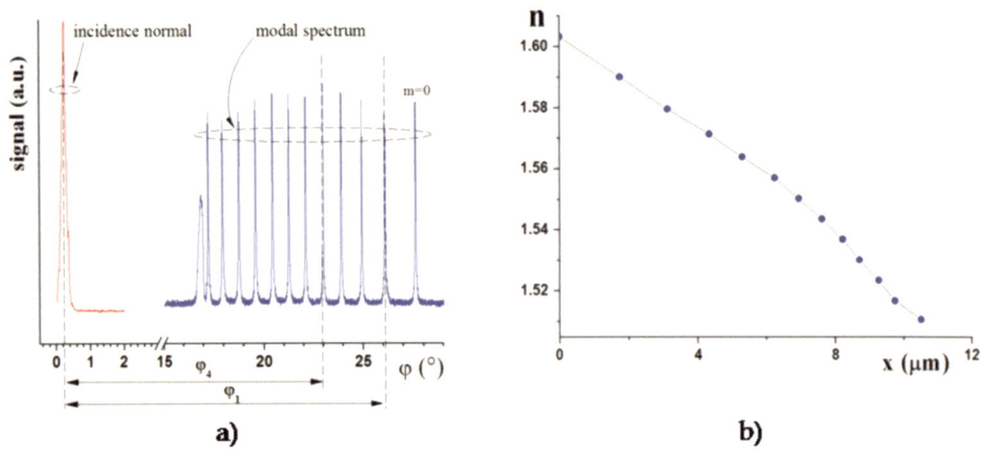

Figure 14. a) The recorded modes spectrum of the waveguide, b) its reconstructed refractive index profile.

5. Mathematical model of binary ion exchange phenomenon

The ion exchange processes are initiated in the glass as a result of the appearance of the admixture ions in its superficial area. This situation occurs when a glass of a sufficiently high temperature is in contact with the liquid phase of the ion source (see Part 3). Then the alkali ions from the liquid phase move into the glass, and mobile modifier ions leave the glass going into the liquid. This keeps the condition of electrical neutrality of the glass. By the appearance of admixture ions at the surface of the glass, with a simultaneous decrease in the concentration of the modifiers, two opposite directed streams of exchanged ions arise in the glass, due to their concentration gradients. Due to the diffusive nature of these phenomena they will continue to be called diffusion processes. The ionic nature of the thermal transport phenomena in the glass makes it possible to stimulate it by an external electric field. These are called the electrodiffusion processes. These two types of processes are used for the initial introduction of admixture ions into the glass area. Their effect is a change of the optical properties of glass (change of the refractive index). This change is proportional to the local concentration of introduced admixture ions (Part 2). The resulting distribution of the refractive index in the glass area may be subject to later modifications in the so-called secondary processes (diffusion, electrodiffusion, heating). This is undoubtedly the biggest advantage of the ion exchange technology, which, despite the restriction on the values of the changes of the refractive index received in the glass, gives wide possibilities of modifying its distribution.

Knowledge of the physical basis of the phenomenon of the ion exchange allows to create a theoretical model describing this process. It can be used to determine the kinetics of the exchange phenomenon for the concrete glass-admixture system.

A model description of the exchange phenomenon is easiest to refer to the preliminary diffusion process, in which the technological parameters are: the temperature and its duration. Temperature determines the kinetics of this process, while time is only an integrating factor. Thus, the kinetics parameters of the ion exchange, determined on the basis of diffusion processes for the selected glass-admixture system, will be dependent only on the temperature of the process.

The kinetics parameters of the ion exchange determined on the basis of diffusion processes should, by the assumed mathematical model, provide the opportunity to describe other cases of the ion exchange processes in the same glass-admixture system. The verification of the correctness of the model description of the ion exchange will be a confrontation of outcomes of specific processes calculated on its basis, with the measurement results of structures produced in the assumed conditions.

For the ion exchange processes, carried out in the multicomponent glasses, various types of modifier ions from the glass can make their share. In order, therefore, to determine their effect on the production of change in the refractive index (see equation (10) and (12)), it is needed to know the weight factors of their participation in the process. Still, a two-component model based on the ion exchange process has been adopted. For this model the following assumptions have been made:

- there are only two types of ions in this process. On the part of the source, these are the cations of the metal introduced into the glass (ions of A type). On the part of the glass substrate, these are the modifier cations (ions of B type), having the lowest activation energy,
- both types of ions have the same valence and diffuse by similar mechanisms,
- for each type of ions the relationship between an individual diffusion constant D_i and the electrochemical mobility μ_i is given by the Einstein equation:

$$\mu_i = \frac{e}{HkT} D_i \qquad i = A, B \qquad \left(m^2 V^{-1} s^{-1}\right), \tag{22}$$

where: e – elementary electrical charge, k – the Boltzmann constant, T – the absolute temperature, H – the correlation coefficient ($0<H\leq 1$) [11]

- the coupling between the streams of exchanged ions occurs only by the electrostatic forces,
- self-diffusion coefficients of both types of exchanged ions can be a function of normalized concentration, given in the form [61]:

$$\begin{aligned} D_A &= D_A(u,T) = D_{0A}(T)\cdot f_A(u) \\ D_B &= D_B(w,T) = D_{0B}(T)\cdot f_B(w), \end{aligned} \tag{23}$$

where: u, w – normalized concentrations of ions, respectively A and B type, T – the temperature on an absolute scale.

The dependencies of the diffusion coefficients on the temperature are given by the Arrhenius equation:

$$D_{0i}(T) = D_{0i}^{*} \cdot \exp\left(-\frac{\Delta Q_i}{RT}\right) \qquad (i = A, B), \tag{24}$$

where: ΔQ_i – activation energy of the ions i-type, R – universal gas constant.

In addition, as outlined in Part 2, there is an assumption of proportionality of the changes in the refractive index of the glass to the concentration of introduced admixture.

To describe the exchange processes in glass the arrangement geometry was assumed, as given in the Fig.15. In the glass, after the introduction of admixture (A), two streams of ions will occur due to their concentration gradients:

$$\begin{aligned} \vec{\Phi}_A &= -D_A \nabla c_A + \mu_A c_A \vec{E}_0 \qquad \left(m^{-2} s^{-1}\right) \\ \vec{\Phi}_B &= -D_B \nabla c_B + \mu_B c_B \vec{E}_0 \qquad \left(m^{-2} s^{-1}\right), \end{aligned} \tag{25}$$

where: D_A, D_B – self-diffusion coefficients of exchanged ions ($m^2 s^{-1}$), c_A, c_B – concentrations of ions A and B type (m^{-3}), μ_A, μ_B – electrical mobility of ions A and B type ($m^2 V^{-1} s^{-1}$).

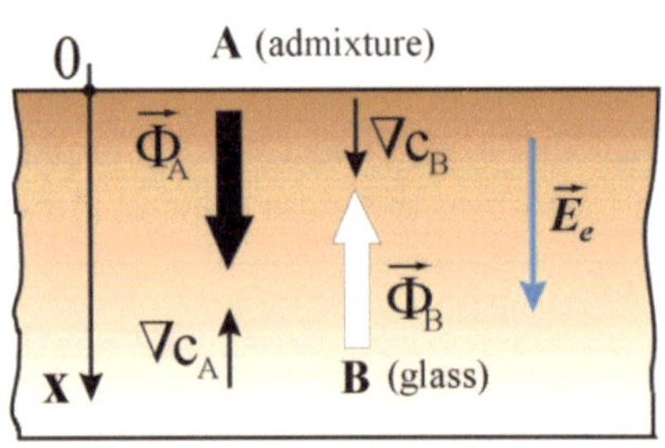

Figure 15. The arrangement geometry

The vector $\vec{E}_0$ (Vm^{-1}) in the equations (25) means the local electric field intensity in the glass:

$$\vec{E}_0=\vec{E}_e+\vec{E}_d, \tag{26}$$

where: $\vec{E}_e$ – external electric field (Vm^{-1}),. $\vec{E}_d$.–diffusion electric field, resulting from the differencies in the mobility of the exchanged ions (Vm^{-1}).

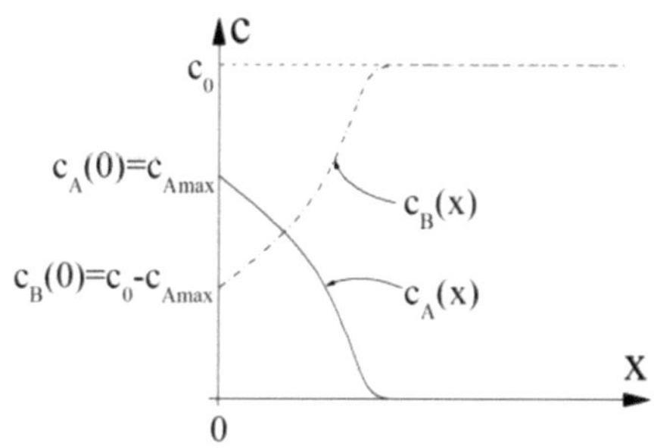

Figure 16. The concentration of the distributions of the admixture ions and modifier ions in the glass during the exchange process

If there is a difference in the diffusion coefficients D_A and D_B of both types of ions involved in the process, then in the glass the so-called diffusion potential occurs, resulting from the difference in their mobility. The associated electric field $\vec{E}_d$ is a coupling factor of the movement of the two streams (slower ions are accelerated and the faster ions are slowed down by this field). Taking into account an additional externally applied electric field $\vec{E}_e$, marked in the Fig.15, allows to generalize the considered ion exchange process to the diffusion with the electric field (electrodiffusion).

In the area of the glass the equilibrium concentration c_0 is determined, defined as the concentration of all the mobile ions. The Fig.16 demonstratively presents the one-dimensional distributions of the concentration of exchanged ions, which in any place of the glass meet the condition:

$$c_A(x)+c_B(x)=c_0 \tag{27}$$

The situation here shows the state, in which the maximum concentration of the admixture introduced into the glass reaches on its surface the value $c_{Amax}<c_0$. This means that at the glass surface only part of the exchangeable ions of the glass has been replaced by the admixture ions. Such a situation may take place when using diluted sources of admixture ions. When normalizing the equation (27) to the value of c_0, it is obtained:

$$\frac{c_A(x)}{c_0}+\frac{c_B(x)}{c_0}=1 \tag{28}$$

The case $c_{Amax}=c_0$ corresponds to the total replacement of ions of its modifiers at the glass surface, by admixture ions. The normalized concentrations of both types of ions u and w are introduced as:

$$u(x)=\frac{c_A(x)}{c_0} \qquad w(x)=\frac{c_B(x)}{c_0} \tag{29}$$

From the above definitions the obvious relations result:

$$\begin{aligned} &u_{max}=u(0)=\frac{c_{Amax}}{c_0} && \lim_{x\to\infty} u(x)=0 \\ &w_{min}=w(0)=1-u(0) && \lim_{x\to\infty} w(x)=c_0 \\ &w=1-u \quad \Rightarrow && \nabla w=-\nabla u \end{aligned} \tag{30}$$

Using the relations (29), the equations (25) can be represented as:

$$\begin{aligned} \vec{\Phi}_u &= \frac{\vec{\Phi}_A}{c_0} = -D_A(u)\nabla u + \mu_A u\vec{E}_0 \quad \left(ms^{-1}\right) \\ \vec{\Phi}_w &= \frac{\vec{\Phi}_B}{c_0} = -D_B(w)\nabla w + \mu_B w\vec{E}_0 \quad \left(ms^{-1}\right) \end{aligned} \tag{31}$$

In the area of glass in which there is no admixture (ions *A* type) $c_A=0 \Rightarrow u=0$ there is no electric diffusion field, so the local electric field is equal to the external field ($\vec{E}_0=\vec{E}_e$). Therefore the requirement of electrical neutrality of the glass puts the condition:

$$\vec{\Phi}_u + \vec{\Phi}_w = \mu_B\vec{E}_e \tag{32}$$

For the processes taking place without an external electric field, this condition reduces to the form:

$$\vec{\Phi}_u + \vec{\Phi}_w = 0 \tag{33}$$

Considering the equations (31), the condition (32) can be expressed as:

$$-D_A(u)\nabla u + \mu_A u\vec{E}_0 - D_B(w)\nabla w + \mu_B w\vec{E}_0 = \mu_B\vec{E}_e \tag{34}$$

By introducing the quantity:

$$\alpha = 1-\frac{D_A(u)}{D_B(w)} = 1-\frac{D_A(u)}{D_B(1-u)} \tag{35}$$

based on the equation (34), using the relations (30), an expression that describes a value of the local electric field in the glass is obtained:

$$\vec{E}_0 = \frac{\vec{E}_e}{1-\alpha u} - \frac{HkT}{e}\cdot\frac{\alpha\nabla u}{1-\alpha u} \tag{36}$$

The first component in (36) is associated with an external electric field, while the second expresses the diffusion field. Attention should be paid to the fact, that the sign of the diffusive component in the above equation is determined by the quantity α, depending on the normalized concentration of admixture ions introduced into the glass.

For the vector of normalized flow of admixture ions $\vec{\Phi}_u$, the continuity equation is valid in the whole volume of the glass:

$$\frac{\partial u}{\partial t}+\nabla\cdot\vec{\Phi}_u=0 \tag{37}$$

Taking into account (31) in equation (37) we obtain:

$$\frac{\partial u}{\partial t}=\nabla\cdot\left[D_A(u)\nabla u-\mu_A u\vec{E}_0\right] \tag{38}$$

For the volume of the glass for the external electric field vector the equality must be fulfilled: $\nabla\cdot\vec{E}_e=0$. Taking into account this fact in the calculation of the operator $\nabla\cdot\vec{E}_0$ in the equation (38), after the transformations the final form of the equation is obtained, describing the spatio-temporal changes in normalized concentration of admixture in the glass:

$$\begin{aligned}\frac{\partial u}{\partial t}&=\frac{D_A(u)}{1-\alpha u}\left[\Delta u+\frac{\alpha}{1-\alpha u}(\nabla u)^2-\frac{e}{HkT}\frac{1}{1-\alpha u}\nabla u\cdot\vec{E}_e\right]+\\&+\frac{1-u}{(1-\alpha u)^2}\left[\nabla u-\frac{e}{HkT}u\vec{E}_e\right]\cdot\nabla D_A+\frac{u(1-\alpha)^2}{(1-\alpha u)^2}\left[\nabla u-\frac{e}{HkT}u\vec{E}_e\right]\cdot\nabla D_B,\end{aligned} \tag{39}$$

where: $\alpha(u)=1-\dfrac{D_A(u)}{D_B(1-u)}$

Still, the dependence of the of diffusion coefficients from their normalized concentration was assumed in the form:

$$D_A(u)=D_{0A}e^{Au},\quad D_B(u)=D_{0B}e^{B(1-u)} \tag{40}$$

In this case, for a one-dimensional diffusion with no external electric field, the equation (39) takes the form:

$$\frac{\partial u}{\partial t}=\frac{D_{0A}e^{Au}}{1-\alpha u}\cdot\frac{\partial^2 u}{\partial x^2}+\frac{D_{0A}e^{Au}\left[\alpha+(1-u)A\right]-u(1-\alpha)^2 D_{0B}Be^{B(1-u)}}{(1-\alpha u)^2}\left(\frac{\partial u}{\partial x}\right)^2 \tag{41}$$

In the above equation:

$$\alpha=1-\frac{D_{0A}}{D_{0B}}\cdot\exp\left[u(A+B)-B\right] \tag{42}$$

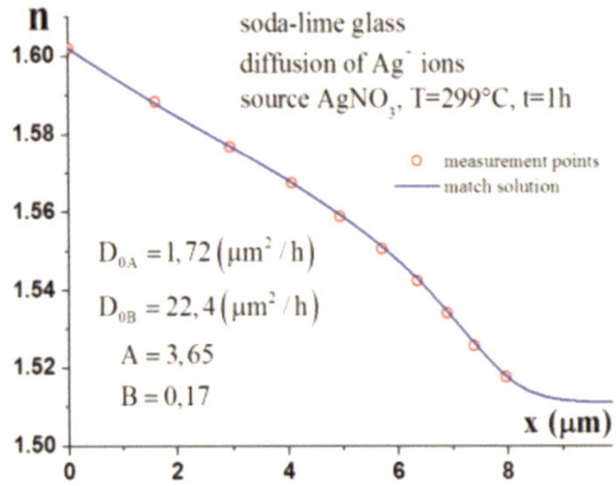

Figure 17. Refractive index profile of the waveguide obtained by fitting the solution of the diffusion equation into measurement points, together with the designated parameters of the diffusion coefficients

From the assumption of proportionality of the changes of refractive index, to the concentration of introduced admixture, results the relationship (one-dimensional case):

$$n(x)=n_b+\Delta n_s \cdot u(x) \tag{43}$$

In the above equation $n(x)$ is the absolute value of the refractive index of glass, n_b is the refractive index of undoped glass, Δn_s is the maximum change in the doping of the glass (for the preliminary diffusion - at its surface), and $u(x)$ is the spatial distribution of normalized concentration of admixture introduced into the glass.

5.1. Determination of diffusion coefficients by solving the diffusion equation

This procedure is based on solving the equation (41) describing the process of thermal diffusion in the one-dimensional case. Determination of the parameters of the diffusion equation reduces to the problem of finding the minimum of the matching function $\Psi(\vec{p})$, defined on the four-dimensional vector of diffusion parameters: $\vec{p}=[D_{0A},D_{0B},A,B]$. The matching function $\Psi(\vec{p})$ is defined as the sum of squared differences between the solution of diffusion equation and the points, set of the measurements of the refractive index profile of the waveguide, then divided by the number of these points. Thus, the determined diffusion parameters are the coordinates of the vector $\vec{p}_{min}$ minimizing the function $\Psi(\vec{p}_{min})$. Fig.17 shows the example fitting of the theoretical profile to the measurement points and the values of the diffusion coefficients: D_{0A}, D_{0B}, A and B designated in this way. The shown example relates to the ion exchange of $Ag^+ \leftrightarrow Na^+$ realized in the soda-lime glass in the diffusion process of the Ag^+ ions from the source of $AgNO_3$. The temperature of the process $T=299°C$, time of the diffusion $t=1h$.

5.2. Determination of the temperature dependences of exchanged ions' coefficients

The way of determining the diffusion coefficients (described in the previous part) is used to determine their dependencies on the temperature of these processes. This issue is very important, because based on the diffusion coefficients the theoretical distributions of admixture concentration in the glass are calculated. The knowledge of the temperature dependencies $D_{0A}(T)$ and $D_{0B}(T)$ of the diffusion coefficients of glass-admixture system allows modeling the technological processes that can be implemented at different temperatures. A very important technological problem, which is the repeatability of produced optical structures' properties, needs also to determine the impact of temperature changes on the final result of the process.

Temperature dependence of the diffusion coefficients is exponential, given by the Arrhenius equation (24). To fully determine the relation of $D_{0A}(T)$ and $D_{0B}(T)$ two factors are needed to be known: D_{0i}^* and ΔQ_i ($i=A, B$).

To determine these factors, one should realize a minimum of two thermal diffusion processes, at different temperatures, and then determine the coefficients of linear regression for dependence $lnD_{0i}(T^{-1})$. The duration of these processes must be chosen so as to produce the multimode waveguides in the glass, because for them the reconstruction procedure of the refractive index profiles ensures high reliability of restoring their shapes. An illustration of such proceedings, for soda-lime glass doped with silver ions, is set out below.

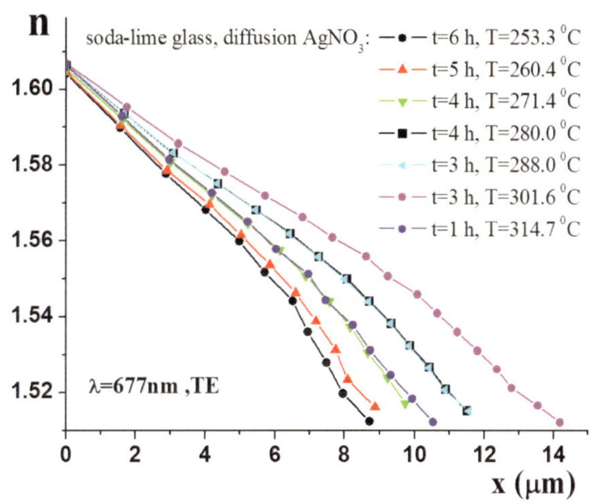

Figure 18. The refractive index profiles of waveguides produced in soda-lime glass in different temperatures diffusion of Ag^+ ions

Fig.18 shows the refractive index profiles of the waveguides produced in soda-lime glass in the processes of diffusion of silver ions realized at different temperatures. Times of diffusion processes have been chosen so that each refractive profile had to at least 10-modes for the wavelength of $\lambda=677$ *nm*. Because, as a result of diffusion of silver ions into this type of glass, a significant stress birefringence in the emerging waveguide structures is not observed, therefore the measurements of the refractive index profiles are referenced only to one TE polarization state. For each refractive profile the parameters D_{0A} and D_{0B} of the diffusion equation were then determined. Results of the calculations of these parameters are summarized in Table 2.

temperature of the process T (℃) / (K)	ΔT (K)	D_{0A} ($\mu m^2/h$)	D_{0B} ($\mu m^2/h$)
253.3 / 526.5	±0.5	0.296	3.77
260.4 / 533.6	±0.5	0.407	4.86
271.4 / 544.6	±0.5	0.658	7.28
280.0 / 553.2	±0.5	0.910	9.22
288.0 / 561.2	±0.5	1.21	11.9
301.6 / 574.8	±0.5	1.71	15.9
314.7 / 587.9	±0.5	2.79	25.1

Table 2. The determined D_{0A} and D_{0B} parameters of the diffusion equation for the ion exchange of $Ag^+ \leftrightarrow Na^+$ in soda-lime glass

The Fig.19 presents the measurement points of the temperature dependencies of the diffusion coefficients D_{0A} and D_{0B}, according to data compiled in Table 2. After taking the logarithm form of equation (24) and matching it to these points, the activation energies of exchanged ions ΔQ_A and ΔQ_B and the values of the logarithms of diffusion coefficients D^*_{0A} and D^*_{0B} are set.

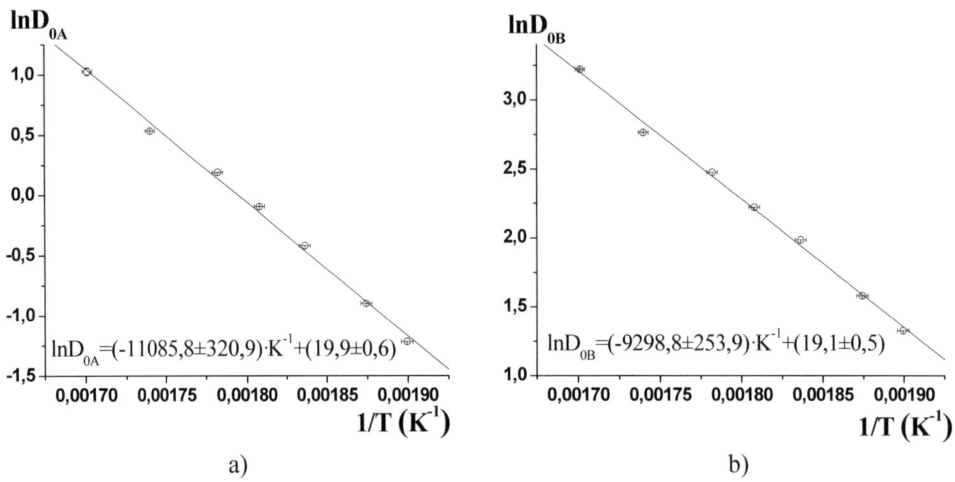

Figure 19. The dependencies of the diffusion parameters: a) D_{0A} and b)D_{0B} from the temperature (according to the Table 2)

The determined values are as follows:

$$\Delta Q_A = (92.2 \pm 2.7)\ kJ/mol \quad \ln D^*_{0A} = 19.9 \pm 0.6 \tag{44}$$

$$\Delta Q_B = (77.3 \pm 2.2)\ kJ/mol \quad \ln D^*_{0B} = 19.1 \pm 0.5$$

The equations describing the temperature dependencies of the diffusion coefficients of exchanged ions have the form:

$$\begin{aligned} D_{0A}(T) &= \exp\left(-\frac{11085.8}{T}+19.9\right) \quad \left(\mu m^2/h\right) \\ D_{0B}(T) &= \exp\left(-\frac{9298.8}{T}+19.1\right) \quad \left(\mu m^2/h\right) \end{aligned} \tag{45}$$

6. The real-time control of diffusion processes in glass

Obtaining a repeatability of effects of diffusion processes (in terms of changes in refractive index of glass) requires a close monitoring of both time and temperature. Ensuring the stability of the diffusion process' temperature of the order of ℃ fraction at the typical for ion exchange $Ag^{+}\leftrightarrow Na^{+}$ level of ~*300* ℃ is very difficult (practically impossible) to meet.

It is however possible to control the diffusion on an ongoing basis. The kinetics of this process is linked to its temperature by the Arrhenius equation (24). In turn, in equation (41) describing the diffusion of admixture ions into the glass, there are (in a transparent manner) the diffusion coefficients of exchanged ions $D_{0A}(T)$ and $D_{0B}(T)$, which are the functions of the temperature. Therefore, it is possible, by monitoring the temperature in the real time during the diffusion process, to determine the current values of these coefficients. Solving the equation (41) describing a one-dimensional diffusion is to integrate it in time. It takes a certain time step Δt. This step can be identified with the actual time, while not giving the upper cut-off time for the process of integrating the diffusion equation. By measuring the temperature in the crucible in certain moments of time t_p, which are the total multiples of the time step Δt, by the known temperature dependence $D_{0A}(T)$ and $D_{0B}(T)$, the current values of diffusion coefficients $D_{0A}(t_p)$ and $D_{0B}(t_p)$ can be calculated.

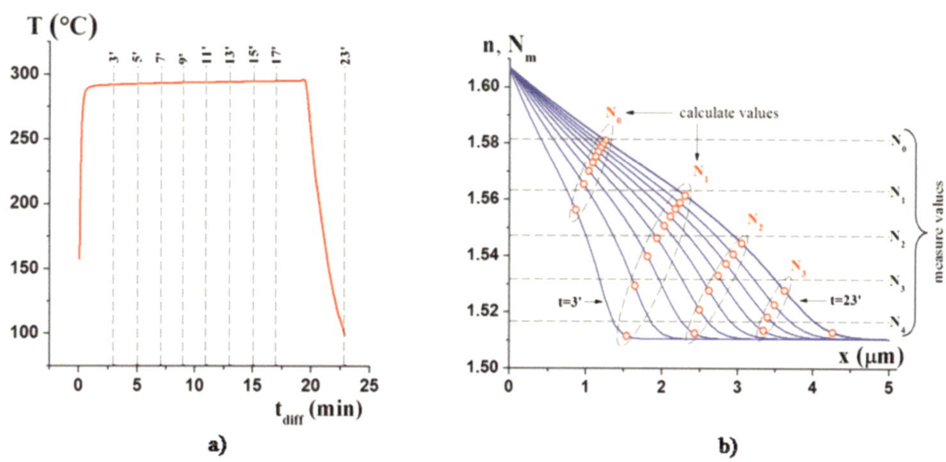

Figure 20. Illustration of the control of the diffusion process. a) Temperature dependence from the time b) Calculated refractive index profiles and the corresponding effective refractive indices of modes (*λ=677nm*, *TE* polarization) for the selected moments of time

In that case, if the duration of the integral equation (41) will run parallel to the actual time of the diffusion process, then the solution of this equation, at the current time t described by the function $u(x,t)$, will present the current distribution of normalized concentration of the admixture introduced into the glass. On the basis of the function $u(x,t)$ and equation (43), in which the disperse values $n_b(\lambda)$ and $\Delta n_s(\lambda)$ are known, the refractive waveguide profile $n(x,\lambda,t)$ is calculated for the current time t corresponding to the wavelength λ. Solving the mode equation (17) for the assumed polarization, which would be propagated in a waveguide with a refractive index $n(x,\lambda,t)$, one can calculate all the effective refractive indices of such modes of the waveguide.

This shows that such a way of monitoring the diffusion process creates the possibility of an ongoing control of the resulting waveguide, both in terms of its own modes and changes during their effective refractive indices. Figure 20 illustrates the example of the control of the diffusion process using this method. Glass substrate was immersed in the molten salt by means of the handle (Fig.21). It provides a direct contact of the thermocouple with the glass substrate. Pressure pulsations allow the continuous stirring of the contents of the crucible. The Figure 20 shows the temperature curve of a selected diffusion process. This method of process control also includes the process of cooling the substrate after removing it from the crucible. Time step of monitoring the process temperature was t_p=2 s. Fig.20b shows the refractive index profiles of the resulting waveguide, calculated at selected moments of time, with the resulting effective indices of modes. The calculated refractive index profiles correspond to the moments of time marked in Fig.20a. The Fig.20b also marks the effective indices obtained from measurements of the resulting waveguide for wavelength λ=677nm with TE polarization. Their measurement uncertainties were ΔN_m=0.0004. From the results shown in Fig.20b it can be seen that the calculated effective indices are slightly smaller than the measured values. This difference is the smallest for the TE_0 mode and is of 0.0006. For higher order modes it increases, reaching the highest value of 0.004 for the TE_4 mode. These differences result from the numerical modeling of the process, and also from the

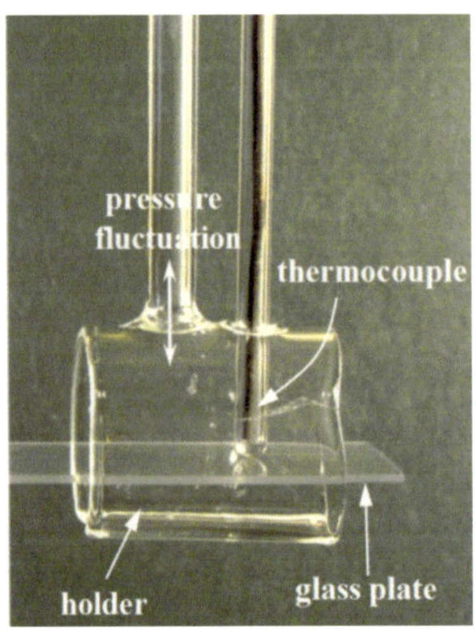

Figure 21. The glass holder

measurement method of the temperature of the substrate. The measurement thermocouple visible in the Fig.21 is placed in a protective glass tube. Therefore it has a rather large thermal inertia. As a result, the temperature recorded with its use does not reflect the actual temperature of a glass substrate, which it has at any given moment of time. The largest differences of the temperature indications occur at stages of heating and cooling of the glass substrate. Therefore the significance of these differences in made calculations will be large, especially for the short periods of diffusion.

Fig.22 presents the results of the seven diffusion processes of Ag^+ ions into the soda-lime glass. The aim of these processes was to obtain uniform waveguide structures in terms of both quantity and value of effective refractive indices of modes. All produced waveguides were 5-modes for the wavelength λ=677 nm, with *TE* polarization. Fig.22a shows the time dependences of each process' temperature. While the Fig.22b shows dispersion ($N_{m,max}$-$N_{m,min}$), of the measured effective refractive indices, shown as a function of the order of the modes. The measurement uncertainties of all effective refractive indices were: $\Delta N_m=0.0004$. The obtained results show good repeatability of the realized processes.

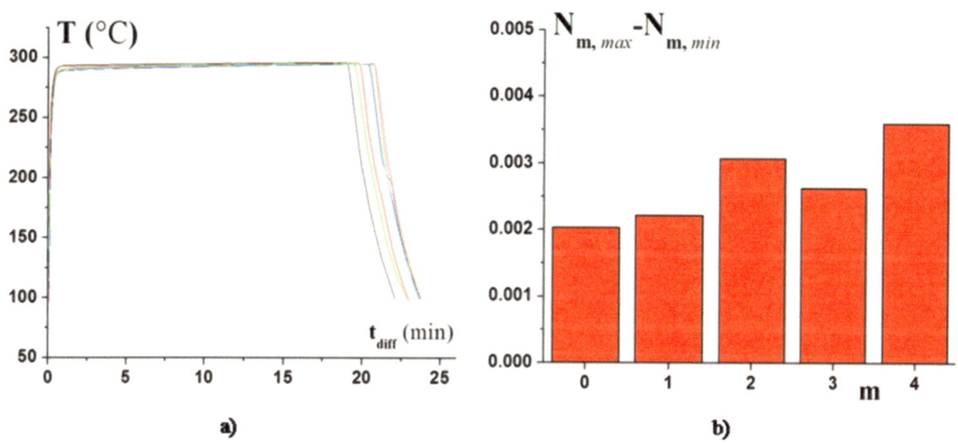

Figure 22. The repeatability of the diffusion processes (soda-lime glass, ion exchange: $Ag^+ \leftrightarrow Na^+$, admixture source: $AgNO_3$) a) time dependences of the temperatures of the processes b) dispersion of the measured effective refractive indices for the individual orders of modes

Fig.23 shows the enlarged fragments of the recorded temperatures of the processes from Fig.22a. This shows that the temperature differences are, in extreme cases, even of a few Celsius degrees. With such a dispersion of temperatures, obtaining the repeatability of effective refractive indices of produced waveguides would not be possible without the use of the proposed process control. The method used to monitor the diffusion process makes its final effect independent from the temperature. The same result is achieved in a shorter time of diffusion process, when the average temperature is higher and vice versa. This regularity is clearly visible in Fig.23.

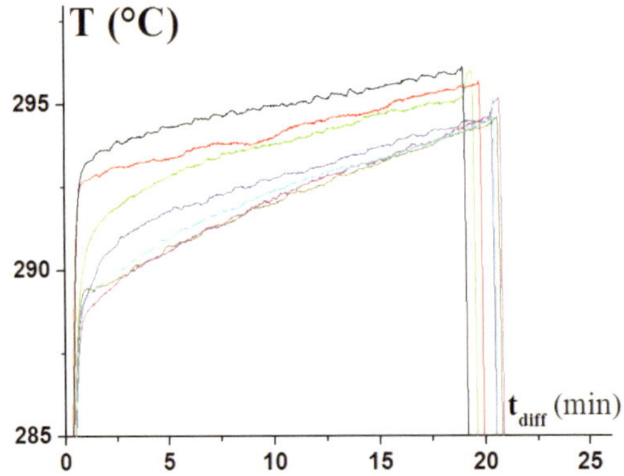

Figure 23. The temperature differences of the diffusion processes

7. Ion exchange in glass application

The method of ion exchange as a technology tool to produce changes in refraction in the glass is still in use. Most of these applications takes place in optoelectronics. There are, of course, limits of its application. They arise from the possibility of physical effects generated by this method. The main limitation is the obtained maximum increment of refractive index of the glass Δn. The value of this increase depends on the type of glass and the type of admixture ions. The maximum reached value of Δn is of the order of *0.1*. However, the big advantage of this method is its low price. Equipment requirements here are much smaller (cheaper) than in the case with other methods, such as vacuum technology. Also the materials used in this method: the glass substrates and ion admixture sources are relatively inexpensive.

A characteristic feature of the effect of changes in refractive index of glass obtained by this method is the gradient nature of those changes. This creates the possibility of implementing gradient micro-optic components. The greatest use of this method in optoelectronics are taking place in the production of waveguide structures. Attractive in particular are application of this method in the manufacture of optical amplifiers in glasses doped with rare earth ions, the production of the structures of planar lasers in glasses doped with rare earth elements Er^{3+} and Yb^{3+} (glass MM-2 of Kigre company [62], glass IOG-1 of Schott company [63]), and the production of the sensor structures. The combination of the ion exchange method and the zol-gel method is also interesting. The zol-gel method enables the production of thin (of the order of a fraction of micrometer) layers whose refractive index can take values in the range of $1.45 \div 2.2$ [64]. These layers are the glasses received in the low-temperature polymerization and gelation processes of the suitable solutions. Large (compared with the ion exchange) values of the refractive index of these layers make this method attractive for integrated optics. The waveguide structures produced in the glass substrates by ion exchange method are homogeneous structures, extending from the glass surface at a depth of several to tens (in some cases even more) μm. For example, for typical values of the refractive index of glass ~ 1.5, with a maximum change of the refractive index of ~ 0.1, the depths of the waveguide areas, providing the conditions for single mode propagation, are of the order of $1\ \mu m$. On the other hand, using diluted admixture sources, which increments of the refractive index can be very small, to obtain single-mode structures requires a depth of doped glass in the range from a few to several μm. Greater depths of doped regions in the glass are formed by the multimode structures. Whereas the homogeneous waveguide layers, produced by the zol-gel method, because of their small thickness (tens to hundreds of *nm*) are single-mode structures with large values of the refractive index as compared to the waveguides produced by ion the exchange method. This combination of small changes in the refractive index with the large depths of glass doping areas at the same time, in the case of ion-exchange method with very small layer thicknesses (obtained by the zol-gel method) with a high refractive index, enables the production of hybrid waveguide structures. In this case, at the glass surface, in which the gradient structure is manufactured, a thin uniform layer of dielectric material is prepared by the zol-gel method. By appropriate selection of the refractive index and thickness of this layer in such a structure, the effect of conducting zero-order mode in the homogeneous layers, while propagating higher order modes in the gradient area of the glass, can be achieved. Such gradient-homogeneous waveguides may constitute a new category of leading light structures, with specific modes properties.

With the use of the zol-gel method the silica layers can also be applied to the glass surface. The use of photolithographic processes of shaping the topology of these layers makes that they can act as a dielectric masking coverings in technological processes of ion exchange. This method can also produce strip waveguide structures of *rib* type [65]. Therefore the combination of ion-exchange method and the zol-gel method creates new possibilities for creating the elements of integrated optics. The combination of both methods is also justified in the case of constructing the structures of amplitude optical sensors, using the evanescent field spectroscopy. In this case, the waveguide path is produced in the glass substrate by the ion exchange method, which refractive index profile can be shaped to a fairly large range, e.g. by the electrodiffusion processes. Then, by the zol-gel method, a sensor layer which covers the gradient waveguide [66, 67] is prepared. Another area of application of the ion exchange method is surface hardening of the glass. When using certain admixture ions (e.g. K^+) in the admixture area of the glass, a significant stress induces arising from the difference in radii of exchanged ions. These stresses are of a clamping character, and their magnitude is of the order of hundreds of N/mm^2. When taking into account the depth of glass doping of tens of μm typical for ion exchange processes, the considerable stress gradients in the superficial area of glass are achieved. This gives the effect of its hardening.

Having regard to those facts, it can be stated that the ion exchange method remains as an attractive technological tool for both: technical and economical reasons. Therefore a further research related to its use for producing the gradient structures in the glass is very useful. Ensuring a high repeatability of processes, guaranteeing the established tolerance of the optical properties of produced structures, is a task that has made this method an effective technological tool.

Author details

Roman Rogoziński
Optoelectronics Department, Silesian University of Technology, Gliwice, Poland

Acknowledgement

This paper was financed from the funds of the National Science Centre, awarded on the basis of the decision DEC-2011/01/B/ST7/06525.

8. References

[1] Kao K.C., Hockman G.A., *Dielectric fibre surface waveguides for optical frequencies*, Proc. IEEE, 1966, Vol.113, No.7, 1151-1158.

[2] Doremus, R.H., *Exchange and Diffusion of Ions in Glass*, The Journal of Physical Chemistry, 1964, Vol.68, No.8, 2212-2218.

[3] Doremus, R.H., *Ion Exchange in Glass*, in Ion-Exchange, Marinsky, J.A., Ed., New York: Marcel Dekker, 1969, Vol.2, 1-42.

[4] Kahnt H., *Ionic transport in glasses*, Journal of Non-Crystalline Solids, 1996, Vol.203, 225-231.

[5] French W.G., Pearson A.D., *Refractive index changes produced in glass by ion-exchange*, Ceramics Bulletin, 1970, Vol.49, No.11, p.474.

[6] Giallorenzi T.G., West E.J., Kirk R., Ginter R., Andrews R.A., *Optical waveguides formed by thermal migration in glass*, Applied Optics, 1973, Vol.12, No.6, 1281-1285.

[7] Izawa T., Nagakome H., *Optical Waveguide Formed by Electrically Induced Migration of Ions in Glass Plates*, Applied Physics Letters, 1972, Vol.21, 584-586.

[8] Findakly T., *Glass Waveguides by Ion Exchange: A Review*, Optical Engineering, 1985, Vol.24, No.2, 244-250.

[9] Najafi S.I., *Introduction to Glass Integrated Optics*, Ed., Boston: Artech House 1992.

[10] Ramaswamy, R.V., Srivastava, R., *Ion-Exchange Glass Waveguides: A Review*, Journal of Lightwave Technology, 1988, Vol.6, No.6, 984-1002.

[11] Terai R, Hayami R., *Ionic Diffusion in Glasses*, Journal of Non-Crystalline Solids, 1975, Vol.18, 217-264.

[12] Garfinkel H. M., *Ion-exchange equilibria between glass and molten salts*, The Journal of Physical Chemistry, 1968, Vol.72, No.12, 4175-4181.

[13] Araujo R.J., Likitvanichkul S., Thibault Y., Allan D.C., *Ion exchange equilibria between glass and molten salt,* Journal of Non-Crystalline Solids, 2003, Vol.318, 262-267.

[14] Zachariasen W.H., *The atomic arrangement in glass*, Journal American Chemistry Society, 1932, Vol.54, 3841-3851.

[15] Warren B.E., Loring A.D., *X-Ray diffraction study of the structure of soda-silica glasses,* Journal American Ceramic Society, 1935,Vol.18, 269-276.

[16] Isard J.O., *The Haven ratio in glasses,* Journal of Non-Crystalline Solids, 1999, Vol.246, 16-26.

[17] Fantone S.D., *Refractive Index and Spectral Models for Gradient-Index Materials*, Applied Optics, 1983, Vol.22, 432-440.

[18] Huggins M. L., *The refractive index of silicate glasses as a function of composition*, Journal Optical Society of America, 1940, Vol.30, No.10, 495-504.

[19] Chartier G.H., Jaussaud P., de Oliveira A.D., Parriaux O., *Fast Fabrication Method for Thick and Highly Multimode Optical Waveguides*, Electronics Letters, 1977, Vol.13, 763-764.

[20] De Bernardi C., Malvicino C., Morasca S., Morra M., *Time and Temperature Influence on Surface Index Change* K^+*-*Na^+ *Ion Exchanged Optical Waveguides*, Journal Applied Physics, 1988, Vol.63, No.1, 234-236.

[21] Gortych J.E., Hall D.G., *Fabrication of Planar Optical Waveguides by* K^+*-Ion Exchange in BK7 and Pyrex Glass*, IEEE Journal of Quantum Electronics, 1986, Vol.22, 892-895.

[22] Honkanen S., Tervonen A., McCourt M., *Control of birefringence in ion-exchanged glass waveguides,* Applied Optics, 1987, Vol.26, No.22, 4710-4711.

[23] Johansson J., Dianta G., Coutaz J.-L., *Optical Waveguides Fabricated by Ion Exchange in High-Index Commercial Glasses*, Applied Optics, 1992, Vol.31, No.15, 2796-2799.

[24] Miliou A.N., Srivastava R., Ramaswamy R.V., *Modeling of the Index Change in* K^+*-*Na^+ *Ion-Exchanged Glass*, Applied Optics, 1991, Vol.30, No.6, 674-681.

[25] Najafi S.I., *Optical Behaviour of Potassium Ion-Exchanged Glass Waveguides,* Applied Optics, 1988, Vol.27, No.17, 3728-3731.

[26] Noutsios P.C., Yip G.L., *Shallow buried waveguides made by purely thermal migration of* K^+ *ions in glass*, Optics Letters, 1990, Vol.15, No.4, 212-214.

[27] Tervonen A., Honkanen S., *Feasibility of potassium-exchanged waveguides in BK7 glass for telecommunication devices,* Applied Optics, 1996, Vol.35, No.33, 6435-6437.

[28] Urnes S., Na^+*-*K^+ *exchange in silicate glasses,* Journal American Ceramics Society, 1973, Vol.56, No.10, 514-517.

[29] Yip G.L., Albert J., *Characterization of Planar Optical Waveguides by* K^+*-Ion Exchange in Glass*, Optics Letters, 1985, Vol.10, No.3, 151-153.

[30] Neuman V., Parriaux O., Walpita L., *Double-Alkali Effect: Influence of Index Profile of Ion-Exchanged Waveguides*, Electronics Letters, 1979, Vol.15, 704-706.

[31] Reichelt A., Clemens P.C., Mthlein H.F., *Single-Mode Waveguides and Components by Two-Step* Cs^+*-*K^+ *Ion-Exchange in Glass*, Proc. SPIE-Int. Soc. Opt. Eng.: "Glasses for Optoelectronics" 1989, Vol.1126, 166-168.

[32] Bogomolova L.D., Ganshin V.V., Jachkin V.A., Kubrinskaya M.E., Petrova V.Z., *EPR and Optical Study of Copper Diffusion Layers Produced by Ion Exchange in Oxide Glasses*, Journal of Non-Crystalline Solids, 1981, Vol.45, 249-255.

[33] Gevorgyan S.S., *Single-step buried waveguides in glass by field-assisted copper ion-exchange*, Electronics Letters, 1990, Vol.26, No.1, 38-39.

[34] Gallagher J.G., de La Rue R.M., *Single-Mode Stripe Optical Waveguides Formed by Silver Ion Exchange*, Electronics Letters, 1976, Vol.12, No.16, 397-398.

[35] Gato L., Srivastava R., *Time Dependent Surface Index Change in Ion Exchanged Waveguides*, Optics Communications, 1996, Vol.123, No.4-6, 483-486.

[36] Ghatak A.K., Khular E., Thyagarajan K., *Modes in Optical Waveguides Formed by Silver Sodium Ion Exchange*, IEEE Journal of Quantum Electron., 1978, Vol.14, No.6, 389-391.

[37] Gonella F., *Stress-induced optical effect in Ag^+-Na^+ion-exchanged glass waveguides*, Optics Letters, 1992, Vol.17, No.23, 1667-1669.

[38] Martin M., Videau J.J., Canioni L., Adamietez F., Sarger L., Le Flem G., *Planar waveguides formed by Ag^+-Na^+ ion exchange in nonlinear optical glasses: diffusion and optical properties*, Applied Optics, 2000, Vol.39, No.3, 435-440.

[39] Millar C.A., Hutchins R.H., *Manufacturing tolerances for silver-sodium ion-exchanged planar optical waveguides*, Journal Physics D; Applied Physics, 1978, Vol.11, 1567-1576.

[40] Najafi S.I., Srivastava R., Ramaswamy R.V., *Wavelength-Dependent Propagation Characteristics of Ag^+-Na^+ Exchanged Planar Glass Waveguides*, Applied Optics, 1986, Vol.25, No.11, 1840-1843.

[41] Ramaswamy R.V., Cheng H.C., Shirastava R., *Process optimization of buried Ag^+-Na^+ ion-exchanged waveguides: theory and experiment*, Applied Optics, 1988, Vol.27, No.9, 1814-1819.

[42] Stewart G., Laybourn P.J.R., *Fabrication of Ion-Exchanged Optical Waveguides From Dilute Silver Nitrate Melts*, IEEE Journal of Quantum Electronics, 1978, Vol.14, No.12, 930-934.

[43] Stewart G., Millar C.A., Layboum P.J.R., Wilkinson C.D.W., De la Rue R.M., *Planar optical waveguides formed by silver-ion migration in glass*, IEEE Journal of Quantum Electronics, 1977, Vol.13,No.4, 192-200.

[44] Camy P., Roman J.E., Willems F.W., Hempstead M., van der Plaats J.C., Pfel C., Beguin A., Koonen A.MJ., Wilkinson J.S., Lerminiaux C., *Ion-Exchanged Planar Lossless Splitter at 1.5 μm*, Electronics Letters, 1996, Vol.32, 321-323.

[45] Bach H., Neuroth N., *The Properties of Optical Glass*, Springer-Verlag Berlin Heidelberg 1995.

[46] Jackel J.L, *Glass waveguides made using low melting point nitrate mixtures*, Applied Optics, 1988, Vol.27, No.3, 472-475.

[47] Rogoziński R., *Refractive index profiles of planar waveguides produced in electrodiffusion processes*, Optica Applicata, 2004,Vol.34, No.4, 489-505.

[48] White J.M., Heidrich P.F., *Optical waveguide refractive index profiles from measurement of mode indices: A simple analysis*, Applied Optics, 1976, Vol.15, No.1, 151-155.

[49] Batchelor S., Oven R., Ashworth D.G., *Reconstruction of refractive index profiles from multiple wavelength mode indices*, Optics Communications, 1996, Vol.131, 31-36.
[50] Albert J., Yip G.L., *Refractive-index Profiles of Planar Waveguides Made by Ion-Exchange in glass*, Applied Optics, 1985, Vol.24, No 10, 3692-3693.
[51] Chiang K.S., *Construction of refractive-index profiles of planar dielectric waveguides from the distribution of effective indexes*, Journal of Lightwave Technology, 1985, Vol.3, No.2, 385-391.
[52] Hertel P., Menzler H.P., *Improved Inverse WKB Procedure To Reconstruct Refractive Index Profiles of Dielectric Planar Waveguides*, Applied Physics, 1987, Vol.B 44, 75-80.
[53] Liñares J., Lipovskii A.A., Tagantsev D.K., Turunen J., *Characterization of ion diffusion process in glasses with simple mode-index measurements*, Optical Materials, 2000, Vol.14, 115-120.
[54] Rogoziński R., *Determination of refractive index profiles of planar buried waveguides on the basis of a set of modal propagation constants*, Optics Communications, 2003, Vol.219, 199-214.
[55] Weiss M.N., Srivastava R., *Determination of ion-exchanged channel waveguide profile parameters by mode-index measurement*, Applied Optics, 1995, Vol.34, No.3, 455-458.
[56] Tien P.K., Ulrich R., *Theory of prism-film coupler and thin film light guides*, Journal Optical Society of America, 1970, Vol.60, 1325-1337.
[57] Ulrich R., Torge R., *Measurement of Thin Film Parameters with a Prism Coupler*, Applied Optics, 1973, Vol.12, No.2, 2901-2908.
[58] Kersten R.T., *The prism-film coupler as a precision instrument*, Optica Acta,1975, Vol.22, No.6, 503-521.
[59] Seligson J., *Prism couplers in guided-wave optics: design considerations*, Applied Optics, 1987, Vol.26, No.13, 2609-2611.
[60] Agan S., Ay F., Kocabas A., Aydinli A., *Stress effect in prism coupling measurements of thin polymer films*, Applied Physics A, 2003, Vol.80, 341-345.
[61] Lupascu A., Kevorkian A., Boudet T., Saint-André F., Persegol D., Levy M., *Modeling ion exchange in glass with concentracion-dependent diffusion coefficients and mobilities*, Optical Engineering, 1996,Vol.35, No.6, 1603-1610.
[62] Kigre,Inc.100 Marschland Road, Hilton Head, SC 29926, www.kigre.com
[63] Schott AG, Hattenbergstrasse 10, 55122 Mainz, Germany, www.schott.com/advanced_optics
[64] Orignac X., Vasconcelos H.C., Du X.M., Almeida R.M., *Influence of Solvent Concentration on the Microstructure of* SiO_2-TiO_2 *Sol-Gel Films*, Journal of Sol-Gel Science and Technology, 1997, Vol.8, 243–248.
[65] Karasiński P., Rogoziński R., Rib waveguides fabricated by means of chemical etching of sol-gel SiO_2:TiO_2 films, Optics Communications, 2005, Vol.245, 237-242.
[66] Karasiński P., Rogoziński R., *Influence of refractive profile shape on the distribution of modal attenuation in planar structures with absorption cover*, Optics Communications 2007, Vol.269, No.1, 76-88.

[67] Karasiński P., *Sol-gel derived sensitive films for evanescent wave ammonia sensors,* Optica Applicata, 2003,Vol.33, No.2-3, 477-487.

Section 3

Applications of Ion Exchangers

Chapter 8

Selective Removal of Heavy Metal Ions from Waters and Waste Waters Using Ion Exchange Methods

Zbigniew Hubicki and Dorota Kołodyńska

Additional information is available at the end of the chapter

http://dx.doi.org/10.5772/51040

1. Introduction

Environmental pollution by toxic metals occurs globally through military, industrial, and agricultural processes and waste disposal (Duffus, 2002). Fuel and power industries generate 2.4 million tons of As, Cd, Cr, Cu, Hg, Ni, Pb, Se, V, and Zn annually The metal industry adds 0.39 million tons/yr of the same metals to the environment, while agriculture contributes 1.4 million tons/yr, manufacturing contributes 0.24 million tons/yr and waste disposal adds 0.72 million tons/yr. Metals, discharged or transported into the environment, may undergo transformations and can have a large environmental, public health, and economic impact (Brower et al. 1997; Nriagu & Pacyna, 1988; Gadd & White, 1993).

Among different techniques used for removal of high concentrations of heavy metals, precipitation-filtration, ion exchange, reverse osmosis, oxidation-reduction, solvent extraction, as well as membrane separation should be mentioned (Hubicki,et al. 1999; Dąbrowski et al. 2004). However, some of the wastes contain substances such as organics, complexing agents and alkaline earth metals that may decrease the metal removal and result in unacceptable concentrations of heavy metals in the effluents. The pollutants of concern include cadmium, lead, mercury, chromium, arsenic, zinc, cobalt and nickel as well as copper. They have a number of applications in basic engineering works, paper and pulp industries, leather tanning, petrochemicals, fertilizers, etc. Moreover, they have also negative impact on human health.

Cadmium is a metal of great toxicological concern. An important source of human exposure to cadmium is food and water, especially for the population living in the vicinity of industrial plants, from which cadmium is emitted to the air. In the case of exposure to occupational cadmium compounds, they are absorbed mainly by inhalation. Through the

INTECH
open science | open minds

© 2012 Hubicki and Kołodyńska, licensee InTech. This is a paper distributed under the terms of the Creative Commons Attribution License (http://creativecommons.org/licenses/by/3.0), which permits unrestricted use, distribution, and reproduction in any medium, provided the original work is properly cited.

gastrointestinal tract less than 10% cadmium is absorbed. An important source of human exposure to cadmium is food and water. In natural water its typical concentration lies below 0.001 mg/dm^3, whereas, the upper limit recommended by EPA (Environmental Protection Agency) is less than 0.003 mg/dm^3. The maximum limit in drinking water is 0.003 mg/dm^3.

Cadmium accumulates in kidneys, pancreas, intestines and glands altering the metabolism of the elements necessary for the body, such as zinc, copper, iron, magnesium, calcium and selenium. Damage to the respiratory tract and kidneys are the main adverse effects in humans exposed to cadmium compounds. In humans exposed to fumes and dusts chronic toxicity of cadmium compounds is usually found after a few years. The main symptom of emphysema is that it often develops without preceding bronchitis. The second basic symptom of chronic metal poisoning is kidney damage. It includes the loss and impairment of smell, pathological changes in the skeletal system (osteoporosis with spontaneous fractures and bone fractures), pain in the extremities and the spine, difficulty in walking, the formation of hypochromic anemia. The most known 'Itai-Itai' disease caused by cadmium exposure is mixed osteomalacia and osteoporosis. However, an important source of cadmium in soils are phosphate fertilizers. Large amounts of cadmium are also introduced to soil together with municipal waste. The high mobility of cadmium in all types of soils is the reason for its rapid integration into the food chain. Daily intake of cadmium from food in most countries of the world is 10-20 mg.

Lead is a toxic metal, which accumulates in the vital organs of men and animals and enters into the body through air, water and food. According to the WHO (World Health Organization) standards, its maximum limit in drinking water is 0.05 mg/dm^3 but the maximum discharge limit for lead in wastewater is 0.5 mg/dm^3. Its cumulative poisoning effects are serious haematological damage, anaemia, kidney malfunctioning, brain damage etc. Chronic exposure to lead causes severe lesions in kidney, liver, lungs and spleen.

Lead is used as industrial raw material in the manufacture of storage batteries, pigments, leaded glass, fuels, photographic materials, matches and explosives. Lead being one of very important pollutants comes from wastewaters from refinery, wastewaters from production of basic compounds containing lead, wastewaters with the remains of after production solvents and paints. Large toxicity of lead requires that its contents are reduced to the minimum (ppb level). To this end there are applied chelating ions with the functional phosphonic and aminophosphonic groups. Also weakly basic anion exchangers in the free base form can be used for selective removal of lead(II) chloride complexes from the solutions of pH in the range 4-6. Also a combined process of cation exchange and precipitation is often applied for lead(II) removal form wastewaters (Pramanik et al. 2009). The average collection of lead by an adult was estimated at 320-440 mg/day. Acute poisoning with inorganic lead compounds occurs rarely. In the case of acute poisoning in man, the symptoms are burning in the mouth, vomiting, abdominal cramps, diarrhea, constipation progressing to systolic, blood pressure and body temperature. At the same time there is hematuria, proteinuria, oliguria, central nervous system damage. Alkyl lead compounds are more toxic than inorganic lead connections. Tetraethyl lead toxicity manifested primarily in lead damage of the nervous system. Toxic effects of lead on the central nervous system are observed more in children. In adults, the effects of lead toxicity occur in the peripheral nervous system. Symptoms of chronic poisoning may vary. The acute form of poisoning known as lead colic is the general state of various spastic internal organs and neurological damage in the peripheral organs. Long-term lead poisoning can lead to organic changes in the central and peripheral nervous systems. Characteristic symptoms include pale gray skin colour and the lead line on the gums (blue-black border).

In nature, natural circulation of **mercury** vapour has a significant influence on the content of the soil and water. Elemental mercury in the rain water creates compounds by oxidation to divalent mercury. Both the chemical reaction, and under the influence of biological factors, and especially the activity of bacteria in the sediments of water bodies methyl and dimethyl mercury compounds are formed. Mercury, a fixed component of the waste water treatment that may be used for soil fertilization is a major threat to the inclusion of the metal in nutritional products. Drinking water may contain up to 300 ng Hg/dm^3, in highly industrialized areas it can reach up to 700 ng/dm^3. Daily consumption of mercury from food in the general population is less than 20 µg/day. 80% of mercury absorbed by the respiratory system is retained in the body. In the case of ingestion of inorganic mercury salts, salivation, burning in the throat, vomiting, bloody diarrhea, necrosis of the intestinal mucosa and kidney damage, leading to anuria and uremii can occur. The concentration of mercury vapour over 1 mg/m^3 damages lung tissue and causes severe pneumonia. The classic symptoms of metallic mercury vapour poisoning are manifested by tremor, mental disorders, inflammation of the gums. Its maximum limit in drinking water is 0.0005 mg/dm^3.

Chromium, occurring as Cr(III) or Cr(VI) in natural environments, is an important material resource, an essential micronutrient or toxic contaminant. Cr(III) is required for normal development of human and animal organisms but Cr(VI) activates teratogenic processes, disturbs DNA synthesis and can give rise to mutagenous changes leading to malignant tumours (WHO, Report 1998). Natural sources of chromium include weathered rocks, volcanic exhalations and biogeochemical processes and, in the man-polluted environment, mainly wastes after processing and utilization of chromium compounds. Chromium is an important and widely applied element in industry. The hexavalent and trivalent chromium is often present in electroplating wastewater. Other sources of chromium pollution are leather tanning, textile, metal processing, paint and pigments, dyeing and steel fabrication. To remove toxic chromium compounds from sewages there are used such methods as: precipitation, coagulation, solvent extraction and various kinds of membrane processes, ion flotation, adsorption and ion exchange (Bajda, 2005). The maximum limit in drinking water is 0.05 mg/dm^3. The Polish drinking groundwater chromium content ranges on the average from 0.07 to 2 mg/dm^3. 0.02 mg/dm^3 is accepted as the permissible content of chromium in groundwater. The daily dose taken by the adult can be 50-200 mg/day (or 60-290 mg/day). Cr(III) cation predominates in most tissues except the liver. Chromium is associated with nucleic acids and is the subject to the concentration in liver cells. It plays an important role in the metabolism of glucose, certain proteins and fats, is part of enzymes and stimulates the activity of others. All compounds of chromium, with the exception of chromate, are rapidly cleared from the blood. Chromium also accumulates in the liver and kidneys. High concentrations of chromium, observed in the lungs of people exposed to this metal, indicate that at least part of chromium is stored in this organ in the form of insoluble compounds. The binding of chromium with the elements of the blood and transport of chromium by the blood depends mainly on its valence. Hexavalent chromium readily crosses the membranes of red blood cells and after reduction to trivalent chromium is bound to hemoglobin. The reduction of hexavalent to trivalent chromium, occurring within cells, considered as the activation of the carcinogenic chromium, increases because the probability of interaction of trivalent chromium on the DNA. Clinical signs of acute toxicity of chromium compounds are characterized by severe abdominal pain, vomiting and bloody diarrhea, severe kidney damage with hematuria leading to anuria, observed gastrointestinal ulceration. Chromium compounds and chromic acid are especially dangerous and cause serious damage to internal organs. Chronic exposure leads to chronic disorders in the body.

Arsenic is present in over 160 minerals. It is readily bioaccumulative and therefore its concentration in polluted waters may reach 430 mg/dm^3 in plants and 2.5 mg/dm^3 in fish. The upper limit of arsenic recommended by US EPA, EU and WHO is 0.01 mg/dm^3. However, many countries have retained the earlier WHO guideline of 0.05 mg/dm^3 as their standard.

Arsenic accumulates in tissues rich in keratin, like hair, nails and skin. Arsenic and its inorganic compounds can cause not only cancer of the respiratory system and skin, but also neoplastic lesions in other organs. Arsenic compounds enter the body from the gastrointestinal tract and through skin and respiratory system. Arsenic compounds have affinity for many enzymes and can block their action, and above all disturb the Krebs cycle. Inorganic arsenic compounds are more harmful than organic and among them AsH_3 and As_2O_3 should be mentioned. 70-300 mg of As_2O_3 is considered to be the average lethal dose for humans. The dose of 10-50 ppb for 1 kg of body weight can cause circulatory problems, resulting in necrosis and gangrene of limbs. The dominant effects of arsenic in humans are changes in the skin and mucous membranes as well as peripheral nerve damage. There are xerosis soles and palms, skin inflammation with ulceration. In addition, there is perforation of the nasal septum. The values of the maximum allowable concentration (NDS) in Poland set for inorganic arsenic compounds are 0.3 mg/m^3 and 0.2 mg/m^3 for AsH_3.

In nature, **zinc** occurs in the form of minerals. An important source of zinc pollution is the burning of coal, petroleum and its products. Incineration of municipal solid waste can introduce about 75% zinc to urban air. Also, municipal wastewater generally contains significant amounts of zinc. The use of municipal and industrial waste in agriculture results in the accumulation of zinc in the surface layers of soil. Another source of this metal in soils are some preparations of plant protection products, as well as phosphatic fertilizers. The degree of toxicity of zinc is not big, but it depends on the ionic form, and changes under the influence of water hardness and pH. The daily average download of zinc by an adult is estimated at about 10-50 mg /day. The toxic dose is 150-600 mg. It is necessary for the proper functioning of living organisms and it is involved in the metabolism of proteins and carbohydrates. High doses of zinc cause damage to many biochemical processes followed by its deposition in the kidneys, liver, gonads. Kidney play an important role in maintaining zinc homeostasis in the body. Zinc is relatively non-toxic to humans and animals. Hazard zinc mainly connected with secondary copper deficiency does not give specific symptoms.

Nickel is a moderately toxic element as compared to other transition metals. It is a natural element of the earth's crust; therefore its small amounts are found in food, water, soil, and air. Nickel occurs naturally in the environment at low levels. Nickel concentrations in the groundwater depend on the soil used, pH, and depth of sampling. The average concentration in the groundwater in the Netherlands ranges from 7.9 μg/dm^3 (urban areas) to 16.6 μg/dm^3 (rural areas). Acid rain increases the mobility of nickel in the soil and thus might increase nickel concentrations in the groundwater. In the groundwater with a pH below 6.2, nickel concentrations up to 0.98 mg/dm^3 have been measured, whereas the upper limit recommended by FAO (Food & Agricultural Organization of the United Nation) for nickel in water is 0.02 mg/dm^3. According to the Polish standards the maximum discharge limit for nickel in waste water is 2-3 mg/dm^3. The maximum limit in drinking water in Europe is 0.01 mg/dm^3. Although it has been suggested that nickel may be essential to plants and some animal species as well as in human nutrition, this metal causes damage to humans. Nickel occurs in seams of coal in the amount of 4-60 mg/kg. Crude oil contains about 50-350 mg/kg of the metal. The most dangerous is tetracarbonyl nickel occurring mostly in nickel refineries. The content of this metal in industrial and municipal wastewater ranges 20-3924 mg/kg. An important source of nickel pollution is its emissions to the air, the combustion of coal and liquid fuels, primarily by diesel engines. It is assumed that the concentration of nickel in the waters of the rivers should be about 1 μg/dm^3, while in most rivers of Europe it is as high as 75 μg/dm^3. Large amounts of nickel are given to surface waters from municipal wastewater in which the concentration exceeds 3000 ppm s.m. The permissible concentration should be 20 μg/dm^3. Nickel readily accumulatives particularly in phytoplankton or other aquatic plants. The daily absorption of nickel by humans ranges 0.3-0.5 mg. In humans, the absorption of nickel from the gastrointestinal tract is less than 10%. Nickel taken with food and water is poorly absorbed and rapidly excreted from the body. It accumulates mainly in bones, heart, skin and various glands. Nickel inhalation of atmospheric air is largely accumulated in the lungs. Practically fatal or acute poisoning with nickel or its salts is not found. The most toxic compound is carbonyl nickel. An excess of inhaled nickel causes damage to the mucous membranes. Moreover, its symptoms are allergic disorders (protein metabolism disorder in plasma, changes in the chromosomes and changes in bone marrow and cancer. It is known that inhalation of nickel and its compounds can lead to serious problems, including, among others, respiratory system cancer. Moreover, nickel can cause skin disorder which is a common occupational disease in workers who handle its large amounts. Also dermatitis is the most common effect of chronic dermal exposure to nickel. Chronic inhalation exposure to nickel in humans also results in detrimental respiratory effects.

Copper is generally found in the earth's crust, usually in the form of sulphides. Municipal and industrial waste waters are an important source of pollution of rivers and water reservoirs. Copper accumulating plants may be the cause of poisoning. Copper is present in all types of water, and its content is subject to large variations (Barceloux, 1999). The natural content of copper in the river water ranges 0.9-20 μg/dm^3 and for saline waters 0.02-0.3 μg/dm^3. Copper is an essential nutritional element being a vital part of several enzymes. It is one of the components of human blood. The estimated adult dietary intakes are between 2 and 4 mg/day. The demand for copper is increased in pregnant women, children and the elderly. Good dietary sources of copper include animal liver, shellfish, dried fruit, nuts and chocolate. In some cases drinking water may also provide significant levels of copper. Copper in the body is involved in oxidation-reduction processes, acts as a stimulant on the amount and activity of hemoglobin, in the process of hardening of collagen, hair keratinization, melanin synthesis as well as affects on lipid metabolism and properties of the myelin sheath of nerve fibers. In animal cells it is mainly concentrated in the mitochondria, DNA, RNA, and the nucleus. Copper readily forms a connection with various proteins, especially those of sulphur. Although copper is an essential metal, it can, in some circumstances, lead to toxic effects including liver damage and gastrointestinal disturbances. Such as Wilson's disease (also known as hepatolenticular degeneration), Indian Childhood Cirrhosis (ICC) which are characterised by an accumulation of copper-containing granules within liver cells. Ingestion of high levels of copper salts is known to cause gastrointestinal upsets. Additionally, absorption of copper compounds by inhalation causes congestion of the nasal mucosa, gastritis, diarrhea and toxic symptoms such as chronic lung damage. Copper compounds act on the intact skin, causing it to itch and inflammation. They can cause conjunctivitis, ulceration and corneal opacity, nasal congestion and as well as sore throat and nasal septum. The upper limit recommended by WHO for copper is less than 1.3 mg/dm^3. The maximum limit in drinking water is 0.05 mg/dm^3 (Fewtrell et al. 1996).

2. Ion exchange (IX)

Ion exchange may be defined as the exchange of ions between the substrate and surrounding medium. The most useful ion exchange reaction is reversible. When the reaction is reversible, the ion exchanger can be reused many times. Generally resins are manufactured in the spherical, stress and strain free form to resist physical degradation. They are stable at high temperatures and applicable over a wide pH range. Ion exchange resins, which are completely insoluble in most aqueous and organic solutions, consist of a cross linked polymer matrix to which charged functional groups are attached by covalent bonding (Sherrington, 1998). The ion exchangers which contain cations or anions as counter-ions are called cation exchangers or anion exchangers, respectively. The usual matrix is polystyrene cross linked for structural stability with 3 to 8 percent of divinylbenzene (3-8 % DVB) (Kunin, 1958; Helfferich, 1962). The resins of higher cross linking (12-16% DVB) are more costly, both to make and to operate and they are specially developed for heavy duty industrial applications. These products are more resistant to degradation by oxidizing agents such as chlorine, and withstand physical stresses that fracture lighter duty materials. Typical ion exchangers are produced with a particle size distribution in the range 20-50 mesh (for separation of anions from cations or of ionic species from nonionic ones). For more difficult separations, materials of smaller particle size or lower degrees of cross linking are necessary. Moreover, when the separation depends solely upon small differences in the affinity of the ions, a particle size of 200-400 mesh is required and when the selectivity is increased by the use of complexing agents, the particle size in the 50-100 mesh is adequate. The ion exchangers finer than 100 mesh are employed for analytical purposes and for practical applications on the commercial scale the materials finer than 50 mesh are used.

Depending on the type of functional groups of exchanging certain ions, the ion exchangers with strongly acidic e.g., sulphonate $-SO_3H$, weakly acidic e.g., carboxylate -COOH, strongly basic e.g., quaternary ammonium $-N^+R_3$ and weakly basic e.g., tertiary and secondary amine $-N^+R_2H$ and $-N^+RH_2$ should be mentioned. The strong acidic cation exchangers are well dissociated over a wide pH range and thus reaching its maximum sorption capacity. On the other hand, weak acidic cation exchangers containing, for example, carboxylic functional groups reach the maximum sorption capacity at pH> 7.0 as presented in Fig.1.

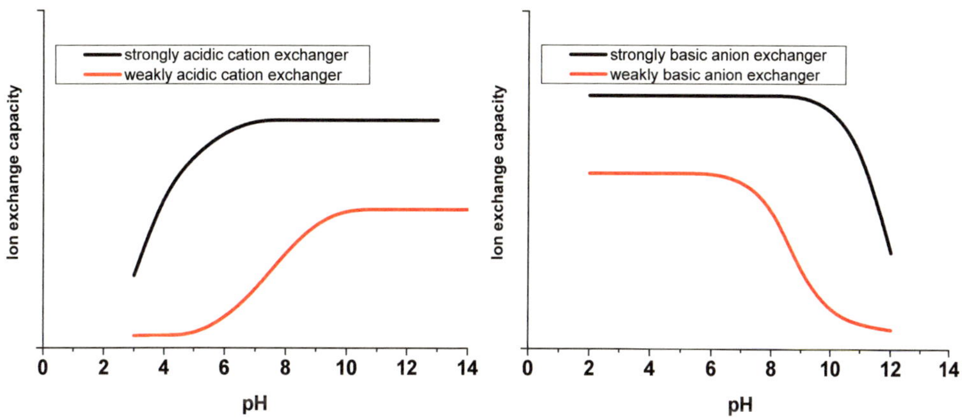

Figure 1. The sorption capacity of ion exchangers depending on pH.

Additionally, ion exchangers possess: the iminodiacetate functional groups ($-N\{CH_2COOH\}_2$), phenol ($-C_6H_4OH$), phosphonic ($-PO_3H_2$) and phosphine ($-PO_2H$) functional groups. These groups are acidic in nature and are dissociated with the exchange of H^+ or Na^+ ions for other cations from the solution. Negative charge of the functional groups is offset by an equivalent number of mobile cations so-called counter ions. Counter ions can be exchanged for other ions from the solution being in the contact with the resin phase.

There are also amphoteric exchangers, which depending on the pH of the solution may exchange either cations or anions. More recently these ion exchangers are called bipolar electrolyte exchange resins (BEE) or zwitterionic ion exchangers (Nesterenko & Haddad, 2000). The aminocarboxylic amphoteric ion exchangers AMF-1T, AMF-2T, AMF-2M, ANKB-35 as well as the carboxylic cation exchanger KB-2T were, for example used for recovery of Ni(II) from the $Mn(NO_3)_2–H_2O$ system (Kononowa et al. 2000).

The individual ions present in the sample are retained in varying degrees depending on their different affinity for the resin phase. The consequence of this phenomenon is the separation of analyte ions, such as metal ions, however, the nature and characteristics of the resin phase determine the effectiveness of this process (Fritz, 2005). The affinity series which for various types of ion exchangers are as follows:

2.1. Cation exchangers with the sulphonic functional groups

It is well known that the affinity of sulphonic acid resins for cations varies with the ionic size and charge of the cation. The affinity towards cation increases with the increasing cation charge:

$$Na^{+} < Ca^{2+} < Al^{3+} < Th^{4+},$$

and in the case of different cations with the same charge the affinity increases with the increasing atomic number:

$$Li^{+} < H^{+} < Na^{+} < NH_4^{+} < K^{+} < Rb^{+} < Cs^{+} < Ag^{+} < Tl^{+}$$

$$Mg^{2+} < Ca^{2+} < Sr^{2+} < Ba^{2+}$$

$$Al^{3+} < Fe^{3+}.$$

Generally, the affinity is greater for large ions with high valency.

For the strong acidic cation exchanger the affinity series can be as follows:

$$Pu^{4+} >> La^{3+} > Ce^{3+} > Pr^{3+} > Nd^{3+} > Sm^{3+} > Eu^{3+} > Gd^{3+} > Tb^{3+} > Dy^{3+} > Ho^{3+} > Er^{3+} > Tm^{3+} > Yb^{3+} > Lu^{3+} > Y^{3+} > Sc^{3+} > Al^{3+} >> Ba^{2+} > Pb^{2+} > Sr^{2+} > Ca^{2+} > Ni^{2+} > Cd^{2+} > Cu^{2+} > Co^{2+} > Zn^{2+} > Mg^{2+} > UO_2^{2+} >> Tl^{+} >> Ag^{+} > Cs^{+} > Rb^{+} > K^{+} > NH_4^{+} > Na^{+} > H^{+} > Li^{+}$$

and for Lewatit SP-112 it is as: $Ba^{2+} > Pb^{2+} > Sr^{2+} > Ca^{2+} > Ni^{2+} > Cd^{2+} > Cu^{2+} > Co^{2+} > Zn^{2+} > Fe^{2+} > Mg^{2+} > K^{+} > NH_4^{+} > Na^{+} > H^{+}$.

2.2. Cation exchangers with the carboxylic functional groups

Cation exchangers with the carboxylic functional groups show the opposite the affinity series for alkali and alkaline earth metal ions. Noteworthy is the fact that the cations exhibit a particularly high affinity for H^{+}. The affinity of this type of cation is therefore as follows:

$$H^{+} > Mg^{2+} > Ca^{2+} > Sr^{2+} > Ba^{2+} > Li^{+} > Na^{+} > K^{+} > Rb^{+} > Cs^{+}.$$

2.3. Anion exchangers with the quaternary ammonium functional groups

The charge of the anion affects its affinity for the anion exchanger in a similar way as for the cation exchangers:

$$\text{citrate} > \text{tartrate} > PO_4^{3-} > AsO_4^{3-} > ClO_4^{-} > SCN^{-} > I^{-} > S_2O_3^{2-} > WO_4^{2-} > MoO_4^{2-} > CrO_4^{2-} > C_2O_4^{2-}$$

$$> SO_4^{2-} > SO_3^{2-} > HSO_4^{-} > HPO_4^{2-} > NO_3^{-} > Br^{-} > NO_2^{-} > CN^{-} > Cl^{-} > HCO_3^{-} > H_2PO_4^{-} > CH_3COO^{-}$$

$$> IO_3^{-} > HCOO^{-} > BrO_3^{-} > ClO_3^{-} > F^{-} > OH^{-}$$

for *Dowex 1* (type 1):

$$ClO_4^- > I^- > HSO_4^- > NO_3^- > Br^- > NO_2^- > Cl^- > HCO_3^- > CH_3COO^- > OH^- > F^-,$$

for *Dowex 2* (type 2):

$$ClO_4^- > I^- > HSO_4^- > NO_3^- > Br^- > NO_2^- > Cl^- > HCO_3^- > OH^- > CH_3COO^- > F^-$$

2.4. Anion exchangers with the tertiary and secondary amine functional groups

Only with the exception of the OH- ion, the affinity of the anion exchangers with the tertiary and secondary functional groups is approximately the same as in the case of anion exchangers with the quaternary ammonium functional groups. These medium and weakly basic anion exchangers show very high affinity for OH^- ions.

Anion exchange materials are classified as either weak base or strong base depending on the type of exchange group. These are two general classes of strong base anion exchangers e.g. types 1 and 2 depending on chemical nature. The synthesis of the weak base anion exchangers with the tertiary amine groups is usually provided by the chloromethylation of PST-DVB followed by the amination by secondary amine (Drăgan & Grigoriu, 1992). Weak base resins act as acid adsorbers, efficiently removing strong acids such as sulphuric and hydrochloric ones. They are used in the systems where strong acids predominate, where silica reduction is not required, and where carbon dioxide is removed in degasifiers. Preceding strong base units in demineralizing processes, weak base resins give more economical removal of sulphates and chlorides. The selectivity for the bivalent ions such as SO_4^{2-} depends strongly on the basicity of the resin, the affinities of various functional groups following the order: primary > secondary > tertiary > quaternary. Therefore among the factors affecting the sorption equilibrium the most important are: first of all nature of functional groups and the concentration of the solution (Boari et al. 1974). At low concentration the resin prefers ions at higher valency and this tendency increases with solution diluting. It should be also mentioned that obtaining resins with the primary amine functional groups is difficult by chemical reactions on polystyrene-divinylbenzene copolymers. Weakly basic anion exchangers can be used, for example for zinc cyanide removal from the alkaline leach solutions in the Merrill Crowe process (Kurama & Çatlsarik, 2000).

2.5. Gel and macroporous resins

The development in polymerization technique has provided novel matrices for a series of new ion exchangers. They differ from the earlier corresponding copolymers that are characterized by being essentially cross linked gels of polyelectrolytes with pore structure defined as the distance between polymeric chains.

It is well known that the fouling of the resin by organic compounds and mechanical stress imposed by plant operating at high flow rates are the most important problems encountered in the use of the ion exchange resins (De Dardel & Arden, 2001). To overcome these problems the ion exchangers with a high degree of cross linking containing artificial open pores in the form of channels with diameters up to 150 nm were introduced (Fig. 2).

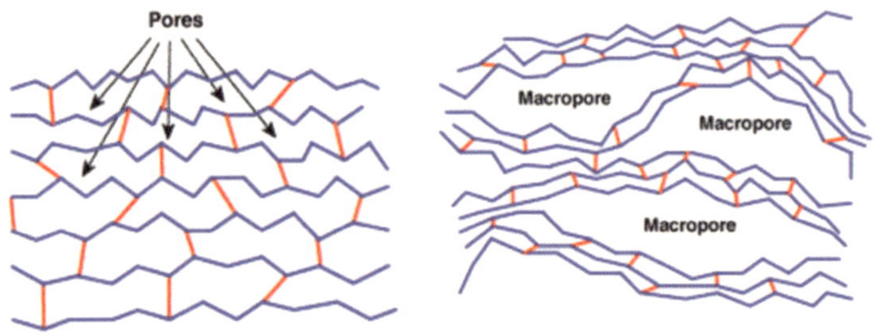

Figure 2. The structure of gel and macroporous ion exchangers (http://dardel.info/IX/index.html)

The first macroporous ion exchanger was a carboxylic resin made by Rohm and Haas, which covered a wide variety of acrylic compounds copolymerized with polyvinyl cross linking agents to make insoluble, infusible weakly acidic resins. By 1948 AmberliteTM IRC-50, made by the copolymerization of methacrylic acid and divinylbenzene was in production and possessed the 'sponge structure' (Abrams & Milk, 1997). According to the definition by Stamberg and Valter (1970) the macroporous resin should be characterised by measurable inner surface by any suitable method resulting from pores 5 nm, even in the completely dried state. In contrast, the gel materials did not show any porosity in the dry state. Then the term 'macroreticular' (sometimes abbreviated to MR) was selected to distinguish resins with a particular type of porosity obtained by application of precipitating diluents such as t-amyl alcohol. In 1979 Amber-Hi-Lites stated that 'macroreticular' resins are those made by a copolymerization technique which brings about precipitation during the polymerization, thus resulting in a product which has two phases, a gel phase in the form of microspheres formed during the phase separation and the pore phase surrounding the microspheres (Kunin, 1979). Later when quantitative porosity measurements were used it was shown that other methods of preparation gave products similar to those declared as 'macroreticular'. Therefore classification of resins should be based on their properties and function (Ion exchange resins and adsorbents, 2006).

During last decades the great progress was made by the development of the macroporous ion exchange resins. It should be mentioned that macroporous resins can also perform as adsorbents because of their pore structure. For organic ion exchange resins based on cross linked polystyrene the porosity was originally selected by the degree of cross linkage. These gel type resins are able to sorb organic substances from water according to their degree of porosity and the molecular weight of the adsorbate. They not only allow for large molecules or ions to enter the sponge like structure but also to be eluted during the regeneration. Therefore they perform two functions: ion exchange by means of the functional groups and the reversible adsorption and elution due to the macroporous structure. They are also resistant against organic fouling which results in a longer resin life compared with the conventional gel type ion exchangers as well as the quality of the treated water is much better because of the adsorption of organic species by the macroporous structure. The SEM scan of the macroporous anion exchanger Lewatit MonoPlus MP 500 is presented in Fig. 3.

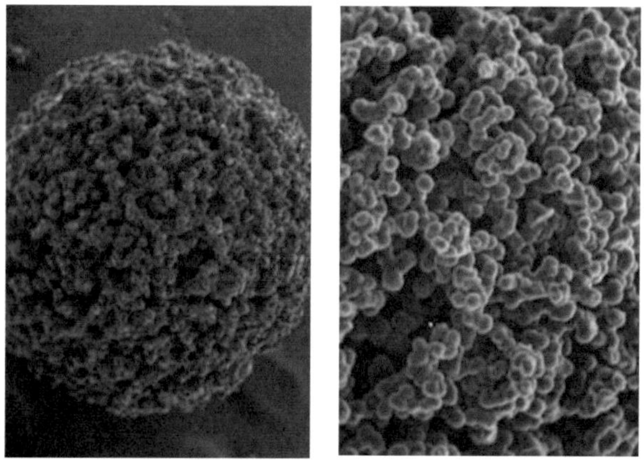

Figure 3. SEM scans of macroporous resins.

Ion exchange applications can be performed by either column (flow continuous) and batch technique. In column operations the ion exchange resin is placed in the vertical column to form a bed. Once the application is completed, the resin can be regenerated to use in another cycle. In batch operations the resin is shacked in a vessel with the solution to be treated. After the application is completed, the resin can be regenerated in place or transferred to a column for regeneration.

While the main aims in the production of conventional ion exchangers were focused on obtaining a high ion exchange capacity and improved chemical resistance and thermal and mechanical strength, in the case of monodisperse ion exchange resins, these efforts directed towards improvement of kinetic parameters. Heterodisperse ion exchangers are usually characterized by a standard grain size of 0.3-1.2 mm and uniformity coefficient (UC) within the limits of 1.5-1.9. In the case of monodisperse ion exchange resins during the manufacturing process the grain size from 0.6 mm and uniformity coefficient within the limits 1.1-1.2 is usually achieved. In addition, monodisperse ion exchangers, due to the uniform packing of the column, show more than 12% higher ion exchange capacity, faster kinetics of exchange and a much higher mechanical strength, which is extremely important from the economical point of view. As the particle size of the ion exchanger material and its uniformity are the most important parameters influencing the hydraulics and kinetics of the ion exchange therefore the monodisperse ion exchangers provided better flow characteristics in column applications in comparison to the conventional heterodisperse ion exchangers (the flow rate decreases with the decreasing particle size, however, smaller particles have larger outer surface, but cause larger head loss in the column processes) (Scheffler, 1996; Krongauz & Kocher, 1997). The visualization of the monodisperse and hetrodisperse ion exchangers is presented in Fig. 4a-b.

For example the research carried out by Zainol & Nicol (2009a) shows that in the the sorption process of Ni(II) and other metal ions the monodisperse resin (Lewatit MonoPlus TP 207) proved to be superior to the conventional heterodisperse ones in terms of loading capacity for Ni(II) and also the kinetics of adsorption. This makes it a preferred choice for different applications.

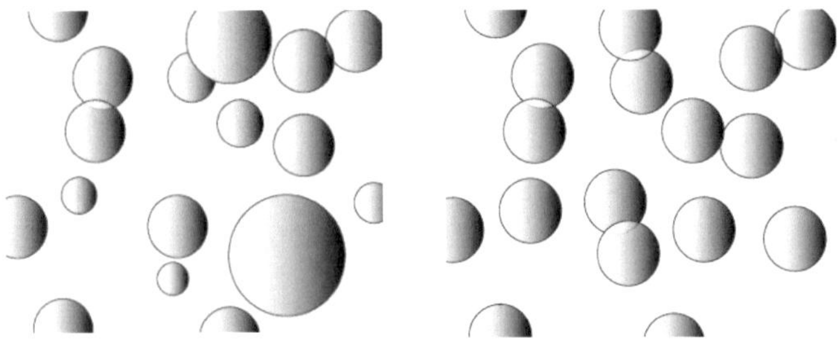

Figure 4. The monodisperse and hetrodisperse ion exchangers.

The influence of temperature on the equilibrium properties of ion exchange resins was studied extensively. The decrease of the capacity of the cation exchange resins based on the polystyrene matrix due to the operation temperature is not a significant problem. However, the relatively slight decomposition gives enough decomposition products to cause significant problems elsewhere. This may be decomposition of the bone polystyrene matrix, resulting in styrene sulphonic acid derivatives or as a substitution of the sulphonic group giving sulphate. Further decomposition of styrene sulphonic acid derivatives will also result in sulphate as one of the end products (desulphonation). The amount of sulphate produced is sometimes high. The information on the stability of the ion exchange resins mainly deals with the anion exchange resins. The mechanism of the degradation of quaternary ammonium salts and tertiary anions is well-known (Reynolds, 1982; Fernandez-Prini, 1982; Fisher, 2002). The effect of temperature on the properties of chelating ion exchangers was also described in the paper by Ivanov (1996).

3. Application of ion exchangers for heavy metal ions removal

Ion exchange technique can remove traces of ion impurities from water and process streams and give a product of desired quality. Ion exchangers are widely used in analytical chemistry, hydrometallurgy, antibiotics, purification and separation of radioisotopes and find large application in water treatment and pollution control (Clifford, 1999; Luca et al. 2009). The list of metals which are recovered and purified on an industrial scale by means of ion exchange include: uranium, thorium, rare earth elements (REEs), gold, silver, platinum metals (PGM), chromium, copper, zinc, nickel, cobalt and tungsten.

In some of these cases, the scale of operations is relatively small, for instance in the rare earth elements or noble metals, but the values of recovered metals are very high. Ion exchange process is particularly suitable for purification of metal ions with a high value and low processing. The alternative is also a process of large-scale recovery of trace amounts of metals from waste streams, such as cadmium and mercury, chromium, or copper and zinc. The use of ion exchange processes in hydrometallurgy is high and every year continues to grow. It is associated mainly with the progress of what is observed in the synthesis of new selective chelating ion exchangers containing complexing ligands (Minczewski, et al. 1982;

4. Chelating and special ion exchangers

Typical disadvantage of lack a of the selectivity towards heavy metal ions and alkali and alkaline earth metal ions of most widely used functionalized ion exchangers such as Chelex 100 is overcome by introducing chelating ligands capable of removing selective metal ions. It exhibits high affinity for heavy metal ions: **$Cu^{2+} > Hg^{2+} > Pb^{2+} > Ni^{2+} > Zn^{2+} > Cd^{2+} > Co^{2+} > Fe^{2+} > Mn^{2+} > Ca^{2+} > Mg^{2+} > Sr^{2+} > Ba^{2+} >>>$ alkali ions $> H^+$**, whereas for sulphonic ones the analogous affinity series can be as presented earlier (for Lewatit SP 112).

Generally, the functional group atoms capable of forming chelate rings usually include oxygen, nitrogen and sulphur. Nitrogen can be present in a primary, secondary or tertiary amine, nitro, nitroso, azo, diazo, nitrile, amide and other groups. Oxygen is usually in the form of phenolic, carbonyl, carboxylic, hydroxyl, ether, phosphoryl and some other groups. Sulphur is in the form of thiol, thioether, thiocarbamate, disulphide groups etc. These groups can be introduced into the polymer surface by copolymerization of suitable monomers, immobilization of preformed ligands, chemical modification of groups originally present on the polymer surface. However, the last two are most often used (Warshawsky, 1987). Chelating resins with such type of ligands are commonly used in analysis and they can be classified according to Fig.1. (Kantipuly et al. 1990). The choice of an effective chelating resin is dictated by the physicochemical properties of the resin materials. These are the acid-base properties of the metal species and the resin materials, the polarizability, selectivity, sorptive capacity, kinetic and stability characteristics of the resin. The sorption capacity of chelating ion exchangers depends mainly on the nature of functional groups and their content as well as solution pH as for their selectivity it depends on the relative position of functional groups, their spatial configuration, steric effects, and sometimes their distance from the matrix and to a lesser extent on the properties of the matrix. Their use allows the recovery of valuable metals from ores and sludge, sea water and industrial effluents. They are used as flotation agents, depressants, flocculants and collectors.

It is worth emphasizing that these resins are invaluable wherever it is necessary to concentrate or remove elements present in very low concentrations.

With a range of well known chelating ion exchangers only a few types are produced on an industrial scale. Among the most important ones these with the functional groups: amidoxime, dithiocarbamate, 8-hydroxychinoline, iminodiacetate, aminophosphonic, bispicolylamine, diphosphonic, sulphonic and carboxylic acid groups, thiol, thiourea as well as isothiourea should be selected (Sahni & Reedijk, 1984; Busche et al. 2009). Among them the chelating ion exchangers possessing methylglucoamine, bis(2pirydylmethyl)amine also known as bispikolilamine, thiol etc. are used for special applications such as removal of precious metal ions, heavy metal ions from the acidic medium, boron and special oxoanions removal. A separate group are ion exchangers of solvent doped type used for In, Zn, Sn, Bi, etc. separation. The advantages of ion exchangers from these groups include good selectivity, preconcentration factor, binding energy and mechanical stability, easy regeneration for multiple sorption-desorption cycles and good reproducibility in the sorption characteristics.

5. Chelating ion exchangers with the hydroxamic and amidoxime functional groups

The choice of hydroxamic acids is based on their application in mineral processing as collectors in flotation of haematite, pyrolusite or bastnaesite ores. The copolymer of malonic acid dihydroximate with styrene-divinylbenzene was used for uranium(VI) removal from sea water (Park & Suh, 1996). In the paper by Ahuja (1996) it was found that glycin hydroximate resin shows maximum adsorption for Fe(III), Cu(II) and Zn(II) at pH 5.5; for W(VI), U(VI), Co(II) and Ni(II) at pH 6.0 as well as for Cd(II) at pH 6.5. It can be recommended for separation of Cu(II) from Co(II) and Ni(II) at pH 5.5. However, the iminodiacetic– dihydroximate resin can be applied for U(VI), Fe(II), Cu(II) separation according to the affinity series: $\mathbf{UO_2^{2+} > Fe^{3+} > Cu^{2+} > Zn^{2+} > Co^{2+} > Cd^{2+} > Ni^{2+} > Zn^{2+}}$. Hydroxamic acids exist in the two tautomeric forms:

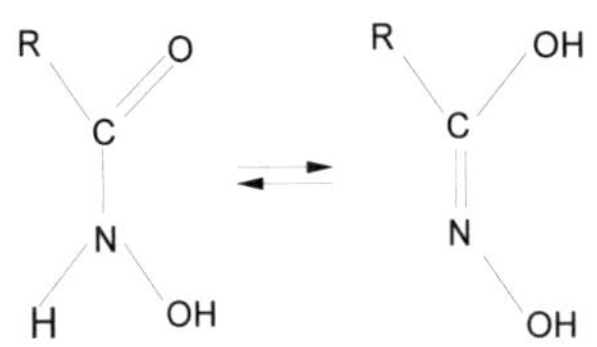

(a) (b)

and metal ions are coordianated by the hydroxamide functional group (a).

Chelating resins with the amidoxime functional groups such as Duolite ES-346, and Chelite N can be applied for the concentration of solutions containing Ag(I), Al(III), Cd(II), Co(II), Cr(III), Cu(II), Fe(III), Hg(II), Mn(II), Ni(II), Mo(VI), Pb(II), Ti(IV), U(VI), V(V) and Zn(II) in the presence of alkali and alkaline earth metal ions (Samczyński & Dybczyński, 1997; Dybczyński et al. 1988). Alkali and alkaline earth metal ions are poorly retained by these resins. Duolite ES-346 is commonly used to extract uranium(VI) from seawater and As(III) from aqueous solutions. It can be also applied for Pd(II) removal (Chajduk-Maleszewska & Dybczyński, 2004). It is characterized by high selectivity towards Cu(II) ions due to the presence of amidoxime groups and small quantities of hydroxamic acid (RCONHOH):

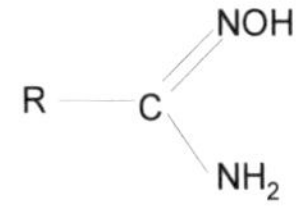

where: R is the resin matrix.

It was found that for the amidoxime resins the selectivity series can be as follows: **Cu(II) > Fe(III) > As(III) > Zn(II) > Ni(II) > Cd(II) > Co(II) > Cr(III) > Pb(II).** Interesting results were obtained by observing the impact of acidity on the behaviour of this ion exchanger. At low pH values (<3) there was a decrease in the chelating ability of Duolite ES 346 for heavy metal ions as well as degradation of its functional groups according to the reactions (Ferriera et al. 1998):

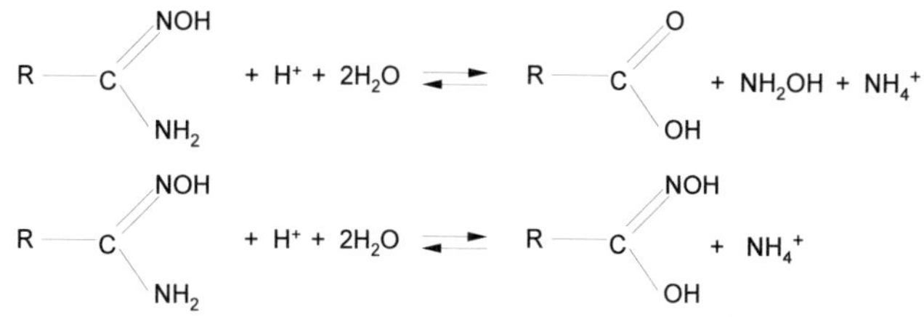

However, the second reaction is the representative of the degradation of amidoxime groups under less acidic conditions (pH < 3.0). The increase in pH causes the weakening of the hydrogen ions competition for active sides resulting in an increase in the complexation of metal ions such as Cu(II). The fact that the degradation of the functional groups of Duolite ES-346 occurs under the influence of strong mineral acids is a serious problem which can significantly reduce the chelating capacity of the resin. However, this effect was made use of the recovery of adsorbed ions on the resin ion exchange. Corella et al. (1984) demonstrated that poly(acrylamidoxime) can be successfully used for the preconcentration of trace metals from aqueous solutions.

Also salicylic acid is a ligand which can selectively complex with Zn(II), Pb(II), Fe(III). Using the salicylic acid loaded resin for preconcentration of Zn(II) and Pb(II), it was proved that the preconcentration factors are much higher than those for bis(2-hydroxyethyl)dithiocarbamate (Saxena et al. 1995). However, the phenol-formaldehyde resin with the salicylaldoxime and salicylaldehyde functional chelating groups shows high selectivity for Cu(II) ions (Ebraheem & Hamdi, 1997).

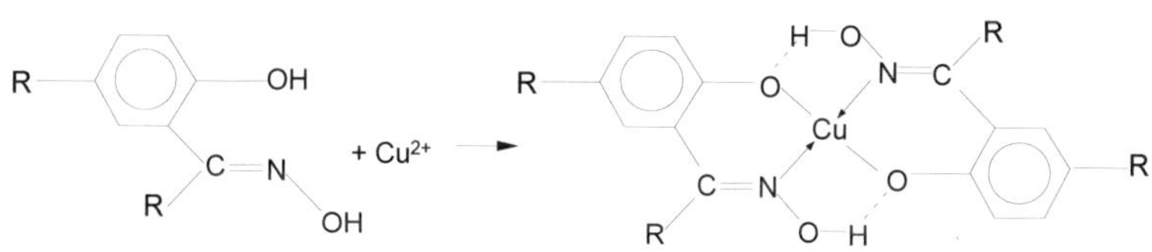

The affinity order of metal complexes of salicylaldoxime is as follows: $Fe^{3+} > Cu^{2+} > Ni^{2+} > Zn^{2+} > Co^{2+}$.

6. Chelating ion exchangers with the dithiocarbamate functional groups

The high affinity for transition metal ions is also exhibited by the classes of ion exchangers with the dithiocarbamate functional groups (including commercially available Nisso ALM 125), in which sulphur is the donor atom. Ion exchangers of this type have high affinity for the Hg(II), Pb(II), Cd(II) ions as well as precious metal ions, however, they do not adsorb alkali and alkaline earth metals. It was shown that dithiocarbamates obtained with the share of primary amines are less stable than those obtained with the share of secondary amines, and the binding of metal ions to the functional group of the donor atom increases in the number of $Fe^{2+} < Ni^{2+} < Cu^{2+}$. The sorption efficiency is dependent on the presence of ions in the solution such as SCN^-. In the paper by McClain and Hsieh (2004) the selective removal of Hg(II), Cd(II) and Pb(II) was presented. This resin is also effective for separation and concentration of Mn(II), Pb(II), Cd(II), Cu(II) , Fe(III) and Zn(II) from complex matrices (Yebra-Biurrun et al. 1992). The copolymer of poly(iminoethylo)dithiocarbamate was used for sorption of VO^{2+}, Fe(II), Fe(III), Co(II), Ni(II) and Cu(II) (Kantipuly et al. 1990)

7. Chelating ion exchangers with the 8-hydroxyquinoline functional groups

A simple method for immobilization of 8-hydroxyquinoline in a silica matrix is described by Lührmann (1985). The sorbent was used in the sorption of Cu(II), Ni(II), Co(II), Fe(III), Cr(III) , Mn(II), Zn(II), Cd(II), Pb(II) and Hg(II) at pH from 4 to 6. It was shown that the sorption capacity varies in the range from 0.2 to 0.7 mM /g, and the partition coefficients from 1×10^3 to 9×10^4. Ryan and Weber showed (1985) that this type of sorbent has better sorption properties with respect to Cu(II) than Chelex 100 with the iminodiacetate functional groups. Th sorbents based on 8-hydroxyquinoline can be used, e.g. for concentration of the trace metal ions Mn(II), Co(II), Ni(II), Cu(II), Zn(II), Cd(II), Pb(II) and Cr(III) from sea water (Pyell & Stork, 1992).

8. Chelating ion exchangers with the iminodiacetic functional groups

Recently the attention has been paid to the ion exchangers with the amino- or iminoacids groups. The presence of two carboxyl groups and the tertiary nitrogen atom provides strong preference for chromium(III) and copper(II) (Marhol & Cheng, 1974). Therefore, for the commercial chelating ion exchangers such as Chelex 100, Dowex A 1, CR-20, Lewatit TP 207, Lewatit TP 208, Purolite S 930, Amberlite IRC 748 (formely Amberlite IRC 718) or Wofatit MC-50 the sorption process of metal ions proceeds according to the order: **Cr^{3+} > Cu^{2+} > Ni^{2+} > Zn^{2+} > Co^{2+} > Cd^{2+} > Fe^{2+} > Mn^{2+} > Ca^{2+} >> Na^{+}**. This type of ion exchangers also exhibis high affinity for Hg(II) and Sb(V) ions. It should be noted that depending on the pH value they may occur in the following forms (Zainol & Nicol. 2009a):

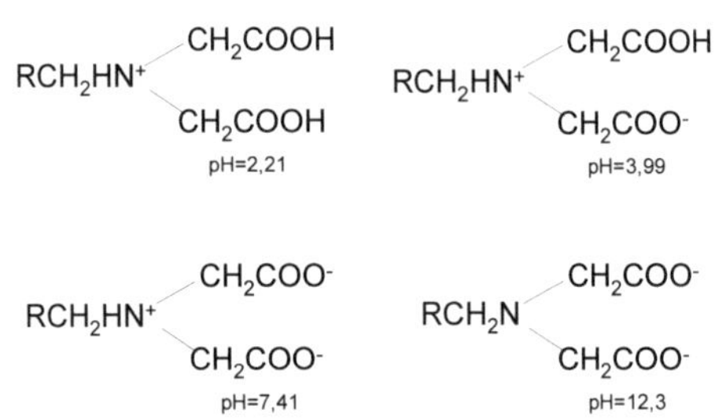

At pH <2.0 the nitrogen atom and the two carboxylic groups are protonated. In this case the chelating ion exchanger with the iminodiacetic functional groups behaves as a weakly basic anion exchanger. At pH ~ 12, the nitrogen atom and the two carboxylic groups undergo deprotonation – the ion exchanger behaves as a typical weakly acidic cation exchanger. For pH medium values, the iminodiacetic resin behaves as an amphoteric ion exchanger. The iminodiacetate groups provide electron pairs so that the binding force for the alkaline earth metals is 5000 times as large as that for alkali metals like Ca(II), which react with divalent metals to form a stable coordination covalent bond. Therefore, the affinity series determined for the iminodiacetic ion exchanger can be presented in the order: **Hg^{2+} > UO_2^{2+} > Cu^{2+} > Pb^{2+} > Ni^{2+} > Cd^{2+} > Zn^{2+} > Co^{2+} > Fe^{2+} >Mn^{2+} > Ca^{2+} > Mg^{2+} > Ba^{2+} > Sr^{2+} >> Li^+ > Na^+ > K^+.**

Amberlite IRC 748 in the K(I) form was also used for removal of Ca(II), Mg(II) from the potassium chromate solution (Yua et al. 2009). The optimum pH obtained for Ca(II) and Mg(II) adsorption onto Amberlite IRC 748 from the potassium chromate solution is 9.8 and 9.5, respectively. It was also noted that an increase of temperature and resin dosage resulted in their higher adsorption and the equilibrium conditions were attained within 480 min. The experimental data were relatively well interpreted by the Langmuir isotherm and the monolayer adsorption capacities of Ca(II) and Mg(II) were equal to 47.21 mg/g and 27.70 mg/g, respectively. This is of great importance because manufacturing of chromium trioxide by electrolyzing chromate salts, as a green process with the zero emission of waste, is studied widely now (Li et. al 2006). It should be also pointed out that separation factors between Mg(II) and Ca(II) and other divalent metal ions on an iminodiacetate resin are much smaller than those expected from the stability constants of their IDA complexes in solutions. Such phenomena were qualitatively described as the 'polymer effect' or operation of ion exchange as well as complexation reactions. Pesavento et al. (1993) gave a quantitative explanation for these anomalies on the basis of the Gibbs-Donnan model. Ca(II) and Mg(II) ions are adsorbed forming the $R(Hida)_2M$ complexes in acidic media and R(ida)M in neutral and alkaline systems whereas Ni(II) or Cu(II) etc. forms the R(ida)M complexes:

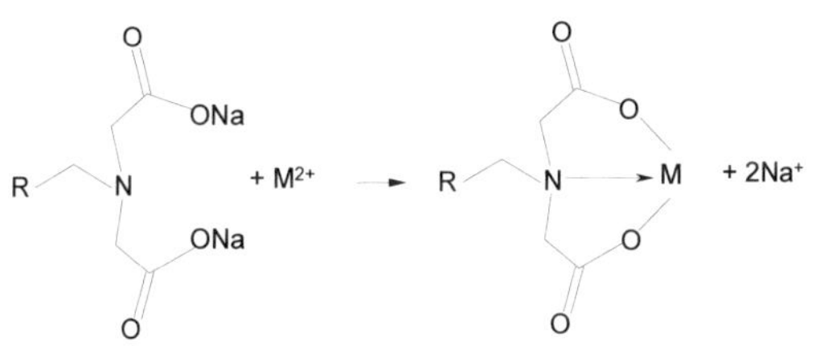

Commercially available chelating resins with the iminodiacetate functional group (Amberlite IRC 748, Lewatit TP 207, Lewatit TP 208, Purolite S 930, Lewatit MonoPlus TP 207) have been evaluated for their suitability for the adsorption of Ni(II) and other metal ions (Al(III), Ca(II), Co(II), Cr(III), Cu(II), Fe(II/III), Mg(II), Mn(II) and Zn(II)) from the tailings of a pressure acid leach process for nickel laterites. The Amberlite IRC 748 and TP MonoPlus 207 resins were found to be the most suitable in terms of loading capacity for nickel and kinetics of adsorption. Although all the five resins studied have the same functional groups their performance is not identical. The observed differences are possibly caused by variations in the synthesis procedure which results in variations in the structure of the matrix, degree of cross linking, density of functional groups, proportion of iminodiacetate groups and also the particle size (Zainol & Nicol, 2009a)

Additionally, the research carried out by Biesuz et al. (1998) shows that in the case of Ni(II) and Cd(II) sorption the structure of formed complexes is different. Ni(II) forms complexes of

R(ida)M type, whereas Cd(II) $R(idaH)_2M$. However, in the paper by Zagorodni & Muhammed (1999) it was stated that the complexes $R(Hida)_2M$ should be extremely weak or even impossible. The adsorption equilibrium of Ni(II), Co(II), Mn(II) and Mg(II) on Amberlite IRC 748 has been discussed in (Zainol & Nicol, 2009b). The resin proves to have high selectivity for Ni(II) and Co(II) which suggests that these metals can be easily separated from Mg(II) and Mn(II) at pH 4 and 5. The following order of selectivity of the resin was also found: Ni(II) > Co(II) > Mn(II) > Mg(II).

The kinetics of Cd(II) sorption from separate solutions and from the mixtures with the nonionic surfactant Lutensol AO-10 (oxyethylated alkohols) in the hydrogen form of chelating iminodiacetic ion exchanger has also been investigated (Kaušpėdienė et al. 2003). It was stated that the sorption of Cd(II) from separate solutions and from the mixture with AO-10 is controlled by the intraparticle diffusion in acidic (pH 5) and alkaline media (pH 7.6). The presence of AO-10 leads to a decrease in the rate of intraparticle diffusion. The iminodiacetate resin has a large collective adsorption with Cr(III) ion. The Cr(III) form bearing waste water can be removed at any pH in the range 3-6 at 2h of the phase contact time. Therefore for treatment of leather tanning, electroplating, textile and dyeing waste water the application of this resin is economical (Gode & Pehlivan, 2003)

Adsorption of trivalent metal ions on iminodiacetate resins was not studied as extensively as that of divalent metal ions. The known selectivity order of trivalent metal ions on an iminodiacetate resin can be presented as: $Sc^{3+} > Ga^{3+} > In^{3+} > Fe^{3+} > Y^{3+} > La^{3+} > Al^{3+}$ (Yuchi et al. 1997).

Also since the end of the 1960s fibrous adsorbents with the iminodiacetic acid groups have been studied. For example, the capacity of a commercially available iminodiacetic acid fiber named Ionex IDA-Na was established to be 0.9-1.1 mmol/g for Cu(II). The fibrous materials containing iminodiacetate groups were developed by the group of Jyo et al. (2004) . Although the metal ion selectivity of the present fiber was close to that of iminodiacetic acid resins, the metal adsorption rate of chloromethylstyrene-grafted polyethylene coated polypropylene filamentary fiber is much higher than that of commercially available granular exchangers of this type having cross-linked polystyrene matrices. In the column mode adsorption of Cu(II), breakthrough capacities of Cu(II) were independent of the flow rates of feeds up to 200-300/h. The main reasons for the extremely fast adsorption rate of sorbent can be ascribed to the diameter of the fiber being much less than those of the resins as well as to the fact that the functional groups were introduced onto non cross linked grafted polymer chains. Their chemical and physical stabilities are comparable to those of commercially available iminodiacetic acid resins.

9. Chelating ion exchangers with the phosphonic and aminophosphonic functional groups

Among various types of ion exchangers with the acidic ligands, those having phosphonate functionality are of particular interest since they are selective towards heavy metal cations.

Development of this type of ion exchangers started in the late 1940s with phosphorylation of poly(vinyl alcohol) using various phosphorylating agents (Trochimczuk & Streat, 1999; Trochimczuk 2000). Besides phosphate, phosphinic and phosphonic resins, containing –$OPO(OH)_2$, –PO(OH) and -$PO(OH)_2$ functional groups, respectively, they also contain methylenediphosphonate, ethylenediphosphonate and carboxyethyl phosphonate ones (Marhol et al. 1974;

Kabay, 1998a; Ogata et al. 2006). In all cases they display good selectivity towards metal ions even at very low pH (except for ethylenediphosphonate and carboxyl containing resins, being less acidic, more selective at the pH value from 1 to 2).

Chelating ion exchangers with the phosphonic functional groups are characterized by extremely high selectivity towards Th(IV) and U(IV,VI) as well as Cu(II), Cd(II), Zn(II), Ni(II), Ag(I), Au(III) and Fe(III) ions. Commercially available resins containing the phosphonic groups are Diaion CRP200 and Diphonix® Resin. In the case of Diphonix® Resin besides the diphosphonic functional groups in the structure of the ion exchanger, there are also carboxylic and sulphonic functional groups whose presence determines better hydrophilic properties. Diphonix® Resin as well as Diphonix A with the functional phopshonic and ammonium (type 1) or pyridyne (type 2) groups have been of significant interest lately (Chiarizia et al. 1993; Horwitz et al. 1993; Chiarizia et al. 1994; Chiarizia et al. 1996, Alexandratos, 2009). Diphonix® Resin was developed by the Argonne National Laboratory and University of Tennessee. It is synthesized by a patented process involving copolymerization of tetraalkylvinylidene diphosphonate with styrene, divinylbenzene, and acrylonitrile followed by sulphonation with concentrated sulphuric acid. Finding a method for effective copolymerization of vinylidene-1,1-diphosphonate (VDPA) ester was a major achievement because of the steric hindrance imposed on the vinylidene group by the diphosphonate group. This difficulty was overcome by using another relatively small monomer, acrylonitrile, as a carrier to induce polymerization of vinylidene-1,1-diphosphonate (Horwitz et al. 1994; Horwitz, et al. 1995). The protonation constants of Diphonix Resin® which are pK_1 and $pK_2 < 2.5$ $pK_3=7.24$ and $pK_4=10.46$ appear almost equal to the protonation constants of the starting material VDPA which are $pK_1=1.27$, $pK_2 = 2.41$, $pK_3=6.67$ and $pK_4=10.04$ (Nash et al. 1994).

In the past few years there were many publications on the separation of lanthanides and actinides on the chelating resins with the phosphonic groups. Lanthanides in minerals occur in small amounts, usually in the form of mixtures, often isomorphic, so that their extraction and separation create many problems. To this end also Diphonix® Resin can be used especially at low pH. It is characterized by high affinity for U(VI), Pu(IV), Np(IV), Th(IV), Am(III) and Eu(III). It was found that from 1 M HNO_3 solutions the distribution coefficient of Diphonix® Resin for U(VI) ions is 70,000 compared to 900 for sulphinic acid resin (Alexandratos, 2007) and the recovery coefficient for Eu(III) under the same conditions is 98.3, whereas for the sulphonic acid resin 44.9 (Ripperger & Alexandratos, 1999). In the paper by Phillips et al. it was demonstrated that Diphonix Resin® can be successfully used for removal of uranium from the solutions of pH > 5 including high concentration of NO_3^- ions as it is less sensitive to interference by such ions as carbonates, nitrates(V), sulphates(VI), Fe(III), Ca(II) and Na(I) (Philips et al. 2008). It can be also used for removal of V(V), Cr(III), Mn(II), Co(II), Ni(II), Zn(II), Cd(II), Hg(II) and Pb(II) form waters and waste waters; V(V), Cr(III), Mn(II), Co(II), Ni(II), Cu(II), Zn(II), Cd(II), Hg(II) and Pb(II) from drinking water; Mn(II), Co(II) and Ni(II) from waste waters of the oil industry; Cr(III) from acidic solutions, Fe(III) from the solutions containing complexing agents in the process of removing scale and radionuclides from radioactive waste waters. Smolik et al. (2009) investigated separation of zirconium(IV) from hafnium(IV) sulphuric acid solutions on Diphonix Resin®. It was found that the best medium for separation of hafnium(IV) and zirconium(IV) is 0.5 M sulphuric acid. A decrease in temperature lowers the degree of metals separation, while lower flow rates through the column increases zirconium(IV) from hafnium(IV) separation. Recent studies have shown that Diphonix Resin® can also be used for removal of Cd(II) and Cr(III) from the phosphoric acid solutions through column tests. Kabay et al. (1998b) found that the acid concentration strongly determines the resin behaviour with respect to the sorption/elution of Cd(II) and Cr(III). In the paper by Cavaco et al. it was pointed but that Diphonix Resin® has strong affinity for Cr(III) ions and high selectivity towards Fe(III) and Ni(II) (Cavaco et al. 2009). The mechanism of sorption on Diphonix Resin® can be written as (Hajiev et al. 1989):

$$R{-}(PO_3H_2)_2^{2-} + M^{2+} \rightleftarrows R{-}(PO_3H_2)^{2-}{\rightarrow}M^{2+}$$

where: R is the resin matrix.

However, according to the literature, Diphonix Resin® has the best selectivity for transition metals such as Fe(III), Cu(II) and Ni(II) over Cr(III). High affinity of Diphonix Resin® for Fe(III) compared to the mono- and divalent ions e.g. Ca(II) was reported in several papers. Owing to its very good separation capability, Diphonix Resin® was also applied in the project FENIX Iron Control System to remove iron from the spent copper electrolyte in Western Metals Copper Ltd. (Queensland, Australia). In this plant, copper(I) sulphate(VI) was used as a reducing agent at the reaction temperature of 85 °C to increase the elution of Fe(III):

$$Fe_2(SO_4)_3 + 6HR \rightleftarrows 2Fe(R)_3 + 3H_2SO_4$$

$$CuSO_4 + Cu \rightleftarrows Cu_2SO_4$$

$$2Fe(R)_3 + 3H_2SO_4 + Cu_2SO_4 \rightleftarrows 2FeSO_4 + 2CuSO4 + 6HR$$

where: R is the resin matrix.

In the paper by Lee & Nicol (2007) it was proved that sorption capacities of Diphonix Resin® for Fe(III) and Co(II) ions in the sulphate(VI) system at pH 2 are equal to 130 mg/g and 90 mg/g, respectively.

The obvious disadvantage of this ion exchanger is therefore the fact that it is difficult to remove Fe(III) ions. To this end 1-hydroxyethane-1,1-diphosphonic acid (HEDP) is used. In the case of Cr(III), Mn(II), Co(II), Ni(II), Cu(II), Zn(II), Cd(II), Hg(II), Sn(II) and Pb(II) ions 2M H_2SO_4 can be also applied.

In the group of chelating ion exchangers containing phosphonic and aminophosphonic functionalities the resins with aminoalkylphosphonic functional groups, such as Duolite C-467, Duolite ES-467, Lewatit OC 1060, Purolite S 940, Purolite S 950 and Chelite P occupy a significant position. In the sorption of heavy metal ions on this kind of chelating ion exchangers the following affinity series is obtained: $Pb^{2+} > Cu^{2+} > UO_2^{2+}$, Zn^{2+}, $Al^{3+} > Mg^{2+} > Sr^{2+} > Ca^{2+} > Cd^{2+} > Ni^{2+} > Co^{2+} > Na^{+} > Ba^{2+}$. These ion exchangers as well as the previously mentioned phosphonic ones exhibit poor affinity for Ca(II) and Mg(II). The effectiveness of sorption of the above mentioned metal ions, however, decreases with the decreasing pH. It is worth mentioning that depending on pH value, the aminoalkylphosphonic groups may occur in the following forms:

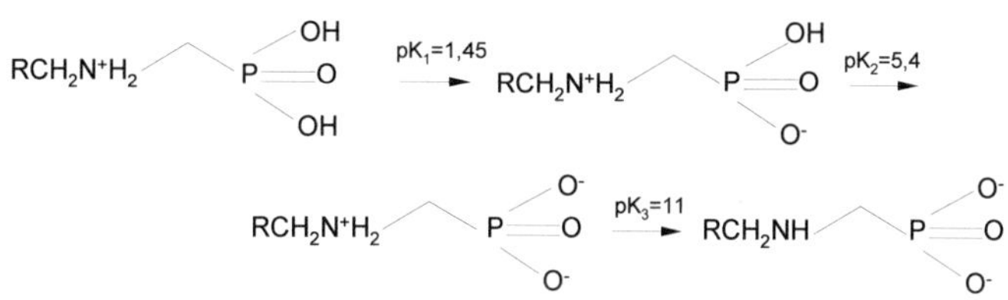

and therefore the selectivity of metal ions sorption depends on the degree of ionization of phosphonic groups. In the case of acidic solutions due to protonation of the nitrogen atom of aminophosphonic group there are formed combinations with the following structure:

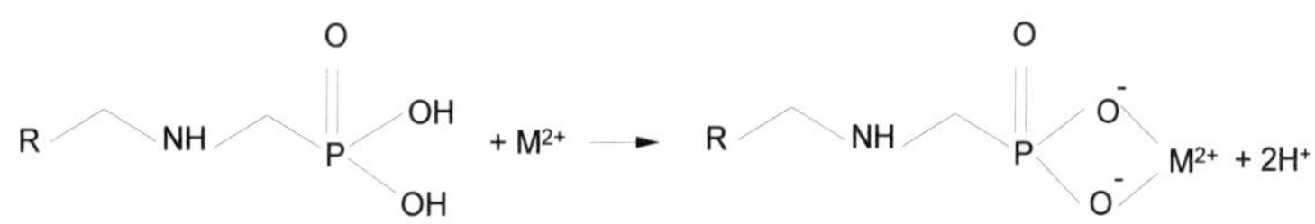

One of the most favourable modes of chelation of the phosphonic acid group is the formation of a four-membered ring through determination of two P-OH groups.

Additionally, in the case of the aminoalkylphosphonic groups, due to the fact that between the aromatic ring of the matrix and the nitrogen atom there is also presented the alkyl group, the increase of the electron density on the nitrogen atom of the amino group is expected. It affects the growth of its protonation. Therefore, this preferred zwitterion form can be as follows:

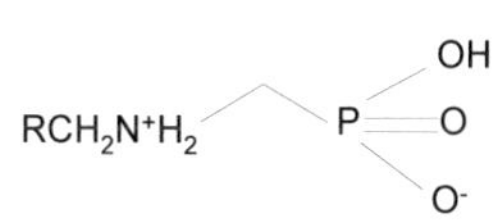

However, the possibility of coordination of the secondary nitrogen atom at lower pH seems to be impossible with respect to its protonated nature and also for steric reasons. Therefore the only potential donor and binding sites of Duolite ES-467 are the oxygen atoms of the phosphonic groups at lower pH values. The chelating, aminomethylphosphonic functional group is also potentially a tridentate ligand having two bonding sites at a phosphonic acid groups and one coordination site at the secondary nitrogen atom (Kertman, 1997; Nesterenko et al. 1999). Formation of a four membered ring through bonding of one of the

OH groups and coordination of the oxygen atom has also been reported. These structures are presented below:

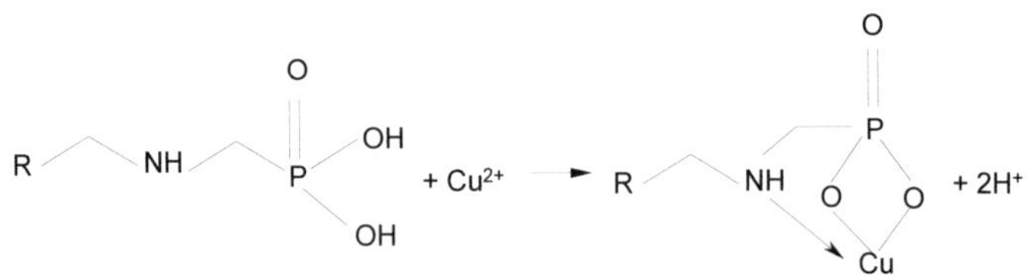

Chelating ion exchangers with the aminoalkylphosphonic functional groups, like picolylamine resins - Dowex M 4195 exhibits moderate selectivity for Cu(II) over Fe(III) in the acidic sulphate(VI) solutions compared to the iminodiacetic acid resins which show no or limited selectivity depending on pH. The stability constants for divalent metal ions with aminomethylphosphonic acid have been found in the order: $Ca^{2+} < Mg^{2+} < Co^{2+} < Ni^{2+} < Cu^{2+} > Zn^{2+}$ (Sahni et al. 1985). In the paper by Milling and West (1984), it was found that Duolite ES-467 possesses a higher capacity for copper(II) ions compared to nickel(II) and iron(III) and that the capacity decreases with the decreasing pH and metal ion concentration in the solution.

Besides Duolite ES-467, Purolite S 950 has been proved to have a high affinity for various heavy metal ions and it is successfully applied in metallurgical and wastewater treatment processes. In the paper by Koivula et al. (2000) Purolitc S-950 was used for purification of effluents from metal plating industry containing Zn(II), Ni(II), Cu(II) and Cd(II) ions. Among others, it was stated that Purolite S-950 showed lower sorption capacity equal to 1.2 eq/dm^3 for zinc chloride compared to zinc solutions containing KCl and NH_4Cl (1.3 eq/dm^3). Under analogous conditions the sorption capacity for Cd(II) was 1.1 eq/dm^3. Recovery of Ni(II) and Co(II) from organic acid complexes using Purolite S 950 was also studied by Deepatana & Valix (2006). They found that sorption capacities for nickel sulphate(VI) for Dowex M4195 (94.51 mg/g), Amberlite IRC 748 (125.03 mg/g) and Ionac SR-5 (79.26 mg/g) are much higher than those for Purolite S-950 in the case of sorption of Ni(II) complexes with citric acid (18.42 mg/g), malic acid (14.45 mg/g) and lactic acid (19.42 mg/g) mainly due to the steric hindrance. For Co(II) ions analogous results were obtained (citric (5.39 mg/g for citric acid; 7.54 mg/g for malic acid and 10.48 mg/g for lactic acid). The elution efficiencies of these complexes from Purolite S-950 resins were high (82–98%) therefore it would appear that the adsorption process involves weak interactions. However, in the case of the sorption of Cu(II) and Zn(II) ions from the sulphate solutions at pH 1.9 on the aminomethylphosphonic resin Lewatit R 252K and the iminodiacetic resin Lewatit TP 207 it was found that separation factors were much lower for Lewatit R 252K (83.0 at 10 °C and 30.0 at 80 °C) than for Lewatit TP 207 (1.67 at 10 °C and 1.4 at 80 °C) (Muraviev et al. 1995).

10. Chelating ion exchangers with the methylglucamine functional groups

Selective ion exchange resins also include chelating ion exchangers containing N-methyl (polyhydroxohexyl)amine functional groups also called methylglucamine. Commercially available ion exchangers of this type are: Amberlite IRA 743, Duolite ES-371, Diaion CRB 02, Dowex BSR 1, Purolite S 108 and Purolite S110.

These ion exchangers show high selectivity for boron (in the form of trioxyboric acid H_3BO_3) (Alexandratos, 2007; Alexandratos, 2009). The boron sorption process proceeds according to the scheme:

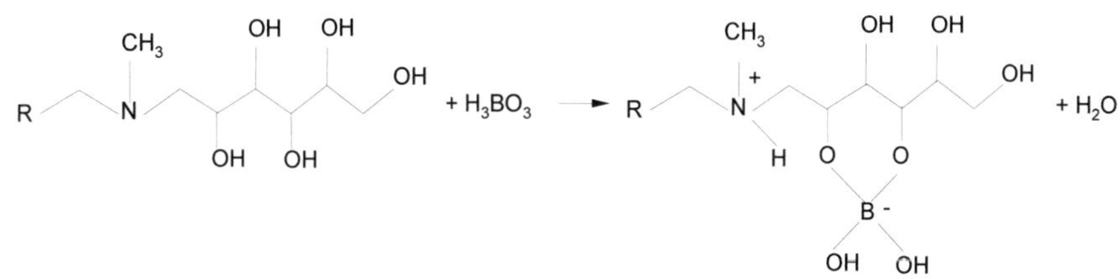

Besides boron the following components of waste water should be also taken into account: Na(I), K(I), Ca(II), Mg(II), Cl^-, SO_4^{2-}, HCO_3^-, CO_3^{2-} and effects should be also considered. Ion exchangers of this type can be used in the removal of Cr(VI) and As(V) (Dambies et al. 2004; Gandhi et al. 2010) although the mechanism of sorption of chromate ions(VI) involves both electrostatic interactions with the protonate amino group and the reduction of Cr(Vl) to Cr(III):

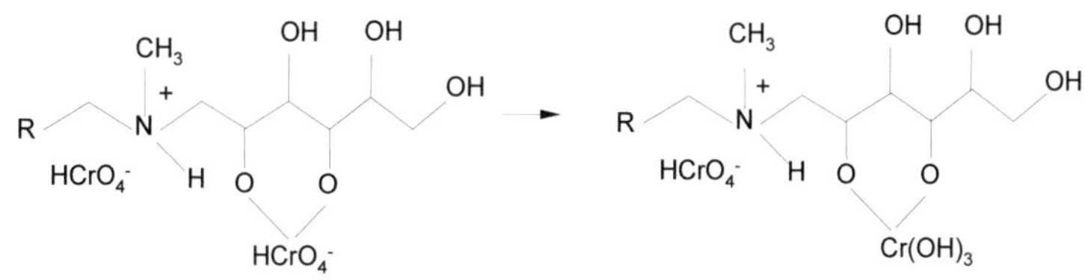

As for arsenate removal the process should be conducted from aqueous solutions at neutral pH. The percent removal of arsenate from the aqueous solution of 100 mg/ dm³ arsenate and 560 mg/dm³ sulphate on NMDG resin is 99% and the reaction is unaffected by the presence of phosphate ions and the solution pH above 9.0, indicating that it can be regenerated with the alkaline solution. It was determined that the key variable in its selectivity is that the resin has to be protonated prior to contact with the aqueous solution (Alexandratos, 2007).

11. Chelating ion exchangers with the bis(2-pyridylmethyl)amine functional groups

The ion exchange resins with the bis(2-pyridylmethyl)amine (bpa) functional groups also known as bispicolylamine are capable of selective sorption of transition metals, particularly Cu(II) ions due to the presence of donor atoms (nitrogen atoms) which are capable of coordination reaction with Cu(II). Due to this fact, such chelating ion exchange resins can combine ion exchange and complexing reactions and then exhibit high selectivity for metal ions. Dowex M 4195 possessing such functional groups is commercially available. It was synthesized in the early 1970s by Dow Chemical Co. and formerly known as Dowex XFS-4195 or DOW 3N. Also two others: Dowex M4196 (formerly Dowex XFS-4196) N-(2-hydroxyethyl)picolylamine or Dowex XFS-43084 (DOW 2N) with N-(2-hydroxypropyl)picolylamine were recognized (Jones & Pyper, 1979; Grinstead, 1984).

Bis(2-pyridylmethyl)amine (bpa) is an uncharged tridentate ligand having the ability to form charged complexes with most divalent metals. The 1:1 complexes with the metal ions of $[M(bpa)]^{2+}$ type are stable (Hirayama & Umehara, 1996). Based on the pK_a values of bis(2-pyridylmethyl)amine (pK_1=0.5, pK_2=2.2, pK_3=3.4), it can be stated that at low pH values three nitrogen atoms would be protonated, while in the middle range of pH only one. For instance, Cu(II) ions (the coordination number is equal to 4) with the bis(2-pyridylmethyl)amine group and water molecule coordinate to it giving a square planar structure. In the next stage the H_2O molecule can be replaced by the anion, which is able to coordinate Cu(II) by a ligand exchange reaction:

$$R–N(bpa)_2 + H^+ \rightleftarrows R–NH^+(bpa)_2$$

$$2R–HN^+(bpa)_2 + M^{2+} \rightleftarrows [R–HN^+(bpa)_2]_2 \rightarrow M^{2+}$$

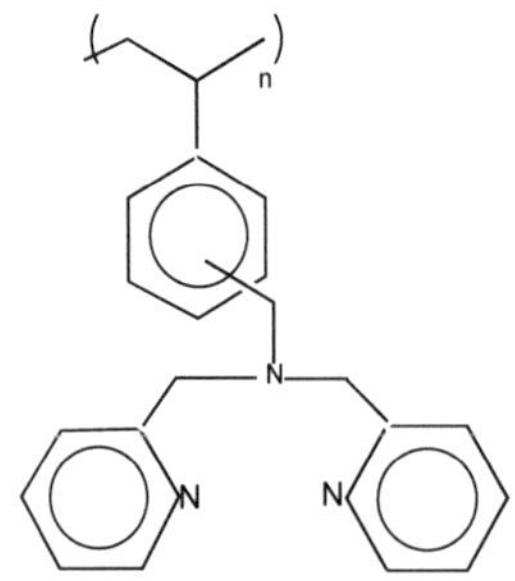

The complexes formed in the resin phase possess the following structure:

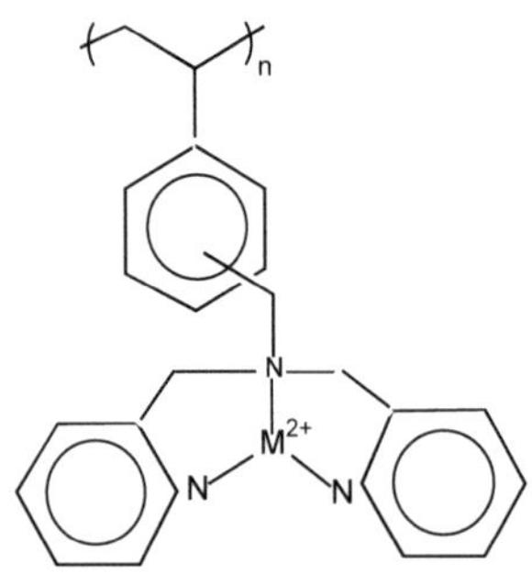

The obtained complex ion exchanger provides a new mode for the recognition of ions in the chromatographic analysis. Dowex M-4195 is a weak base ion exchanger and 1 M H_2SO_4 is in the protonated form (pKa = 3.2). It is also resistant to osmotic shock. Diniz et al. (2000, 2002, 2005) showed that the affinity series of metal ions determined in the one-component system for Dowex M 4195 is as follows: Cu(II) > Ni(II) > Co(II) > Pb(II) > Fe(III) > Mn(II) and it is slightly different from that in the multicomponent system: Cu(II) > Ni(II) > Pb(II) > Fe(III) > Co(II) > Mn(II). The affinity of the transition metal cations for Dowex M 4195 in most cases was in agreement with the Irving-Williams order (Irving and Williams, 1953): Fe(II) < Co(II) < Fe(III) < Ni(II) < Cu(II) > Zn(II). It can be used for purification of chloride solutions after leaching of Mn(II) containing trace amounts of Co(II), Pb(II), Ni(II) and Cu(II). It can be also used for gold recovery (Tuzen, 2008).

On a commercial scale Dowex M 4195 has been used, among others, for separation of Ni(II) ions in the presence of Co(II) at INCO's Port Colborne refinery in Canada and Zambia Chambishi Cobalt Plant (Diniz et al. 2005) for purifying cobalt electrolytes. The efficiency of sorption of both ions is affected not only by pH, but by also by the concentration of sulphate(VI) ions and temperature. It is worth mentioning that separation of the twin pair Co(II)-Ni(II) is one of the most difficult tasks in inorganic chemistry. Contrary to Lewatit TP 207 and Amberlite IRC 718 with the iminodiacetate functional groups, Dowex M-4195 is characterized by the maximum sorption capacity towards Cu(II) ions in the pH range 1-4 (Melling and West, 1984). The sorption process can be presented in the following reaction:

$$RH_n + Cu^{2+} \rightleftarrows R\text{–}Cu^{2+} + nH^+$$

where: R is the resin matrix, n is the stoichiometric ratio, for n≠2 the SO_4^{2-} ions sorption occurs.

Partial washing out of copper(II) ions proceeds by means of 4M H_2SO_4, whereas the total one by means of $NH_3{\cdot}H_2O$. In the case of sorption of Cu(II) ions sorption in the presence of Fe(III) ions, the ion exchange mechanism must be assumed. Sorption of both ions is affected not only by pH, but also by concentration of sulphate(VI) ions and temperature. Fe(III) ions sorption increases significantly with the temperature rise from 293 K to 303 K, whereas it does not change for Cu(II) ions.

The ion exchangers with the picolylamine functional groups can be the basis for obtaining the polymeric ligand exchanger (PLE) with the structure presented above (Zhao et al. 1998; Kołodyńska 2009c):

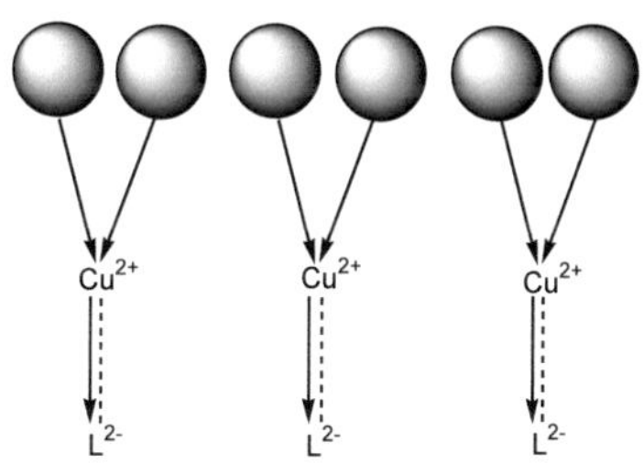

Such kind of ion exchanger consists of the cross linked polystyrene-divinylbenzene matrix, covalently attached bispicolylamine functional groups and Lewis acid cations (such as Cu^{2+}, Ni^{2+}, Fe^{3+}, Co^{2+} according to the series: $Cu^{2+} > Ni^{2+} > Fe^{3+} > Co^{2+}$) coordinated to the functional groups and without neutralization of their positive charge so that the anion exchanger is obtained. It is expected to show high affinity not only for phosphates(V) HPO_4^{2-}, arsenate(V) $HAsO_4^{2-}$ and chromate(VI) CrO_4^{2-} ions but also oxalates ox^{2-}, perchlorates ClO_4^-, tartaric acid as well as simultaneous and selective removal of heavy metal and chromate ions (contrary to other ion exchangers) (Saygi et al. 2008; Dimick, 2008; Du et al. 2008). It was found that for phosphates(V) HPO_4^{2-} removal, the sorption efficiency is much higher than that for the strongly basic macroporous anion exchanger Amberlite IRA 958 and proceeds according to the reaction:

$$R–M^{2+}(2Cl^-) + HPO_4^{2-} \rightleftarrows R–M^{2+}(HPO_4^{2-}) + 2Cl^-$$

The desorption process can be written as follows:

$$R–M^{2+}(HPO_4^{2-}) + 2Cl^- + H^+ \rightleftarrows R–M^{2+}(2Cl^-) + H_2PO_4^-$$

$$R–M^{2+}(HPO_4^{2-}) + 2Cl^- \rightleftarrows R–M^{2+}(2Cl^-) + HPO_4^{2-}$$

Four chelating ion exchange polymeric resins were tested to remove Ni(II) and Co(II) from synthetic solutions on the commercially available ion exchangers Dowex M4195, Amberlite IRC 748, Ionac SR-5 and Purolite S930. Among the selected resins, Dowex M4195 showed the best results for Ni(II) and Co(II) selective sorption from acid liquors in the whole pH range and with small influence of other elements. Even at pH 1 Dowex M4195 was the most effective (Mendes & Martins, 2004).

12. Chelating ion exchangers with the thiol, thiourea and isothiourea functional groups

In the group of ion exchangers with the thiol functional groups (Chelite S, Duolite ES-465, Imac GT 73) Imac TMR resin is very important. It is the macroporous ion exchanger with the PS-DVB matrix, which besides the thiol ones, also possesses the sulphone groups. Imac TMR is used for selective sorption of Hg(II) ions from the process solutions as well as Ag(I), Au(III), Pt(IV) and Pd(II). The sorption process of Hg(II) with the saline solution proceeds according to the reaction:

$$R—SH + Hg^{2+} \rightarrow (R—S)_2Hg^{2+} + 2H^+$$

$$R—SH + HgCl^+ \rightarrow R—SHgCl + H^+$$

Also ion exchangers with the isothiourea functional groups (Ionac SR 3, Lewatit TP 214, Purolite S 920, Srafion NMRR) exhibit high affinity for Hg(II) ions. They are also selective for noble metal ions. Depending on the pH, the isothiourea groups occur in the following forms:

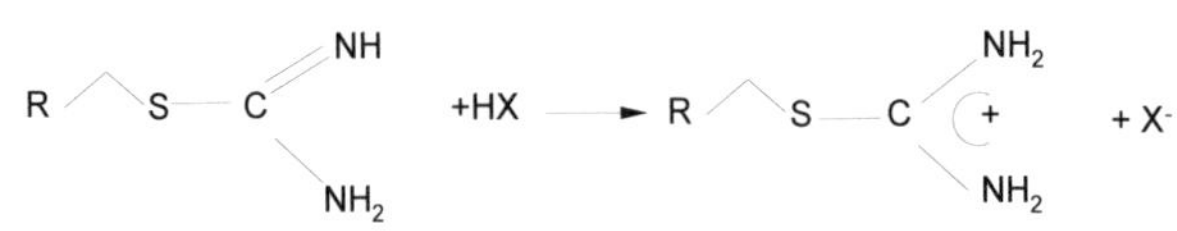

For the first form, coordination bond formation is possible, whereas for the second one the sorption process proceeds in accordance with the anion exchange mechanism:

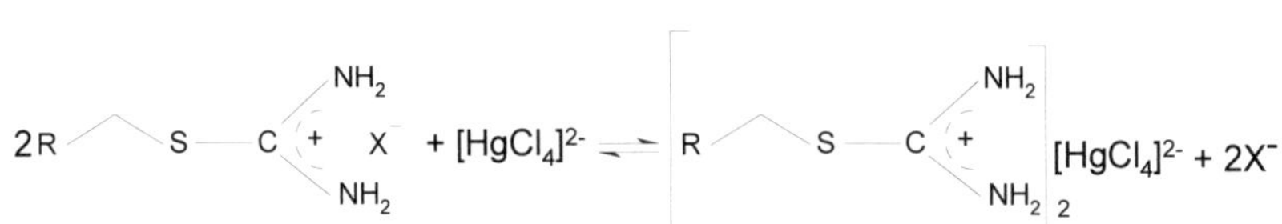

13. Retardion 11A8

Dowex Retardion 11A8 is an example of a very interesting ion exchanger of the type 'snake in a cage' with the quaternary ammonium and carboxylic functional groups (amphotheric resin) (Dybczyński, 1987). It is produced by polymerizing acrylic acid monomer inside an anion exchange resin. Polyacrylic chains are (snake) alternate with the PS-DVB matrix (cage) and therefore they are trapped inside the cross linked ion exchange resin and cannot diffuse out. As a result, cationic and anionic sites are so closely associated that they partially neutralize their electric charges. Mobile ions, such as chlorides, nitrates(V) are attracted and retained on these unique sites until they are eluted with hot water.

Dowex Retardion 11A8 can be used for selective separation of Cd(II) ions in the presence of other heavy metal ions (Samczyński & Dybczyński, 2002). Cd(II) ions are sorbed from 2 M HCI and 2 M NH_4OH with 0.1 M NH_4Cl systems according to the reactions:

$$(n\text{-}2)RCl + [CdCl_n]^{2-n} \rightarrow R([CdCl_n]^{2-n}) + (n\text{-}2)Cl^-$$

$$2RCOONH_4 + [Cd(NH_3)_n]^{2+} \rightarrow (RCOO)_2[Cd(NH_3)_{n-x}] + 2NH_4^+ + nNH_3$$

In the case of separation of Ga(III), In(III), Tl(III), Pt(II), Pd(II) and Na(I), Ni(II), Cu(II) and Zn(II) mixtures in acidic media the resin acts mainly as an anion exchanger. For the elements that can exist as both cations and anions in solution (e.g., Ni, Co, Cu and Zn), the amphoteric properties of Retardion11A8 permit more specific isolation of certain elements from complex mixtures than would be possible with the use of monofunctional ion exchangers (Dybczyński & Sterlińska, 1974).

14. Impregnated ion exchangers

The impregnated resins obtained by physically loaded organic reagents on a solid inert support material such as Amberlite XAD resins are an attractive material for separation and preconcentration of heavy metal ions (Prabhakaran & Subramanian, 2003). They are characterized by good porosity, uniform pore size distribution, high surface area as a chemical homogeneous, non-ionic structure. For instance, it was found that Amberlite XAD-2 functionalized with dithiocarbamate ligand, 1,8-dihydroxynaphthalene-3,6-disulphonic acid (chromotropic acid), 2(2-thiazolylazo)-*p*-cresol, l-(2-pyridylazo)-2-naphthol, calmagite, xylenol orange (Abollino et al. 1998; Ferreira & Brito, 1999, Ferreira et al. 1999; Ferreira et al. 2000a; Ferreira et al. 2000b; Tewari & Singh, 1999; Tewari & Singh, 2000; Tewari & Singh, 2001; Tewari & Singh, 2002) can be used for selective sorption and preconcentration of heavy metal ions. Amberlite XAD-4 loaded with sodium diethyl dithiocarbamate; 2,3-dihydroxy benzoic acid (DHBA), ammonium pyrrolidine dithiocarbamate (APDC) and piperidine dithiocarbamate (pipDTC) were used for preconcentration and determination of metal ions in various matrices (Uzun et al. 2001; Hosseini et al. 2006; Ramesh et al. 2002). However, the most promising polymeric support with a larger surface area is Amberlite XAD 16. Amberlite XAD-16 loaded with quercitin (Sharma & Pant, 2009) is characterized by good adsorbent properties for large amounts of uncharged compounds (Tokalıoğlu et al. 2010). The solid phase extraction

(SPE) process with the application of such materials is characterized by important advantages such simplicity, flexibility, economical, rapid, higher enrichment factors, absence of emulsion and low cost because of lower consumption of reagents.

In general, sorption selectivity of a resin can be affected by both sorbate-sorbent and sorbate-solvent interactions. It has been well recognized that resin matrix and functional groups can strongly affect ion exchange capacity and selectivity. Therefore in the presented paper the chelating ion exchangers Diphonix Resin® containing diphosphonic, sulphonic and carboxylic acid groups and Dowex M 4195 with the bis(2-pyridylmethyl) amine functional group were used for the sorption of Cu(II), Zn(II), Co(II), Pb(II) complexes with Baypure CX 100 (IDS) and Cu(II), Zn(II), Cd(II), Pb(II) complexes with Trilon M (MGDA). The presence of the sulphonic functional groups determines better hydrophilic properties of Diphonix Resin® compared to the traditional monofunctional ion exchangers.

15. Experimental

In the paper the results of the sorption of heavy metal ions such as Cu(II), Zn(II), Cd(II) and Pb(II) in the presence of the complexing agents of a new generation Baypure CX 100 (IDS) and Trilon M (MGDA) on commercially available chelating ion exchangers are presented.

The essential physicochemical properties of these chelating agents are given in Table 1.

16. Characteristics of the chelating ion exchange resins

The chelating ion exchange resins Dowex M 4195 and Diphonix Resin® were tested. Their short characteristics are presented in Table 2.

Properties	IDS	MGDA
Structure	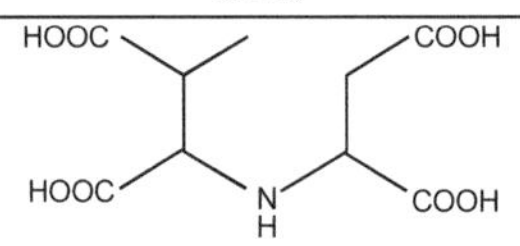	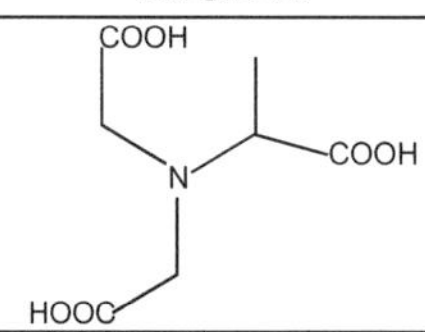
Form supplied	liquid	liquid
Molecular weight	337.1	271.0
Appearance	colourless to light yellow	clear yellowish
pH	10.3-11.4	11.0
Density [g mL^{-1}]	1.32-1.35 g/mL	1.31 g/mL
Solubility in H_2O	in any ratio	in any ratio
Solubility in NaOH	in any ratio	in any ratio
Biodegradability [%]	> 80%	> 68%
Termal stability	in any range	in any range

Table 1. Physicochemical properties of IDA and MGDA.

Properties	Dowex M 4195	Diphonix Resin®
Matrix	PS-DVB	PS-DVB
Structure	macroporous	gel
Functional groups	bis(2-pyridylmethyl) amine bis-picolylamine	diphosphonic sulphonic carboxylic
Commercial form	weak base, partially H_2SO_4 salt	H^+
Appearance	brown to green, opaque	beige, opaque
Total capacity	1.3 [eq/dm³]	5,6 [mol/kg]
Moisture content	40-60 %	58.3 %
Bead size	0.300-1.200 [mm]	0.074-0.150 [mm]
Density	0.67 [g/cm³]	1.05-1.11 [g/cm³]
Max temp. range	353 K	313 K
Operating pH range	2 - 6	0 - 12

Table 2. Physicochemical properties of Dowex M 4195 and Diphonix Resin®.

Before the experiments, the resins were washed with hydrochloric acid (0.1 M) or sulphuric acid (0.5 M) to remove impurities from their synthesis. After pre-treatment they were washed with deionised water.

The solutions of Cu(II), Zn(II), Cd(II) and Pb(II) complexes with Baypure CX 100 and Trilon M with the desirable concentrations were prepared by mixing appropriate metal chlorides or nitrates with the complexing agents solutions, respectively. For the studies the obtained solutions were used without pH adjustment. The pH values of the solutions of Cu(II), Zn(II), Cd(II) and Pb(II) complexes with IDS were as follows: 6.7, 6.5, 69 and 7.3, respectively. For the Cu(II), Zn(II), Cd(II) and Pb(II) complexes with MGDA these values were equal to 8.3, 9.8, 10.5 and 10.4. The other chemicals used were of analytical grade.

In batch experiments, 50 cm^3 of sample solution and ion exchanger (0.5 g) were put into a conical flask and shaken at different time intervals using the laboratory shaker Elpin type 357, (Elpin-Plus, Poland). After the pH of solutions was stabilized and equilibrated, the ion exchangers were filtered. The experiments were conducted in three parallel series. The reproducibility of the measurements was within 5%. Adsorption isotherms were obtained with different initial concentrations varying from 1×10^{-3} M to 2.5×10^{-2} M of metal ions and ligands while keeping the constant amount of resins at room temperature (295 K). The equilibrium between the solid and liquid phases was modelled by the Langmuir and Freundlich equations as presented earlier (Kołodyńska, 2010a; Kołodyńska 2010b; Kołodyńska 2010c). Kinetic studies were carried out at different time intervals varying from 1 to 120 min keeping the constant amount of resins at room temperature (295 K). The shaking speed was 180 rpm to maintain resin particles in suspension.

The amount of heavy metal complexes sorbed onto the resins was calculated by the difference between the amounts added and already present in the solution and that left in the solution after equilibrium.

The pH values were measured with a PHM 84 pH meter (Radiometer, Copenhagen) with the glass REF 451 and calomel pHG 201-8 electrodes. The concentrations of heavy metals were measured with the AAS spectrometer Spectra 240 FS (Varian, Australia).

17. Results

As for the removal of toxic metal ions many different methods are available. Among them, the most commonly used are ion exchange, adsorption, reduction and precipitation. In many cases, the environmentally most compatible and cost-effective solutions include combination of two or more of these processes. From different waste waters those containing heavy metal ions and complexing agents require special attention.

18. Complexing agents

For over fifty years synthetic chelating agents from the group of aminopolycarboxylic acids (APCAs) have been the basis in many technological processes. Ethylenediaminetetraacetic acid (EDTA), nitrilotriacetic acid (NTA) and diethylenetriaminepentaacetic acid (DTPA) are the best known traditional complexing agents. They are commonly applied in many branches of industry forming stable, water soluble complexes with various metal cations or as a masking agent. Nowadays there are a number of alternative products on the market which claim to be as effective as EDTA and NTA. Among them, IDS and MGDA should be listed.

Iminodisuccinic acid (IDS) also known as Baypure CX 100 is a medium-strong chelator consisting of: iminodisuccinic acid sodium salt > 32%, aspartic acid sodium salt < 7%, fumaric acid sodium salt < 3.5 %, hydroxysuccinic acid < 0.9 %, maleic acid sodium salt < 0.9 % (IDS Na-salz, 1998; Vasilev et al. 1996; Vasilev et al. 1998; Reinecke et al. 2000, Kołodyńska et al. 2009; Kołodyńska, 2009a). Iminodisuccinic acid sodium salt can form quintuple-bonded complexes with metal ions. In this case, complexing occurs via the nitrogen and all four carboxyl groups. As a result of the octahedric structure of the complete complex, a water molecule is required for the sixth coordination point (Kołodyńska, 2009b). In the paper by Hyvönen it was found that for low pH conditions (less than 3), the tendency for M(II)/M(III) ions to form complexes with IDS may be assumed as: Cu(II)>Fe(III)>Zn(II)>Mn(II), whereas for pH >7 it can be as follows: Cu(II)>Zn(II)> Mn(II)>Fe(III) (Hyvönen et al. 2003; Hyvönen & Aksela, 2010). IDS is able to replace EDTA when rather moderate chelating agents are sufficient for masking alkaline earth or heavy metal ions. As a substitute for EDTA it is used in a variety of applications, including detergent formulations, corrosion inhibitors, production of pulp and paper, textiles, ceramics, photochemical processes, and as trace nutrient fertilizers in agriculture.

Methylglycinediacetic acid (MGDA) was patented by BASF and marketed under the brand name Trilon M. The active ingredient contained in Trilon M is the trisodium salt of methylglycinediacetic acid. The acid dissociation constants pK_a of MGDA are as follows: pK_1=1.6, pK_2=2.5 pK_3=10.5 (Jachuła et al. 2011; Jachuła et al. 2012). The most important property of Trilon M is the ability to form complexes (MGDA is a tetradentate chelating ligand where chelation involves three carboxylate groups and nitrogen atom) with metal ions, soluble in water in the large pH range 2-13. These complexes remain stable, especially in alkaline media and even at temperatures of up to 373 K. It is worth mentioning that MGDA chelating capacity was investigated by Tandy et al. (2004) in soil washing. It was found that 89-100% of MGDA can be degraded in 14 days, 90% of EDDS in 20 days while no EDTA was degraded in 30 days.

Fig. 5a-b shows the comparison of the logarithmic stability constants (log K) for the complexes of IDS and MGDA and selected metal ions with the stability constants for EDTA.

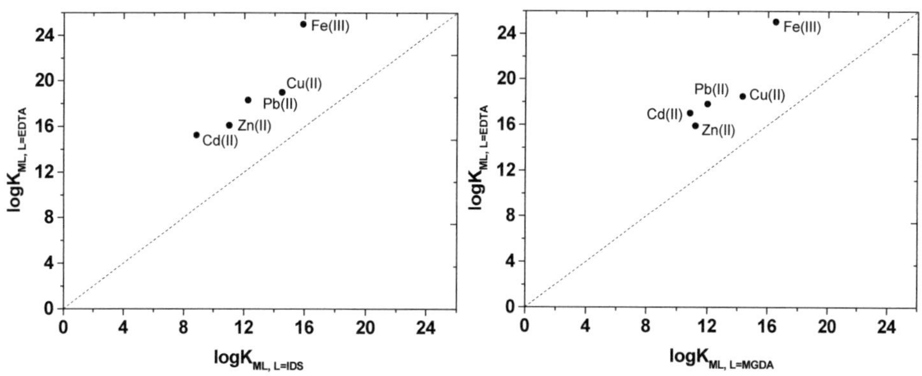

Figure 5. a-b. Comparison of conditional stability constants values of some complexes of metals with EDTA and IDS (a) as well as EDTA and MGDA (b).

A high or moderately high value for log K of Cu(II), Zn(II), Cd(II) and Pb(II) and first of all Fe(III) with IDS and MGDA indicates that these chelating agents have a high affinity for particular metal ions and they provide a preliminary indication of whether the chelating agent is suitable for the specific application.

As these complexing agents are widely applied, removal of their complexes with heavy metals is essential, especially when typical chemical precipitation methods are ineffective, even if solutions with high metal concentrations are treated. Therefore, more advanced techniques are required for cleaning up such contaminants and retardation of heavy metal ions mobility. Among these, the ion exchange with application of selective resins appears to be a more promising method for the treatment of such solutions.

Generally, chelating properties and selectivity of ion exchangers have been enhanced by: (i) immobilization of ligands with multiple coordinating sites such as bifunctional polymers or polyfunctional polymers possessing different functional groups, (ii) immobilization of low molecular weight complexing agents, (iii) by preparation of ion imprinted polymers (IIP), (iv) preparation of reactive ion exchangers (RIEX), (vi) immobilization of specific donor groups through application of Pearson's hard soft acid base theory, (vii) immobilization of macrocycles e.g. crown ethers, calixarenes, resorcinarenes etc. These approaches correspond to both chelating ion exchangers Dowex M 4195 and Diphonix Resin®. Additionally, their sorption selectivity can be affected by sorbate-sorbent and sorbate-solvent interactions. It has been well recognized that the resin matrix and the functional groups can strongly affect ion exchange capacity and selectivity (Clifford & Weber, 1983; Barron & Fritz, 1984; Li et al. 1998). Therefore, in the case of chelating ion exchangers, where the formation of coordination bonds is the basis of the sorption process, besides the parameters related to physicochemical properties of the resins, the effect of the presence of complexing agents should be also taken into account.

In the presence of the complexing agents, IDS and MGDA, there are formed:

$$M^{2+} + H_n ids^{n-4} \rightleftarrows [M(H_n ids)]^{n-2} \text{ where } n=1,2,3$$

and

$$M^{2+} + H_n mgda^{n-3} \rightleftarrows [M(H_n mgda)]^{n-2}, \text{ where } n=1,2.$$

Therefore using selective chelating ion exchangers the sorption effectiveness will be dependent on the decomposition of neutral or anionic species of $[MH_2L]$, $[MHL]^-$ and $[ML]^{2-}$ type, where L=ids^{4-}, $mgda^{3-}$. Additionally, the 'sieve effect' is also important (Kołodyńska, 2010b; Kołodyńska 2010c; Kołodyńska 2011). In the case of the chelating resin Dowex M 4195 possessing the bis(2-pyridylmethyl)amine (bpa) functional groups, depending on the pH value the mechanism of sorption can be as presented earlier. Additionally, the ionic interaction mechanism between the protonated amines and the anionic complexes of the $[ML]^{2-}$ and $[ML]^-$ is also possible (Kołodyńska 2011). Therefore, appropriate reactions can be as follows:

$$2R{-}HN^+(bpa)_2\,Cl^- + [ML]^{2-} \rightleftarrows [R{-}HN^+(bpa)_2]_2[ML]^{2-} + 2Cl^-$$

$$R{-}HN^+(bpa)_2\,Cl^- + [MHL]^- \rightleftarrows [R{-}HN^+(bpa)_2][MHL]^- + Cl^-$$

or

$$R{-}HN^+(bpa)_2\,Cl^- + [ML]^- \rightleftarrows [R{-}HN^+(bpa)_2][ML]^- + Cl^-$$

where: R is the Dowex M 4195 skeleton (PS-DVB), L is the ids^{4-} or $mgda^{3-}$ ligand.

The analogous mechanism of sorption in the case of Diphonix chelating ion exchanger should be considered.

Kinetic studies

For the kinetic data, a simple kinetic analysis was performed using the pseudo first order and the pseudo second order equations:

$$\log(q_e - q_t) = \log(q_e) - \frac{k_1 t}{2.303}$$

$$\frac{t}{q_t} = \frac{t}{q_e} + \frac{1}{k_2 q_e^2}$$

where: q_e is the amount of metal complexes sorbed at equilibrium (for the pseudo first order model also denoted as q_1 and q_2 for the pseudo second order model) (mg/g), qt is the amount of metal complexes sorbed at time t (mg/g), k_1, k_2 are the equilibrium rate constants (1/min), respectively.

The sorption of Cu(II), Zn(II), Cd(II) and Pb(II) complexes with IDS on Dowex M 4195 in the M(II)-L=1:1 system is presented in Fig.6a. The analogous data for the Cu(II), Zn(II), Cd(II) and Pb(II) complexes with MGDA sorption on Dowex M 4195 are presented in Fig.6b and for Diphonix Resin® in Figs.6c and 6d.

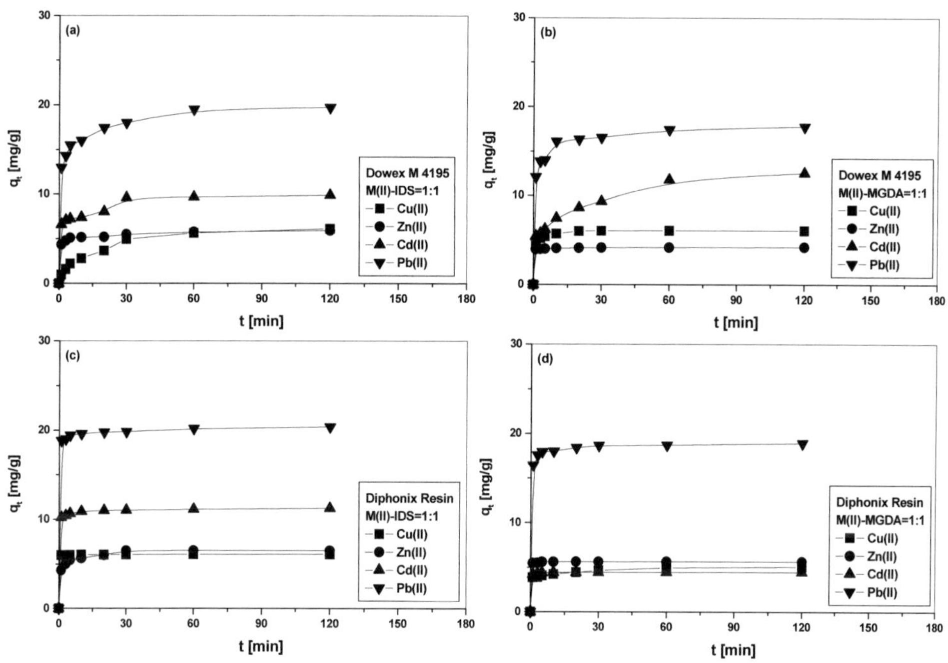

Figure 6. a-b. The effect of the phase contact time on the sorption capacities of Cu(II), Zn(II), Cd(II) and Pb(II) complexes with IDS on Dowex M 4195 (a) and Diphonix Resin® (c) as well as the Cu(II), Zn(II), Cd(II) and Pb(II) complexes with MGDA on Dowex M 4195 (b) and Diphonix Resin® (d) (c_0 1×10^{-3} mol/dm^3, shaking speed 180 rpm, shaking time 1-120 min, room temperature).

The straight lines of t/q_t vs. t suggest the applicability of the pseudo second kinetic model to determine the q_e, k_2 and h parameters (from the intercept and the slope of the plots). These kinetic parameters are presented in Tables 3 and 4.

It was shown that the equilibrium was reached very quickly. More than 90% of metal ions were bound to Dowex M 4195 and Diphonix Resin® within 10-20 min of the phase contact time and therefore a slight increase until a plateau was reached after about 2 h was observed. The values of the theoretical q_e for the studied resins were in good agreement with those obtained experimentally ($q_{e,exp}$). On Dowex M 4195 about 95 %, 100 %, 99 % and

System	$q_{e.exp}$ [mg/g]	q_2 [mg/g]	k_2	h	R^2
Cu(II)-IDS=1:1	5.63	5.61	1.012	5.789	0.9987
Zn(II)-IDS=1:1	5.91	5.88	1.007	4.897	0.9988
Cd(II)-IDS=1:1	9.81	9.89	0.987	12.456	0.9999
Pb(II)-IDS=1:1	19.10	19.00	0.845	16.789	0.9992
Cu(II)-MGDA=1:1	6.05	5.98	2.335	9.237	0.9999
Zn(II)-MGDA=1:1	4.17	4.03	1.017	16.783	0.9996
Cd(II)-MGDA=1:1	12.55	12.23	0.924	10.123	0.9999
Pb(II)-MGDA=1:1	17.77	17.46	0.688	7.525	0.9999

Table 3. The pseudo second order kinetic parameters for the sorption of Cu(II), Zn(II), Cd(II) and Pb(II) complexes with IDS and MGDA on Dowex M 4195.

System	$q_{e.exp}$ [mg/g]	q_2 [mg/g]	k_2	h	R^2
Cu(II)-IDS=1:1	6.12	6.21	2.211	10.207	0.9999
Zn(II)-IDS=1:1	6.01	6.09	1.345	7.123	0.9991
Cd(II)-IDS=1:1	10.21	10.11	0.988	23.434	0.9999
Pb(II)-IDS=1:1	20.39	20.26	0.876	37.551	0.9998
Cu(II)-MGDA=1:1	5.66	5.61	3.469	11.111	0.9999
Zn(II)-MGDA=1:1	4.48	4.48	2.395	48.077	0.9999
Cd(II)-MGDA=1:1	10.23	10.24	0.024	2.475	0.9999
Pb(II)-MGDA=1:1	18.94	18.93	0.188	67.568	0.9999

Table 4. The pseudo second order kinetic parameters for the sorption of Cu(II), Zn(II), Cd(II) and Pb(II) complexes with IDS and MGDA on Diphonix Resin®.

97.5 % of the Cu(II), Zn(II), Cd(II) and Pb(II) complexes with IDS and 94 %, 98 %, 96 % and 95 % complexes with MGDA are sorbed at this time, respectively. On Diphonix Resin® for the Cu(II), Zn(II), Cd(II) and Pb(II) complexes with IDS and MGDA the adequate values are as follows: 94 %, 89 %, 97 % and 98 % as well as 97 %, 86 %, 99 % and 96 %. These results indicate that the sorption process of metal ions in the presence of IDS and MGDA on Dowex M 4195 and Diphonix Resin® followed a pseudo second order kinetics, which meant that both the external mass transfer and intraparticle diffusion together were involved in the sorption process. The correlation coefficients (R2) obtained for the pseudo second order kinetic model are in the range 0.9991 -1.000 for all metal complexes. The pseudo first order parameters were not shown because the correlation coefficients for this model are low (0.7438-0.8745 for the IDS complexes and 0.919-0.986 for the MGDA complexes on Diphonix Resin®.

The breakthrough curves for Cu(II) ions in the presence of MGDA on Dowex M4195 from single metal ion solutions of a concentration $1x10^{-3}$ M are shown in Fig. 7. Typical 'S' shaped curves were obtained in the experiments. Analogous results were obtained on Diphonix Resin®. It should be mentioned the UV exposition does not have a significant effect on the decomposition of the complexes in the resin phase.

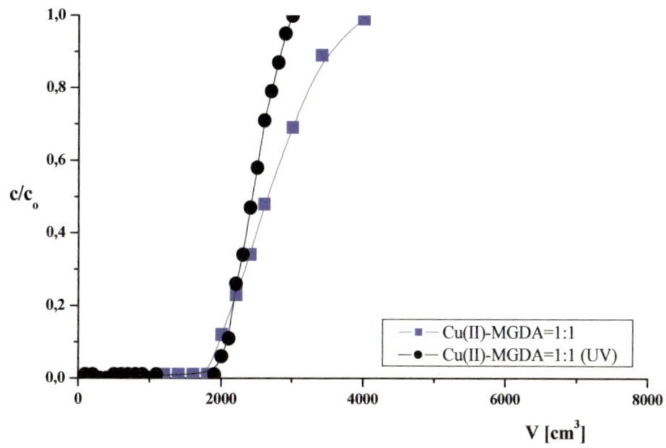

Figure 7. The breakthrough curves of Cu(II) complexes with MGDA on Dowex M 4195 without and with UV exposition (c_0 1×10^{-3} mol/dm^3, bed volume 10 cm^3, flow rate 0.6 cm^3/min)

It is well known that the particle size of ion exchange resins influences the time required to establish equilibrium conditions and two types of diffusion must be considered in an ion exchange equilibrium e.g. the film diffusion (the movement of ions from a surrounding solution to the surface of an ion exchange particle) and the internal diffusion (the movement of ions from the surface to the interior of an ion exchange particle). Film diffusion is usually the controlling reaction in dilute solutions whereas the internal diffusion is controlling in more concentrated solutions. The particle size of an ion exchange resin affects both the film diffusion and the internal diffusion (Kołodyńska, 2011).

According to the manufacturer data the particle size of Dowex M 4195 is 0.300-1.200 mm. However, Diphonix Resin® available on the commercial scale is in the range 0.30-0.85 mm, 0.15-0.30 mm and 0.075-0.15 mm.

In the presented paper Diphonix Resin® with the particle size 0.075-0.150 mm was used to study the sorption process of Cu(II), Zn(II), CdII) and Pb(II) in the presence of IDS and MGDA. In the paper by Cavaco et al. (2009) it was found that for the range 0.15-0.30 mm, 50 % of the particles have diameters less than 0.223 mm. As follows from the obtained results, the bead size of the used chelating ion exchangers has also approximately the Gaussian distribution (Fig. 8 a-b). It was found that with the increase of bead dimensions, the volume fractions of disc-similar beads decrease and the beads are more spherical (Kołodyńska, 2011).

A decrease in the particle size thus shortens the time required for equilibration of particle size and pore characteristics have an effect on equilibrium concentration and influence sorption kinetics. Therefore this factor is essential, especially when the sorption of metal complexes, not metal ions is taken into account. In the case of large complexes the sieve effect is observed.

Kinetic sorption experiments were also carried out with the increased complexes concentrations from 1×10^{-3} mol/dm^3 to 2×10^{-2} mol/dm^3 and these results were presented in

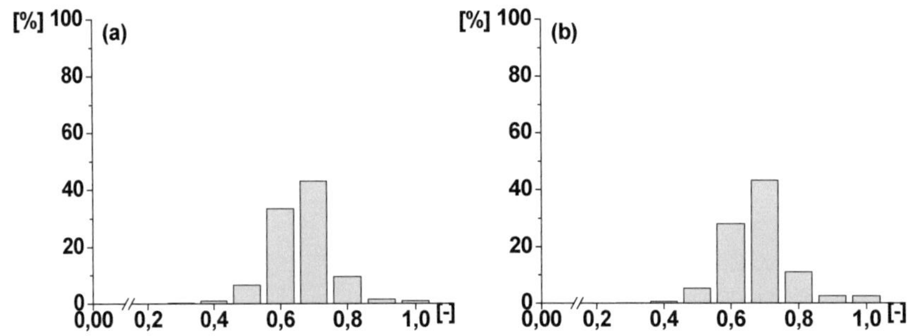

Figure 8. a-b. Comparison of the distribution of the bead size of Dowex M 4195 (a) and Diphonix Resin® (b) based on the Zingg classification.

(Kołodyńska, 2011). It was found that with an increase of metal complexes concentrations a continuous increase in the amount adsorbed per unit mass of ion exchanger was observed till the equilibrium was achieved. For the pseudo second order kinetic model, the rate k_2 values decrease with the increasing initial concentrations, while h increases.

19. pH effects

The effect of pH was studied for the Cu(II), Zn(II), Cd(II) and Pb(II) in the M(II)-IDS=1:1 and M(II)-MGDA=1:1 systems at the pH varied from 2 to 12. The optimal sorption range of the Cu(II), Zn(II), Cd(II) and Pb(II) complexes with IDS practically does not change in the pH range from 4 to 10 both on Dowex M 4195 and Diphonix Resin® whereas, at high pH values, decrease in removal efficiency is observed. In the case of the Cu(II), Zn(II), Cd(II) and Pb(II) complexes with MGDA a slight decrease in sorption efficiency with the increasing pH was also shown.

20. Adsorption studies

The Langmuir equation was applicable to the homogeneous adsorption system, while the Freundlich equation was the non-empirical one employed to describe the heterogeneous systems and was not restricted to the formation of the monolayer. The well-known Langmuir equation was represented as:

$$\frac{1}{q_e} = \frac{1}{b q_0 c_e} + \frac{1}{q_0}$$

where: q_e is the equilibrium M(II) ions concentration on the ion exchanger, (mg/g), c_e is the equilibrium M(II) ions concentration in solution (mg/dm^3), q_0 is the monolayer capacity of ion exchanger (mg/g), b is the Langmuir adsorption constant (L/g) related to the free energy of adsorption.

The values of q_0 and b were calculated from the slope and the intercept of the linear plots c_e/q_e vs. c_e. On the other hand, the Freundlich equation was represented as:

$$q_e = K_F c_e^{bF}$$

where: KF and 1/n are the Freundlich constants corresponding to the adsorption capacity and the adsorption intensity.

The plot of ln q_e vs. ln c_e was employed to generate the intercept KF and the slope 1/n.

The exemplary results presented in Fig.9a-b indicate that for the studied range of concentration of Cu(II) complexes with MGDA (1×10^{-3} M - 2×10^{-2} M) the sorption capacity of Dowex 4195 and Diphonix Resin® increases.

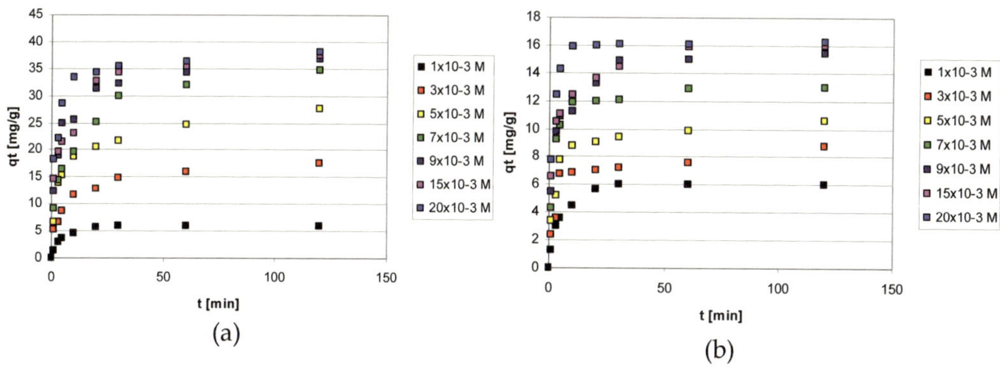

Figure 9. a-b. The effect of the concentration on the sorption capacities of Cu(II) complexes with MGDA on Dowex M 4195 (a) and Diphonix (b) (c_0 1×10^{-3} -20×10^{-3} mol/dm^3, shaking speed 180 rpm, shaking time 1-120 min, room temperature).

The experimental data obtained for the sorption of Cu(II), Zn(II), Cd(II) and Pb(II) in the presence of IDS and MGDA on Dowex M 4195 and Diphonix Resin® were well represented by the Langmuir isotherm model (Table 5). The correlation coefficients of the linear plot of c_e/q_e vs. c_e obtained from them were high, ranging from 0.9512 to 0.9999 (Kołodyńska, 2011). The highest values of the Langmuir parameter q_0 were obtained in the case of Pb(II)

System	Langmuir				Freundlich		
	$q_{e,\,exp}$	q_0	K_L	R^2	K_F	n	R^2
Cu(II)-IDS=1:1	**38.23**	**37.56**	0.023	0.9876	9.54	3.21	0.9865
Zn(II)-IDS=1:1	**21.07**	**20.87**	0.046	0.9923	6.23	2.45	0.9456
Cd(II)-IDS=1:1	**88.00**	**87.65**	0.008	0.9998	4.23	4.58	0.9687
Pb(II)-IDS=1:1	**122.37**	**121.58**	0.012	0.9989	1.23	5.69	0.9623
Cu(II)-MGDA=1:1	**37.99**	**38.05**	0.026	0.9932	12.48	4.23	0.9758
Zn(II)-MGDA=1:1	**16.78**	**17.01**	0.052	0.9983	7.56	3.69	0.9823
Cd(II)-MGDA=1:1	**65.78**	**63.21**	0.061	0.9996	6.23	6.11	0.9877
Pb(II)-MGDA=1:1	**98.78**	**97.64**	0.042	0.9999	2.48	9.25	0.9837

Table 5. The Langmuir and Freundlich isotherm parameter values for the sorption of Cu(II), Zn(II), Cd(II) and Pb(II) ions in the presence of IDS and MGDA on Dowex M4195.

complexes with IDS and MGDA on Dowex M 4195 and Diphonix Resin®. They are equal to 121.58 mg/g and 97.64 mg/g on Dowex M 4195 and 112.37 mg/g and 100.20 mg/g on Diphonix Resin®, respectively.

For the studied systems regeneration tests were conducted using HCl, HNO_3, H_2SO_4 and NaCl at 1M and 2M concentrations. Based on the series of five experiments using known amounts of Cu(II) complexes with IDS and MGDA sorbed, it was established that the overall recoveries of Cu(II) eluted from Dowex M 4195 and Diphonix Resin® by 2M HCl and H_2SO_4 were above 98 %, suggesting that the recovery is quantitative.

21. Conclusions

The presence of biodegradable complexing agents of a new generation that is IDS and MGDA affects the sorption process of Cu(II), Zn(II), Cd(II) and Pb(II) ions on Dowex M4195 and Diphonix Resin®. The effectiveness of sorption depends on the type of complexes and their stability that facilitates their decomposition in the resin phase. The batch equilibrium was relatively fast and reached equilibrium after about 10-20 min of the contact. The experimental data have been analyzed using the Langmuir and Freundlich models. The sorption of studied metal ions in the presence of IDS and MGDA Dowex M 4195 and Diphonix Resin® followed the pseudo second order kinetics. As follows from the experiment pH does not have a significant effect on the sorption of Cu(II), Zn(II), Cd(II) and Pb(II) ions in the presence of IDS and MGDA on the chelating ion exchanger under consideration. The affinity of the above analyzed heavy metal complexes with IDS and MGDA Dowex M 4195 and Diphonix Resin® were found to be as follows: Pb(II) > Cd(II) > Cu(II) > Zn(II) for IDS and MGDA. The studied complexing agents can be proposed as alternative chelating agents to EDTA or NTA for the removal of heavy metal ions from waters and wastewaters.

Author details

Zbigniew Hubicki and Dorota Kołodyńska
Maria Curie-Skłodowska University, Poland

22. References

Abollino, O., Aceto, M., Bruzzoniti, M.C., Mentasti, E., Sarzanini, C. (1998) Determination of metals in highly saline matrices by solid-phase extraction and slurry-sampling inductively coupled plasma-atomic emission spectrometry, *Analytica Chimica Acta*, Vol. 375, pp. 293-298.

Abrams, I.M., Milk, J.R. (1997) A history of the origin and development of macroporous ion-exchange resins, *Reactive and Functional Polymers*, Vol. 35, 7-22.

Ahuja, M., Rai, A.K., Mathur P.N. (1999) Adsorption behaviour of metal ions on hydroximate resins, *Talanta*, Vol. 43, pp. 1955-1963.

Alexandratos, S.D. (2007) New polymer-supported ion-complexing agents: Design, preparation and metal ion affinities of immobilized ligands, *Journal of Hazardous Materials*, Vol. 139, pp. 467–470

Alexandratos, S.D. (2009) Ion exchange resins. A retrospective from Industrial and Engineering Chemistry Research, *Industrial and Engineering Chemistry Research*, Vol. 48, pp. 388-398.

Bajda, T. (2005) Chromatite Ca[CrO_4] in soil polluted with electroplating effluents (Zabierzów, Poland), *Science of the Total Environment*, Vol. 336, pp. 269-274.

Barceloux, D.G. (1999) Copper, *Journal of Toxicology – Clinical Toxicology*, Vol. 37, pp. 217-230.

Barron, R. E.; Fritz, J. S. (1984) Effect of functional group structure on the selectivity of low-capacity anion exchangers for monovalent anions, *Journal of Chromatography*, Vol. 284, pp. 13-25.

Biesuz, R., Pesavento, M., Gonzalo, A., Valiente, M. (1998) Sorption of proton and heavy metal ions on a macroporous chelating resin with an iminodiacetate active group as a function of temperature, *Talanta*, Vol. 47, pp. 127-136

Boari G., Liberti, L., Merli, C., Passino, R. (1974) Exchange equilibria on anion resins, *Desalination*, Vol. 15, pp. 145-166.

Brower, J.B., Ryan, R.L., Pazirandeh, M. (1997) Comparison of ion-exchange resins and biosorbents for the removal of heavy metals from plating factory, *Environmental Science and Technology*, Vol. 31, pp. 2910-2914.

Busche, B., Wiacek, R., Davidson, J., Koonsiripaiboon, V., Yantasee W., Addleman, R.S., Fryxell, G.E. (2009) Synthesis of nanoporous iminodiacetic acid sorbents for binding transition metals, *Inorganic Chemistry Communications*, Vol. 12, pp. 312-315.

Cavaco, S.A., Fernandes, S., Augusto, C.M., Quina, M.J., Gando-Ferreira, L.M. (2009) Evaluation of chelating ion-exchange resins for separating Cr(III) from industrial effluents, *Journal of Hazardous Materials*, Vol. 169, pp. 516-523.

Clifford, D., Weber, W.J., Jr. (1983) The determinants of divalent/monovalent selectivity in anion exchangers, *Reactive Polymers, Ion Exchangers, Sorbents*, Vol. 1, pp. 77-89.

Clifford, D.A. (1999) *Ion exchange and inorganic adsorption*, In: Water Quality and Treatment Fifth Edition, Ray Lettermam, ed., McGraw Hill, Inc., New York, pp. 9.1-9.91.

Chajduk-Maleszewska, E., Dybczyński, R. (2004) Effective separation and preconcentration of trace amounts of Pd on Duolite ES 346 resin and its use for the determination of Pd by NAA, *Analytical Chemistry*, Vol. 49, pp. 281-297.

Chiarizia, R., Horwitz, E.P., Alexandratos, S.D., Gula, M.J. (1997) Diphonix resin: a review of its properties and applications, *Separation Science and Technology*, Vol. 32, pp. 1-35.

Chiarizia, R. Horwitz, E.P., Alexandratos, S.D. (1994) Uptake of metal ions by a new chelating ion-exchange resin. Part 4. Kinetics, *Solvent Extraction and Ion Exchange*, Vol. 12, pp. 211-237.

Chiarizia, R. D'Arcy, K.A. Horowitz, E.P. Alexandratos, S.D., Trochimczuk, A.W. (1996) Uptake of metal ions by a new chelating ion exchange resin. Part 8. Simultaneous uptake of cation and anion species, *Solvent Extraction and Ion Exchange*, Vol. 14, pp. 519-542.

Corella, M.B., Siggia, S., Barnes, R.M. (1984) Synthesis and characterization of a poly(acryloamidoxime) metal chelating resin, *Analytical Chemistry*, Vol. 56, pp. 967-972.

Dambies, L., Salinaro, R., Alexandratos, S.D. (2004) Immobilized N-methyl-D-glucamine as an arsenate-selective resin, *Environmental Science and Technology*, Vol. 38, pp. 6139-6146.

Dąbrowski, A., Hubicki, Z., Podkościelny, P., Robens, E. (2004) Selective removal of the heavy metal ions from waters and industrial wastewaters by ion-exchange method, *Chemosphere*, Vol. 56, pp. 91-106.

De Dardel, F., Arden, T.V. (2001) *Ion exchangers*. Principles and applications, Ullmann's Encyclopaedia of Industrial Chemistry, Sixth Edition, Wiley-VCH Verlag GmbH, pp. 1-70.

Deepatana A., Valix M. (2006) Recovery of nickel and cobalt from organic acid complexes: Adsorption mechanisms of metal-organic complexes onto aminophosphonate chelating resin, *Journal of Hazardous Materials*, Vol. 137, pp. 925-933.

Dimick, P.D., Kney, A., Tavakoli, J., Mylon, S.E., Zhao, D. (2008) A comparison of metal-loaded DOW3N ion exchanger for removal of perchlorate from water, *Separation Science and Technology*, Vol. 43, pp. 2343-2362.

Ding, X., Mou, S.F., Liu, K., Yan, Y. (2000) Improved scheme of chelation ion chromatography with a mixed eluent for the simultaneous analysis of transition metals at mg l^{-1} levels, *Journal of Chromatography*, Vol. 883, pp. 127-136.

Ding, X., Mou, S. (2001) Retention behavior of transition metals on a bifunctional ion-exchange column with oxalic acid as eluent, *Journal of Chromatography*, Vol. 920, pp. 101-107.

Diniz, C.V., Doyle, F.M., Martins A.H. (2000) Uptake of heavy metals by chelating resins from acidic manganese chloride solution, *Minerals Metallurgy Processing*, Vol. 17, pp. 217-222.

Diniz, C.V., Doyle, F.M., Ciminelli, V.S.T. (2002) Effect of pH on the adsorption of selected heavy metal ions from concentrated chloride solutions by the chelating resin Dowex M-4195, *Separation Science and Technology*, Vol. 37, pp. 3169-3185.

Diniz, C.V., Ciminelli, V.S.T., Doyle, F.M. (2005) The use of the chelating resin Dowex M-4195 in the adsorption of selected heavy metal ions from manganese solutions, *Hydrometallurgy*, Vol. 78, pp. 147-155.

Drăgan, S., Grigoriu, G. (1992) Ion exchangers, I. Anion exchanges with tertiary amine groups on poly(acrylonitrile-co-divinylbenzene) network, *Die Angewandte Macromolekulare Chemie*, Vol. 200, pp. 27-36.

Du, W., Pan, B., Jiang, P., Zhang, Q., Zhang, W., Pan, B., Zhang, Q., Zhang, Q. (2008) Selective sorption and preconcentration of tartaric acid using a copper(II)-bound polymeric ligand exchanger, *Chemical Engineering Journal*, Vol. 139, pp. 63-68.

Duffus, J.H. (2002) „Heavy metals" – a meaningless term? *Pure Applied Chemistry*, Vol. 74, pp. 793-807.

Dybczyński, R. (1987) Selective separation of zinc from other elements on the amphoteric resin Retardion 11A8 and its use for the determination of zinc in biological materials by neutron activation analysis, *Analyst*, Vol. 112, pp. 449-453

Dybczyński R., Hubicki Z., Kulisa K. (1988) Ion exchange behaviour of 23 elements and amphoteric properties of chelating resin Duolite ES 346 containing amidoxime groups, *Solvent Extraction and Ion Exchange*, Vol. 6, 699-724.

Dybczyński R., Sterlińska E. (1974) The use of the amphoteric ion-exchange resin retardion11A8 for inorganic separations, *Journal of Chromatography*, Vol.102, pp. 263-271.

Ebraheem, K.A.K. Hamdi, S.T. (1997) Synthesis and properties of a copper selective chelating resin containing a salicylaldoxime group, *Reactive and Functional Polymers*, Vol. 34, pp. 5-10.

Fernandez-Prini, R. (1982) Hydrothermal decomposition of ion-exchange resins, *Power Industry Research*, Vol. 2., pp.101-108.

Ferriera, L.M., Loureiro, J.M., Rodrigues, A.E. (1998) Sorption of metals by an amidoxime chelating resin. Part I: Equilibrium, *Separation Science and Technology*, Vol. 33, pp. 1585-1604.

Fisher, S. (2002) Spotlight on the cation resin, *Power Plant Chemistry*, Vol. 4, pp. 407-410

Ferreira, S.L.C. de Brito, C.F. (1999) Separation and preconcentration of cobalt after sorption onto Amberlite XAD-2 loaded with 2-(2-thiazolylazo)-*p*-cresol, *Analytical Science*, Vol. 15, p.189.

Ferreira, S.L.C., De Brito, C.F., Dantas, A.F., Lopo De Araújo, N.M., Costa, A.C.S. (1999) Nickel determination in saline matrices by ICP-AES after sorption on Amberlite XAD-2 loaded with PAN, *Talanta*, Vol. 48, pp. 1173-1177.

Ferreira, S.L.C., Lemos, V.A., Moreira, B.C., Costa, A.C.S., Santelli, R.E. (2000a) An on-line continuous flow system for copper enrichment and determination by flame atomic absorption spectroscopy, *Analytica Chimica Acta*, Vol. 403, pp. 259-264.

Ferreira, S.L.C., Ferreira, J.R., Dantas, A.F., Lemos, V.A., Araújo, N.M.L., Costa, A.C.S. (2000b) Copper determination in natural water samples by using FAAS after preconcentration onto Amberlite XAD-2 loaded with calmagite, *Talanta*, Vol. 50, pp. 1253-1259.

Fewtrell, L., Kay, D., Jones, F., Baker, A., Mowat, A. (1996) Copper in drinking water. An investigation into possible health effects, Public Healh, Vol. 110, pp. 175-178.

Fritz J.S. (2005) Factors affecting selectivity in ion chromatography, Journal of Chromatography, Vol. 1085, pp. 8-17.

Gadd, G.M., White, C. (1993) Microbial treatment of metal pollution – a working biotechnology? *Trends in Biotechnology*, Vol. 11, pp. 353-359.

Gandhi, M.R., Viswanathan, N., Meenakshi, S. (2010) Adsorption mechanism of hexavalent chromium removal using Amberlite IRA 743 resin, *Ion Exchange Letters*, Vol. 3, pp. 25-35.

Gode, F., Pehlivan, E. (2003) A comparative study of two chelating ion-exchange resins for the removal of chromium(III) from aqueous solution, *Journal of Hazardous Materials*, Vol. 100, pp. 231-243.

Grinstead, R.R. (1984) *New developments in the chemistry of XFS 4195 and XFS 43084 chelating ion exchange resins*, In: Ion Exchange Technology, Society of Chemical Industry, New York, pp. 509-518.

Hajiev, S.N., Kertman, S.V., Leykin, U.A., Amelin, A.N. (1989) Thermochemical study of ion echange processess. V. Sorption of copper ions in complex-forming resins, *Thermochimica Acta*, Vol. 139, pp. 327-332.

Helfferich F. (1962) Ion Exchange Resins, Mc-Grow Hill:New York.

Hirayama, N., Umehara, W. (1996) Novel separation of inorganic anions using a charged complex ion-exchanger, *Analytica Chimica Acta*, Vol. 334, pp. 1-4

Horwitz, E.P., Chiarizia, R., Diamond, H., Gatrone, R.C., Alexandratos, S.D., Trochimczuk, A.Q., Crick, D.W. (1993) Uptake of metal ions by a new chelating ion-exchange resin. Part 1. Acid dependencies of actinide ions, *Solvent Extraction and Ion Exchange*, Vol. 11, pp. 943-966.

Horwitz, E.P., Alexandratos, S.D., Gatrone, R.C., Chiarizia, R. (1994) Phosphonic acid based ion exchangers, US Patent 5281631.

Horwitz, E.P., Alexandratos, S.D., Gatrone, R.C., Chiarizia, R. (1995) Phosphonic acid based ion exchange resins. US Patent 5449462.

Hosseini, M.S., Raissi, H., Madarshahian, S. (2006) Synthesis and application of a new chelating resin functionalized with 2,3-dihydroxy benzoic acid for Fe(III) determination in water samples by flame atomic absorption spectrometry, *Reactive and Functional Polymers*, Vol. 66, pp. 1539-1545.

Hubicki, Z., Jakowicz, A., Łodyga, A. (1999) *Application of the ions from waters and sewages*, In: Adsorption and its applications in industry and environmental protection. Studies in surface science and catalysis, ed. A. Dąbrowski, Elsevier, Amsterdam, New York.

Hyvönen, H., Orama, M., Saarinen, H., Aksela, R. (2003) Studies on biodegradable chelating ligands: complexation of iminodisuccinic acid (ISA) with Cu(II), Zn(II), Mn(II) and Fe(III) ions in aqueous solution, *Green Chemistry*, Vol. 5, pp. 410-414.

Hyvönen, H., Aksela, R. (2010) Complexation of 3-hydroxy-2,2'-iminodisuccinic acid (HIDS) with Mg^{2+}, Ca^{2+}, Mn^{2+}, Fe^{3+}, Fe^{2+}, Co^{2+}, Ni^{2+}, Cu^{2+}, and Zn^{2+} ions in aqueous solution, *Journal of Coordination Chemistry*, Vol. 63, pp. 2013-2025.

IDS Na-salz (1998) Eine neue umweltfreundliche alternative zu klassischen komplexierungsmitteln. Bayer AG brochure, Leverkusen.

Ion exchange resins and adsorbents (2006) Dow Chemical, Co. brochure.

Irving, H., Williams, R.J.P. (1953) The stability of transition-metal complexes. *Journal of Chemical Society*, Vol. 162, pp. 3192-3210.

Ivanov, V.A., Timofeevskaya, V.D., Gorshkov, V.I., Drozdova, N.V. (1996) The role of temperature in ion exchange processesof separation and purification, *Journal of Radioanalytical and Nuclear Chemistry*, Vol. 208, pp. 23-45.

Jeyakumar, S., Mishra, V.G. Das, M.K., Raut, V.V., Sawant, R.M., Ramakumar K.L. (2011) Separation behavior of U(VI) and Th(IV) on a cation exchange column using 2,6-pyridine dicarboxylic acid as a complexing agent and its application for the rapid separation and determination of U and Th by ion chromatography, *Journal of Separation Science*, Vol. 34, pp. 1-8.

Jones, K.C., Pyper, R.A. (1979) Copper recovery from acidic leach liquors by continuous ion-exchange and electrowinning, *Journal of Metals*, Vol. 4, pp. 19-25.

Kabay, N., Demircioğlu, M., Yaylı, S., Günay, E., Yüksel, M., Sağlam, M., Streat, M. (1998a) Recovery of uranium from phosphoric acid solutions using chelating ion-exchange resins, *Industrial and Engineering Chemistry Research*, Vol. 37, pp. 1983-1990.

Kabay, N., Demircioğlu, M., Ekinci, H., Yüksel, M., Sağlam, M., Akçay, M., Streat, M. (1998b) Removal of metal pollutants (Cd(II) and Cr(III)) from phosphoric acid solutions by chelating resins containing phosphonic or diphosphonic groups, *Industrial and Engineering Chemistry Research*, Vol. 37, pp. 2541-2547.

Kantipuly, C., Katragadda, S., Chow, A. Gesser, H.D. (1990) Chelating polymers and related supports for separation and preconcentration of trace metals, *Talanta,* Vol. 37, pp. 491-517.

Kaušpėdienė, D., Snukiškis, J., Gefenienė, A. (2003) Kinetics of cadmium(II) sorption by an iminodiacetic ion exchanger in the presence of a nonionic surfactant, *Desalination,* Vol. 154, pp. 67-77.

Kertman, S.V., Kertman, G.M., Amelin, A.N., Leykin, Yu.A. (1997) Heats of the immersion of Co^{2+} and Cu^{2+} contained chelating resins, *Thermochimica Acta,* Vol. 297, pp. 49-56.

Koivula, R., Lehto, J., Pajo, L., Gale, T., Leinonen, H. (2000) Purification of metal plating rinse waters with chelating ion exchangers, *Hydrometallurgy,* Vol.56, pp. 93-108.

Kołodyńska, D.; Hubicka, H. & Hubicki, Z. (2009). Studies of application of monodisperse anion exchangers in sorption of heavy metal complexes with IDS, *Desalination*, Vol. 239, pp. 216-228.

Kołodyńska, D. (2009a). Polyacrylate anion exchangers in sorption of heavy metal ions with the biodegradable complexing agent, *Chemical Engineering Journal*, Vol. 150, pp. 280-288.

Kołodyńska, D. (2009b). Iminodisuccinic acid as a new complexing agent for removal of heavy metal ions from industrial effluents, *Chemical Engineering Journal*, Vol. 152, pp. 277-288.

Kołodyńska D. (2009c) Chelating ion exchange resins in removal of heavy metal ions from waters and wastewaters in presence of a complexing agent, *Przemysł Chemiczny*, Vol. 88, pp. 182-189 (in Polish).

Kołodyńska, D. (2010a). Biodegradable complexing agents as an alternative to chelators in sorption of heavy metal ions, *Desalination and Water Treatment. Science and Engineering*, Vol. 16, pp. 146-155.

Kołodyńska, D. (2010b). Diphonix Resin® in sorption of heavy metal ions in the presence of biodegradable complexing agents of a new generation, *Chemical Engineering Journal*, Vol. 159, pp. 27-36

Kołodyńska, D. (2010c). The effect of the treatment conditions on metal ions removal in the presence of complexing agents of a new generation, *Desalination*, Vol. 263, pp. 1595-169.

Kołodyńska, D. (2011) The chelating agents of a new generation as an alternative to conventional chelators for heavy metal ions removal from different waste waters, In: Expanding issues in desalination (ed., Robert Y. Ning) InTech, Publishers 2011, pp. 339-371.

Kołodyńska, D., Jachuła, J., Hubicki, Z. (2012) Removal of heavy metal complexes with MGDA by synthetic resin Diphonix, *Environmental Engineering and Management Journal*, in press.

Jachuła, J., Kołodyńska, D., Hubicki, Z. (2011) Sorption of Cu(II) and Ni(II) ions in presence of novel chelating agent methylglycinediacetic acid by microporous ion exchangers and sorbents from aqueous solutions, *Central European Journal of Chemistry*, Vol. 9, pp. 52-65.

Jachuła, J., Kołodyńska, D., Hubicki, Z. (2012) Methylglycinediacetic acid as a new complexing agent for removal of heavy metal ions from industrial wastewater, *Solvent Extraction and Ion Exchange*, in press.

Kononova, O.N., Kholmogorov, A.G., Kachin, S.V., Mytykh, O.V., Kononov, Y.S., Kalyakina, O.P., Pashkov, G.L. (2000) Ion exchange recovery of nickel from manganese nitrate solutions, *Hydrometallurgy*, Vol. 54, pp. 107-115.

Krongauz, V.V., Kocher, C.W. (1997) Kinetics of ion exchange in monodisperse resin, *Journal of Applied Polymer Science*, Vol. 59, pp. 1271-1283.

Kunin R. (1958) *Ion Exchange Resins*, 3rd Ed; Wiley: New York.

Kunin, R. (1979) Amber-Hi-Lites 161, Rohm and Haas Co.

Kurama, H., Çatlsarik, T. (2000) Removal of zinc cyanide from a leach solution by an anionic ion-exchange resin, *Desalination*, Vol. 129, pp. 1-6.

Lee, M.S., Nicol, M.J. (2007) Removal of iron from cobalt sulphate solutions by ion exchange with Diphonix resin and enhancement of iron elution with titanium(III), *Hydrometallurgy*, Vol. 86, pp. 6-12.

Li, P.; SenGupta, A.K. (1998) Genesis of selectivity and reversibility for sorption of synthetic aromatic anions onto polymeric sorbents, *Environmental Science and Technology*, Vol. 32, pp. 3756-3766.

Li, C.W., Qi, T., Wang, F.A., Zhang, Y., Yu, Z.H. (2006) Variation of cell voltage with reaction time in electrochemical synthesis process of sodium dichromate, *Chemical Engineering and Technology*, Vol. 29, pp. 481-486.

Luca, C., Vlad, C.D., Bunia, I. (2009) Trends in weak base anion exchangers resins, *Revue Roumaine de Chimie*, Vol. 54, pp. 107-117.

Lührmann, LM., Stelter, N., Kettrup, A. (1985) Synthesis and properties of metal collecting phases with silica immobilized 8-hydroxyquinoline, *Fresenius Journal of Analytical Chemistry*, Vol. 322, pp. 47-52.

Marhol, M., Beranová, H., Cheng, K.L. (1974) Selective ion-exchangers containing phosphorus in their functional groups. I. Sorption and separation of some bivalent and trivalent ions, *Journal of Radioanalytical and Nuclear Chemistry*, Vol. 21, pp. 177-186.

Marhol, M., Cheng, K.L. (1974) Some chelating ion-exchange resins containing ketoiminocarboxylic acids as functional groups, *Talanta*, Vol. 21, pp. 751-762.

McClain, A., Hsieh, Y.L. (2004) Synthesis of polystyrene-supported dithiocarbamates and their complexation with metal ions, *Journal of Applied Polymer Sciences*, 92 (2004) 218-225.

Mendes, F.D., Martins A.H. (2004) Selective sorption of nickel and cobalt from sulphate solutions using chelating resins, *International Journal of Mineral Processing*, Vol. 74, pp. 359-371.

Melling J., West, D.W. (1984) Proceedings of the International Conference on Ion-Exchange, Society of Chemical Industry, Cambridge, England, p. 724.

Minczewski, J., Chwastowska, J., Dybczyński, R. (1982) *Separation and preconcentration methods in inorganic trace analysis*, Wiley-VCH, New York.

Muraviev, D., Gonzalo, A., Valiente, M. (1995) Ion exchange on the resin with temperature-responsive selectivity. 1. ion exchange equilibrium of Cu^{2+} and Zn^{2+} on iminodiacetic and aminomethylphosphonic resin, *Analytical Chemistry*, Vol. 67, pp. 3028-3035.

Nash, K.L., Rickert, P.G., Muntean, J.V. (1994) Uptake of metal ions by a new chelating ion exchange resin. Part 3: Protonation constants via potentiometric titration and solid state 31P NMR spectroscopy, *Solvent Extraction and Ion Exchange*, Vol. 12, pp. 193-209.

Nesterenko, P.N., Shaw, M.J., Hill, S.J., Jones, P. (1999) Aminophosphonate-functionalized silica: A versatile chromatographic stationary phase for high performance chelation ion chromatography, *Microchemical Journal*, Vol. 62, pp. 55-69.

Nesterenko, P.N., Haddad, P.R. (2000) Zwitterionic ion-exchangers in liquid chromatography, *Analytical Sciences*, Vol. 16, pp. 565-574.

Nriagu, J.O., Pacyna, J.M. (1988) Quantitative assessment of worldwide contamination of air, water and soils by trace metals, *Nature*, Vol. 333, pp. 134-139.

Ogata, T., Nagayoshi, K., Nagasako, T., Kurihara, S., Nonaka, T. (2006) Synthesis of hydrogel beads having phosphinic acid groups and its adsorption ability for lanthanide ions, *Reactive and Functional Polymers*, Vol. 66, pp. 625-633.

Park, I.H., Suh, J.M. (1996) Preparation of uranyl ion adsorptivity of macroreticular chelating resins containing a pair of neighboring amidoxime groups in a monomeric styrene units, *Angewandte Makromolekulare Chemie*, Vol. 239, pp. 121-132.

Pesavento, M., Biesuz, R., Gallorini, M., Profumo, A. (1993) Sorption mechanism of trace amounts of divalent metal ions on a chelating resin containing iminodiacetate groups, *Analytical Chemistry*, Vol. 65, pp. 2522-2527.

Phillips, D.H., Gu B., Watson, D.B., Parmele, C.S. (2008) Uranium removal from contaminated groundwater by synthetic resins, *Water Research*, Vol. 42, 260-268.

Prabhakaran, D., Subramanian, M.S. (2003) A new chelating sorbent for metal ion extraction under high saline conditions, *Talanta*, Vol. 59, pp. 1227-1236.

Pramanik, S., Sarkar, S., Paul, H., Chattopadhyay, P. (2009) A new polymer with 2-methoxy-1-imidazolylazobenzene functionality for determination of lead(II) and iron(III), *Indian Journal of Chemistry*, Vol. 48A, pp. 30-37.

Pyell U., Stork, G. (1992) Preparation and properties of an 8-hydroxyquinoline silica gel synthesized via Mannich reaction, *Fresenius Journal of Analytical Chemistry*, Vol. 342, pp. 281-286.

Ryan, D.K., Weber, J.H. (1985) Comparison of chelating agents immobilized on glass with Chelex 100 for removal and preconcentration of trace copper(II), *Talanta*, 32 (1985) 859-863

Ramesh, A., Mohan, K.R., Seshaiah, K. (2002) Preconcentration of trace metals on Amberlite XAD-4 resin coated with dithiocarbamates and determination by inductively coupled plasma-atomic emission spectrometry in saline matrices, *Talanta*, Vol. 57, pp. 243-252.

Reinecke, F., Groth, T., Heise, K.P., Joentgen, W., Müller, N., Steinbüchel, A. (2000) Isolation and characterization of an *Achromobacter xylosoxidans* strain B3 and other bacteria capable to degrade the synthetic chelating agent iminodisuccinate, FEMS Microbiological Letters, Vol. 188, pp. 41-46.

Reynolds T.D. (1982) Unit operations and processes in environmental engineering, BC Engineering Division, Edition II, Boston, MA.

Ripperger K.P., Alexandratos, S.D. (1999) Polymer-supported phosphorus-containing ligands for selective metal ion complexation. In: Adsorption and its applications in industry and environmental protection. Studies in surface science and catalysis, ed. A. Dąbrowski, Elsevier, Amsterdam, New York.

Saygi, K.O., Tuzen, M., Soylak, M., Elci, L. (2008) Chromium speciation by solid phase extraction on Dowex M 4195 chelating resin and determination by atomic absorption spectrometry, *Journal of Hazardous Materials*, Vol. 153, pp. 1009-1014.

Sahni, S.K., Reedijk, J. (1984) Coordination chemistry of chelating resins and ion exchangers, *Coordination Chemistry Reviews*, Vol. 59, pp. 1-139.

Sahni, S.K., Van Bennekom, R., Reedijk, J. (1985) A spectral study of transition-metal complexes on chelating ion exchange resin containing aminophosphonic acid groups, *Polyhedron*, Vol. 4, pp. 1643-1658.

Samczyński, Z., Dybczyński, R. (1997) Some examples of the use of amphoteric ion exchange resins for inorganic separations, *Journal of Chromatography*, Vol. 789, pp. 157-167.

Samczyński, Z., Dybczyński, R. (2002) The use of Retardion 11A8 amphoteric ion exchange resin for the separation and determination of cadmium and zinc in geological and environmental materials by neutron activation analysis, *Journal of Radioanalytical and Nuclear Chemistry*, 254 (2002) 335-341.

Saxena, R., Singh, A.K., Rathore, D.P.S. (1995) Salicylic acid functionalized polystyrene sorbent Amberlite XAD-2. Synthesis and applications as a preconcentrator in the determination of zinc(II) and lead(II) by using atomic absorption spectrometry, *Analyst*, Vol. 120, pp. 403-405.

Scheffler, A. (1996) Lewatit-MonoPlus. The latest generation of monodisperse ion exchange resin with outstanding properties for optimising water treatment system, Technical Bulletin, Bayer AG.

Sharma, R.K., Pant, P. (2009) Solid phase extraction and determination of metal ions in aqueous samples using Quercetin modified Amberlite XAD-16 chelating polymer as metal extractant, *International Journal of Environmental Analytical Chemistry*, Vol. 89, pp. 503-514.

Sherrington, D.C. (1998) Preparation, structure and morphology of polymer supports, *Chemical Communications*, pp. 2275-2286.

Smolik, M., Jakóbik-Kolon, A., Porański, M. (2009) Separation of zirconium and hafnium using Diphonix® chelating ion-exchange resin, *Hydrometallurg*, Vol. 95, pp. 350-353.

Stamberg, J., Valter, V. (1970) Entfärbungsharze - Anwendungsprinzipien, charakterisierung und verwendung (Decolourizing resins - principles of application, characterization and use). Akademie-Verlag, Berlin, Akadbmie-Verlag, Berlin, p. 63.

Tandy, S., Bossart, K., Mueller, R., Ritschel, J., Hauser, L., Schulin, R., Nowack, B. (2004) Extraction of heavy metals from soils using biodegradable chelating agents, Environmental Science and Technology, Vol. 38, pp. 937-944.

Tewari, P.K., Singh, A.K. (1999) Amberlite XAD-2 functionalized with chromotropic acid: Synthesis of a new polymer matrix and its applications in metal ion enrichment for their determination by flame atomic absorption spectrometry, *Analyst*, Vol. 124, pp. 1847-1851.

Tewari, P.K., Singh, A.K. (2000) Amberlite XAD-7 impregnated with Xylenol Orange: a chelating collector for preconcentration of Cd(II), Co(II), Cu(II), Ni(II), Zn(II) and Fe(III) ions prior to their determination by flame AAS, *Fresenius Journal of Analytical Chemistry*, Vol. 367, pp. 562-567.

Tewari, P.K., Singh, A.K. (2001) Synthesis, characterization and applications of pyrocatechol modified Amberlite XAD-2 resin for preconcentration and determination of metal ions in water samples by flame atomic absorption spectrometry (FAAS), *Talanta*, Vol. 53, pp. 823-833.

Tewari, P.K., Singh, A.K. (2002) Preconcentration of lead with Amberlite XAD-2 and Amberlite XAD-7 based chelating resins for its determination by flame atomic absorption spectrometry, *Talanta*, Vol. 56, pp. 735-744.

Tokalıoğlu, S., Ergün H., Çukurovalı, A. (2010) Preconcentration and determination of Fe(III) from water and food samples by newly synthesized chelating reagent impregnated Amberlite XAD-16 resin, *Bulletin of Korean Chemical Society*, Vol. 31, pp. 1976-1980.

Trochimczuk, A.W., Streat, M. (1999) Novel chelating resins with aminothiophosphonate ligands, *Reactive & Functional Polymers*, Vol. 40, pp. 205-213.

Trochimczuk, A.W. (2000) Synthesis of functionalized phenylphosphinic acid resins through Michael reaction and their ion-exchange properties, *Reactive and Functional Polymers*, Vol. 44, pp.9-19.

Tuzen, M., Saygi, K.O., Soylak M. (2008) Novel solid phase extraction procedure for gold(III) on Dowex M 4195 prior to its flame atomic absorption spectrometric determination, *Journal of Hazardous Materials*, Vol. 156, pp. 591-595.

Uzun, A., Soylak, M., Elçi, L. (2001) Preconcentration and separation with Amberlite XAD-4 resin; determination of Cu, Fe, Pb, Ni, Cd and Bi at trace levels in waste water samples by flame atomic absorption spectrometry, *Talanta*, Vol. 54, pp. 197-202

Warshawsky, A. (1987) Chelating ion exchangers, in: Ion exchange and sorption processes in hydrometallurgy, M. Streat, D.A. Naden (Eds.), Wiley VCH, Chichester, pp. 166-225

Vasilev, V.P., Katrovtseva, A.V. Gorelov, I.P. Tukumova, N.V. (1996) Stability of complexes of Ni(II) with iminodisuccinic acid, *Zhurnal Neorganic Khimii*, Vol. 41, pp. 1320-1323.

Vasilev, V.P., Katrovtseva, A.V., Bychkova, S. A., Tukumova, N. V. (1998) Stability of complexes of Co(II) and Cu(II) with iminodisuccinic acid, *Zhurnal Neorganic Khimii*, Vol. 43, pp. 808-809.

Yebra-Biurrun, M.C., García-Dopazo, Bermejo-Barrera, A., Bermejo-Barrera, M.P. (1992) Preconcentration of trace amounts of manganese from natural waters by means of a macroreticulap poly(dithiocarbamate) resin, *Talanta*, Vol. 39, pp. 671-674.

Yua, Z., Qia, T., Qua, J., Wanga, L., Chu, J. (2009) Removal of Ca(II) and Mg(II) from potassium chromate solution on Amberlite IRC 748 synthetic resin by ion exchange, *Journal of Hazardous Materials*, Vol. 167, pp. 406-417.

Yuchi, A., Sato, T., Morimoto, Y., Mizuno, H., Wada, H. (1997) Adsorption mechanism of trivalent metal ions on chelating resins containing iminodiacetic acid groups with reference to selectivity, *Analytical Chemistry,* Vol. 69, pp. 2941-2944.

Zagorodni, A.A., Muhammed, M. (1999) Explanation of the Zn/Cu dual temperature separation on Amberlite IRC-718 ion exchange resin, *Separation Science and Technology,* Vol. 34, pp. 2013-2021.

Zainol, Z., Nicol, M. (2009b) Ion-exchange equilibria of Ni^{2+}, Co^{2+}, Mn^{2+} and Mg^{2+} with iminodiacetic acid chelating resin Amberlite IRC 748, *Hydrometallurgy,* Vol. 99, pp.175-180.

Zainol, Z., Nicol, M. (2009a) Comparative study of chelating ion exchange resins for the recovery of nickel and cobalt from laterite leach tailings, *Hydrometallurgy,* Vol. 96, pp. 283-287.

Zhao, D., SenGupta, A.K., Stewart, L. (1998) Selective removal of Cr(VI) oxyanions with a new anion exchanger, *Industrial and Engineering Chemistry Research,* Vol. 37, pp. 4383-4387.

Chapter 9

Ion Exchange and Application of Layered Silicate

Kyeong-Won Park

Additional information is available at the end of the chapter

http://dx.doi.org/10.5772/51564

1. Introduction

Layered Na^+-titanosilicate ($Na_4Ti_2Si_8O_{22}·4H_2O$) [1-5] has interesting potential applications. It contains titanosilicate layers composed of tetrahedral SiO_4 units and square pyramidal TiO_5 polyhedral [3]. The interlayer surfaces are composed of five-coordinated titanium (IV), and thus, are less coordinated than the octahedrally-coordinated microporous and layered titanosilicates [5]. The existence of five-coordinated titanium (IV) in interlayer surface makes Na^+-titanosilicate a promising material for oxidation catalysis. Each layer is separated by water molecules solvated around interlayer Na^+ ions [4]. Furthermore, this layer structure allows the intercalation of large organic and inorganic molecules.

Roberts et al. [3] synthesized Na^+-titanosilicate using Ti-alkoxide, and Ferdov and co-workers [5] described an effective means of producing Na^+-titanosilicate using $TiCl_4$ as a titanium source without an organic template. Kostov-kytin and co-workers [6] also investigated on the phase transition of Na^+-titanosilicate during heating at 300-700 °C. However, studies of Na^+-titanosilicate are still in the early stage. For industrial applications, in which Na^+-titanosilicate is used as a catalytic support, studies on surface chemistry, surface area, and porosity are needed, and on the relations between its surface properties and basic intercalation chemistry.

The galleries are normally occupied by exchangeable cations such as Na^+, Ca^{2+}, and Mg^{2+}. They happen easily an ion exchange reaction with quaternary ammonium ions or other organic cation in water [7]. The acid treatment of Na^+-titanosilicate can also produce Si-OH groups in interlayer surface by an ion exchange of exchangeable cations for H^+. Si-OH groups can offer an excellent bonding site for an organic base or alkoxysilane compound.

Pillaring of metal oxides in layered silicate, such as, natural and synthetic layered silicate is being increasingly studied [8–13]. In general, pillaring is achieved by the direct introduction of bulk inorganic (polyoxocations) or organic precursors (metal alkoxides) between the interlayers of layered silicate. Pillaring processes that use metal alkoxide are facilitated by a

INTECH
open science | open minds

© 2012 Park, licensee InTech. This is a paper distributed under the terms of the Creative Commons Attribution License (http://creativecommons.org/licenses/by/3.0), which permits unrestricted use, distribution, and reproduction in any medium, provided the original work is properly cited.

preswelling step where by interlayer regions are exposed to quaternary ammonium [8,9,13]. However, preswelling procedures are problematic because they are complex, non-quantitative and require reagents. Recently, we reported a method of introducing metal alkoxides or organic precursors into H^+-layered silicates without a preswelling step [14–18].

Layered materials have been often used to design and construct organic–inorganic nanomaterials because of the ease and variety of modifications possible by the introduction of organic and inorganic compounds into the interlayer space [19–27]. In recent years, the chemical modification into the interlayer surface of layered materials has become the focus of increased research. Ruiz-Hitzky and Rojo [28,29] grafted trimethylsilyl groups to the interlayer silanol groups of H^+-magadiite starting with intercalation compounds from polar organic molecules. Shimojima et al.[30], Okutomo et al. [31], Ogawa et al. [32] and Yanagisawa et al. [33] reported on the trimethylsilylation, diphenylmethylsilylation and octyldimethylsilylation of magadiite, kenyaite and kanemite using the quaternary ammonium-exchanged form of silicates as intermediates. Thiesen et al. [34] also reported on the silylation of H^+-kenyaite using alkylamines together with a silylating agent.

We now report on the silylation of organic functional groups in the interlayer surface of a layered material. Functional organo-layered silicates with a functional group silylate in the interlayer surface can offer new opportunities for designing nanocomposites with the desired function because of their highly accessible interactions with various chemicals. In polymer–clay nanocomposite, physical properties depend on the interaction between exfoliated clay surfaces and the polymer. Interlayer surfaces bonded chemically by functional groups can interact with active groups in the polymer [35]. In particular, attached amine groups in the surface can chemically bond with epoxy [36], nylon [37], and urethane polymers [38], creating a bridge between the exfoliated layered surface and the polymer [35]. Their expanded gallery of functional groups can also store expensive reagents, such as drugs or enzymes. Layered materials with attached amine groups may be used to adsorb heavy metal ions for photosensitive species.

Here, we found that the H^+-titanosilicate (formed by proton exchange of Na^+-titanosilicate) produced H^+-titanosilicate/DDA (dodecylamine)/TEOS (tetraethylorthosilicate) intercalation compounds in DDA–TEOS solution. In these intercalation compounds, the long chain amine DDA appears to act as a gallery height expander and as a base catalyst during TEOS hydrolysis. Furthermore, it also appears to act as a liquid crystal template that forms surfactant-like molecular assemblies in galleries. The physical properties of the SPT derivatives produced were investigated by XRD (powder X-ray diffraction), BET (Brunauer, Emmett and Teller)-surface area, and SEM (scanning electron micrographs), and these investigations confirmed that SPT derivatives are mesoporous materials with large surface areas, highly ordered gallery structures, and high thermal resistances.

Intercalation and silylation were easily achieved by entropy differences between the interlayer gallery and the outside, which were caused by vaporizing the relatively more volatile ethanol. The ethanol on the outside vaporizes more rapidly than that in the interlayer gallery. DDA can also have a role as gallery expander and silylation catalyst.

Ethanol can quantitatively control the amount of OTES (octyltriethoxysilane) and DDA needed for gallery silylation, and the residual water from vaporization catalyzes the silylation reaction. This process was achieved by a quantitative procedure under atmospheric conditions without consumption of expensive reagents or an effluence of waste liquid. Our method is a promising route in leading to the preparation of new functional nanomaterials to bond with a variety of functional groups in the interlayer surface of layered materials.

2. Experimental section

2.1. Synthesis of Na^+-titanosilicate and H^+-titanosilicate

Na^+-tanosilicate was prepared by the method described by Ferdov et al.[5]. SiO_2 (14.8 g; particle size 200 μm, Merck) and 11.0 g NaOH (Junsei, Japan) were added to 200 ml of distilled water and completely dissolved by heating to boiling. Subsequently, 3.3 ml of $TiCl_4$ (Yakuri Pure Chemicals Co., Japan) hydrolyzed in 100 ml distilled water was added to the above solution. The gel obtained was then transferred into 1000 ml Teflon-lined autoclaves. The crystallization was performed under static conditions at 180 °C for 50 h. The reaction product was filtered and washed with deionized water and dried at 40 °C. H^+-titanosilicate was obtained by the ion exchange of Na^+ in Na^+-titanosilicate by H^+ using 0.1 N HCl solution. A suspension of Na^+-titanosilicate (10 g) in deionized water (200 ml) was titrated slowly with 0.1 N HCl solution to a final pH 2.0 and then maintained at this value for 24 h. H^+-titanosilicate was recovered by filtering, washing with deionized water (until Cl^- free), and drying in air at 40 °C.

2.2. Silica-pillared H^+-titanosilicate (SPT)

Silica-pillared H^+-titanosilicate derivatives were prepared using methods similar to those reported by Kwon et al. [14,15], which introduce TEOS and DDA into the interlayer regions of H^+-layered silicates without a separate preswelling step. Mixtures of H^+-titanosilicate, DDA (Aldrich), and TEOS (Aldrich) at molar ratios in the range 1:10:14–20 were allowed to react for 1 h at room temperature. Here, DDA molecules intercalated into H^+-titanosilicate interlayers by forming hydrogen bonds with interlayer surface Ti–OH groups. TEOS also intercalated with DDA by solvation. Mixtures at this stage were composed of DDA/TEOS co-intercalated H^+-titanosilicate gels. Unreacted DDA and TEOS were removed by vacuum filtration, which resulted in the isolating of DDA/TEOS co-intercalated H^+-titanosilicate gels. The hydrolysis of TEOS in interlayer spaces was conducted in pure water. The reaction was conducted by dispersing DDA/TEOS intercalated H^+-titanosilicate gels in deionized water at room temperature, when the viscous gray gels became white solids. Bubbles and heat were also produced at 5 min into this reaction. After soaking for 30 min, the solid products obtained were filtered, washed three times with ethanol, and oven dried at 90 °C. Resultant powders were siloxane-pillared H^+-titanosilicates. These powders were then heated for 5 h at 500 °C in air to remove template DDA, and organic by-products resulting from the hydrolysis of TEOS, to produce silica-pillared H^+-titanosilicate (SPT) derivatives.

The SPT derivatives so obtained were also then heated for 5 h at 600 or 700 °C in air to examine their thermal resistances.

2.3. Silylation of H^+-titanosilicate

DDA–OTES mixed solutions were prepared by dissolving OTES and DDA in ethanol (95%). The concentration of OTES in solution was between 0.01 and 0.1 M, and the DDA concentration was held constant at 0.1 M. H^+-titanosilicate(DDA–OTES–TS) with intercalated OTES and DDA preparation H^+-titanosilicate (0.321 g, 0.5 mmol) was then dispersed in 6 mL (0.01, 0.05, 0.1 M OTES 3 mL + 0.1 M DDA 3 mL) of these solutions (ultrasound, 20 min, at room temperature) and evaporated and dried for 24 h at 50 °C. OTES–TS was recovered by filtering, washed with ethanol until free of DDA using DDA–OTES–TS, and then dried in the air at 50 °C.

3. Characterization

X-ray diffraction data were recorded using a Bruker diffractometer using Cu K*a* radiation. The chemical compositions of Na^+-titanosilicate and H^+-titanosilicate were analyzed by energy dispersive spectroscopy (EDS; Link system AS1000-85S) and by thermo-gravimetric analysis (TGA; Dupont 9900 thermoanalyser, 10 °C/min to 900 °C, 100 ml/min N_2 purge). Scanning electron micrograph (SEM) measurements were carried out using a JEOL JSM-840A SEM. Samples, were stuck onto adhesive tape, sputter coated with gold and morphological variations were examined. Transmission electron micrographs (TEM) were obtained with a JEOL JEM-200 CX transmission electron microscope operating at 200 kV, using a thin-section technique. Powdered samples were embedded in Epoxy resin and then sectioned with a diamond knife. Microtome sectioned samples were examined. Solid-state ^{29}Si MAS NMR experiments were performed on a Bruker DSX-400 spectrometer (KBSI Daegu center, Korea) operating at 79.5 MHz. ^{29}Si MAS NMR spectra were obtained using 30-40° pulse width at a spinning rate of 6 kHz. Cross-polarization experiments were carried out with delay times of 3 s, 90° pulse of 5 s, and contact times of 2000 μs. The chemical compositions of Na^+-titanosilicate and H^+-titanosilicate were analyzed by energy-dispersive X-ray spectrometer (EDS, Link system AS1000–85S) and thermogravimetric analysis (TGA, 10 °C/min to 900 °C, 100 cm^3/min N_2 purge). The composition of the product was determined by CHN analysis (CHNS-932 (Leco)). The FT-IR spectrum was recorded on a Digilab FTS-40 FT-IR unit using the KBr disk method. Raman spectra were obtained using a Jobin Yvon/Horiba LabRAM spectrometer equipped with an integral microscope (Olympus BX 41). The 632.8 nm line of an air-cooled He/Ne laser was used as an excitation source.

4. Results and discussion

4.1. Synthesis of Na^+-titanosilicate and H^+-titanosilicate

The X-ray powder diffraction pattern of air-dried Na^+-titanosilicate, as shown in Fig. 1(a), exhibited several 001 reflections corresponding to a basal spacing of 1.08 nm. The peak positions of this synthetic sample agreed closely with previously reported values [3,5]. Slow titration of Na^+-titanosilicate with 0.1 N HCl produced H^+-titanosilicate by exchanging Na^+ for H^+ in interlayers. The X-ray powder diffraction pattern of air-dried H^+-titanosilicate (Fig. 1(b)), exhibited 001 reflections corresponding to a basal spacing of 0.92 nm. This decrease in basal spacing relative to the Na^+ form indicates a loss of interlayer H_2O due to H^+ for Na^+ exchange. Also, the general broadening of diffraction peaks indicated that stacking disorder had occurred during this proton exchange. SEM morphologies of Na^+-titanosilicate and H^+-titanosilicate are shown in Fig. 2. Na^+-titanosilicate particles were composed of plates, and H^+-titanosilicate exhibited a similar morphology. The chemical compositions of Na^+-titanosilicate and H^+-titanosilicate were determined by TGA and EDS analysis. The TGA curve (Fig. 3(a)) indicated that air-dried Na^+-titanosilicate lost 8.5 wt% of its total initial weight as water below 200 °C. An additional 1.5 wt% was lost between 200 and 1000 °C, which was assigned to the dehydration loss of surface bound -OH groups. The chemical compositions of Na^+-titanosilicate and H^+-titanosilicate are shown in Table 1. By combining Na_2O, SiO_2, TiO_2 and weight loss, we obtained an empirical composition for synthetic Na^+-titanosilicate of $Na_4Ti_2Si_8O_{22}{\cdot}4H_2O$, which compares well with those reported previously [3-6]. H^+-titanosilicate, as shown in Fig. 3(b), showed an initial weight loss of 7.5% below 250 °C due to H_2O desorption, and a 6 wt% weight loss above 300 °C, which was attributed to the elimination of -OH groups from the structure. Based on water loss and the absence of sodium, we obtained an empirical unit cell composition of $H_4Ti_2Si_8O_{22}{\cdot}4H_2O$ or H^+-titanosilicate.

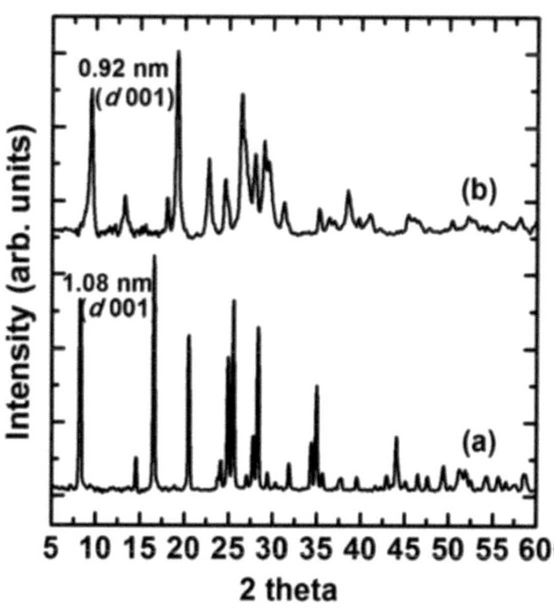

Figure 1. X-ray diffraction patterns of (a) Na^+-titanosilicate and (b) H+-titanosilicate.

Figure 2. SEM micrographs of (a) Na^{+}-titanosilicate and (b) H^{+}-titanosilicate.

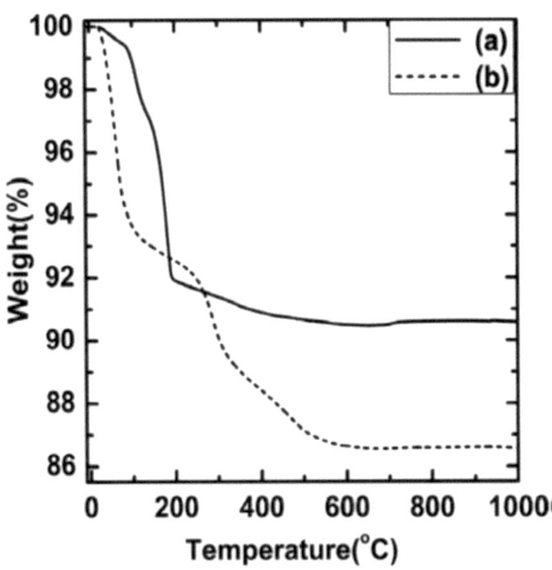

Figure 3. TGA curves of (a) Na^{+}- titanosilicate and (b) H^{+}-titanosilicate.

	Weight percent				Total	Atomic ratio			
Samples	Na_2O^a	SiO_2^a	TiO_2	H_2O		Na^a	Si^a	Ti^a	H_2O
Na^+-titanosilicate	14.80	58.50	18.20	8.50	100	3.92	8.00	1.90	3.89
HT	-	69.05	23.15	7.80	100	-	8.00	2.01	3.02

[a] EDS data. HT = H^+-titanosilicate.

Table 1. Composition of Synthetic Na^+-titanosilicate and H^+-titanosilicate

4.2. Silica-pillaring

Mixtures of H^+-titanosilicate, DDA and TEOS produced intercalation compounds in which DDA and TEOS were simultaneously intercalated into the interlayer regions. Intercalation compounds of the gel type were recovered by vacuum filtration, which allowed unreacted DDA and TEOS to be removed. The recovered gels were DDA/TEOS co-intercalated H^+-titanosilicates. In these compounds, DDA molecules between layers are solvated by TEOS molecules, which results in an additional interlayer expansion. Addition of water to the gels caused the rapid hydrolysis of TEOS in interlayer spaces. This hydrolysis occurred rapidly (within 5 min) in pure water because of the catalytic effect of DDA, and resulted in the formation of siloxane-pillared H^+-titanosilicates. The calcination of siloxane-pillared H^+-titanosilicate derivatives for 5 h at 500 °C resulted in the formation of silica-pillared H^+-titanosilicate (SPT) derivatives. Fig. 4 shows X-ray diffraction patterns of SPT derivatives prepared at H^+-titanosilicate:DDA:TEOS reaction stoichiometries of 1:10:14 (SPT-1), 1:10:16 (SPT-2), 1:10:18 (SPT-3) and 1:10:20 (SPT-4). SPT derivatives exhibited refractions corresponding to a basal spacing of 4.16–4.32 nm, which depended on the molar ratio of DDA to TEOS (with the exception of SPT-1). These results indicate that the molar ratio of DDA to TEOS is an important factor for successful pillaring. If the DDA/TEOS ratio is too high, the interlayer space is maintained by template DDA, and gallery TEOS cannot form a pillar firm enough between layers. Accordingly, in this situation the removal of DDA by calcination can cause the disordered collapse of interlayer spaces, which results in severe peak broadening or peak disappearance, as observed in SPT-1. Since the H^+-titanosilicate layer sheet thickness is 0.92 nm, the corresponding gallery heights are 3.24–3.40 nm, which are nearly double the chain length of DDA (1.65 nm). This implies that DDA molecules in the gallery are arranged as molecular assemblies as lamellar bilayers and that TEOS produce firm enough silica-pillars to prop the expanded gallery after the removal of DDA. Moreover, gallery height was found to increase slightly as the molar ratio of TEOS was increased. This implies that pillar size and strength depends on the quantity of gallery TEOS. The basal spacings and gallery heights of SPT derivatives are summarized in Table 2.

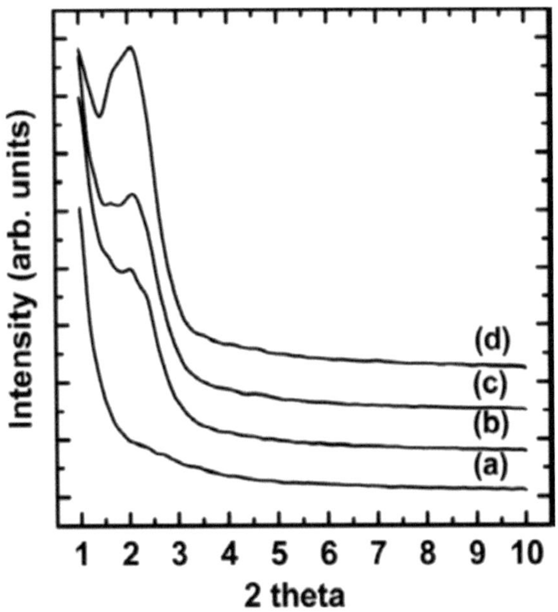

Figure 4. X-ray diffraction patterns for calcined silica-pillared H^+-titanosilicate prepared at H^+-titanosilicate:DDA:TEOS reaction stoichiometries of: (a) 1:10:14 (SPT-1), (b) 1:10:16 (SPT-2), (c) 1:10:18 (SPT-3) and (d) 1:10:20 (SPT-4).

Samples	d-spacing(nm)	S_{BET}	S_{mic}	S	H-K pore size (nm)
SPT-1	-	618	511	106	1.4
SPT-2	4.16 (3.24)	535	406	129	2.8
SPT-3	4.25 (3.33)	538	415	123	3.0
SPT-4	4.32 (3.40)	570	420	150	3.4
HT	0.92	5	-	-	

SPT = silica-pillared H^+-titanosilicate; ( ) = gallery height; gallery height = basal spacing – 0.92 nm (thickness of H^+-titanosilicate); HT = H^+-titanosilicate.

Table 2. Basal spacing, gallery heights and surface area analyses (m^2/g) for SPT products.

Fig. 5 also exhibits XRD peaks for SPT derivatives heated for 5 h at 600 or 700 °C in air. XRD peaks of SPT samples were well preserved after heat treating to 700 oC. Basal spacings and gallery heights are shown in Table 3. Basal spacing was also not altered by exposure to high temperatures. This indicates that control of DDA/TEOS ratio and the rapid hydrolysis of TEOS in water contribute to the formation of firm silica-pillars. Here, TEOS is not lost from interlayer spaces during hydrolysis, because of its waterinsolubility and rapid hydrolysis.

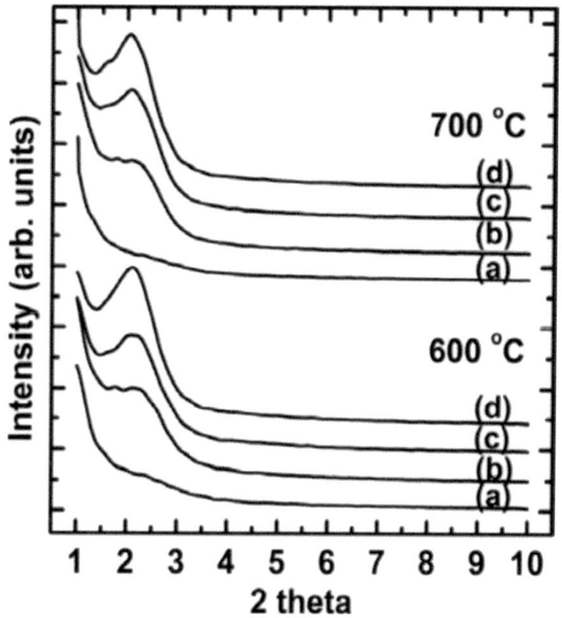

Figure 5. X-ray diffraction patterns of SPT derivatives heated for 5 h at 600 or 700 °C in air.

Typical SEM images of SPT are exhibited in Fig. 6. Most of the platelets in samples were unaffected by hydrolysis, though they swelled slightly more than H^+-titanosilicate. The small particles observed around platelets were probably broken platelets and amorphous SiO_2 caused by the hydrolysis of surface TEOS. The amount of platelet destruction was found to be related to the hydrolysis conditions used, e.g., reaction time pH of solution, and polarity of the solvent. However, this platelet destruction did not reflect SPT gallery structure destruction, probably because the gallery structure within broken platelet particles is preserved.

Figure 6. SEM micrographs of SPT derivatives: (a) SPT-1, (b) SPT-2, (c) SPT-3, and (d) SPT-4.

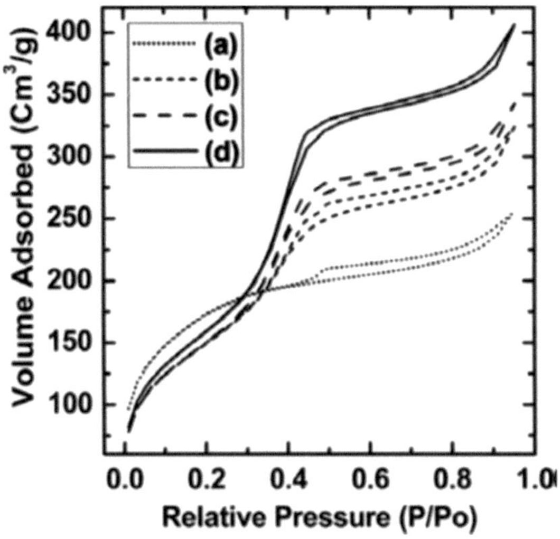

Figure 7. Nitrogen adsorption isotherms of SPT derivatives.

Fig. 7 illustrates typical N_2 adsorption/desorption isotherms for SPT. All isotherms, except for SPT-1, showed a well-defined step indicative of framework-confined mesoporosity similar to MCM-41 related materials [39]. The nearly linear portions of the adsorption curves in the partial pressure region 0.03–0.35 are also indicative of small mesopores (~0.2–3.0 nm diameter) [40]. BET surface areas and microporous and non-microporous surface areas for SPT products are summarized in Table 3. Surface areas were obtained by fitting adsorption data below $P/P_0 = 0.1$ to the BET equation [40]. H^+-titanosilicate had a total surface area of 5.0 m^2/g due to adsorption at non-porous external surfaces, but SPT products had dramatically larger total surface areas of between 535 and 618 m^2/g, which depended on the DDA/TEOS molar ratio used. Furthermore, most this surface area was attributable to the presence of micropores of <2 nm in diameter.

Samples	Basal spacing (nm)		S_{BET}	S_{mic}	S	H-K pore size (nm)
	600 °C	700 °C				
SPT-1	-	-				
SPT-2	4.14 (3.22)	4.07 (3.15)				
SPT-3	4.23 (3.31)	4.14 (3.22)				
SPT-4	4.28 (3.36)	4.22 (3.30)	551[a] 505[b]	429[a] 394[b]	122[a] 111[b]	3.2[a] 3.0[b]

S_{BET} is the N_2 BET surface area; S_{mic} and S are the microporous and nonmicroporous surface areas, respectively, obtained from *t*-plots of the nitrogen adsorption data. H-K pore size = Horvath and Kawazoe pore size (obtained from nitrogen adsorption data). [a,b]data heated at 600 °C and 700 °C, respectively.

Table 3. Basal spacing, gallery height and surface area analyses (m^2/g) for SPT derivatives heated for 5 h at 600 °C and 700 °C.

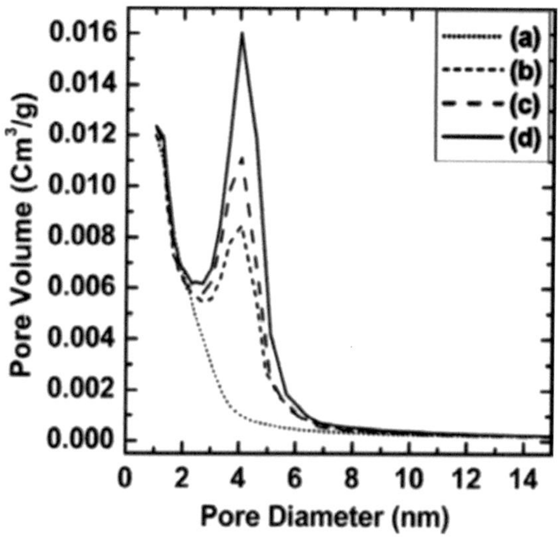

Figure 8. Pore size distributions of SPT derivatives.

Horvath and Kawazoe analysis [41] of the N_2 adsorption data (Fig. 8), yielded pore sizes of 2.8–3.4 nm, which is similar to the gallery height (3.24–3.40 nm). This indicates that DDA plays an important structure-directing role during silica-pillaring. Furthermore, pore size distributions depended to some extent on DDA/TEOS molar ratio, which reflects differences in DDA molecular assemblies in the interlayer space caused by changing DDA/TEOS molar ratios.

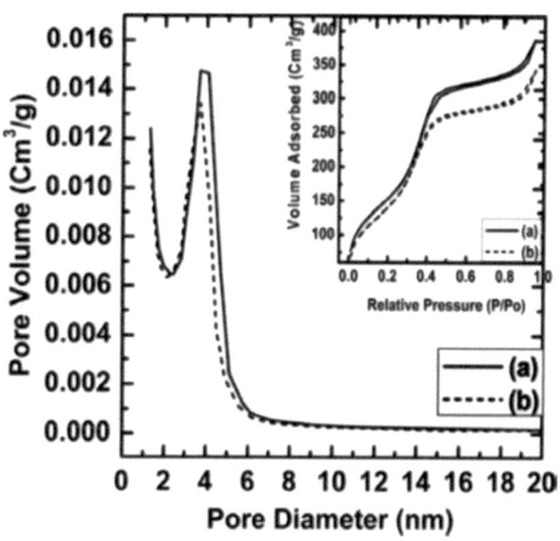

Figure 9. Pore size distributions and nitrogen adsorption isotherms of SPT-4 heated at (a) 600 °C or (b) 700 °C.

Fig. 9 shows typical isotherm and pore size distributions for SPT-4 heat treated at 600 or 700 °C. Surface areas and pore sizes are also shown in Table 3. Heat treatment hardly affected the pore size distribution and surface area of SPT-4, demonstrating the thermally stability of SPT derivatives.

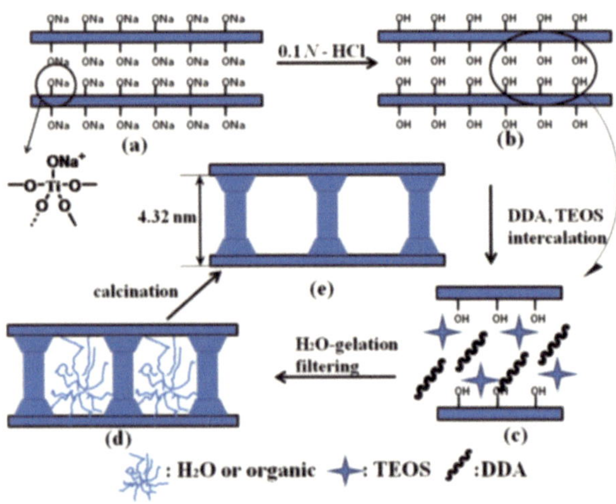

Figure 10. Schematic for the preparation of mesoporous silica-pillared H^+-titanosilicate.

Fig. 10 presents a schematic representation of mesoporous silica-pillared layered titanosilicate.

The XRD patterns of DDA–OTES–TS treated with 0.01, 0.05, 0.1 M OTES-0.1 M DDA-ethanol solution are shown in Fig. 11. Treatment of H^+-titanosilicate with the OTES–DDA mixed solution ensured successful silylation by OTES. Dried OTES–DDA–TS powder consists of H^+-titanosilicate with intercalated OTES and DDA. In the gallery, DDA changes into dodecylammonium cations [$CH_3(CH_2)_{11}NH_3^+$] as a result of the interaction with acidic silanol groups. Here, condensation between adjacent layers can not occur because of the large expansion of the interlayer space, with basal spacing between ~3.87 nm (Fig. 11(b-d)) by DDA. However, increasing concentrations of OTES broaden the reflections and decrease the basal spacing because of the high degree of grafting that reduces the amount of intercalated DDA. The XRD patterns of DDA–OTES–TS after washing with ethanol to remove the DDA showed that an increasing concentration of OTES broadened the reflections and decreased the basal spacing, because a high degree of grafting reduces the amount of intercalated OTES (Fig. 11(e-g)). The interlayer space had a basal spacing of ~2.95 nm (Fig. 11(e-g)). This small increase in basal spacing indicates that OTES molecules arrange with paraffin-type in the gallery.

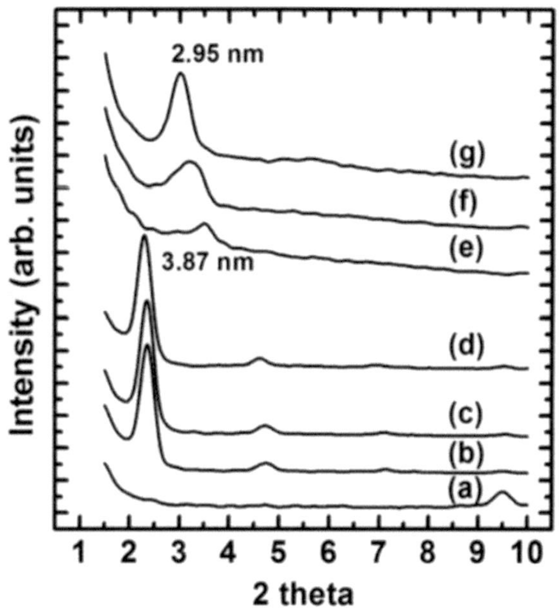

OTES–TS, (c) 0.1 M DDA-0.05 M OTES–TS, (d) 0.1 M DDA-0.1 M OTES–TS, (e) 0.01 M OTES–TS, (f) 0.05 M OTES–TS and (g) 0.1 M OTES–TS.

Figure 11. XRD patterns of (a) as-prepared H^+-titanosilicate, (b) 0.1 M DDA-0.01 M

The FT-IR spectrum of H^+-titanosilicate and OTES–TS is shown in Fig. 12. The band at 1469 cm^{-1} has been assigned to the asymmetric $-CH_3$ deformation of $-Si-CH_3$ [42,43], while the 2852, 2922 and 2960 cm^{-1} bands corresponded to $-CH_2CH_3$. The –OH stretching band is seen at 3208 cm^{-1} in –OH form and the H_2O bending vibration is seen at 1619 and 1645 cm cm^{-1}. In the OTES–TS spectrum, the absorption band from the Si–O–Si type of motion overlaps with strong Si–O–Ti absorption bands at 586 and 866 cm^{-1}. However, additional bands were observed at 631, 699, 776, 909, 1054, 1149 cm^{-1} in the OTES–TS spectra. These bands were also observed in the spectrum of titanosilicate [44–47] and probably arose from Ti–O–Si linkages formed by silylation. The bonding of the alkylsilyl groups with the silicate layers was revealed by the Raman spectra (Fig. 13). Titanosilicate is composed of $-(Si-O)_4-Ti-O-$ units with their Si-O-Si environments observed at around 600 cm^{-1}, Ti–O–Si: 960 cm^{-1} and –OH: 970, 1100, 3600 cm^{-1} [48–51]. The reaction products exhibit Raman bands at 223, 328, 385, 429, 477, 600, 920, 1029, 1078 and 1301 cm^{-1}, respectively. The spectra also showed a substantial increase in the relative intensity of Ti–O–Si signals due to the increase in formed Ti–O–Si units by the silylation of interlayer Ti–OH groups.

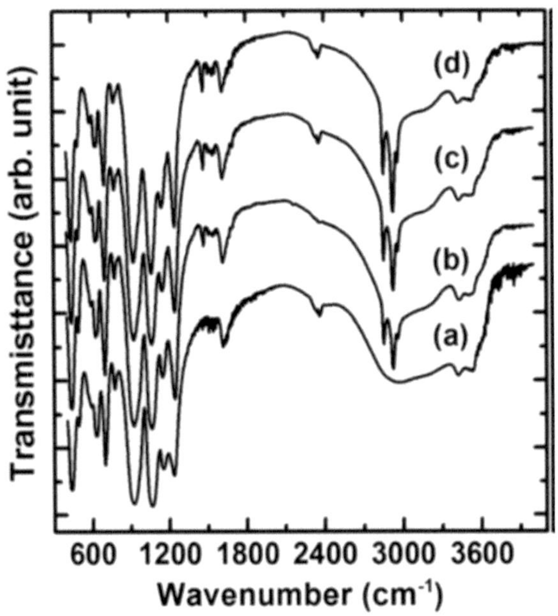

Figure 12. FT-IR spectra of (a) as-prepared H^+-titanosilicate, (b) 0.01 M OTES–TS, (c) 0.05 M OTES–TS and (d) 0.1 M OTES–TS.

TGA data from DDA–OTES–TS and insert of OTES–TS treated in N_2 at a heating rate of 10 °C/min are shown in Fig. 14. The weight loss of DDA–OTES–TS in the temperature range from 100 to 250 °C was ~23.0 wt.% and was probably caused by water and DDA contained between the interlayers. A weight loss of ~25 wt.% was recorded between 250 and 650 °C, because the TS surface organic chains group was thermally cracked in this temperature range (Fig. 14). These values represent the loadings of organic functions on OTES–TS and are in agreement with the C, H, N contents of 9.4–20.6 wt.% obtained by chemical analysis. The composition of the products determined by the C, H, N analysis is summarized in Table 1. The C, H, N content in OTES–TS increased with increasing OTES concentration. The C content in DDA–OTES–TS decreased with increasing OTES concentration because DDA adsorption decreases at higher OTES content. The almost constant N content indicates that the sum of OTES and DDA remain nearly independent from the degree of silylation.

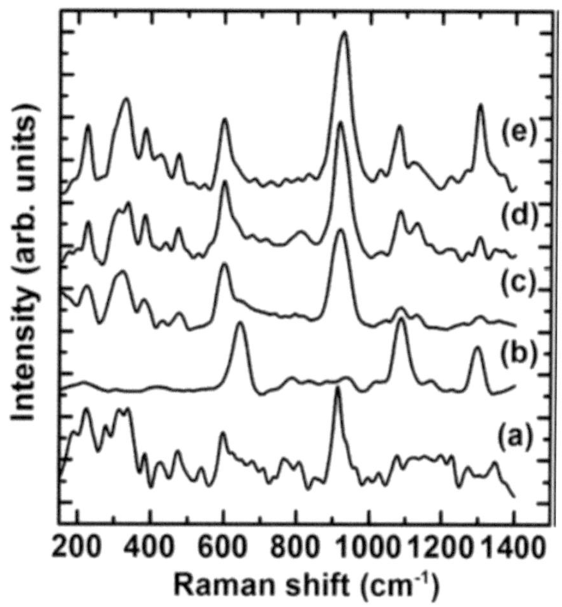

Figure 13. Raman spectra of (a) as-prepared H^+-titanosilicate, (b) OTES, (c) 0.01 M OTES–TS, (d) 0.05 M OTES–TS and (e) 0.1 M OTES–TS.

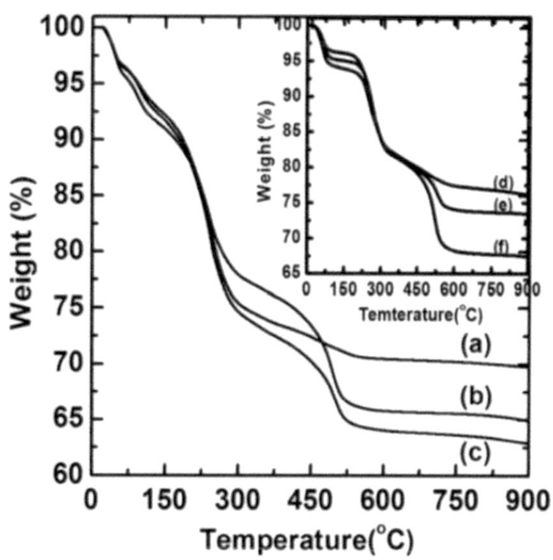

Figure 14. TGA of (a) 0.1 M DDA-0.01 M OTES–TS, (b) 0.1 M DDA-0.05 M OTES–TS, (c) 0.1 M DDA-0.1 M OTES–TS, (d) 0.01 M OTES–TS, (e) 0.05 M OTES–TS, (f) 0.1 M OTES–TS.

Silylated-TS	Composition Weight Percent				
	C [1]	H [1]	N [1]	CHN [2]	SiO_2, TiO_2 (%)
0.10 M DDA-0.01 M OTES	16.80	3.55	1.75	20.1	77.9
0.10 M DDA-0.05 M OTES	22.08	4.27	1.92	25.2	71.73
0.10 M DDA-0.10 M OTES	23.05	4.80	1.67	29.0	70.48
0.01 M OTES [3]	9.38	2.65	0.81	1.4	87.16
0.05 M OTES [3]	14.53	3.45	0.99	7.1	81.03
0.10 M OTES [3]	20.60	4.00	1.08	14.2	74.32

[1] Evaluated by CHN analysis
[2] Evaluated by CHN calculation (Contains CHN of 0.1M DDA, 0.01, 0.05, 0,1M-OTES)
[3] Washed with EtOH

Table 4. Chemical composition of silylated titanosilicate.

The solid-state ^{29}Si MAS NMR spectra for silylated H^+-titanosilicate show Q^3 (-105.7 ppm), Q^4 (-110 to -112 ppm), and T^2 (-50 to -60 ppm) signals with diverse environments of silicon (Fig. 15). The increase of the Q^3/Q^4 ratio compared with H^+-titanosilicate clearly indicates the grafting of OTES molecules to the surface silanol groups. The new peaks ($T^?$) near -59.5 ppm, in 0.1 M OTES–TS were attributed to the Si atoms of the grafted OTES molecules. In general, Si atoms of the silicate network exhibited signals (Q^3 and Q^4) in the range of -102 and -112 ppm, whereas Si atoms in the silylating agent grafted to the surface showed signals (T^2) at -50 to -60 ppm. Caravajal et al. [52], Impens et al. [53], Lin et al. [43] and D'Amore and Schwarz [54] all reported the mechanism and solid NMR data for the silylation of amorphous silica, alumina and Ti-MCM-41 with alcoxysilane. The Si signals for OTES silylated on silica was in the range of -49 to -68 ppm, depending upon the bonding type. Caravajal et al. [55] and Shimojima et al.[30] also showed that Si signals for alkyltrichlorosilane grafted on kanemite appeared between -56 and -65 ppm. The signals from 0.5 M DDA-0.1 M OTES–TS were similar to signals of kanemite silylated by alkyltrichlorosilane. The TEM image (Fig. 16) explains the reason why OTES–TS samples, in spite of partial destruction of platelets in the outer appearance in the SEM, exhibit uniform distance distributions and ordered basal spacing. The crosssection of an OTES–TS platelet shows several silicate sheets with uniform spacing of ~2.95 nm. A hypothetical diagram for the silylation by OTES and the intercalation of DDA during the evaporation of the solvent is shown in Fig. 17. The evaporation of ethanol starts at the outside of the particles, resulting in a higher concentration of DDA and OTES on the external surface and subsequently promoting the intercalation of OTES and DDA (Fig. 17(b and c)). The evaporation of residual ethanol in the gallery results in the silylation of OTES (Fig. 17d). The water molecules contained in ethanol catalyze the interlayer surface silylation of OTES during the evaporation of ethanol. The condensation of alkoxysilane is known to be catalyzed by bases. DDA as a gallery expander can also play a role as catalyst for the grafting of OTES into the surface silanol groups.

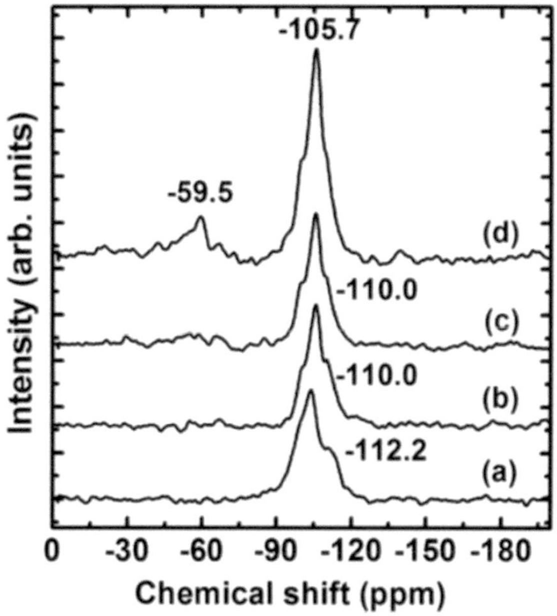

Figure 15. ^{29}Si MAS NMR spectra of (a) as-prepared H^+-titanosilicate, (b) 0.01 M OTES– TS, (c) 0.05 M OTES–TS and (d) 0.1 M OTES–TS.

Figure 16. Transmission electron micrographs for OTES–TS.

The narrow pore size distributions (~0.7 nm peak width at half peak maximum) of SPT derivatives are similar to those of the related MCM-41 (~0.5 nm peak width), which is produced using a surfactant template [39]. This indicates that DDA molecules play a decisive role in pore formation in SPT derivatives, and that they form molecular assemblies similar to those formed by surfactant micelles. An optimal DDA/TEOS molar ratio in interlayer regions could afford near ideal conditions for the formation of micellar neutral amine assemblies, and such conditions favor the DDA-templated hydrolysis of TEOS. Tanev and Pinnavaia [56] demonstrated that the assembly of hexagonal mesoporous metal oxides can also be achieved by hydrogen bonding between neutral amine and TEOS. In porous materials formed by this gallery-templated reaction, specific surface areas are composed of the surfaces of micropores, where pore walls act as pillars and mesopores. Therefore, the lateral spacing between pillars in conventional pillared layered materials is not the sole contributor to increased microporosity.

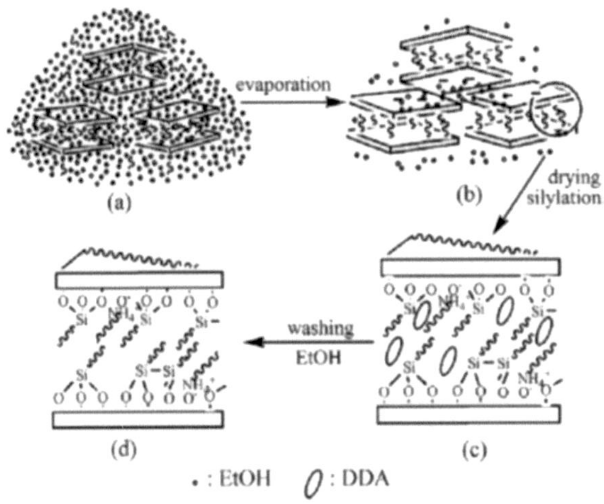

Figure 17. Schematic representation of the intercalation and silylation of DDA and OTES for H⁺-titanosilicate.

SPT derivatives exhibit a broad pore size distribution as compared with MCM-41. Furthermore, they offer new opportunities for the rational design of heterogeneous catalyst systems because of their complementary chemical functionality, thermal stability, and stable pore size distribution to the small mesopore range (1.0–2.0 nm). In addition, the existence of five-coordinated titanium(V) in the interlayer surfaces of SPT derivatives makes then promising potential oxidation catalysts.

5. Conclusion

The simultaneous intercalation of DDA and TEOS into H^+-titanosilicate interlayers and subsequent intragallery DDA-catalyzed hydrolysis of TEOS resulted in mesoporous silica-pillared H^+-titanosilicate (SPT) derivatives. These derivatives exhibited refractions corresponding to a basal spacing of 4.16–4.32 nm, a uniform pore size of 2.8–3.4 nm and large surface areas of 535–618 m^2/g. The structural and physical properties of SPT derivatives exhibited excellent thermal resistance, i.e., they remained stable after heating for 5 h at 700 °C in air. Our results indicate that DDA plays a decisive role in pore formation, because it acts as a base catalyst and as a micelle-like template during the hydrolysis of TEOS. In particular, the rapid hydrolysis of TEOS in water controls TEOS outflow from interlayers and contributes to the formation of firm silica-pillars.

Our results show that intercalation and silylation proceeded more effectively by the evaporation of solvents than by filtration and drying. This evaporation process should be a very efficient method for the intercalation and silylation of silane coupling agents into layered compounds and organic–inorganic materials.

Author details

Kyeong-Won Park
Department of Chemistry and Research Institute of Natural Science,
Gyeongsang National University, Republic of Korea

6. References

[1] J. Rocha, M.W. Anderson, Eur. J. Inorg. Chem. (2000) 801.
[2] Z. Lin, J. Rocha, P. Brandao, A. Ferrira, A.P. Esculcas, J.D. Pedrosa de Jesus, A. Philippou, M.W. Anderson, J. Phys. Chem. 101 (1997) 7114.
[3] H. Du, M. Fang, J. Chen, W. Pang, J. Mater. Chem. 6 (1996) 1827.
[4] M.A. Roberts, G. Sankar, J.M. Thomas, R.H. Jones, H. Du, M. Fang, J. Chen, W. Pang, R. Xu, Nature 381 (1996) 401.
[5] R. Murugavel, H.W. Roesky, Angew. Chem., Int. Ed. Engl. 36 (5) (1997) 477.
[6] S. Ferdov, V. Kostov-kytin, O. Petrov, Chem. Commun. (2002) 1786.
[7] V. Kostov-kytin, B. Mihailova, S. Ferdov, O. Petrov, Solid State Sci. 6 (2004) 967.
[8] M.E. Landis, B.A. Aufdembrink, P. Chu, I.D. Johnson, G.W. Kirker, M.K.J. Rubin, J. Am. Chem. Soc. 113 (1991) 3189.
[9] J.S. Daily, T.J. Pinnavaia, Chem. Mater. 4 (1992) 855.
[10] E. Ruitz-Hitzky, J.M. Rojo, Nature 287 (1980) 28.
[11] S.Y. Jeong, O.Y. Kwon, J.K. Seo, H. Jin, J.M. Lee, J. Colloid Interf. Sci. 175 (1995) 253.
[12] O.Y. Kwon, S.Y. Jeong, J.K. Seo, B.H. Ryu, J.M. Lee, J. Colloid Interf. Sci. 177 (1996) 677.
[13] A. Galarneau, A. Barodawalla, T.J. Pinnavaia, Nature 374 (1995) 529.
[14] O.Y. Kwon, J. Ind. Eng. Chem. 5 (1999) 314.
[15] O.Y. Kwon, H.S. Shin, S.W. Choi, Chem. Mater. 12 (2000) 1273.
[16] K.W. Park, S.Y. Jeong, O.Y. Kwon, Appl. Clay Sci. 27 (2004) 21.
[17] K.W. Park, O.Y. Kwon, J. Ind. Eng. Chem. 10 (2004) 252.
[18] K.W. Park, O.Y. Kwon, Bull. Korean Chem. Soc. 25 (2004) 965.
[19] Michael E. Landis, B.A. Aufdembrink, P. Chu, I.D. Johnson, G.W. Kirker, M.K. Rubin, J. Am. Chem. Soc. 113 (1991) 3189.
[20] J.S. Daily, T.J. Pinnavaia, Chem. Mater. 4 (1992) 855.
[21] M. Ogawa, K. Kuroda, Bull. Chem. Soc. Jpn. 70 (1997) 2593.
[22] Soon-Yong Jeong, Oh-Yun Kwon, Jeong-Kwon Suh, Hangkyo Jin, Jung Min Lee, J. Colloid Interface Sci. 175 (1995) 253.
[23] Oh-Yun Kwon, Soon-Yong Jeong, Jeong-Kwon Suh, Beyong-Hwan Ryu, Jung-Min Lee, J. Colloid Interface Sci. 177 (1996) 677.
[24] O.Y. Kwon, S.W. Choi, Bull. Korean Chem. Soc. 20 (1999) 69.

[25] O.Y. Kwon, H.S. Shin, Chem. Mater. 12 (2000) 1273.
[26] O.Y. Kwon, K.W. Park, J. Ind. Eng. Chem. 7 (1) (2001) 44.
[27] A. Galarneau, A. Barodawalla, T.J. Pinnavaia, Nature 374 (1995) 529.
[28] E. Ruiz-Hitzky, J.M. Rojo, Nature 287 (1980) 28.
[29] E. Ruiz-Hitzky, J.M. Rojo, Colloid Polym. Sci. 287 (1985) 28.
[30] A. Shimojima, D. Mochizuki, K. Kuroda, Chem. Mater. 13 (2001) 3603.
[31] S. Okutomo, K. Kuroda, M. Ogawa, Appl. Clay Sci. 15 (1999) 253.
[32] M. Ogawa, S. Okutomo, K. Kuroda, J. Am. Chem. Soc. 120 (1998) 7361.
[33] T. Yanagisawa, K. Kuroda, C. Kato, React. Solid 5 (1988) 167.
[34] P.H. Thiesen, K. Beneke, G. Lagaly, J. Mater. Chem. 12 (2002) 3010.
[35] M.W. Weimer, H. Chen, E.P. Giannelis, D.Y. Sogah, J. Am. Chem. Soc. 121 (1999) 1615.
[36] Kostas S. Triantafyllidis, Peter C. LeBaron, In Park, Thomas J. Pinnavaia, Chem. Mater. 18 (2006) 4393.
[37] Hao Fong, Richard A. Vaia, Jeffrey H. Sanders, Derek Lincoln, Peter J. John, Andrew J. Vreugdenhil, John Bultman, Clifford A. Cerbus, Hong G. Jeon, Polymer Prepr. 42 (1) (2001) 354.
[38] S. Solarski, S. Benali, M. Rochery, E. Devaux, M. Alexandre, F. Monteverde, P. Dubois, J. Appl. Polymer Sci. 86 (2004) 238.
[39] C.T. Kresge, M.E. Leonowicz, W.J. Roth, J.C. Vartuli, J.S. Beck, Nature 359 (1992) 710.
[40] S. Brunauer, P.H. Emmett, E. Teller, J. Am. Chem. Soc. 60 (1938) 309.
[41] G. Horvath, K.J. Kawazoe, Chem. Eng. Jpn. 16 (1983) 470.
[42] Jie Bu, Hyun-Ku Rhee, Catal. Lett. 65 (2000) 141.
[43] K. Lin, L. Wang, F. Meng, Z. Sun, Q. Yang, Y. Cui, D. Jiang, F-S. Xiao, J. Catal. 235 (2005) 423.
[44] Damodara M. Poojary, Roy A. Cahill, Abraham Clearfield, Chem. Mater. 6 (1994) 2364.
[45] Yunling Liu, Hongbin Du, Feng-Shou Xiao, Guangshan Zhu, Wenqin Pang, Chem. Mater. 12 (2000) 665.
[46] Elizabeth A. Behrens, Damodara M. Poojary, Abraham Clearfield, Chem. Mater. 8 (1996) 1236.
[47] Yining Huang, Zhimei Jiang, Wilhelm Schwieger, Chem. Mater. 11 (1999) 1210.
[48] Yali Su, Mari Lou Balmer, Bruce C. Bunker, J. Phys. Chem. B 104 (2000) 8160.
[49] Goutam Deo, Andrzej M. Turek, Israel E. Wachs, Zeolites 13 (1993) 365.
[50] R.H. Stolen, G.E. Walrafen, J. Chem. Phys. 64 (1976) 2623.
[51] Vladislav Kostov-Kytin, Boriana Mihailova, Stanislav Ferdova, Ognyan Petrov, Solid State Sci. 6 (2004) 967.
[52] G. Stephen Caravajal, Donald E. Leyden, Gregory R. Quinting, Gary E. Maciel, Anal. Chem. 60 (1988) 1776.
[53] N.R.E.N. Impens, P. van der Voort, E.F. Vansant, Micropor. Mesopor. Mater. 28 (1999) 217.
[54] M.B. D'Amore, S. Schwarz, Chem. Commun. (1999) 121
[55] G.S. Caravajal, D.E. Lyden, G.R. Quinting, G.E. Maciel, Anal. Chem. 60 (1988) 1776.
[56] P.T. Tanev, T.J. Pinnavaia, Science 267 (1995) 865.

Chapter 10

Structural and Ion-Exchange Properties of Natural Zeolite

Aiymgul M. Akimkhan

Additional information is available at the end of the chapter

http://dx.doi.org/10.5772/51682

1. Introduction

Intensive studies of natural and synthetic zeolites properties are being held from the middle of 60-70-ies. Among the various examples of zeolites application it is sufficient to mention emission and purification of normal paraffin hydrocarbons, catalytic reactions of hydrocarbons, extraction of radioactive isotopes, obtaining carriers for catalysts, release of enzymes and removal of impurities polluting the atmosphere [1-3]. The interest of researchers towards such alumosilicates is connected with unique properties of zeolites: an extremely high adsorption capacity, catalyzing action, thermal stability and resistance in different chemical environments. The relative simplicity along with relatively low cost thereof stipulates for a significant availability for mass application.

Zeolitic rocks are widely distributed in the territory of Kazakhstan and are economically and ecologically cost-efficient raw material for creating unique ion-exchange sorbents with selective properties. Development of natural zeolites modification methods opens a possibility of purposeful construction of chemically modified inorganic sorbents able to substitute synthetic analogues and of obtaining new compositions with a prospect of widening the range of ion-exchange materials. The results of the study of structural, physicochemical and ion-exchange properties of natural alumosilicates allow developing the theoretical bases for the directed alternation of the complex of useful properties of natural minerals, meeting the main criteria required for sorption materials.

2. Investigation of physico-chemical properties of natural zeolite

Natural zeolites fields explored in Kazakhstan determine the intensive development of studies in different application areas. In spite of the numerous publications, the physicochemical properties of natural zeolites are still not sufficiently studied.

INTECH
open science | open minds

© 2012 Akimkhan, licensee InTech. This is a paper distributed under the terms of the Creative Commons Attribution License (http://creativecommons.org/licenses/by/3.0), which permits unrestricted use, distribution, and reproduction in any medium, provided the original work is properly cited.

A quick diagnosis of natural zeolites can be significantly enhanced through a combination of such methods of analysis as X-ray phase and thermographic study. Recording the diffraction pattern makes it possible to state a presence of zeolites isostructural to heulandite even in the mixtures [4]. Analysis of thermogram allows determining the species within this group qualitatively. To that effect you can use the low-temperature area up to 400⁰C. According to XRF data, the main phase of zeolite tuff studied is clinoptilolite-heulandite (Figure 1). The distinct peaks 9,58; 8,68; 7,42; 5,69; 5,07; 4,58; 4,37; 3,69; 3,62; 3,45; 3,23 ; 3,03 Å typical for clinoptilolite appear on natural zeolite's diffraction pattern. There are montmorillonite, plagioclase and quartz among other phase components. Comparison of the results with literature data obtained for monophasic clinoptilolite shows that the material in question is close to clinoptilolite according to the composition but it has higher sodium, calcium and iron contents (Table 1). The study thereof showed that it was characterized by Si/Al ratio equal to 3,83.

As a result of the studies held it was stated that zeolite tuff of Shankhanai field of Kazakhstan contained up to 70% of the plate-like shape main product with high ratio between silica and aluminum oxides (heulandite – clinoptilolite). Brown-red colouring thereof specifies to the presence of fine iron oxides.

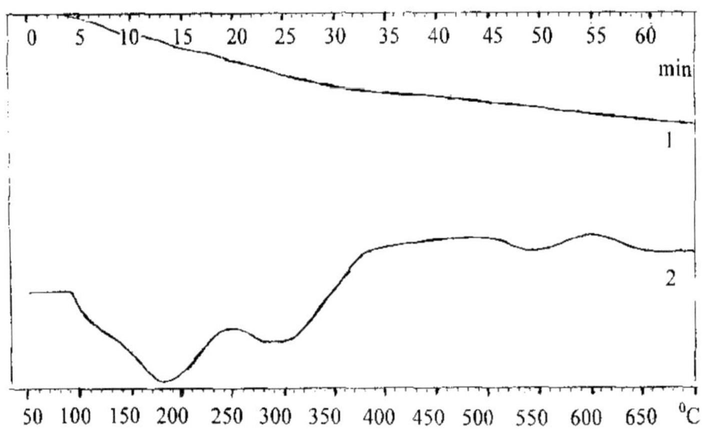

Figure 1. Derivatogram of natural zeolite. 1 – TG; 2 – DTA.

№	HCl, N	*The composition of clinoptilolite, %*							$SiO_2/$
		SiO_2	Al_2O_3	Na_2O	K_2O	CaO	MgO	Fe_2O_3	Al_2O_3
1	---	58,08	15,16	2,30	1,42	4,92	2,53	5,95	3,83
2	3 (20-22⁰C, 24 h)	58,19	15,09	1,75	1,27	4,02	2,28	5,60	3,86
3	12 (20-22⁰C, 24h)	58,86	15,05	1,77	1,25	4,05	2,34	5,54	3,91
4	1,5 (96-98⁰C, 6h)	69,16	15,43	1,67	1,18	2,24	1,41	4,95	4,48
5	Concentration (96-98⁰C, 6h)	79,10	6,20	1,24	1,01	0,58	0,29	0,70	12,80

Table 1. Changing the chemical composition of clinoptilolite.

The results of thermographic studies are given in the Figure 2. Differential thermal analysis (DTA) data show that the dehydration of clinoptilolite-containing tuff begins from the lowest temperatures, approximately from 60°C, and proceeds continuously over a broad temperature range up to 400°C. The mass loss at this temperature reaches 3,3%. The total mass loss while heating zeolite up to 700°C is 6,7%. Water release in zeolite proceeds continuously and smoothly, as evidenced by the mass loss (TG) curve. The DTA curve of the raw zeolite exhibits several endothermic effects with minimums at 100; 180; 280; 550 and 660°C. The character of DTA and TG curves over the entire temperature range of 50–700°C is indicative of several forms of bound water in minerals [4]. The presence of two endothermic effects due to the release of adsorbed water indicates that active sites in clinoptilolite microcavities are inhomogeneous in energy. The natural zeolite heating curve shows the loss of weakly bound adsorption water and capillary-condensation water at 100°C. The energy of interaction of water molecules with potassium ions is lower, and the endothermic effect at 170°C is due to the desorption of water molecules bound as K^+, as well as due to zeolite oxygen framework. Weakly expressed endothermic effects displayed by zeolitic rock thermograms are caused by foreign impurities. At 550°C it is stipulated by the detachment of hydroxyl groups. As follows from the given experimental data, the initial natural zeolite exhibits a fairly high thermal stability (above 700°C).

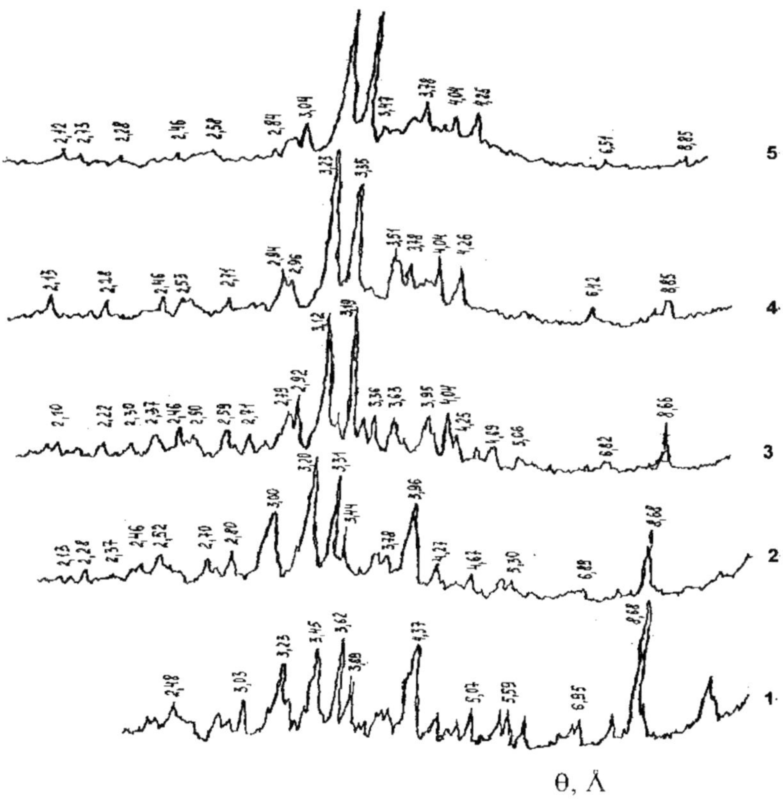

1 –natural zeolite (№1); after treatment: 2 – 3 N HCl solution, 22-25°C (№2); 3 – 12 N HCl solution, 22-25°C (№3); 4 – 1,5 N HCl solution, 96-98°C (№4); 5 – concentrated solution of HCl, 96-98°C (№5)

Figure 2. The diffractograms of the zeolite samples

It is known that in the process of natural mineral sorbents treatment with acids their pore structure develops. Therefore, we tested different ways of acid processing of natural zeolite with hydrochloric acid. From the standpoint of environmental friendliness of the process (simplicity of purification and regeneration of spent solutions), HCl was used as a modifying acid for sorbent treatment, since the wastes of dilute hydrochloric acid are much easier to purify of leaching products and to neutralize.

In the experiments, we varied the acid concentration, temperature and duration of processing. The effectiveness of activation was evaluated in terms of the total porosity of a sorbent and specific surface area thereof.

Activation typically involves three stages: removal of exchangeable cations, framework dealumination and formation of an amorphous silicon-oxygen phase. The sequence and intensity of the stages are determined by processing conditions and the specific features of zeolites. Acid modification of the investigated zeolite with 3 and 12 N HCl solution within the day (№ 2 and №3) at room temperature showed that the value of silica modulus (Table 1) was not changing essentially. More stringent processing conditions lead to a sharp increase in the silica modulus (Si / Al). At modification by 1,5 N HCl solution by heating (№ 4) clinoptilolite is subject to dealumination and removal of cations without significant destruction of the crystal lattice in comparison with the treatment thereof under the same conditions with a concentrated acid (№ 5). In the latter case the amorphous phase of zeolite forms. These results confirm the XRF data (Figure 1). Comparison of natural and modified samples radiographs indicates to a decrease in the intensity of main reflection characteristics of the mineral up to complete disappearance due to the amorphization of the structure and changes in zeolite cation composition after mineral acid treatment.

The obtained results show that removal of cations and dealumination are insignificant under milder conditions of acid processing. More severe activation conditions result in intensification of these processes. The removal of exchangeable zeolite cations and occupation of the free sites by protons have an effect on the cationic density of the framework and can change the sizes of channel windows. The removal of aluminum from mineral framework also results in partial degradation of zeolite crystal structure.

Location and exploration of large deposits of natural zeolites in different regions of Kazakhstan creates significant scientific-technical prospects for the complex use of natural sorbents, especially for the creation of nanomaterials on their basis.

For the study of morphology of the porous structure of natural zeolites and those ones, activated by acidic treatment, adsorption-desorption measurements have been carried out (Table 2). Natural zeolite (sample №1) is characterized, by the isotherm data, by mesoporosity. It also contains some micropores, because nitrogen vapors are absorbed to some degree at low pressures (p/p_s less than 0,1). Its treatment with a 3 N HCl solution (№2) pracically does not change the surface characteristics of the sorbent, and the treatment by a 12 N HCl solution (№3) results in an increase in the total pore volume by a factor of about 2. Porocity increases considerably with the activation with a 1,5 N HCl upon heating (№4). For the initial sample the total pore volume constitutes 0,019574, whereas for the activated ones – 0,19611 (№2);

0,031348 (№3); 0,064101 (№4); 0,122297 (№5) cm^3/g. Chemical treatment increases porosity of zeolite by a factor of 6. Differential curves of the pore size distribution have shown that the studied samples possess mainly mono- and bi-dispersion structures. Natural zeolite possesses the mono-dispersion structure with a large maximum in the area of mesopores d ~ 30 Å. For the sample №3, besides, an increase in the field of d~ 20 Å is observed, i.e. the activation of the mineral with 12 N acid in the usual conditions results in the development of micropores, the volume of which increases more than 5 times (Table 2). Differential dimensional spectra of such samples are characterized by the narrow distribution. The sample №4 is characterized by bioporosity, the volume of mesopores increases by a factor of 1,5 and microporosity – by a factor of 217. The further activation of zeolite also leads to an increase in the micro- and mesoporous structure with an extension of the range of the mesopores (№5).

№ sample	$S_{БЭТ}$, m^2/g	St-diagram, m^2/g	Vmicropores, cm^3/g	Vmesopores, cm^3/g	ΣVpores, cm^3/g
1	5,8714	7,8482	0,000162	0,019412	0,01957
2	6,8487	8,1363	0,000379	0,019232	0,019611
3	9,8917	9,9675	0,000841	0,030507	0,031348
4	108,5298	21,8786	0,035250	0,028851	0,064101
5	181,8055	47,9273	0,050992	0,071305	0,122297

Table 2. Surface area and porosity of the zeolite samples

In the NMR spectra of the zeolite samples № 4 and №5 alongside with the signals from aluminum atoms in the tetrahedral coordination (56 ppm), show the signals with the size of chemical shift of 3 and 11 ppm., characteristic for the atoms of aluminum in the octahedral medium by oxygen (Figure 3). With an increase in decationation their intensity increases. An increase in decationation of zeolite leads to an increase in the intensity of the signal from the atoms of alluminum in the octahedral coordination. On the whole the predominant part of aluminum atoms remains in the structure of the carcass also after removal of cations of zeolite.

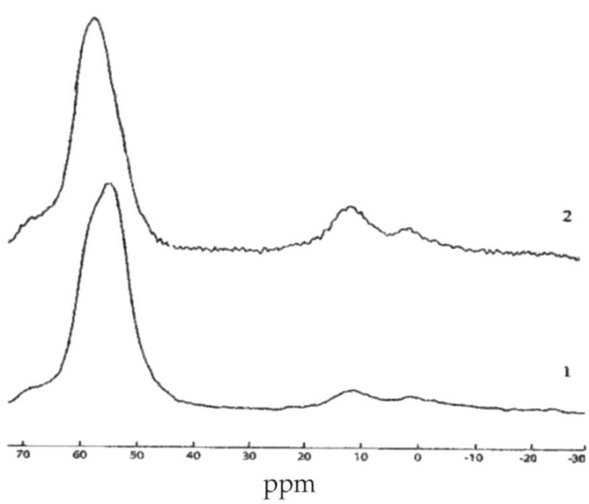

Figure 3. ^{27}Al NMR spectra of zeolite after acid treatment. 1 - sample №4 (1,5 N solution of HCl, 96-98⁰C); 2 - sample № 5 (conc. solution of HCl, 96-98⁰C)

3. Investigation of ion-exchange properties of natural zeolite

The intensive use of zeolites in various fields of science and practice is based largely on their ion exchange properties, which are one of the main parameters that characterize their sorption and technological properties. The use of zeolites as selective sorbents can solve a number of important problems of technology and environment (Table 3). In this regard, commercial development of deposits of natural zeolites, which due to low cost can be applied in various fields where the use of synthetic analogues is marginal, is required.

Method of application	Zeolite	Fields of application
Without regeneration	clinoptilolite mordenite	Cleaning of waste water from the radioactive ions ^{137}Cs, ^{90}Sr, followed by burial of the zeolites. Treatment of domestic wastewater from ammonium nitrogen with followed by the use of zeolite as ammonium fertilizer
	clinoptilolite	Purification of wastewater from the nonferrous metals using zeolites as a flux
With the self refresh	clinoptilolite mordenite	The use of zeolites as filter material for water treatment
	clinoptilolite	The introduction of zeolites in the soil to increase the duration of fertilizer
Regeneration with solutions	clinoptilolite mordenite	Concentration and separation of alkali metals from technological solutions and natural waters
	clinoptilolite	Concentration of Sr from the waste and natural waters
	clinoptilolite mordenite	Concentration of non-ferrous metals from technological solutions, waste and natural waters

Table 3. Use of natural zeolites in ion-exchange process

Acid resistance is crucial to the technological application of clinoptilolite. The study of the conditions for maintaining its crystallinity allows creating a basis to obtain the samples intermediate in the degree of cations removal and dealumination which have acidic properties.

It is established [5] that two stages of acid activated cation removal - ion exchange and dealumination – are observed as a result of zeolite reaction with dilute solutions. The former may be connected with polycationicity of mineral. In the latter case, the divalent cations Ca^{2+} are harder replaced by hydrogen than monovalent ones. In processing zeolites with more concentrated acid solutions there is a shift to the solution of aluminum from tetrahedron with the formation of high-silicon core. The crystal structure thereof can be preserved up to a very high degree of cations removal and dealumination.

In practice, in most cases, clinoptilolite containing material is rarely used in its natural form and it is usually subjected to additional chemical pre-treatment to improve its sorption, mainly ion-exchange properties. The most common method of zeolite adsorbent targeted modification is a cation-exchange modification, consisting in the transfer of polycationic natural material into one of monocationic form (H-, Na, Ca, NH_4-, etc.).

When processing high-silicon zeolites with strong acids a sequential substitution of cations hydroxonium ions [6] occurs. Hydrogen forms of zeolites can be regarded as solid crystalline alumosilicic polyacids. Therefore, the method of potentiometric titration with alkaline solutions which allows evaluating the active groups' acidity usual for ion exchanger may be used for hydrogen forms of zeolites. For high-silicon natural zeolites (clinoptilolite, mordenite) with enhanced stability in acidic solutions in comparison with other zeolites, the hydrogen forms can be obtained directly by treatment of acid solutions. The high stability of clinoptilolite to heating and acids can be explained by an increase in the stability of skeleton associated with a decrease in the number of weaker bonds Al-O (1,66 Å) and an increase in the number of strong bonds Si-O (1,62 Å).

To study the acid-base properties of the investigated zeolite air-dried sample, previously modified with 3 N hydrochloric acid was being treated with 0,1 N alkali solutions within one day at room temperature. Analysis of the potentiometric titration curve (Figure 4) shows that the hydrogen forms of zeolite obtained by treatment with acid are titrated with KOH solution. At alkali charge 0,5 mg-eq / g there is a clear inflection point.

Consequently, the hydrogen forms of zeolites obtained by treatment with HCl have a more acidic hydroxyl grouping. The other exchange centers are subacid and uniformly distributed in the mineral structure. The static exchange capacity of the activated zeolite is 2,7 mg-eq /g.

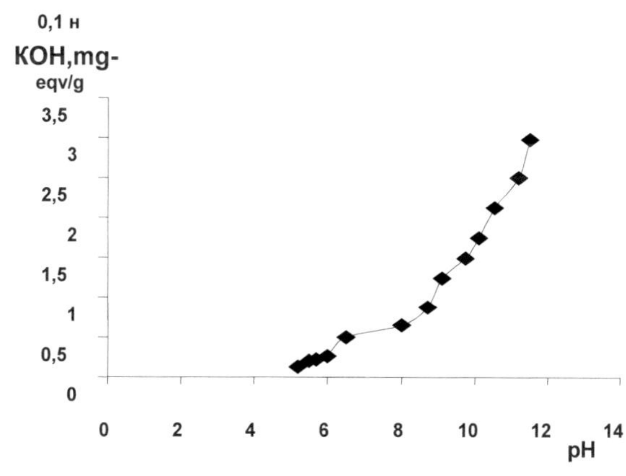

Figure 4. Potentiometric titration curve of the H^+ form of zeolite.

The capacity for chemical modification of natural zeolites gives the possibility of obtaining a large variety of adsorbents, catalysts and other fine general-purpose bodies. The results of these studies contribute to the establishment of limits of clinoptilolite in the presence of acidic reagents, and in addition, the identification of opportunities for obtaining the modified forms with novel properties. The successful implementation of technology of sorption metal extraction is realized by a comprehensive study of the processes occurring in a complex system ion exchanger-solution. The study of adsorption on decationated and dealuminated clinoptilolite samples revealed their ability to selective adsorption of large cations [7,8]. This is apparently due to the proximity of the ions sizes and the kinetic diameters of clinoptilolite channels, which defines a strong Coulomb interaction with structure of the zeolite. In addition, it may be due to the presence of stronger acid sites and stabilizing role of the major cations in conditions of a significant reduction in their number in the internal space stipulated by a decrease in tetrahedral aluminum content.

Analysis of the kinetic curves of lead ions sorption in zeolite under study showed that the time to reach equilibrium decreased as the concentration of original solution increased and the grain size [8] decreased. The obtained results indicate that the rate of exchange of lead ions on the H-form of clinoptilolite is controlled by internal diffusion mechanism of the process. The effective diffusion coefficient depends linearly on the radius of the incoming ion and does not depend on the concentration of external solution.

The problem of waste water treatment is largely due to the lack of low-cost multifunctional sorbents stable during operation. Existing industrial sorbents with high sorption capacity are economically unfeasible due to their high cost. The results of experiments with investigated zeolite show that sorption of phenol on the natural sample and aminated form thereof happens with high degree of extraction from dilute solutions [9]. Modification of the PAV sorbent allows adsorbing phenol from aqueous solutions to the level of MPC and increasing the efficiency of the process in the presence of organic impurity compounds.

The presence of commercial deposits of clinoptilolite and experience of its application in various sorption processes make it possible to use it in the process flow sheets designated to extract a number of metal cations from solutions and wastewater. Significant amounts of the latter require the methods of treatment to have low cost and high efficiency. Addressing these issues is also associated with an additional yield of these metals which are being irretrievably lost currently.

One of the criteria of the waste water condition is nonferrous and heavy metal ions content in it. The latter are contained in waste water of nonferrous and ferrous metallurgy enterprises, concentration plants, galvanizing rooms of plants. There are different methods either physical or chemical which enable to retrieve ions of metals from solutions and industrial waste. The treatment of this sewage by ions isolation in the form of hydroxides cannot provide the required drop of concentration of the latter in sewage. Metals deposition in a form of sulphides is connected with the use of toxic reagents such as Na_2S and H_2S. The effective sewage treatment can be obtained by the sorption process which is easy to control and allows reducing relative detection limits.

The sorption of copper ions (II) under static conditions at different concentrations of mother solution, pH medium and in presence of various metals ions was conducted on investigated natural zeolite.

The effectiveness of sorption in many respects is determined by kinetics of the process. The curves of copper ions sorption are presented in the Figure 5. The equilibrium in distribution of metal ions among the sorbent and solution at room temperature is established within short period of time (30 min). Short time of process is determined by the presence of chemically active groups in surface layers and high selectivity of sorbent toward metal ions. The results of conducted studies showed that natural zeolite has high activity at sorption of Cu^{2+} even in lower concentrations (0,125 g/l). Within 30 minutes the degree of recovery/extraction is 92% and it reaches 95% with increase of content of metal ions up to 0,5 g/l. At sorption of Cu^{2+} in areas of ions metal concentrations in natural waters 0,001-0,0005 mg/l high degree of recovery/extraction (90%) for 30 minutes is also reached, that shows the effectiveness of used sorbent. Hyperactivity of the latter at low contents stipulates for the suitability of use of metal ions for extraction of trace contaminants.

Due to the fact that waste outlet waters may contain both different composition and variable acidity, there is a need to study the influence of pH degree on sorption of copper ions. The curves of recovery degree dependence on acidity of solution under constancy of concentration Cu^{2+} (0,25 g/l) show (Figure 6) that the degree of metal ions recovery in studied conditions decreases with increase of acidity of solution due to competitive effect of hydrogen ions. Thus, in pH<4 area the extracting ability of zeolite falls to 10%. With increase of pH medium to more than 4-5 the sediments of basic salts and hydroxides delayed by ionite are generated and sorption occurs additionally due to adhesion of generated sediments (recovery is 90-91%). Under collective concentration of Cu^{2+}, Pb^{2+}, Co^{2+}, Ni^{2+} with cumulative content of metals ions 0,5 g/l (concentration of each ion is 0,125 g/l) sorption of copper ions occurs with delayed kinetics (Figure 7). The maximum degree of recovery is 67% for 120 minutes. According to the obtained data under accepted conditions natural zeolite equally sorbs copper and lead ions.

It is determined that ratio of the solution volume to sorbent mass influences recovering ability thereof. Growth of zeolite content in the solution from 0,5 g to 2 g (V=50 ml) leads to increase of recovery effectiveness by 3%. Degree of sorption reaches 97% within 30 minutes.

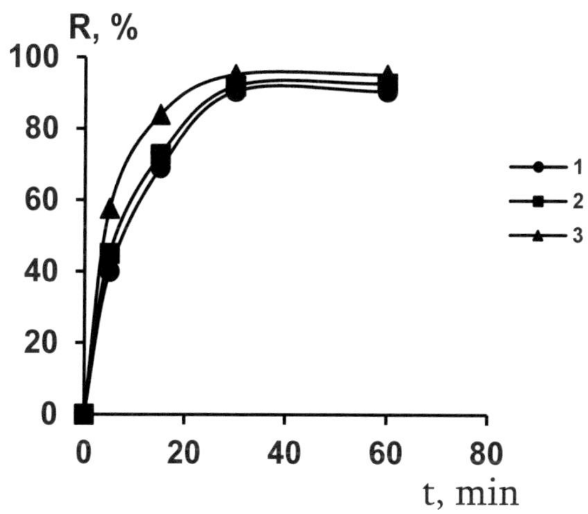

Figure 5. Degree Cu^{2+} by natural zeolite as a function of the process duration t. Cs: 0,125 (1); 0,25 (2); 0,5 (3); $C_{Cu^{2+}}$=0,5 g/l; V=50 ml; m=0,5 g.

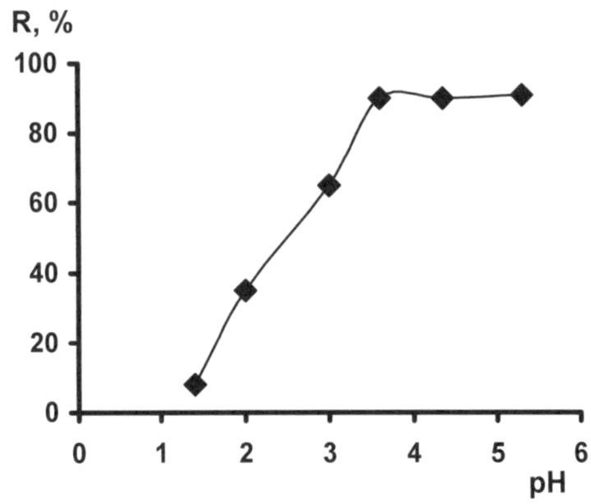

Figure 6. Degree of Cu^{2+} by natural zeolite as a function of the pH. $C_{Cu^{2+}}$=0,25 g/l; V = 50 ml; m = 0,25 g / l; τ= 30 min.

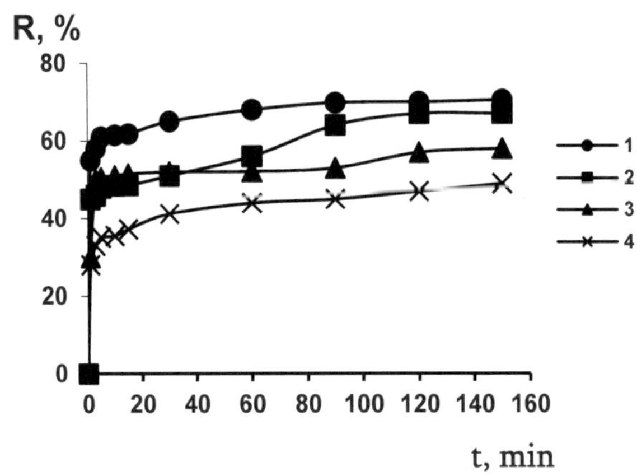

Figure 7. Degree Pb^{2+} (1); Cu^{2+} (2); Co^{2+} (3); Ni^{2+} (4) by natural zeolite as a function of the process duration t. ΣC_{Me2+}=0,5g/l; V=50 ml; m=0,5 g.

Thus, the suggested sorbent effectively recovers copper ions from model solutions within short periods of time. Cheap natural zeolites may be used in the processes which do not provide regeneration of ionites. The opportunity of its one-time use with subsequent burial provides application thereof in such spheres where synthetic zeolites or ion-exchange polymers cannot be used under economical reasons.

4. Features of the sorption of various substances vapors by activated forms of natural zeolite

The comparative ease of zeolites chemical modification opens up wide possibilities for controlled changes in their structure and properties. This makes them very convenient objects for studying the nature of sorption interactions, molecular-sieve effects. The vast majority of these studies were conducted with synthetic zeolites. The natural forms of such minerals which are multicomponent systems complex and variable according to composition are much less known. Their physicochemical properties essentially depend on the content of zeolitic phase in the rock, such as cation exchange form and nature of impurities. Even with the same content of the zeolitic phase in the rock a mismatch in the properties of individual samples may be observed. Natural zeolites are biporous systems; the primary porosity of them is determined by the micropores of crystals, and the secondary one – by the transitional pores and macropores. The latter are the main transport arteries for the supply of a substance; they determine the absorption of relatively large molecules and play an important role for a number of sorption and catalytic processes.

Hydrophilic inorganic sorbents, which include zeolites, are practically useless for adsorption of many organic substances from aqueous solutions since water molecules interact with the OH - groups to form strong hydrogen bonds (5-10 kcal/mol) energy of which exceeds the energy of adsorption of most organic molecules [10]. In this connection, it is important to study the sorption of vapours of various substances on the surface of such minerals.

We used acidoactive (processing for twenty four hours by 3 and 12 N hydrochloric acid solution) air-dried and dehydrated (600ºC) natural zeolite samples as the adsorbate. Sorption was determined by the increase in mass of the sorbent (initial sample 1 g) placed in open weighing cup over liquid adsorbate in a desiccator under normal conditions.

Porous crystals of zeolites are of great interest as highly specific adsorbents both on the molecular sieve effect and also on the nature of their channels' surface. Adsorption peculiarities are connected with the fact the delicacy of the crystal structure creates a large adsorption volume (up to 0,54 cm^3/g for faujasites), and its geometry determines the molecular sieve properties. The presence of acceptor centers (cations, the Lewis acid centers) firmly holding the electron donors, or OH - groups, strongly binding bases (Bronsted centers), causes strong interaction of adsorbed molecules with the adsorbent.

Depending on the nature of these molecules, various types of interactions may occur between them and zeolites. The specificity of zeolites with respect to the molecules of this adsorbate is determined by the values of the interaction energies of different types. In case of the heteropolar adsorbent, the polarization of the adsorbate molecules occurs by its electrostatic field. High adsorption energy of zeolite by molecules capable for specific interaction with cations having a peripheral dipoles (water, aether), the large quadrupoles and π-electron bonds (nitrogen, olefins, benzene) are typical for zeolites. The energy of dispersive attraction makes decisive contribution to the specificity of the process. For zeolites, the prevalence of the energy may be due to the dense environment of adsorbate molecules by the atoms of oxygen frame. Therefore, it is very high energy of adsorption of organic molecules with peripheral functional groups having atoms with free electron pairs and π-bonds for them.

The ratio of Si/Al in natural zeolites can be improved by chemical treatment of the crystals, resulting to the removal of part of aluminum from the frame. Herewith, dealumination can occur without significant changes in their characteristic structural features. A simple option thereof is the treatment of crystals with acid solutions. The result of this chemical action is the artificial increase of the adsorption volume and the effective pores' size.

The curves of the kinetics organic solvents vapours sorption on dehydrated and non-dehydrated zeolite samples are given in the Figure 8. For each of them, they are close to each other, and according to the value of weight gain of the solvent they are in the following order: acetone>toluene>carbon tetrachloride. This range correlates with the size of Vander Waals molecules. Not less important role is played by well-known sorption activity of the polar solvent (acetone). For molecules of hydrophobic solvents water sorbed and bound with cations in the crystallographic positions of the mineral blocks the entrance windows in its cavity. It is shown [11] that the sorption of carbon tetrachloride vapours occurs only on the outer surface of both natural and activated clinoptilolite, and sorption of acetone – in the micropores of the latter. If we assume that water molecules are completely desorbed in zeolite pores dehydrated at 600^{0}C, then the process of sorption of these compounds on the active sites of sorbent runs with displacement of pre-sorbed water.

The results of acetic acid and water vapours sorption on zeolites treated with 3 and 12 N hydrochloric acid are given in the Figure 9. A slight difference in the values of weight gain of the substance on the original forms of mineral is shown in this Figure; a small difference in weight gain of water and organic acid on non-dehydrated and dehydrated zeolites treated with 3 N solution of mineral acid is due to insufficiently developed micro porosity of the mineral. Specific adsorption of water molecules by zeolite is mainly represented by interaction of free electron pairs of oxygen atoms with cations of zeolite cavities surface. The polar molecules of water penetrating into its micropores are sorbed, including as a result of ion-dipole interaction with the active centers (cations, compensating the excess charge of zeolite frame). Water molecules bond with the cations and the frame is predominant in the case of natural zeolites [12]. The adsorbate-adsorbent interaction completely determines the process.

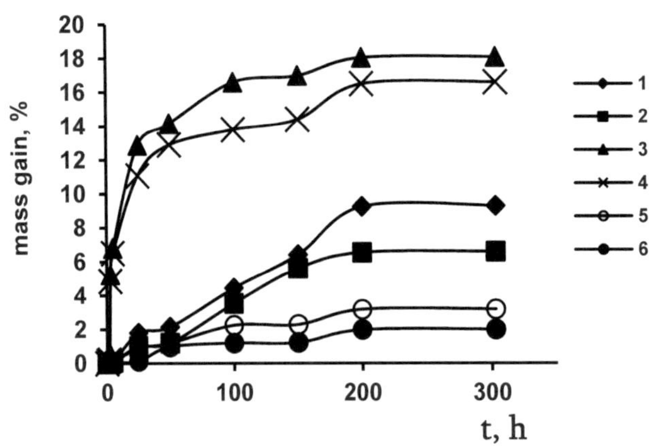

Figure 8. Degree of the mass gain of solutions as a function of the process duration t. sample №2: **dehydrated zeolite 1** – toluene; 3- acetone; 5- carbon tetrachloride. **air-dry zeolite** 2- toluene; 4- acetone; 6- carbon tetrachloride.

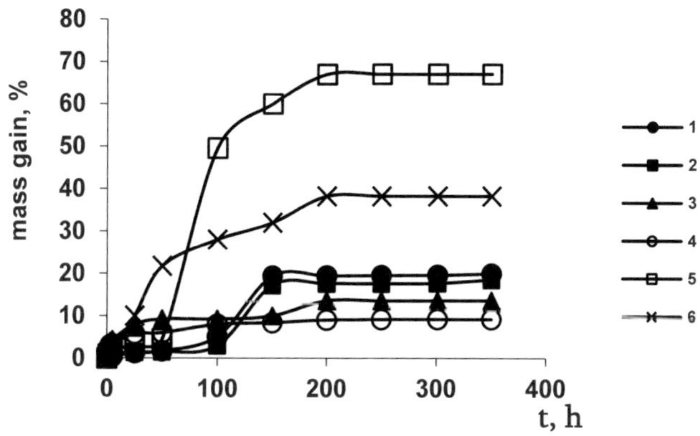

Figure 9. Degree of the mass gain of acetic acid and water as a function of the process duration t. **acetic acid:** (sample №2) 1 - dehydrated zeolite; 2 - air-dry zeolite; (sample №3) 5 - dehydrated zeolite; 6 - air-dry zeolite; **water:** (sample №2) 3 - dehydrated zeolite; 4 - air-dry zeolite

Large scale polarizability (in comparison with water) and low dielectric permeability of the organic acid that increase the binding energy of the cation-fixed ion create preconditions for more effective development of the dispersion forces which play an important role in the mechanism of weak organic acid absorption by ionites.

Acid treatment of zeolite samples with 3 N hydrochloric acid under normal conditions practically does not lead to significant changes in the structural properties of the sorbents, as it is shown above (Table 2). The values of specific surface area for them according to BET and t-graphic have small differences [4]. However, increase of acid concentration under comparable conditions leads to an increase of micro- and mesopores volume. Therefore, zeolite activated by 12 N acid has a higher sorption capacity. Such treatment fosters the expansion of the input windows of mineral channels, and water allocation on ignition contributes to the displacement of cations inside and improvement of primary porosity which depends on the nature of zeolite. This effect seems to be the result of changes in the chemical nature of adsorption sites and porous structure of the mineral in the process of activation. This condition is caused by changes in relative positions of aluminum-oxygen and silicate-oxygen tetrahedrons of the core.

It is shown by example of acetic acid (Figure 9, 10) that there is an evident divergency between the curves of sorption of vapours of the studied substance on non-dehydrated and dehydrated zeolite. Study of the kinetics process showed that for 200 h there is a saturation of natural mineral and sorption volume, is 47 and 89 mmol/g, accordingly.

Increasing of activating acid concentration leads to development of micro-cavities and channels interconnected within the frame of zeolite free of exchangeable cations that promotes the growth of the volume of adsorbed acetic acid. The latter circumstance indicates that the activation process in comparison with the amorphization of zeolite prevails under these conditions; this characterizes the studied mineral as an acid-resistant.

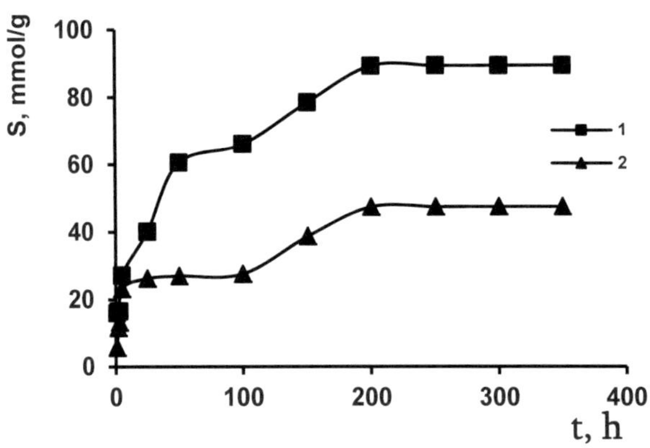

Figure 10. Degree of the sorbtion S of acetic acid vapors by natural zeolite as a function of the process duration t. 1- dehydrated zeolite; 2- air-dry zeolite

The results of spectroscopic (Figure 11) and X-ray studies of natural zeolite activated by 3 and 12 N hydrochloric acid solutions give evidence of preservation of mineral crystal structure due to acid resistance thereof.

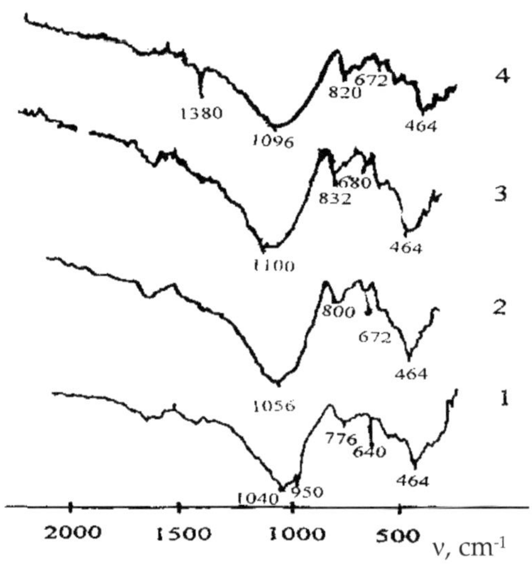

Figure 11. IR spectra of zeolites. 1 - natural zeolite;2 - activated 3 N hydrochloric acid; 3 - activated 12 N hydrochloric acid; 4 - after the sorption of acetic acid vapors.

The absorption bands at 464, 640, 776, 950 cm^{-1} typical for Si-O-Al intratetrahedral deformative and symmetric valence vibrations are prescribed in original zeolite spectrum. When treating it by 3 and 12 N hydrochloric acid, all the characteristic frequencies become more apparent. The position of the peak at 464 cm^{-1} is kept and there is a shift of the absorption bands at 640 and 776 to the long-wave area (672; 680 and 800;832 cm^{-1}, accordingly) without changing the intensity of the spectrums. This gives evidence of preservation of samples crystallinity degree after activation. In the spectrums of zeolite saturated by acetic acid vapours frequencies of adsorbate molecules methyl groups vibrations appear at 1380 cm^{-1} and the absorption band broadens at 1640 cm^{-1} due to overlap of the deformation vibrations of water molecules and asymmetric COO^- valent groups.

Dealumination is accompanied by rupture of Si-O-Al bonds and the resulting shift of asymmetric valent intratetrahedral vibrations frequency from 1040 to 1056 and 1100 cm^{-1}.

The studies showed that an increase in acid concentration leads to an increase in sorption ability of mineral connected with an increase of mineral porosity in the cation removal process. Thus, depending on the nature of adsorbate and adsorbent type, the appropriate modes of activation can be selected.

5. Investigation of mechanochemical modification of natural zeolite with various substances

Diversity of centers formed in zeolites explains the fact that zeolites can be used in a wide range of chemical and technological processes. One of the effective ways of intensification of processing different types of mineral raw materials is mechanical activation, which determines the formation of a significant number of defects in the structure of minerals, which should lead to a change in surface properties of the constituent particles [13, 14].

Mechanochemical modification allows, first of all, intensifying the process of solids dispersing. As a result, the chemisorbed polymer layer passivates the surface of dispersed particles and lyophilized highly dispersed filler is formed. At the same time, the conditions for emergence of chemical interaction at the interface of filler-polymer are created, that leads to a deeper change in the surface.

Environmental cleanness and the possibility of simplifying the process flow sheet determine the prospects for such studies. In this regard, we conducted mechanochemical modification of natural zeolite with aminoacetic acid and epoxy resin for the first time ever.

5.1. Solid-phase interaction of natural zeolite with aminoacetic acid

Layered silicates and zeolites are characterized by the presence of both basic and acid active sites on their surface [4,15]. The solid-phase interaction of these sites with amino acids is of interest [16]. In the sorption on solid supports, amino acids containing two functional groups can form different surface complexes. The structure and chemical properties of the surface and also theoretical and practical aspects of adsorption and catalysis on zeolites are studied by various methods. Widely used spectral methods provide valuable information on the structure and properties of supports [16].

The aim of this work was to study the interaction of α-aminoacetic acid (α-glycine) with natural zeolite during their mechanochemical activation. Natural aluminosilicate, zeolite from Shankhanai deposit (Kazakhstan), was used. To remove nonstructural impurities readily washed out with acids, zeolite was preliminarily activated with a 10% HCl solution for a day. Aluminosilicate samples were dried at 130°C for 2 h; α-glycine was used without additional purification. The mechanochemical activation of initial compounds (the amount of glycine was 30% of that of the zeolite) was performed in an agate mortar for 30 min. The IR absorption spectra of powders pressed into pellets with optically pure KBr were recorded on a UR-20 spectrophotometer.

α-Glycine has a layered crystal structure [14]. Glycine molecules are present in crystals as zwitterions, which are bound with each other by relatively short (therefore strong) N–H···O hydrogen bonds and form antiparallel layers,

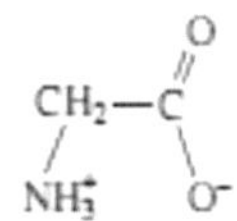

Scheme 1. Scheme 1

The spectra of the samples are presented in the Figure 12. Carboxylic acids form very strong hydrogen bonds; therefore, OH group stretching vibration frequencies are characteristic frequencies. A broad absorption band and several weaker peaks are observed over the range 3000–2500 cm^{-1}. In the spectra of aminoacetic acid and its activated mixture with zeolite, the main peak is observed at 3000 cm^{-1}, and weaker peaks, at 2720, 2616 and 2700, 2600 cm^{-1}, respectively; these peaks can be assigned as combination tones corresponding to C–O (1420 cm^{-1}) and O–H (1300 cm^{-1}) vibrations [17]. The band of stretching vibrations shifts toward lower frequencies, 3488; 3424 cm^{-1}. A pronounced secondary band at 2175 cm^{-1} in the spectrum of α-glycine is evidence of the presence of a bipolar ionic structure; in the modified sample, this band is observed at 2170 cm^{-1}.

Both samples exhibit absorption in the range 1420–1340 cm^{-1}, which is assigned to C–O vibrations closely related to planar bending vibrations of OH groups. In the spectrum of the modified sample, vë=é and antisymmetric stretching vibrations in α-glycine(1720–1540 cm^{-1}) and Si–O stretching vibrations in zeolite (1060 cm^{-1}) are shifted toward longer wavelengths (1700–1500 and 1040 cm^{-1}, respectively). Absorption in the range 1700–1500 cm^{-1} can also be assigned to hydrogen bonds of aminoacetic acid with OH surface groups. This assignment is based on the results obtained for the intra- and intermolecular association of carboxylic acids [18].

Thus, the NH^{3+} stretching vibration and COO^- antisymmetric vibration frequencies shift toward lower frequencies. Similar results were obtained in the mechanochemical activation of caolinite and the amino acid with the formation of a salt with the $NH_2CH_2COO^-$ [14].

It is known [16] that, along with the absorption band of the ionized carboxyl group (1600–1550 cm^{-1}), all amino acids containing the group exhibit two characteristic bands at 1600–1500 cm^{-1}. One of these is observed at 1640–1610 cm^{-1}. For the amino acid under study, bending vibrations occur at 1640 and 1608 cm^{-1}. These bands are also present in the spectrum of the modified zeolite sample; the band at ~1640 cm^{-1} is, however, broadened. This band is a superposition of antisymmetric stretching vibrations of COO^- groups and bending vibrations of water molecules and NH_2 groups. The other characteristic band of

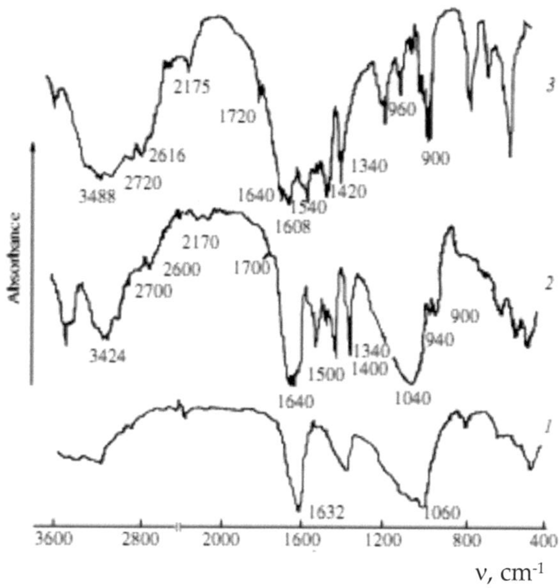

Figure 12. IR spectra of (1) natural zeolite, (2) activated mixture, and (3) aminoacetic acid.

amino acids (1540–1420 cm^{-1}) is, as a rule, more intense than the first characteristic band. In the zeolite activated with the amino acid, this band is shifted toward lower frequencies (1500–1400 cm^{-1}). A similar result was obtained in [14]; however, in the spectra of activated caolinite samples, the absorption band of NH_2 groups was observed at 1550–1450 cm^{-1} and had a low intensity or even disappeared against the background of the maximum at 1640 cm^{-1}.

The absorption band in the frequency range 960–900 cm^{-1} is assigned to OH group bending vibrations. In the spectrum of the modified zeolite sample, this band is narrowed and is observed at 940–900 cm^{-1}.

To summarize, the solid-phase neutralization reaction in the mechanochemical activation of zeolite and the amino acid results in the formation of chemical bonds between carboxyl groups of the acid and basic sites of the aluminosilicate and also coordination bonds between aluminum atoms in the zeolite lattice and nitrogen atoms of amino acid molecules.

5.2. Mechanochemical modification of natural zeolite with epoxy resin

The joint dispersion of fillers with polymers of different nature leads to the breakdown of macromolecules with formation of free radicals, which react with the active centers of minerals to form surface chemical compounds [19].

The specific features of zeolite structure, its high affinity to polar groups, the developed surface area, adsorption capacity determine the possibility of combining the filler with high-molecular compounds. Herewith, the additional intermolecular bonds form and the ability to structure the macromolecule as a result of adsorption at the interface appear. In this regard, the results of the study of the joint dispersion of natural aluminosilicate with an industrial epoxy resin ED-20 [20] were considered.

The results of the study of dependence of the value of epoxy resin weight gain on the surface of zeolite on its concentration showed that epoxy cycles open and the chemisorption of free radicals (Figure 13) takes place. Resin content change from 1 to 20 mass % in the process of dispersion system for 0,5 h leads to an increase in the value of the polymer grafting from 3,2 to 5,7%.

Modification of zeolite for 1 h contributes to a sharp increase in polymer weight gain on the surface, which reaches 7,5% at 1:10 mass parts components ratio (Figure 11). The maximum for these conditions dispersion of zeolite in the presence of ED-20 occurs within the first hour of grinding (Figure 14).

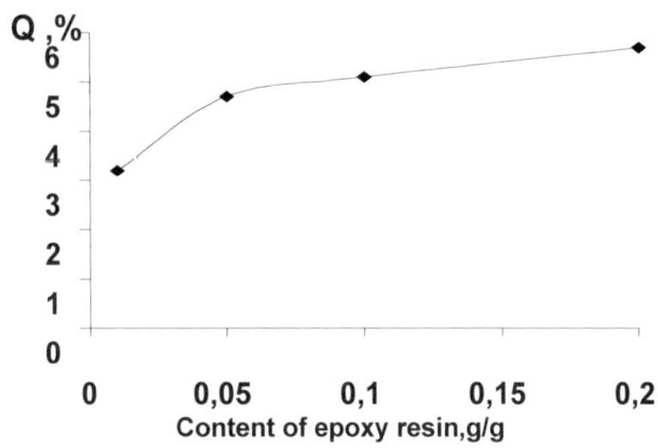

Figure 13. Degree of the grafting of polymer (Q) on the surace of zeolite) as a function of the content of epoxy resin (dispersion for 0,5 h)

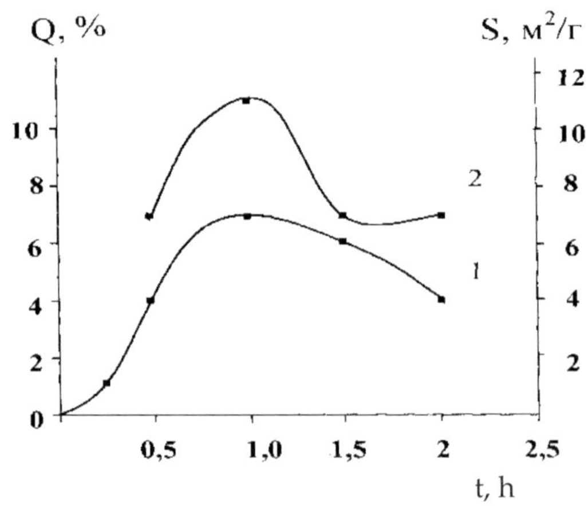

Figure 14. Degree of the grafting of epoxy resin (Q) and the specific surface area (S) minerals (2) as a function of the process duration t. (Resin: Zeolite = 1:10 mass part by weight)

The data obtained shows that at mechanochemical interaction of natural aluminosilicate with an epoxy resin a number of transformations with the components of the system take place. Spectroscopic studies found the significant changes in the chemical structure of polymer as a result of the modification. It is observed that with the increase of process duration a decrease in the intensity of absorption band at 832 cm^{-1} and disappearance of frequency at 916 cm^{-1} are observed, characterizing the epoxy groups' vibrations (Figure 15). This indicates the mechanochemical activation of the latter. Free radicals formed as a result of disclosure thereof contribute to the dispersion and intensify the process of grafting, which affects certain increase in polymer weight gain. Thus, with increasing of grinding time up to 1 h, zeolite specific surface increases and the grafting of polymer (Figure 14) takes place. The process is intensified due to the interaction of freshly formed surface with products of mechanodestruction of macromolecules and its modification as a result of epoxy resin free radical grafting. A further increase in the duration of system dispersion up to 2 h leads to a decrease in the specific surface of modified aluminum silicate stipulated by the grafted layer of polymer the amount of which at this time is the maximum.

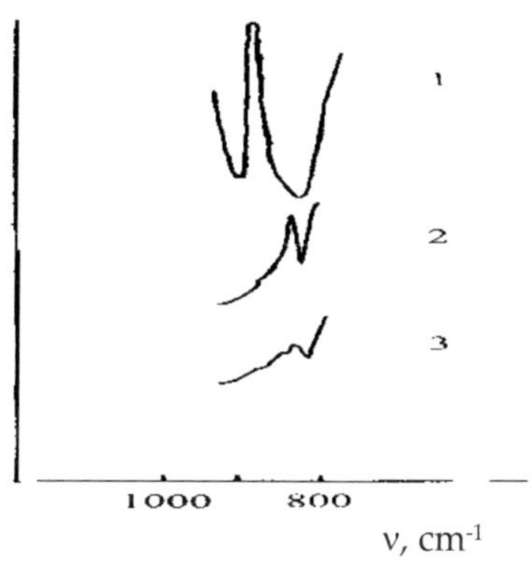

Figure 15. IR spectra of epoxy resin. 1 - in the absence of dispersion; 2 - dispersion for 0,5 h; 3 - dispersion for 1 h.

Extraction data of polymer-zeolite system showed the physical and chemical nature of the adsorption bonds between molecules of resin and aluminum silicate surface. Herewith, the contribution of chemisorption bonds varies depending on the ratio of the components and process duration.

Thus, a modified aluminum silicate which can be used effectively to create a variety of organo-mineral systems with a complex of valuable properties was obtained by combined dispersion of natural zeolite and epoxy resin.

6. Conclusion

The study of sorption activity of natural and modified sorbents has made it possible to substantiate the decisive role of the nature of the acid-base properties of a mineral in the process of extraction, as well as the influence of the porous structure upon sorption of organic compounds (acetone, toluene, acetic acid) and inorganic ions (the ions of nonferrous metals).

Physico-chemical study of the composition and thermal stability of natural zeolite allowed to determine the ratio of Si / Al as a 3,8, a high thermal stability, and to find optimal conditions for acid activation of the zeolite with the aim further modification of the samples with epoxy and amine reagents. It is shown that, depending on the nature of adsorbate and adsorbent type can be selected appropriate modes of activation of natural zeolite as a result of vapor sorption studies of various organic solvents, water and acetic acid.

It has been shown for the first time by the method of IR-spectroscopy that a chemical interaction takes place between the main centers of natural zeolite and carboxylic groups of aminoacetic acid in the process of their mechanic-chemical activation, as well as the formation of a coordination bond between the aluminum atoms of its lattice and nitrogen of aminoacid. Stereoscopic studies have established considerable changes in the chemical structure of the polymer as a result of mechanic-chemical modification of natural zeolite by epoxy resin. The extraction data of the system polymer-zeolite testify to the physical and chemical nature of adsorption bonds between the molecules of resin and the surface of alumosilicate. Incidentally, the contribution of chemosorption bonds is different in dependence of the ratio of the components and duration of the process.

Development of methods for modification of natural zeolite by various oligomers and functional materials opens up the possibility of synthesis of chemically modified materials for ion-exchange technology. Therefore, an interest is evoked by a profound research of the processes, proceeding on the surface of the modified natural sorbents, the state of engrafted compounds and creation of methods of purposeful alternation of the structure of the engrafted layer and the properties of the modified sorbents, connected with this structure, as well as establishing of a possibility to use them in the modern sorption processes, creation of new competitive technologies for the extraction of metal ions from technological solution.

The profound theoretical and practical studies of the methods of modification of natural sorbents will result in the development of available methods of obtaining of organomineral sorbents as the most problematic aspect of the development of the modern ion-exchange technology and related fields. And the achievements in this field should be connected, first of all, with the physical-chemical approach to the assessment and generalization of the existing vast experimental and theoretical material.

Author details

Aiymgul M. Akimkhan
Laboratory of Ion Exchange Resins, A.B. Bekturov Institute of Chemical Sciences, Ministry of Education and Science, Kazakhstan

7. References

[1] Akimbaeva A.M. The development of nanomaterials on the basis of natural zeolite. In: Proceedings of the XVIII Mendeleev congress on general and applied chemistry. Section 2. 23-28 September 2007, Moscow, Russia.

[2] Choudary N.V., Newalkar B.L. Use of zeolites in petroleum refining and petrochemical processes: recent advances. Journal of porous materials. 2011;18(6)685-692.

[3] Macedo-Miranda M.G., Olguín M.T. Arsenic sorption by modified clinoptilolite–heulandite rich tuffs. Journal of Inclusion Phenomena and Macrocyclic Chemistry. 2007;59(1-2) 131-142.

[4] Akimbaeva A. M., Ergozhin E. E. Blumich V., Demko D. Diagnostics of the structure and evaluation of catalytic activity of natural ceolite. Petroleum Chemistry.2006; 46(3)204-213.

[5] Cicishvili G.V., editor. Nature zeolites. Moscow: Chemy; 1985.

[6] Chelichev N.F., Volodin V.F., Krjukov V.L. Ion exchange properties of natural high-silicon zeolites. Moskau: Nauka; 1988.

[7] Tretyakov S.Ja. Study of sorption of radionuclides ^{90}Sr and ^{137}Cs on Natural Sorbents in Model Ecosystems. Radiochemistry 2002; 44(1)89-91.

[8] Akimbaeva A. M., Ergozhin E. E. The Kinetics of Sorption of Lead Ions on Clinoptilolite in the H-Form. Russian Journal of Physical Chemistry A. 2008; 82(3)397–400.

[9] Ergozhin E. E., Akimbaeva A. M. Purification of Phenol-Containing Aqueous Solutions by Natural Zeolite and Its Aminated Form. Petroleum Chemistry. 2005;45(4,)292-295.

[10] Treppel B. Chemisorption. Foreign Literature Publishing House; 1969.

[11] Korneytchuk G.K., Tsherbatyuk N.E., Froer Y.F. Adsorption of acetone and carbon tetrachloride and liquid mixtures of natural and modified clinoptilolite. Russian Journal of Physical Chemistry A. 1973;47(9)2391-2393.

[12] Bykov V.T., Tsherbatyuk N.E. Sorption of water vapor in natural zeolites and the effect of the mobility of water molecules in zeolites in the desorption process. In.: Nature sorbents. Moskow: Nauka; 1967.p156-163.

[13] Dudkin B.N., Louchina I.V., Yusupov V.P. (et al). Mechanochemical activation of kaolinite in the presence of concentrated sulfuric acid. Russian Journal of Applied Chemistry. 2005;78(1)36-40.

[14] Grigorieva T.F., Vorsina I.A., Vorsina I.A. (et al). Interaction in the kaolinite-aminoacetic acid during mechanochemical activation. Inorganic materials. 1997; 33(8)998-1000.

[15] Akimbaeva A. M., Ergozhin E. E. Estimation of Structural and Sorption Characteristics of Activated Bentonite. Colloid Journal. 2007;69(4)401-407.

[16] Akimbaeva A. M. Solid-Phase Interaction of Natural Zeolite with Aminoacetic Acid. Russian Journal of Physical Chemistry A. 2008; 82(2)315–317.

[17] Bellamy L. J. The Infra-Red Spectra of Complex Molecules. London: Methuen; 1960(Moscow: Foreign Literature Publishing House; 1963).

[18] Basyuk V. A., Prikl Zh. IR spectra of surface compounds of carboxic acids on silica in the 1500 - 1800 cm^{-1}. Spectroskopy. 1994; 60(1–2) 39-43.

[19] Lipatov Yu.S. Physico-chemical basis of filling polymers. Moscow: Chemy; 1991.

[20] Ergozhin E.E., Akimbaeva A. M. Organomineral system based on oligomers. In: Proceedings of the IX International Conference on the chemistry and physical chemistry of oligomers, 13–16 september 2005, Odessa,Ukraine.

Chapter 11

Thermodynamic Study of the Synthesis of Zeolites from Coal Ash and Its Use as Sorbents for Heavy Metals

Marisa Nascimento, Patrícia F. Prado, Paulo Sérgio M. Soares and Vicente P. de Souza

Additional information is available at the end of the chapter

http://dx.doi.org/10.5772/51704

1. Introduction

Species of toxic heavy metals cause serious damage to the ecosystem and as a result of this fact there is an increase in research on processes for wastewater treatment. Many of these processes are based on adsorptive properties or ion exchange some of these materials which immobilize the heavy metal species. Recently, various materials of natural or synthetic origin, such as bagasse, coal ash, carbonates, phosphates and zeolites, have been tested for its sorption capacity. Zeolites are commonly used for sorption of heavy metals due to their physical and chemical properties (thermal stability, defined molecular structure and ion exchange capacity).

The synthesis of zeolites from some mineral phases has been investigated by several researchers. It is well established that the coal ash composed mainly of phases such as quartz (SiO_2), mullite ($Al_{4+2x}Si_{2-2x}O_{10-x}$), hematite ($Fe_2O_3$) and magnetite ($Fe_3O_4$) can produce zeolites from alkaline hydrothermal treatment. But the mechanism of hydrothermal synthesis is still not very detailed. In general, increasing the temperature tends to favor the formation of hydrated aluminosilicate phase.

Zeolites synthesized from the coal ash, its adsorptive capacity are dependent on certain reaction conditions, for example, the concentration of metal cation in solution and the temperature at which sorption occurs. However, temperature and reaction time in the stage of the synthesis zeolite are also considered important variables affecting the behavior of the adsorptive materials such as zeolites of heavy metals in solution.

This chapter presents a study based on the sorption of cations of heavy metals by zeolites synthesized from hydrothermal treatment of coal ashes produced in Brazil.

INTECH
open science | open minds

© 2012 Nascimento et al., licensee InTech. This is a paper distributed under the terms of the Creative Commons Attribution License (http://creativecommons.org/licenses/by/3.0), which permits unrestricted use, distribution, and reproduction in any medium, provided the original work is properly cited.

2. Literture review

2.1. Coal

With the progress of humanity and its new challenges, we started to research the various types of energy production coupled with environmental preservation as a tool for sustainable development. Nowadays, the search for new processes using renewable energy sources is increasing, but the non-renewable energy sources occupy yet a large portion of production[1].

Although the use of coal represents a small share in electricity generation, its use is extremely important because it's the non-renewable energy resource most abundant of the globe.

The main environmental problems related to use of coal in power generation are: changes in vegetation and geomorphology of mining activities, formation of acidic water in the handling of coal due to the presence of pyrite, a large amount of waste in the process of improvement, gaseous emissions after combustion and the generation of potentially toxic waste.

One way to reduce the environmental impacts of disposal of these wastes in the environment involves the expansion of the potential for their use. An alternative use of these residues is the processing of coal ash in an sorbent low cost. The ashes of coal are composed mainly of silica and alumina, and you can convert them into zeolitic material after hydrothermal treatment in alkaline medium. The zeolitic material is characterized by high cation exchange capacity and good sorption allowing numerous potential applications.

The ashes of coal are composed mainly of silica and alumina, and you can convert them into zeolitic material after hydrothermal treatment in alkaline media. The zeolitic material is characterized by high cation exchange capacity and good sorption allowing numerous potential applications [2].

2.2. Coal ashes

Coal ashes ares inorganic solid wastes generated after burning coal in power generation processes in power plants worldwide. The formation of ash is achieved by the direct combustion of lignite, which is a solid raw material, consisting of two intimately mixed fractions, an organic (volatile matter more fixed carbon) and a fraction (mineral clay, quartz, pyrite, carbonates, etc.). By the action of heat, generates volatile organic fraction and coke, while the mineral fraction becomes gray with a mineralogy modified in view, the water loss of clays, carbonates and decomposition of sulfide oxidation [3].

The combustion of pulverized coal occurs at high temperatures, between 1200 and 1600º C in an oxidizing gaseous environment and for total or partial melting of matter [4].Three main wastes are generated from coal combustion in power plants, which are:

- Light Gray or steering wheel: they are extremely fine particle size of less than 0.15 mm, is collected in cyclones mechanical or released into the atmosphere;
- Heavy ash, are coarser particle size, is typically agglomerated at the bottom of the furnace being removed by a water flow;
- Slag or ash thick, coarse grain size and has high content of unburned carbon from 10 to 20%;

The fly ashes are usually composed of 30-60% SiO_2, 10-20% Al_2O_3, 5-10% Fe_2O_3, 5-10% MgO and 2-4% CaO, and other compounds.

In the process of coal combustion, these minerals are partially fused particles form ashes and light in which the crystalline phases such as quartz (SiO2) and mullite (3 Al2O3.2SiO2) remain in the core, while the aluminosilicate (Al2O5 Si) remains the surface [5].

The ash generated from burning coal in power plants in Brazil are comprised 65-80% fly ash and bottom ash of 15-35% [6].

In general, mineral coal ashes are aluminosilicates formed by amorphous and crystalline phases and the pH of ash ranges from 4.5 to 12 depending on the characteristics of the carbon precursor geochemical [7].

From the standpoint of generating electric power, coal ash waste are considered as residues, however, to evaluate its characteristics and possible uses, these can be considered as a resource that can be widely used and exploited [8].

The storage of the dry coal ash has the major advantage that its physicochemical properties of the ash remain unchanged, which constitutes a buffer for further use [9].

Several potential applications of coal ash have been developed or are under development at research centers all over the world. These applications are directed primarily to remove toxic metals present in industrial wastes [10] in the building materials industry [11], and the synthesis of zeolites that have been studied by many researchers [12].

2.3. Zeolites

Zeolites are hydrated aluminosilicates formed by three-dimensional crystal structures of SiO_4 tetrahedra and AlO_4, linked together by four oxygen vertices.In this configuration, the negative charges of AlO_4 tetrahedra are compensated by interstitial cations (Na^+, K^+, Ca^{2+} and Ba^{2+}) and form an open structure with large channels, where water and other molecules can stay and have considerable freedom of movement, permitting ion exchange and reversible hydration [13] .

Zeolites include a wide range of mineral and synthetic products, whose structure has an electrical imbalance, as it has an excess negative charge on each atom of aluminum present.Therefore, atoms of alkali metals and alkaline earth metals cause the balancing of charge, can easily be exchanged for other ions. This transfer of material between the spaces intracristalinos is limited by the pore diameter that ranges from one to another zeolite.

In nature, are usually found in deposits associated with the activation of alkaline volcanic rocks. As naturally occurring normally have a high content of impurities and / or do not possess the properties required for their use, therefore, be synthesized zeolites started [14] .

Zeolites exhibit features as high selectivity, high stability and high exchange capacity, which together with the properties of sorption and catalysis, provide their technological application in various sectors [15] .

One advantage of synthetic zeolites is to provide uniformity in size and shape of the channel, another is its chemical composition pre-defined in relation to the purposes intended. Considering, however, its high cost, synthetic zeolites are reserved for applications that require features more uniform structure and composition, as in the catalysis of hydrocarbons and in the detergent industry. The zeolites type A, X and Y are the predominant for commercial use as ion exchangers and adsorbers [16] .

Among various uses, natural zeolites can be applied to wastewater treatment for removal of toxic metals, in removing odors in the air purification and in the conditioning of soil [17].

The fly ash are sources of aluminum and silicon, which are the main elements in the composition of the zeolites, because of this, the high content of reactive stages of ash and fine particle size thereof, are considered excellent feedstock for the synthesis of zeolites[12] .

2.4. Hydrothermal synthesis of zeolites

The synthesis of zeolites from some mineral phases has been investigated by several researchers. It is well established that the coal ash composed mainly of phases such as quartz (SiO_2), mullite ($2SiO_2.3Al_2O_3$), hematite (Fe_2O_3) and magnetite (Fe_3O_4) can produce zeolites from alkaline hydrothermal treatment. By doing the necessary adjustments in the Al / Si, the ashes can be used for the synthesis of zeolites such as Zeolite Na-A, faujazita, zeolite Na-P and philipsita [18] . [19] report the synthesis of zeolite-type Na from coal ash. The synthesis methodology is focused on so-called "activation" of gray, or was more popularly known as "hydrothermal synthesis" in closed systems for solutions of NaOH or KOH.

Pure zeolites are synthesized from supersaturated solutions of silicates and aluminates under stringent experimental conditions [20] .

It is therefore possible to synthesize zeolite from coal ash by the hydrothermal treatment or in aqueous solution at elevated pressure and temperature [14] .Several articles have proposed various methods for the hydrothermal activation of coal ash in zeolite synthesis, using for this variation of several parameters involved in this synthesis [12,21-24].

The zeolites that have been obtained by hydrothermal treatment of the coal ash with sodium hydroxide (NaOH) are analcime, zeolite A, zeolite P (gismondina), zeolite X and Y (faujasite), sodalite, chabazite, gmelinite and cancrinite .

All methods are developed based on the dissolution of the phases of fly ash supported by Al and Si with alkaline solutions and subsequent precipitation of the zeolitic material [12].

According to [25], the hydrothermal reaction mechanism involves stages of dissolution, condensation or gelation and crystallization.

The dissolution of the ash is between 20-120° C. The amorphous silica, quartz and mullite, which are the phases of Al and Si are present in the ash dissolved. The kinetics is very dependent on the concentration of OH^-, which is responsible for dissolving the reagents and keeps them in solution, providing its saturation to occur the formation of zeolites.

The next step is the condensation or gelation which is the phase where the reaction occurs between aluminate and silicate ions with the formation of aluminosilicate gel (hydrogel). In the final stage of crystallization the hydrogel is transformed into a zeolite crystal (the reaction speed is dependent on the amount of Na^+). The crystallization of zeolites generally results in a final product containing 40-60% of zeolite.

The zeolite sinthesis from the coal ash is therefore an equilibrium reaction between the alkaline solution and solid phase. By raising the temperature, the solubility of silica and alumina ions increases and condensation reactions for the formation of crystal nuclei are initiated. The crystal growth leads to a complete dissolution of the original material for the formation of amorphous phases zeolite.

Depending on the experimental conditions and the chemical composition of the ash used, one obtains different zeolites. The most important parameters in the processes of zeolitização by hydrothermal process are the chemical composition of ash concentration and type of agent activation, the ratio liquid / solid, temperature, reaction time and intensity of agitation [9].

Each zeolitic material obtained has application in accordance with its characteristics and properties. At least 15 different types of zeolites can be produced from a gray by varying the activation parameters. The activation conditions may be optimized to maximize the adsorptive capacity of the product obtained and the production costs [12] .

Most studies on the use of zeolites derived from coal ash describes a major potential applications of this zeolites as the sorption of toxic metal ions from solutions of pollutants under laboratory conditions [12].

2.5. Sorption

The sorption process is the separation of components of a mixture through a mass transfer phenomenon when a component in the mixture which may liquid or gaseous called adsorbate is in contact with a solid called sorbent. When the two phases are in contact, the dilute component in the mixture is transfered the liquid phase to the surface of the sorbent [2] .The sorption process depends on several factors such as nature and concentration of the adsorbate, the characteristics of sorbent and the sorption conditions (pH, temperature, ratio solid\liquid).

The sorption phenomena are classified into two types: physical sorption and chemisorption. In physical sorption occurring weak interactions like Van der Waals, featuring a reversible process . By not having any sort of change in the nature of the species involved, the physical sorption is a non-specific, and can occur for different adsorbates [26] .In the chemical sorption or chemisorption, the chemical species involved are altered, as there is an effective exchange of electrons between the solid and the adsorbed molecule, resulting in the formation of a monolayer on the solid surface [27] .

If the sorption of one or more ionic species is accompanied by simultaneous desorption of an equivalent amount of ionic species, this process is regarded as ion exchange. The shapes of sorption usually encountered in environmental studies are the physical sorption, chemical sorption and ion exchange [2] .Being essentially a phenomenon of sorption surface, that has an sorption capacity of sorbent significant, must have large surface area, which implies a highly porous structure.

The sorbent materials, in turn, are natural substances or synthetic crystalline structure whose inner surface of the pores is accessible by a selective combination between the solid and solution [26] .

The sorption isotherms are the main way of studying the capacity to remove toxic metals from different solid sorbents [28] .

2.6. Sorption isotherms

To obtain the sorption isotherms is the first step for quantitative evaluation of the sorption mechanisms. The data obtained from the isotherms can be used to project and define the operating conditions of industrial equipment that are based on the principle of sorption [1].

Isotherms relates the concentration of the adsorbate fluid phase and solid phase at a certain temperature and are represented in graphic form, presenting in various ways that reflect the behavior of the mechanism of sorption. In general, the isotherms convex represent usually the solid microporous sorbent. The more complex shapes may be associated with multilayer sorption or the varying sizes of pores of the sorbent material.

We can cite as examples, the isotherms of Henry, Langmuir, Freundlich, BET and Radke-Prausnitz. The Langmuir and Freundlich isotherms are the most commonly used to demonstrate the balance of removing a metal ion sorption, and are the most used models to describe the mechanisms of sorption on zeolites [29].

To provide us with data on the mechanisms of sorption isotherms can be linearized and applied to mathematical models.

3. Materials and methods

The fly ash used in this study originated from a coal thermo power station located in southern Brazil that operates with pulverized Brazilian coal. Ashes and synthetic zeolites produced by hydrothermal treatment were previously characterized chemically and mineralogically and subsequently used in sorption experiments. These experiments were performed with synthetic solutions of sulphate salts of Pb (II), Cu(II), Zn(II), and Mn(II) prepared with analytical grade reagents in deionized water.

3.1. Zeolite preparation

The coal fly ash was submitted to hydrothermal treatment at different temperatures, reaction times, NaOH concentrations, solid/ liquid ratios (S/L) and aluminum/silicon (Al/Si) ratios as shown in Table 1. A Plackett–Burmann experimental design (2^{5-2} fractional factorial design) was used in order to identify the most important variables in a preliminary analysis [30]. The tests were performed in duplicate. The molar ratio Al/Si was modified by adding analytical grade Al_2O_3 in the reaction media.

The hydrothermal treatment was carried out in a 450 ml reactor PARR-4562, made of Nickel-200 equipped with a turbine impeller and stirred constantly at 300 rpm. The reaction products were filtered and washed with water to remove the excess of sodium hydroxide.

The mineralogical characterization of the zeolites as well as of the ashes was carried out by X-ray diffraction (XRD) in a Bruker – AXS D5005 powder diffractometer with Goebel mirror and CoKa (35 kV/40 mA) radiation.

Tests	Concentration (mol\l) NaOH	Temperature (ºC)	Time (h)	Ratio Al\Si	Ratio Solid\Liquid (S\L)
1	2.0	100	0.5	1.00	1\6
2	5.0	100	0.5	0.51	1\8
3	2.0	150	0.5	0.51	1\6
4	5.0	150	0.5	1.00	1\8
5	2.0	100	6.0	1.00	1\8
6	5.0	100	6.0	0.51	1\6
7	2.0	150	6.0	0.51	1\8
8	5.0	150	6.0	1.00	1\6

Table 1. Experimental conditions of the hydrothermal experiments

3.2. Preliminary evaluation of the response variable

The zeolites as well as the coal fly ash were washed up to pH 9 and subsequently dried at 60ºC for 24 h before the sorption experiments. The sorption capacity was determined by contacting, 50 ml of Mn and Cu solutions (100 mg/l) with 0.5 g of zeolites in plastic bottles. The solution pH of each test was chosen conveniently, between 4 and 5 to avoid cation precipitation.

The bottles were shaken for 2 h at 180 rpm in a KS501 IKA shaker and the solids were filtered with Whatman filter paper. The concentrations of metal ions of all tests were determined in a Varian Atomic Absorption Spectrometer - model Spectra 50B.

The sorption capacity was defined as a percentage and calculated by the equation (1):

$$\text{Sorption capacity (\%)} = ((C_i - C_e) \backslash C_i) \times 100 \tag{1}$$

C_i and C_e are respectively initial and final concentrations of the metal ion in solution.

3.3. Sorption experiments

The sorption experiments for the heavy metals, Cu^{2+}, Pb^{2+}, Zn^{2+}, and Mn^{2+} with synthesized zeolite were carried out using the shaking device and methodology described previously in Section 3.2. A concentration range from 100 to 3000 mg/l was used for each cation. Only the zeolitic material from test 8 (Table 1) was used as sorbent.

The obtained data were plotted and adjusted with isotherm sorption models to analyze the cations sorption onto the sorbents.

3.4. Investigation of temperature effect – Thermodynamic study

Samples of 30 g ash were mixed with 150 ml of a 3 mol/L NaOH solution. This mixture was reacted in an autoclave Parr - 4562, made of nickel alloy 200, equipped with a turbine impeller for 2 hours and 350 rpm agitation and temperatures of 50, 100, 150, 200 and 250 ° C. The product formed after cooling was filtered, washed with 2 liters of distilled water and dried at 60 ° C for 24 hours.

4. Experimental results

4.1. Characterization and preparation

The XRD analysis of the ashes used for zeolite preparation is shown in figure 1. In according to the results of X ray diffraction, the sample ash is mainly formed by mullite phases ($Al_6Si_2O_{13}$) and quartz (SiO_2) with some traces magnetite (Fe_2O_3), shown in Figure 1. Chemical analysis carried content was 56.8% of SiO_2, 24.5% of Al_2O_3 and SiO_2/ Al_2O_3 ratio of 2.32.

Figure 2 shows the XRD of all products of the hydrothermal treatments. It can be observed that treatments resulted in formation of zeolitic phases as zeolite A, cubic analcime, philipsite, hydroxycancrinite and $Na_8(AlSiO_4)_6(OH)_2$ $.2H_2O$. Table 2 shows the sorption capacity obtained for the zeolitic products. The results show that a hydrothermal treatment can increase 10–25 times the % sorption when comparing with the value of the original ashes. The sorption of Mn^{+2} is higher than 85% and the absorption of Cu^{+2} reaches up to 99%. In the case of a solution with 100 ppm of Mn the sorption increased about 25 times when compared with the results of the test for the same Mn concentration performed with ashes with no hydrothermal treatment. The increasing was about 16 times for Cu^{+2}. The tests 1 and 2 showed small sorption capacity for cations tested. In figure 2, these tests did not show zeolitic phases.

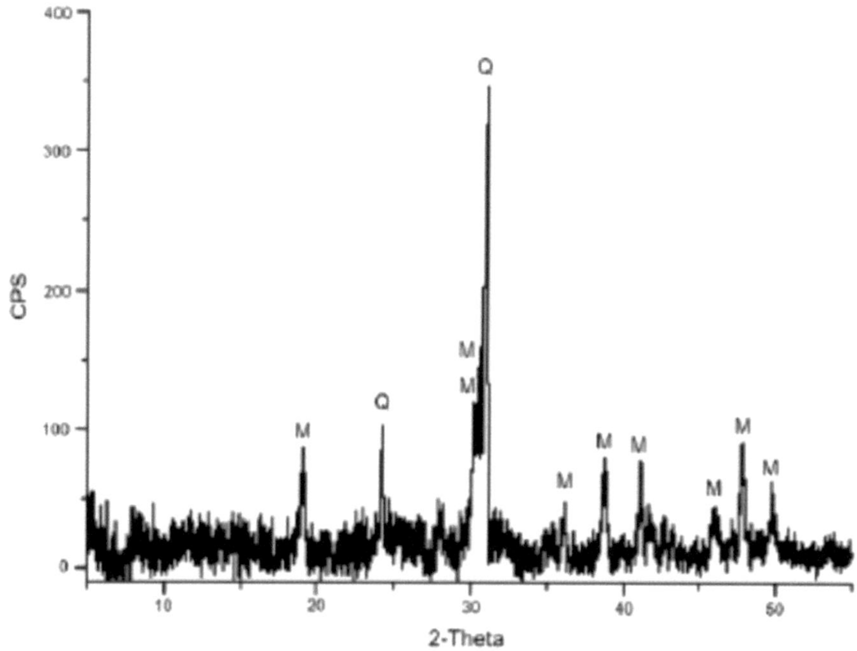

Figure 1. XRD pattern of coal fly ash. Mineral abbreviations: M, mullite; Q , Quartz.

Test	Cu^{+2}	Cu^{+2}	Mn^{+2}	Mn^{+2}
	C_e (mg/l)	Sorption capacity %	C_e (mg/l)	Sorption capacity %
1	37.43	62.57	42.80	57.20
2	34.36	65.64	66.19	33.81
3	1.66	98.34	15.98	84.02
4	0.37	99.63	26.48	73.52
5	1.11	98.89	24.19	75.81
6	0.02	99.98	21.77	78.23
7	0.11	99.89	15.25	84.75
8	0.22	99.78	14.40	85.60
Coal fly ash	93.95	6.05	96.60	3.40

Table 2. Results of Cu^{+2} and Mn^{+2} sorption

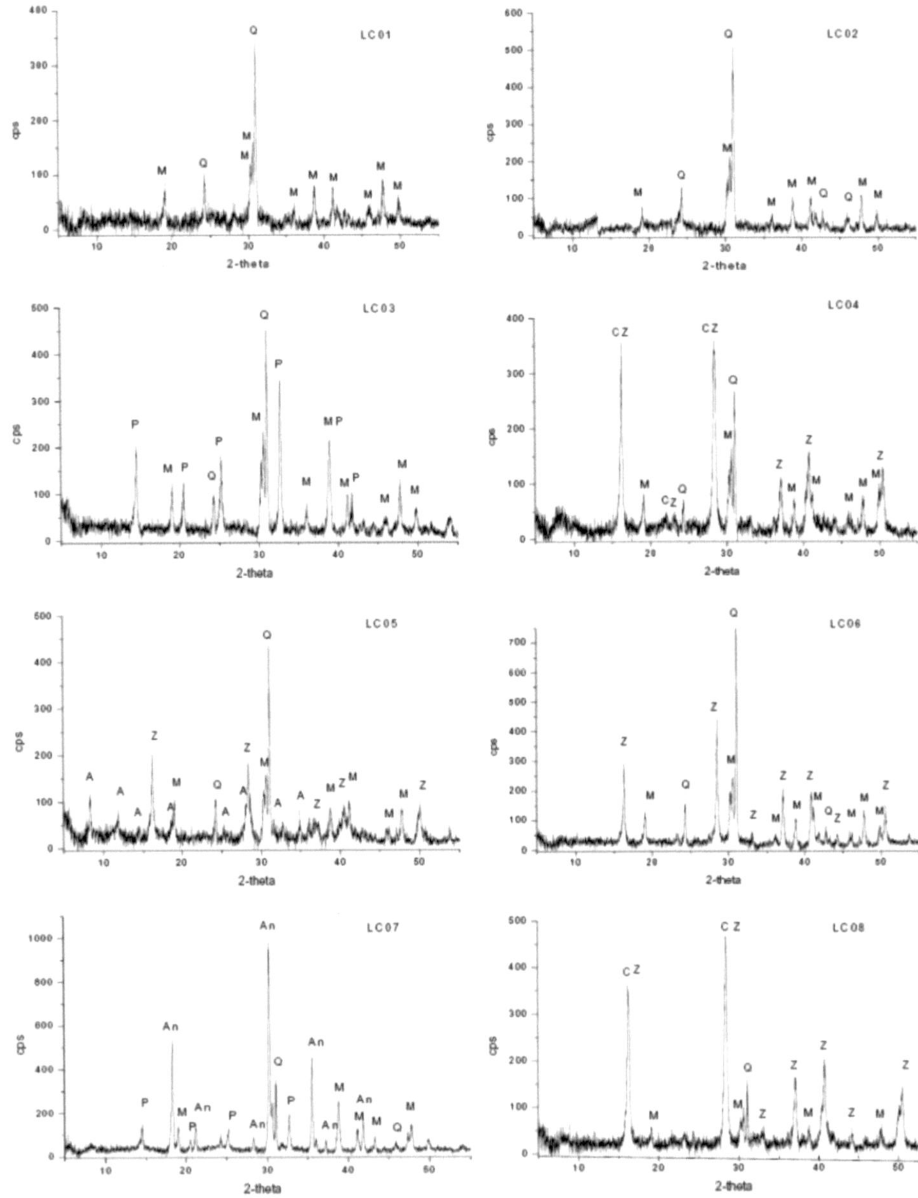

Figure 2. XRD pattern of modified coal fly ash. Phases abbreviations: M, mullite; Q, quartz; Z, zeolite $Na_8(AlSiO_4)_6(OH)_2 . 2H_2O$; C, hydroxycancrinite; A, zeolite A; An, analcime; P, zeolite P. the legends LC01 to LC08 are related with tests 1-8 from Table 1.

Figure 3 and 4 show the results of preliminary statistical analysis and the effects of the variables. The *F*-test was used to identify the most significant variables in the hydrothermal process. The significance level (p-value) adopted was 0.05. Temperature and time were the most significant variables in the synthesis of the zeolites. An increase of the level of these variables tends to increase the sorption capacity for Mn and Cu.

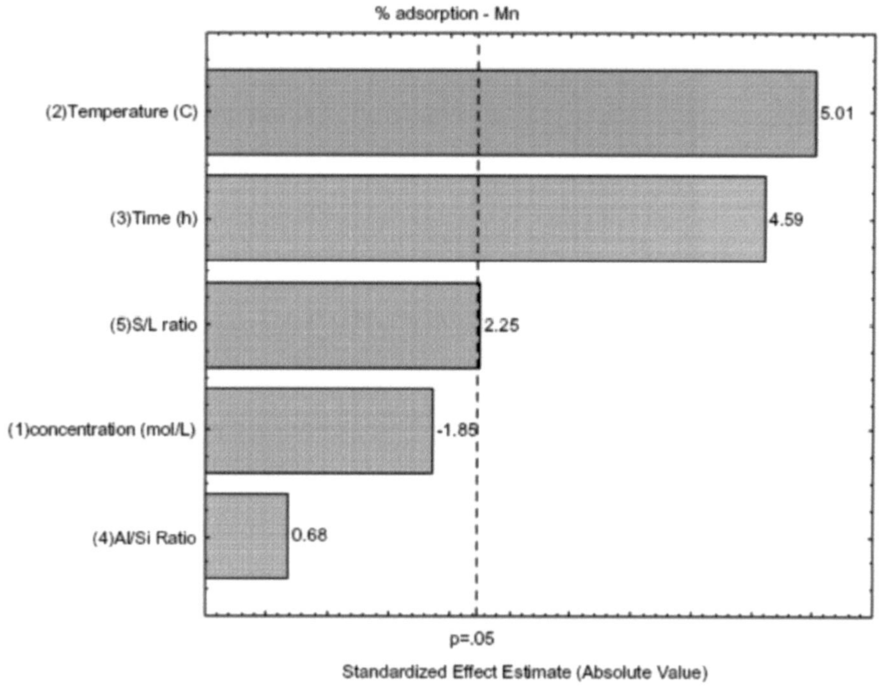

Figure 3. Pareto chart of standardized effects to fatorial planning - % Mn sorption (p-level = 0.05)

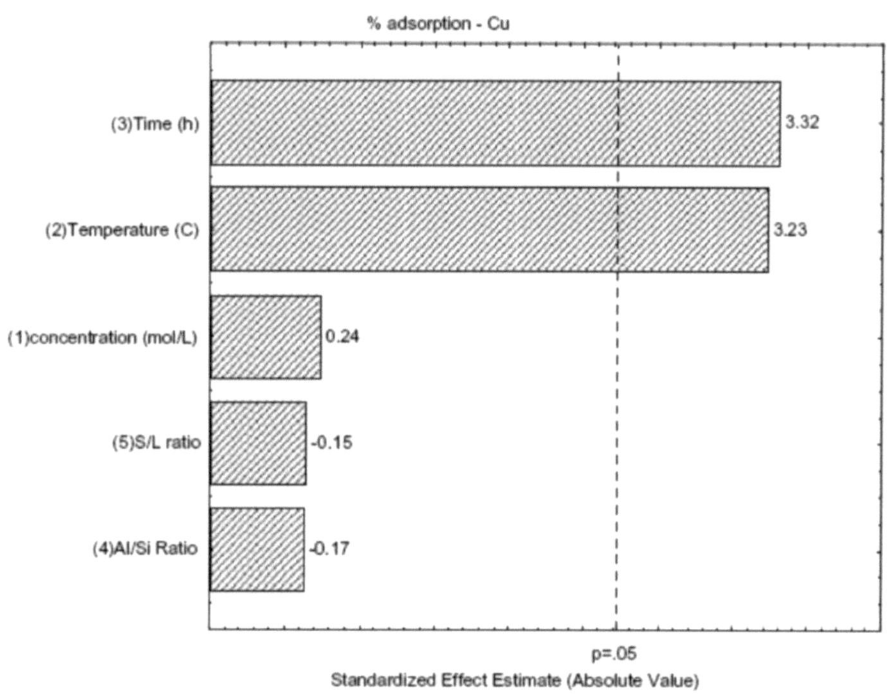

Figure 4. Pareto chart of standardized effects to fatorial planning - % Mn sorption (p-level = 0.05)

Several authors suggest that changes in synthesis temperatures lead to different zeolitic phases. The differences among sorption capacities of the various zeolites synthesized may be credited to the different zeolitic phases present in the products of the hydrothermal synthesis [25,31-32]. An increase in the reaction time tends to promote a better crystallization of the phases formed, which it might also explain the increase in the of sorption capacity [33].

4.2. Sorption studies

The sorption of different metal ion concentrations onto synthetic zeolite at 25^0 C was studied for Cu^{+2} , Pb^{+2}, Zn^{+2}, and Mn^{+2} in the range 100-3000 mg\L keeping all other variables constant. The results are shown in figure 5. The sorption for Cu^{+2}, Pb^{+2}, Zn^{+2}, and Mn^{+2} increases with increasing metal concentration in aqueous solutions. These results indicate that energetically less favorable sites become involved when the concentration of metal in solutions increases. The metal uptake can be credited to different mechanisms of both ion-exchange and sorption. During the ion-exchange process, metal ions have to move through the pores of the zeolite, but also through channels of the lattice, and they have to replace exchangeable cations (mainly sodium and calcium). Difusion is faster through the pores and is retarded when ions move through channels of small diameter. In this case the metal ion uptake can mainly be credited to ion- exchange reactions in the porous of the zeolitic samples [34].

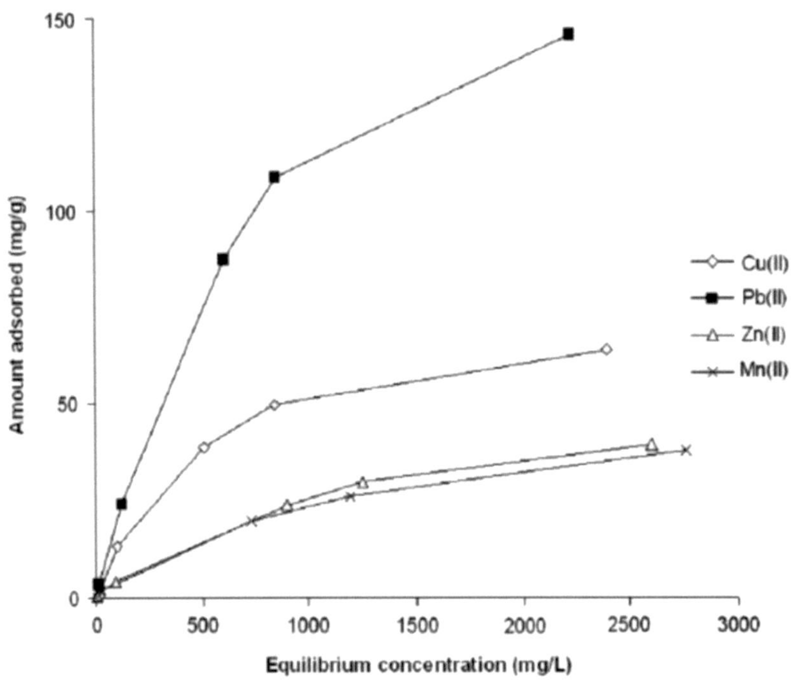

Figure 5. Sorption isotherms of de Mn^{+2} ,Cu^{+2} ,Zn^{+2} , Pb^{+2} .

The preferred order observed for sorption was Pb>Cu>Zn>Mn. In the literature, similar results were obtained when the sorption capacity of a large variety of zeolite minerals for cadmium, copper and zinc and revealed that zinc had the lowest sorption for all zeolites synthesized [31,35-36]. Zeolites obtained under same conditions as philipsite and chabazite had limited sorption capacity (CA) for zinc as compared to copper [37]. The sorption characteristics of Zn (II) onto pure fly ash showed that the solution pH was the key factor affecting the sorption characteristics [38-39].

4.3. Investigation of temperature results – Thermodynamic study

According to figure 6, near the temperature of 100°C, the X-ray diffraction indicates that no phase transformation occurs. According to literature, the reaction kinetics is highly temperature dependent, and it is believed that in this case, a larger reaction time would be required.

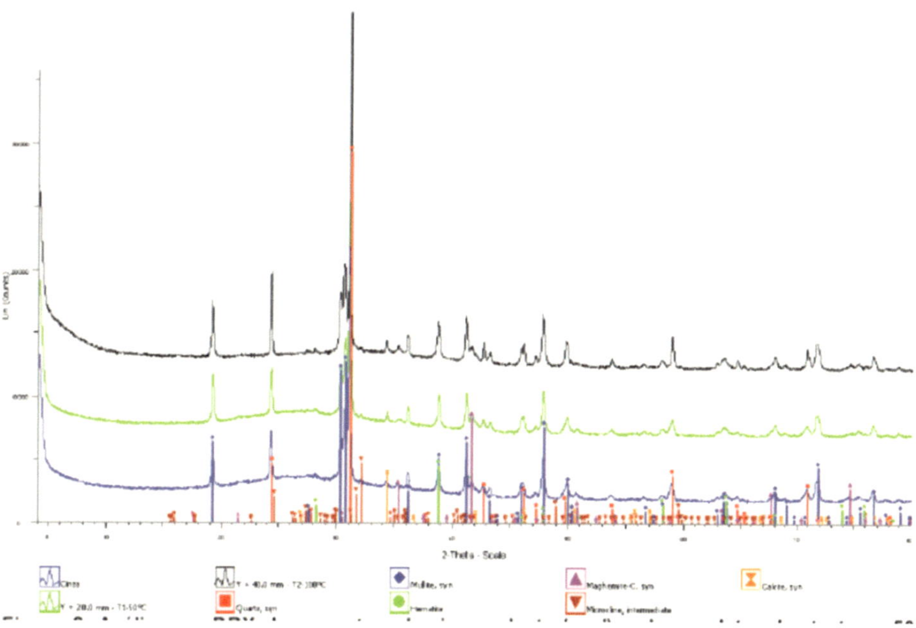

Figure 6. XRD analysis of samples of gray (blue) and the products of the tests at 50 and 100°C (green and black, respectively)

From a temperature of 150ºC (Figure 7), zeolite P was identified in the X-ray diffractograms have been replaced by analcime 200°C (Figure 8), which was later replaced by hydroxycancrinite phase at 250° C (figure 9).

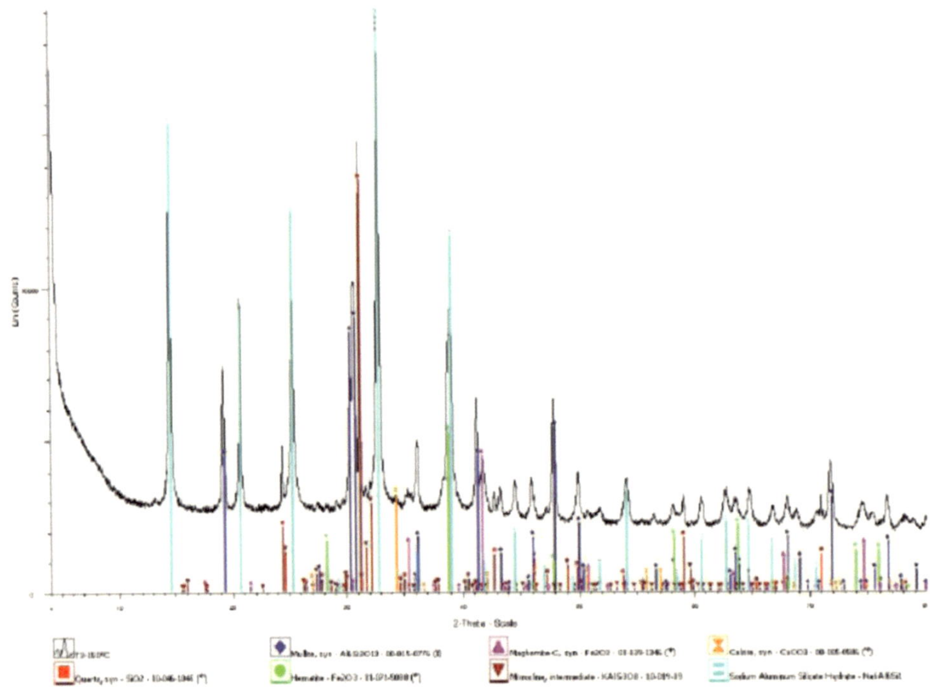

Figure 7. XRD analysis of the product for testing at 150 ° C

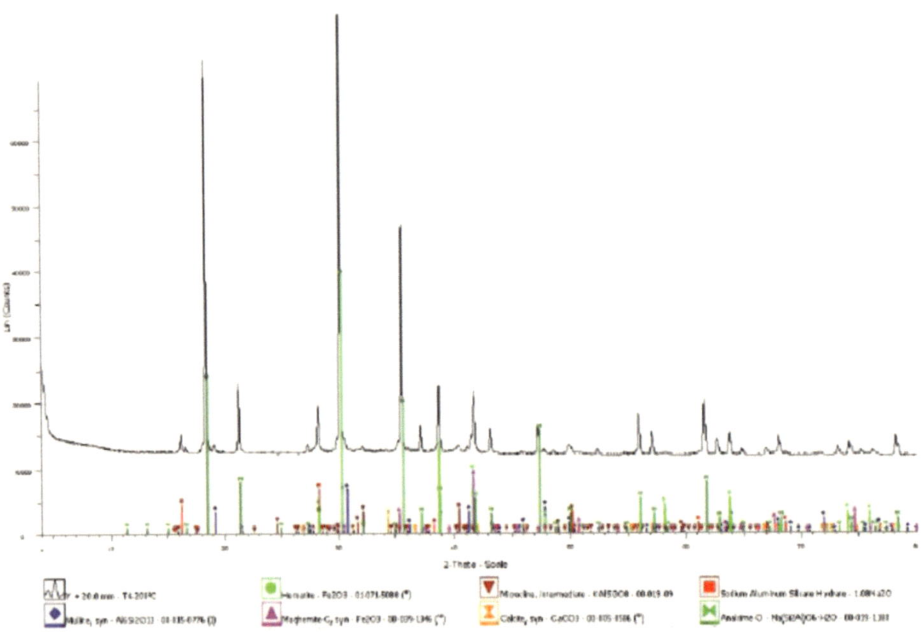

Figure 8. XRD analysis of the test product at 200 ° C

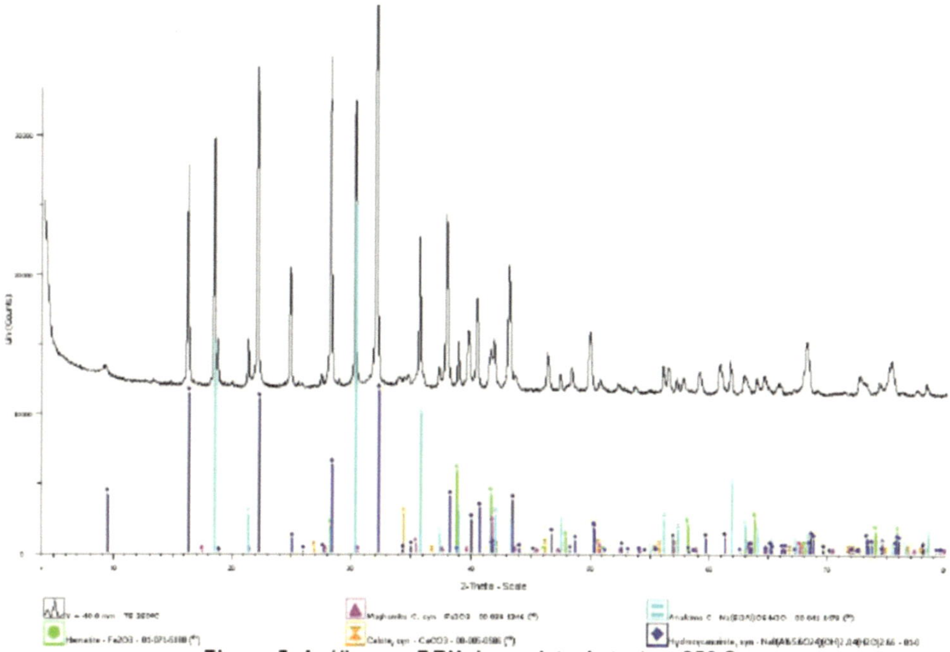

Figure 9. XRD analysis of the product for testing at 250 ° C

From the experimental results, we performed a thermodynamic study, to investigate the effect of temperature on the synthesis of these zeolite phases. For the species treated zeolite in this work, whose thermodynamic properties were not known, estimates have been proposed. Was used for this study a variation of the equation developed by [40] for estimating enthalpies of formation for the species treated in this work, to enable the calculation of free energies at several temperatures.

The proposed equation for calculating the enthalpy of formation of species (kJ/mol) is shown in equation 2, in which β_1 is a parameter calculated by nonlinear regression of data made from enthalpy of known species.

$$\begin{aligned} DH^{o}_{f} &= n_{NA} DH^{o}_{f}\left[NaOH\right] + n_{Al} DH^{o}_{f}\left[Al(OH)_3\right] + \\ &+ n_{Si} DH^{o}_{f}\left[Si(OH)_4\right] - (y-x) DH^{o}_{f}\left[H_2O\right] - b_1 n_{Na} R_{Na} \end{aligned} \tag{2}$$

The magnitudes n_{Na}, n_{Al}, N_{Si}, are the stoichiometric coefficients of the respective Na, Al and Si present in the molecular formula of each species. The quantity y is the coefficient of oxygen stoichiometric, x the stoichiometric coefficient of water incorporated and R_{Na} the radius of sodium ion (0.102 nm). The values of enthalpies of formation ($\Delta H^o f$) and the standard molar entropies ($S^o f$) species used in the calculations were taken from the database program HSC Chemistry 7.0

The calculation of nonlinear regression allowed us to estimate β_1 equal to 613.231, with a correlation coefficient of 0.999. In Table 3 are the values estimated by the model and the waste from the values taken from the database. One can observe that the maximum residue was 3.08% for the species andalusite.

Species	ΔH^o_f (kJ/mol)	ΔH^o_f (model) (kJ/mol)	Residues %	Residues
Cianite	-2581.097	-2604.19	0.894	23.091
Caulinite	-4119.599	-4085.29	0.832	-34.313
Pirofilite	-5637.900	-5618.33	0.347	-19.568
Paragonite	-5944.209	-5901.83	0.712	-42.374
Analcime 1	-3282.348	-3286.31	0.120	3.958
Jadeita	-3032.760	-3011.82	0.690	-20.943
Albita	-3927.659	-3921.25	0.163	-6.404
Analcime 2	-3309.841	-3297.65	0.368	-12.194
Metasilicate de sodium	-1561.511	-1600.57	2.501	39.058
Andalusite	-2686.965	-2604.19	3.080	-82.777

Table 3. First estimation results for the model enthalpy of formation

The estimation of standard molar entropy (S^of) was performed using a simple regression, taking into account only the stoichiometric coefficients of each species (Equation 3). The best results were achieved with a correlation coefficient of 0.97 for the model.

$$S^o_f = 27.148 + 46.608\, n_{Na} - 11.321 n_{Al} + 5.111 n_{Si} + 16.147\, n_O + 43.487\, n_{H2O} \qquad (3)$$

The thermodynamic data used for the calculation and the values estimated by the model S^of and residues, are shown in Table 4.

Species	Na	Al	Si	O	H_2O	S^o_f (J/molK)	S^o_f (model) (J/molK)	Residues	Residues %
Cianite	0	2	1	5	0	86.680	90.353	-7.667	8.845
Caulinite	0	2	2	7	2	205.016	214.744	-6.993	3.411
Pirofilite	0	2	4	11	1	239.171	246.059	-8.905	3.723
Paragonite	1	3	3	11	1	276.833	276.235	0.379	0.137
Analcime 1	0.96	0.96	2.04	6	1	226.776	211.822	11.577	5.105
Jadeite	1	1	2	6	0	133.499	169.542	-19.060	14.277
Albite	1	1	3	8	0	226.400	206.948	11.397	5.034
Analcime 2	1	1	2	6	1	223.802	213.029	10.129	4.526
Andalusite	0	2	1	5	0	104.600	90.354	-24.169	23.106
Sillimannita	0	2	1	5	0	95.790	90.354	10.778	11.251
Nefeline	1	1	1	4	0	123.000	132.136	2.572	2.091

Table 4. Results estimated of S^o_f by linear regression model.

To check the validity of the models surveyed, we compared the values of the standard energies of formation ($\Delta G^o{}_f$) of some species sodium zeolite, whose values of $\Delta G^o{}_f$ were already known. Using the values of $\Delta H^o{}_f$ estimated by the model developed $S^o f$ the value at 298 K calculated by linear regression is calculated to $\Delta G^o{}_f$ values for each species to be compared (Table 5).

Species	ΔG_f^o	ΔG_f^o (calculated)	Residues (%)
Analcime 1	-6144.3	-6138.166	0.100
Analcime 2	-6172.2	-6116.414	0.904
Analcime DH1	-5648.3	-5679.663	0.555
Analcime DH2	-5625.2	-5662.739	0.667
Na-clinoptilolite	-10270.1	-10218.262	0.505
Natrolite	-5330.7	-5384.351	1.006
Na-phillipsite	-7444.2	-7693.379	3.347
Zeolite 4A	-5214.3	-5174.898	0.756
Zeolite 13X	-5767.7	-5711.523	0.974

Table 5. Comparison between the values of $\Delta G^o{}_f$ at 298K (in kJ/mol) calculated from the models.

These values of $\Delta G^o{}_f$ were estimated using the program HSC Chemistry 7.0, using the values of enthalpy of formation and standard molar entropy of the species included in the required database program, and using the balanced chemical equation as the model reaction represented by Equation 4:

$$aNa + bAl + cSi + dO_2(g) + eH_2(g) \Rightarrow \text{estimated species} \quad (4)$$

Where a, b, c, d and e are the respective stoichiometric coefficients. The model of the reacting species (more stable forms of each element) were already in the database program HSC Chemistry 7.0. The maximum error obtained was slightly higher than 3%, showing that the estimate used seems appropriate.

4.4. Calculation of equilibrium

Using the estimated values of enthalpies of formation and standard molar entropy of analcime and hydroxycancrinite species, we calculated the values of free energies as a function of temperature reaction for the formation of analcime reactions (reaction A), hydroxycancrinita (reaction B) and zeolite P (reaction C) from the reaction of mullite with sodium hydroxide. Furthermore, it has been provided in the processing of analcime to hydroxycancrinite (reaction D), Zeolite P to hydroxycancrinite (reaction E) and analcime to zeolite P (reaction F).For each of these reactions studied were calculated the values of ΔG reaction and equilibrium constant at each temperature studied.

Reaction A)

$$Al_6Si_2O_{13} + Na^+_{(aq)} + OH^-_{(aq)} + 8H_2O = NaAlSi_2O_6^* + 5Al\ (OH)_3$$

Reaction B)

$$3Al_6Si_2O_{13}+8Na^+_{(aq)} + 8OH^-_{(aq)}+17.7H_2O = Na_8Al_6Si_6O_{24}(OH)_2{*}2.7H_2O +12Al(OH)_3$$

Reaction C)

$$5Al_6Si_2O_{13} + 6Na(+a) + 6OH(-a) + 45H_2O = Na_6Al_6Si_{10}O_{32}{*}12H_2O + 24Al(OH)_3$$

Reaction D)

$$3\ NaAlSi_2O_6{*}H_2O + 5Na(+a) + 5OH(-a) + 3Al(OH)_3 = Na_8Al_6Si_6O_{24}(OH)_2{*}2.7H_2O + 6.3H_2O$$

Reaction E)

$$3\ Na_6Al_6Si_{10}O_{32}{*}12H_2O + 12Al(OH)_3 + 22Na(+a) + 22OH(-a) = 5Na_8Al_6Si_6O_{24}(OH)_2{*}2.7H_2O + 46.5H_2O$$

Reaction F)

$$Na_6Al_6Si_{10}O_{32}{*}12H_2O = 5\ NaAlSi_2O_6{*}H_2O + Al(OH)_3 + Na(+a) + OH(-a) + 5H_2O$$

It was considered the activities of solid phases equal to 1 and that the activities of Na^+ and $OH^-$are equal.

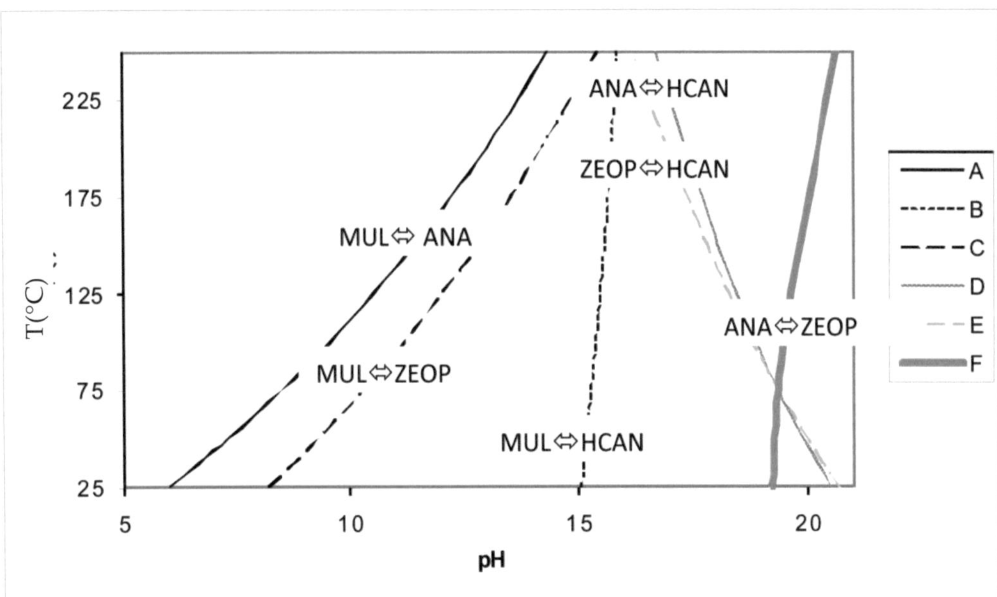

Figure 10. Phase diagram of stability of zeolite formed from the coal ash. (Abreviations: ANA – Analcime; MUL – Mulite ; ZEOP – Zeolite P ; HCAN- Hidroxycancrinite)

It was possible to make a survey of pH versus temperature diagram as shown in Figure 10. It is possible to check, at 25 ° C, a large region of stability of the zeolite P type, ie the reaction of the mullite forming zeolite P, at pH values between 8 and 15. This region decreases as the temperature increases.Starting pH 15, the reaction hydroxycancrinite form, rather than zeolite P. The stability region analcime phase (pH between 6 and 8 to 25 ° C) maintains a constant-width around two pH units, but with increasing temperature is shifted to pH also higher. Higher values of pH, therefore higher concentrations of NaOH, promote the transformation of phases formed in hydroxycancrinite. The remaining balance for the reactions D, E and F are present in high pH values, and in all cases the phase stability hydroxycancrinite prevails.

5. Conclusions

The study demonstrates the application of Brazilian fly ash for zeolite synthesis by hydrothermal treatment. Statistical analysis shown that an increase of time and temperature in zeolite synthesis tends to increase the capacity of sorption of ions tested onto products.

The products obtained are effective sorbent for removal of Pb^{2+}, Zn^{2+}, Mn^{2+} and Cu^{2+} ions from aqueous solutions. The performance of sorption increased up to 25 times when compared with the original ashes.

From the results obtained and discussed, it was concluded that the formation of zeolitic phases from coal ash is thermodynamically possible. The thermodynamic values obtained for the models in the literature and the proposed in this study had similar, and allowed the establishment of equilibrium diagrams that define the regions of thermodynamic stability of the species present in the reaction medium.

The analysis of these diagrams shows that the temperature and concentration of alkali present in the medium are the most important factors affecting the stability of species to be formed during the hydrothermal treatment of coal ash with alkaline solution of NaOH. Among the equilibrium reaction studied, we found that the phase hydroxycancrinite phase tends to be more stable zeolite with increasing reaction temperature and the NaOH concentration, rather than formation of zeolite P and analcime phases. Therefore, to obtain phase zeolite P and zeolite such as analcime, are required low reaction temperatures and low concentrations of NaOH.

These results show the existence of a potential reduction of manufacturing costs of these products zeolites which may eventually enable the production of zeolites from an industrial waste such as coal ash.

Author details

Marisa Nascimento, Paulo Sérgio M. Soares and Vicente P. de Souza
Centre for Minearl Tecnology – CETEM - Ministry of Science, Tecnology and Inovation, MCTI, Brazil

Patrícia F. Prado
Chemistry College, Federal University of Rio de Janeiro – UFRJ, Brazil

6. References

[1] Izidoro J.C., Fungaro D.A.,(2007) Utilização de resíduos de usinas terméletricas a carvão na síntese de zeólitas e sua aplicação na remoção de Zn +2 e Cd+2 em água. Rev. Bras. Pesq. Des.9: 101-106.

[2] Bruno M., Utilização de zeólitas sintetizadas a partir de cinzas de carvão para remoção de corantes em água .(2008) Dissertação de Mestrado. Ipen – Universidade de São Paulo .

[3] Rohde G.M., Zownok O., Chies F., Silva (2006) N.I.W. Cinzas de carvão fóssil no Brasil – Aspectos técnicos e ambientais. Porto Alegre: CIENTEC, v.1 , 202 p.

[4] Kunz A, Peralta- Zamora P, Moraes S.G, Duran N. (2002) Novas tendências no tratamento de efluentes. Química Nova. 25 (1):78-82.

[5] Ojha K., Pradhan N. C , Samanta A. M. (2004) Zeolite from fly ash: synthesis and characterization. Bull. Mater. Sci. 27: 555-564.

[6] Kalkreuth, W. D. (2009) Avaliação dos impactos ambientais das cinzas pesadas e leves provenientes da Usina Termelétrica de Figueira, Paraná. Universidade Federal do Rio Grande do Sul. Available: <http://www1.ufrgs.br/pesquisa/forms/form_dadosProjetoPesquisa.php?Cod=10672>. Accessed: 2012 Fev 27

[7] Ferret L.S., Zeólitas de cinzas de carvão: síntese e uso.(2004) Tese de Doutorado. Universidade Federal do Rio Grande do Sul.

[8] Wang S., Soudi M., Li L., Zhu Z.H.(2006) Coal ash conversion into effective adsorbents for removal of heavy metals and dyes from wastewater. Journal Hazardous Materials ,133: 243–251.

[9] Umanã-Penã, L.C.,Síntesis de zeolitas a partir de cenizas volantes de centrales termoeléctricas de carbón.(2002) Tese de Doutorado. Universidad Politécnica de Catalunya, Barcelona , Espanha.

[10] Banerjee S.S., Joshi M.V., Jayaram R.V.(2004) Removal of Cr(VI) and Hg(II) from aqueous solutions using fly ash and impregnated fly ash. Sep. Sci. Tchenol., 39: 1611-1629.

[11] Chies F., Silva N.I.W, Zwonok O.(2003) Desenvolvimento de blocos e tijolos a partir de cinzas de fundo de carvão. CIPECAL. In: Rocha J.C., John U.M. Utilização de resíduos na construção habitacional, Coleção Habitare ,4: 218-239p.

[12] querol X., Moreno N., Umanã J.C., Alastuey A., Hernández E., López-Soler A., Plana F., (2002) Syntesis of zeolite from coal ash: an overview. International Journal of Coal Geology, 50: 413-423.

[13] Dana J.D (1981). Manual de Mineralogia (Dana- Hurlbut). São Paulo: Livros Técnicos e Científicos Editora S.A., 642 p.

[14] Ferret L.S., Fernandes I.D., Khahl C.A., Endres J.C.T., Maegawa A. (1999) Zeolification of ashes obtained from the combustion of southern's Brazil Candiota coal. Lexington-Kentucky, USA In: International ash utilization symposium. 247-252 p .

[15] Ferreira K D. (1998) Uso de zeólitas na reduçaão do teor de cátion níquel de efluentes galvânicos. Dissertação de Mestrado. Escola Politécnica da Universidade de São Paulo.

[16] Wang Y, Guo Y, Yang Z, Cai H, Querol X .(2003) Synthesis of zeolites resing fly ash and their applications in removing heavy metals from Waters. Sci. in China. 46:967-976.
[17] Luz A., (1995) Zeólitas: Propriedades e Usos Industriais. Rio de Janeiro: CETEM/MCT
[18] NaSCIMENTO M., Soares P.S. Souza V.P., (2009) Adsorption of heavy metals cations using coal fly ash modified. Fuel. 88: 1714-1719.
[19] Querol X., Plana F., Alastuey A., López-Soler A. (1997) Synthesis of Na-zeolite from fly ash. Fuel. 76: 793-799.
[20] Izidoro J.C., Estudo sobre a reamoção de íons metálicos em água usando zeólitas sintetizadas a partir de cinzas de carvão. (2008) Dissertação de Mestrado. Ipen – Universidade de São Paulo .
[21] Moreno N, Querol X, Ayora C, Alastuey A, Fernández-Pereira C (2002) Potential environmental applications of pure zeolitic material synthesized from fly ash. J Environ Eng. 127:994–1002.
[22] Lee D. B, Matsue N., Henmi T (2001) Influence of NaOH concentrations dissolved in seawater and hydrothermal temperatures on the synthesis of artificial zeolite from coal fly ash. Clay Science.11: 451-463.
[23] Wang G. H, Zhang Q., Song C , Liu F (2001) Synthesis of zeolites by alkaline activations of fly ash. Journal of University of Science and Technology Beijing. 8: 161-163.
[24] Querol X, Umanã J. C, Plana F, Alastuey A, López- Soler A ,Medinaceli A, Valero A, Domingo M. J, Garcia-Rojo E (2001) Synthesis of zeolites from fly ash at pilot plant scale. Examples of potential applications.Fuel. 80: 857 - 865.
[25] Murayama N, Yamamoto H, Shibata J (2002) Mechanism of zeolite synthesis from coal fly ash by alkali hydrothermal reaction. Int J Miner Process. 64:1–17.
[26] Ortiz N (2000) Estudo da utilização de magnetita como material adsorvedor dos metais Cu+2, Pb+2, N+2 e Cr+3, em solução. Tese de Doutorado – Instituto de Pesquisas Energéticas e Nucleares, São Paulo.
[27] Costa A. E (2005) Adsorção e purificação de corantes naturais com sílica amorfa. Dissertação de Mestrado - Universidade Federal de Santa Catarina,Florianópolis.
[28] Zambon G. A (2003) Remoção de chumbo (Pb+2) utilizando zeólita natural clinoptilolita. Dissertação de mestrado - Universidade Estadual de Campinas, São Paulo.
[29] Peruzzo L. C (2003) Influência de agentes auxiliares na adsorção de corantes de efluentes da indústria têxtil em colunas de leito fixo. Dissertação de mestrado. Universidade Federal de Santa Catarina, Florianópolis.
[30] Montgomery D, Calado V (2003) Planejamento de experimentos usando o Statistica. Rio de Janeiro: e-papers.
[31] Querol X., Alastuey A., Moreno N. (2006) Immobilization of heavy metals in polluted soils by the addition of zeolitic material synthesized from coal fly ash. Chemosphere. 62: 171-180.
[32] Molina A, Poole C. (2004) A comparative study using two methods to produce zeolites from fly ash. Miner Eng.17:167–173.
[33] Hernández S, Juan R., Andrés JM, Ruiz C.(2007) Synthesis of granular zeolitic materials with high cation exchange capacity from agglomerated coal fly ash. Fuel. 86:1811–1821.

[34] Erdem E, Karapinar N, Donat R.(2004) The removal of heavy metal cations by natural zeolites. J Colloid Interf Sci . 14: 280-309.

[35] Hsu T, Yu C, Yeh C.(2008) Adsorption of Cu+2 from water using raw and modified coal fly ashes. Fuel. 87:1355–1359.

[36] Fungaro D A, Izidoro J C, Almeida R S.(2005) Remoção de compostos tóxicos de solução aquosa por adsorção com zeólita sintetizada a partir de cinzas de carvão. Ecética Química 30:31–35.

[37] Colella C, Gennaro M, Langella A, Pansini M.(1998) Evaluation of natural philipsite and chabazite as cation exchangers for copper and zinc. Sep Sci Technol. 33:467–481.

[38] Hui K S, Chao C Y H. (2006) Synthesis of MCM-41 from coal fly ash by a green approach: influence of synthesis pH. J Hazard Mater.137:1135–1148.

[39] Bayat B. (2002) Comparative study of adsorption properties of Turkish fly ashes. The case of nickel (II), copper(II) and zinc(II). J. Hazardous Mater. 95:251–273.

[40] Mattigod S.V, Rai D, Eary L.E, Ainsworth C.C, (1990) Geochemical factors controlling the immobilization of inorganic constituints from fossil fuel combustion residues I. Review of the major elements. Journal Envior. Qual. 19:188-201.

Chapter 12

Influence of KNO_3 Bath Composition on Ion Exchange Process of Commercial Soda Lime Silicate Float Glass

Vincenzo M. Sglavo

Additional information is available at the end of the chapter

http://dx.doi.org/10.5772/52064

1. Introduction

Ion-exchange has been used exactly for one century to modify the surface properties of glass; Schülze in 1913 [1] was in fact the first to demonstrate that monovalent cations contained in glass could be exchanged when a soda lime silicate glass was immersed into a bath of molten silver nitrate. During the years, ion-exchange processes were developed and industrialized for numerous applications, such as chemical strengthening of glass articles [2, 3], gradient index (GRIN) lenses [4, 5] and planar waveguides [6]; the ion-exchange properties of glass have been also used to explain the functioning of the pH glass electrode and the chemical durability of glass [7].

In the last years, ion-exchange of glass has re-acquired great interest due to the possibility to increase the strength also of three-times and to work on articles of different shapes (even curved, wavy or hollow) and with limited thickness, avoiding many of the problems associated to thermal tempering such as optical distortions of the surface and premature failure due to NiS inclusions. Several contemporary technologies and applications, which regard the modern everyday life, took fundamental advantages from the application of ion-exchange to glass components such as pharmaceutical packaging, transparent lightweight armor, transparencies for private vehicles, trains and aircrafts, 3D / touch / flexible displays, photovoltaic modules [8].

Typically, an industrial ion-exchange process is performed by placing the glass components for several hours (from about 4 h even to 120 h) in a vessel containing a molten salt; in the most typical configuration sodium (or lithium) containing silicate glass is immersed into a molten potassium nitrate bath at temperatures ranging from 400°C to 500°C [9-11]. During the ion-exchange process the glass can be considered as a matrix of immobile negative

INTECH
open science | open minds

© 2012 Sglavo, licensee InTech. This is a paper distributed under the terms of the Creative Commons Attribution License (http://creativecommons.org/licenses/by/3.0), which permits unrestricted use, distribution, and reproduction in any medium, provided the original work is properly cited.

groups with associated mobile cations (Na^+ or Li^+); the contact with liquid monovalent ion-containing salt induces interdiffusion that can be treated with Nerst-Plank equations. After the ion-exchange process is completed, the articles are rinsed with water and dried, being immediately ready for successive use. For a given glass composition, the overall strengthening is a function of the type of invading ion (salt bath composition), bath temperature and immersion time.

In the past, no much attention has been devoted to the variation of chemical composition of the salt bath, to its influence on the ion-exchange process and resulting chemical, physical and mechanical performances. As a matter of fact, since interdiffusion depends on the concentration of the ions, one should expect that any variation in the liquid bath composition modifies the ion-exchange process. In addition, some recent results [12] pointed out a more effective ion-exchange process when the glass article is immersed in a non-pure liquid bath containing significant amount of the ion to be exchanged present in the glass.

There is therefore an increasing interest in the analysis and understanding of the influence of salt bath composition on ion-exchange process and consequent properties for silicate glasses.

In the present work soda lime silicate float glass from commercial source was considered and were subjected to ion-exchange in different KNO_3 salts. The treated samples are then characterized in terms of potassium penetration profile and mechanical strength to point out the influence of the bath composition on final performances.

2. Experimental procedure

Float soda lime silicate glass plates (nominal thickness = 4 mm) from commercial source was used in the present work. The composition of the glass is reported in Table 1. The glass transition temperature measured by Differential Scanning Calorimeter (DSC) (DSC2010, TA Instruments, USA) method [13] is equal to 577°C.

SiO_2	Al_2O_3	Na_2O	CaO	MgO	other
72	1	13	9	4	1

Table 1. Composition (wt%) of the float glass

The original sheets were manually cut into 50 mm x 50 mm square plates whose edges were carefully rectified and polished by using SiC abrasive paper; care was also used to avoid any damage especially on the surface of the samples.

The samples were subjected to ion-exchange treatment by using a lab-scale furnace that allows the treatment of 19 small square plates contemporaneously. Three different potassium nitrate salt bath were used: (A) chemically pure (>99.9%) KNO_3 from commercial source, (B) pure (>99%) KNO_3 for industrial use from commercial source and (C) KNO_3 from different sources used for at least 1000 h in chemical strengthening process in the cited furnace.

The specimens were initially cleaned with water, gently swabbed and placed in the stainless steel holder. This latter was then positioned within the semi-automatic chemical strengthening furnace. The following conditions were used for the ion exchange process: pre-heating within the furnace above the salt bath surface = 20 min, duration of the ion exchange process = 4 h or 24 h, post cooling above the salt bath surface = 20 min, temperature of the salt bath = 450°C. At the end of each cycle the samples were carefully cleaned with water and gently swabbed before successive mechanical testing. At least 30 specimens were treated by using the same bath/time treatment condition.

Before starting the ion-exchange process the salts were subjected to chemical analysis. Small amounts (few grams) of salt were collected randomly from each solidified bath and dissolved in specific volumes of bi-distilled water; the obtained solutions were then analyzed by Inductively Coupled Plasma – Optical Emission Spectroscopy (ICP-OES) (Spectro-Ciros, Germany) and a multi-element standard (Sigma Aldrich) was used for the quantification of the dissolved ions. The melting point of the salts was also measured by DSC.

The ion-exchanged samples were then subjected to mechanical testing for the measurement of bi-axial flexural strength [14, 15]. Bi-axial flexure was carried out with a ring-on-ring configuration, the upper loading ring and lower support ring having a diameter of 20 mm and 40 mm, respectively. The tests were performed in lab air (temperature ≈ 22°C, relative humidity ≈ 40%) with an actuator speed of 1 mm/min. The failure stress was evaluated from the measurement of the maximum load, F_{max}, by using the following equation:

$$\sigma_F = K \frac{F_{\max}}{t^2} \tag{1}$$

where t is the thickness of the sample and

$$K = \frac{3(1+\mu)}{2\pi}\left[\ln\frac{r_2}{r_1} + \frac{(1-\mu)}{(1+\mu)}\frac{\left(r_2^{1} - r_1^{2}\right)}{0.72\,L^2}\right] \tag{2}$$

μ being the glass Poisson's ratio (equal to 0.2 [16]), L the specimen size (50 mm), r_1 and r_2 the upper and lower support ring radius, respectively. The tests were performed also on one set of as-cut glass plates for comparison.

Fragments of the samples subjected to mechanical testing were used to measure the potassium penetration profile in the surface layers on the fracture surface. The specimens were initially attached to an aluminum disk by using conducting adhesive tape and then coated by sputtering with Au-Pd alloy. Once placed within a Scanning Electro Microscope (SEM) (JSM5500, Jeol, Japan), clean and flat portions of the fracture surface were analyzed and the potassium Kα signal was recorded on specific paths (about 40 μm long) by using the Energy Dispersion X-ray Spectroscopy (EDXS) (EDS2000, IXRF System, USA) probe. By using a similar procedure, the chemical composition of the external surface of the glasses after the ion-exchange process was determined on clean regions of about 0.5 mm^2.

3. Results and discussion

The results of the mechanical tests are summarized in Table 2. One can observe the general increase of the mechanical resistance after the ion-exchange process although the scatter of the measurements does not allow to point out specific differences among the different treatments. There is anyway an increase in the average failure stress in excess to 60-70 MPa that is associated to the development of a compressive residual stress on the surface of the samples.

For a better understanding of the resistance data, strength values were analyzed by using the Weibull statistics. Figure 1 shows the failure stress distributions where the failure probability associated to each single measurement was calculated as:

$$F = \frac{j}{N+1} \tag{3}$$

where j is the rank in the ascending ordered strength distribution and N the total number of specimens. Failure probability is typically related to the tensile stress through the relationship:

$$F = 1 - \exp\left[k\, S \left(\frac{\sigma}{\sigma_0} \right)^m \right] \tag{4}$$

where S is the surface of the sample subjected to tension, k the loading factor, m the Weibull modulus and σ_0 the normalizing stress [17]: it is reminded here that the Weibull modulus is the parameter that represents the scatter of the distribution. Equation (4) can be linearized by calculating twice the natural logarithm, thus obtaining:

$$\ln\left(\ln\left(\frac{1}{1-F} \right) \right) = m \ln\sigma + \ln\frac{k\, S}{\sigma_0^m} \tag{5}$$

Fitting of the data shown in Figure 1, using linear regression, allows the calculation of the Weibull modulus corresponding to each strength distribution. The results are shown in Table 3: it is interesting to observe that the strengthening process is associated to a sensible decrease of the Weibull modulus, *i.e.* to an increase of the strength scatter.

		average	standard deviation	minimum	maximum
as-cut		257	47	140	346
A	4 h	331	71	158	465
	24 h	312	62	155	476
B	4 h	341	103	114	558
	24 h	304	63	151	437
C	4 h	441	155	131	505
	24 h	593	148	185	701

Table 2. Summary of the strength values (MPa) measured by biaxial flexure test on as-cut samples and on ion-exchanged glass (A, B, C).

The effect of ion-exchange appears clearer in Figure 1, although for samples treated in baths A and B the effect of time is not so evident. In any case, glasses exchanged in bath C show the best performances. One interesting aspect regards the different strengthening effect generated on glasses with different initial failure stress. As shown in Figure 1 (and reported quantitatively in Table 3) the tail of the distributions corresponding to low failure probability and limited strength are almost overlapped, regardless the ion-exchange process. This means that specimens characterized by large critical defects are substantially indifferent to the creation of a surface compressive stress whose extension is probably small compared to the depth of the critical flaw.

	A	B	C	as-cut
4 h	5.0	3.5	3.3	5.8
24 h	5.3	5.0	3.9	

Table 3. Weibull modulus of the strength distributions reported in Figure 1.

Some exemplary potassium concentration profiles measured by EDXS on the fracture surface of ion-exchange samples, chosen from very similar others, are shown in Figure 2. In this case, the potassium concentration recorded by line scan was scaled with respect to the intensity of the peak associated to the Kα line measured from surface analysis (Figure 3); as a matter of fact a ratio ≈0.25:0.4:1 was always recorded among the intensities of potassium Kα peak corresponding to specimens treated in bath A, B and C, respectively; in addition, some residual sodium was also detected on the surface of glasses treated in baths A and B.

Figure 2 shows that the depth involved in the ion-exchange process varies between ≈12 μm and ≈20-30 μm if the duration increases from 4 h to 24 h. On the basis of the failure stress measured in as-cut samples, the size of the critical flaw can be estimated using linear-elastic fracture mechanics from the expression:

$$c_F = \left(\frac{K_c}{\psi \, \sigma_F} \right)^2 \tag{6}$$

where K_c is the fracture toughness of the material and ψ the crack shape factor. Considering that biaxial tests were performed in laboratory air and that some sub-critical growth could have occurred before final failure, one can assume $K_c \approx 0.6$ MPa $m^{0.5}$ [18]; in addition, if semi-circular surface flaws are assumed, $\psi \approx 1.26$ [19]. Therefore, from the data in Table 2, values ranging from 3 μm to 12 μm can be calculated for the depth of the critical defect. It is consequently evident that the strengthening process, whose intensity decreases rapidly moving away from the surface, has very different effect on pre-existing flaws, the largest ones being almost unaffected by the ion-exchange. Moreover, the strengthening effect appears to be stronger on glasses treated in bath C that is responsible for a higher surface potassium concentration (and, therefore, for a more effective Na^+-K^+ interdiffusion) and for a deeper profile, especially at 24 h duration.

It is not easy to explain the discrepancies observed among glasses strengthened by using the three considered potassium nitrate baths. The differences among them are in fact relatively limited. Table 4 reports the major impurities in terms of metallic elements contained in the baths: salt A corresponds to substantially pure KNO_3; a larger amount of sodium is present

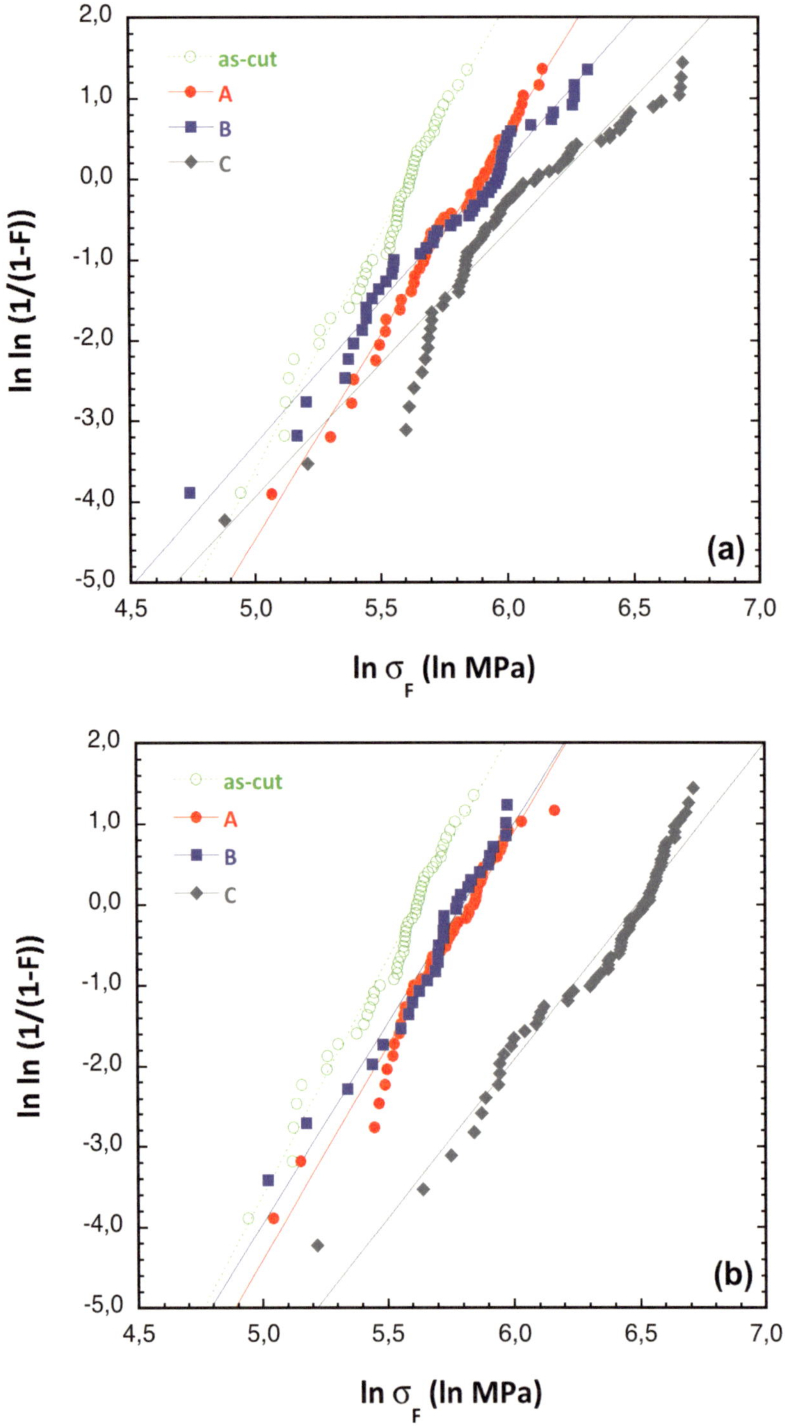

Figure 1. Weibull distributions of the strength for as-cut and ion-exchanged specimens: (a) 4 h, (b) 24 h. Straight lines represent fitting curves.

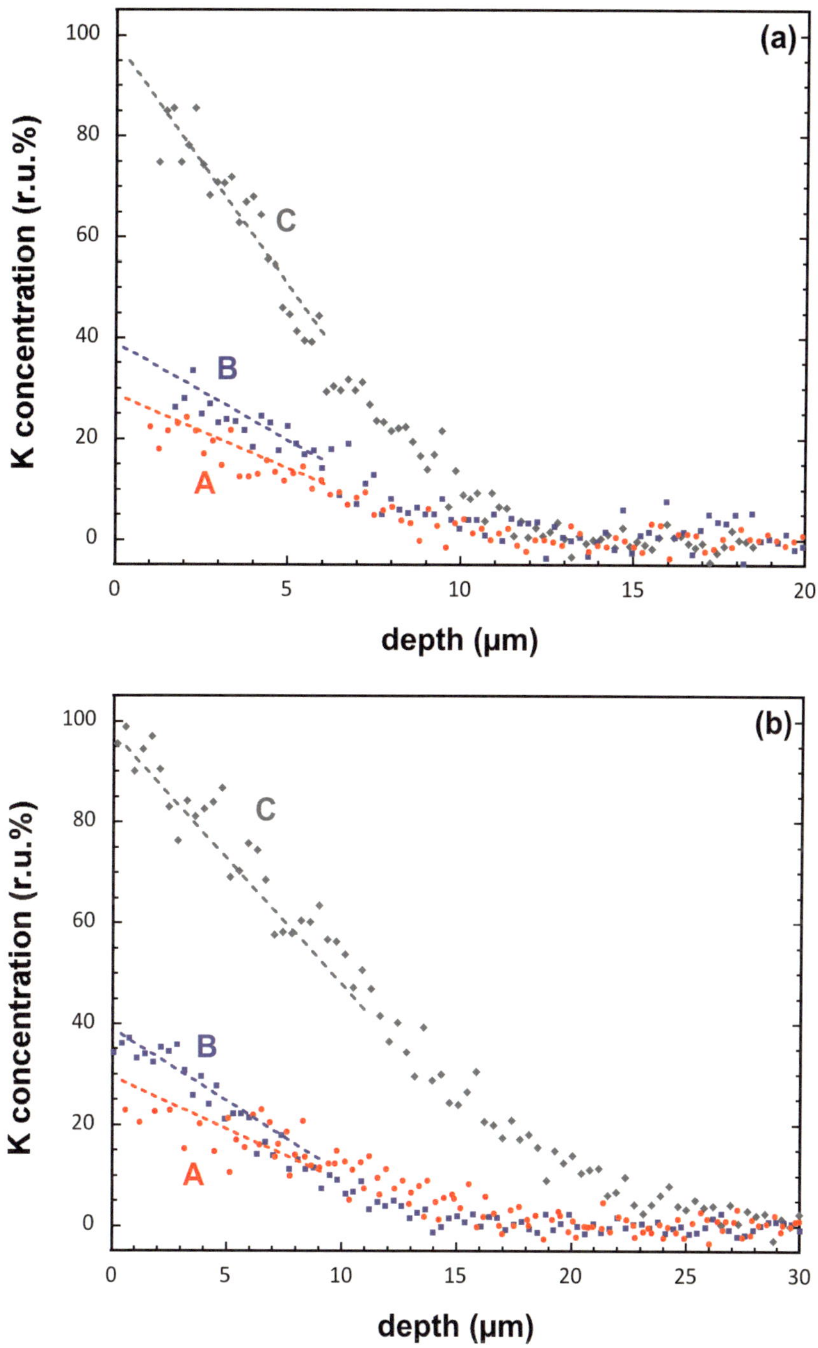

Figure 2. Potassium concentration profile as a function of the depth from the sample surface: (a) 4 h, (b) 24 h. Dashed lines represent tendency curves.

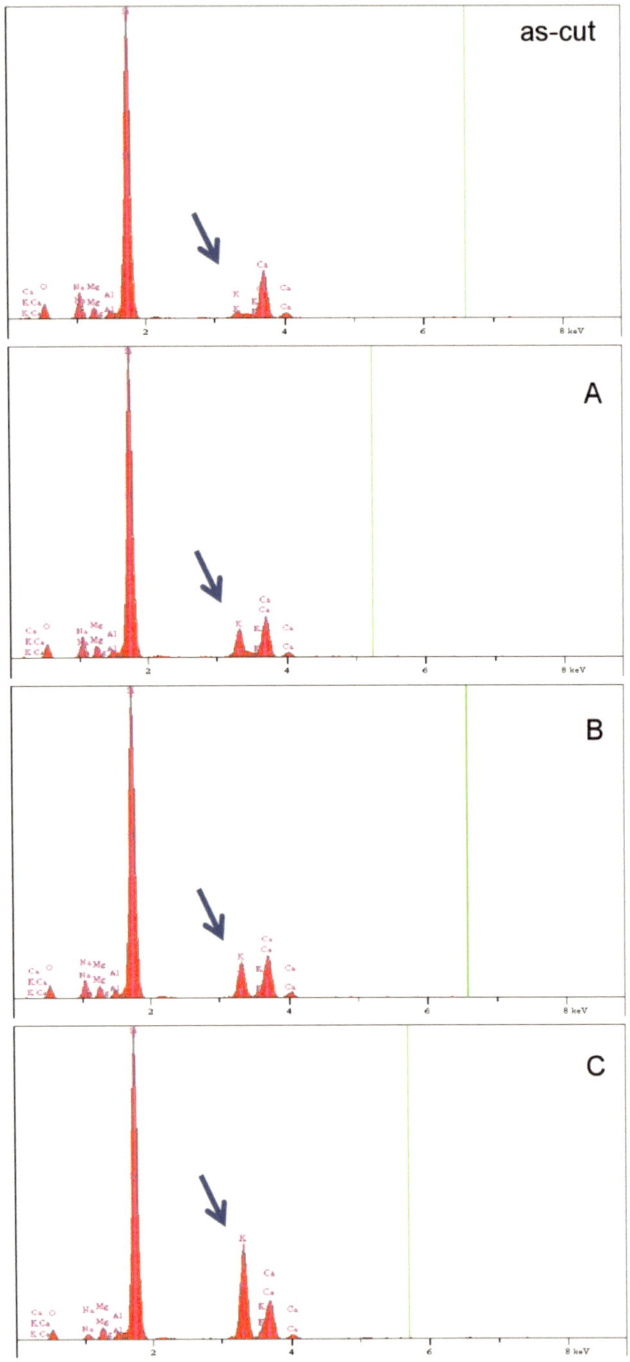

Figure 3. EDXS spectra recorded on the surface of samples subjected to ion exchanged for 24 h. The spectrum recorded on as-cut glass is shown for comparison. The potassium Kα peak is indicated by an arrow.

especially in bath C (around 0.5%) which corresponds to a $NaNO_3$ load of about 1.8%. Similarly, also the melting point of the three salts does not change substantially, varying from 335.5°C (bath A) to 334.6°C (bath B) to 331.0°C (bath C). On the basis of the results shown in Figure 2, it is possible to hypothesize that the slight difference among the three salts has substantial effect on the effective composition or activity of the K^+ ions on the glass surface, the interdiffusion coefficient being affected only to a limited extent. More precisely, the presence of Na^+ ion in the KNO_3 salt bath makes the exchanging process more effective.

The results obtained in the present work can be fitted in the more general framework regarding the effect of mixed salt baths on the chemical strengthening of silicate glass [12, 20-23]. Other previous works reported a sensible increase of the reinforcing effect when ion-exchange is performed in non-pure melted salts although a clear explanation of the physical/chemical mechanisms has not been given yet.

A certain number of different parameters has to be considered such as "impurities" in the glass composition, presence of non-metallic elements in the salt bath, interface interaction between glass and bath, effect of "tin side", exchanging pair of ions, structural and viscous relaxation. Additional analyses and tests are required for a deeper understanding of the problem; nevertheless, the results reported here point out interesting aspects regarding the salt bath composition for higher efficiency chemical tempering processes.

	A	B	C
Na	40±5	1100±50	5200±50
Ca	<1	<1	15±5
Cr			30±5
Zn	<1	<1	2±1

Table 4. Metallic elements content (ppm) in the KNO_3 baths.

4. Conclusions

The composition of the potassium nitrate salt bath in terms of impurities, mostly Na, seems to have an important role in the strengthening process of soda lime silicate glass by ion exchange. It has been found that chemical reinforcement is more effective when it is carried out in the bath containing the highest amount of sodium (≈ 0.5%). The results point out interesting practical and industrial aspects regarding chemical tempering processes with higher efficiency.

Author details

Vincenzo M. Sglavo

Department of Materials Engineering and Industrial Technologies, University of Trento, Trento, Italy

5. References

[1] Schulze G. Versuche über die Diffusion von Silber in Glas. Annalen der Physik 1913; 345 335-367.

[2] Varshneya AK. Chemical Strengthening of Glass: Lessons Learned and Yet To Be Learned. International Journal of Applied Glass Science 2010; 1[2] 131–142.
[3] Gy R. Ion exchange for glass strengthening. Mater. Sci. Eng. B 2008; 149 159–165.
[4] Moore DT. Gradient-index optics: a review. Appl. Optics 1980; 19 1035-1038.
[5] Yip GL, Albert J. Characterization of planar optical waveguides by K^+-ion exchange in glass. Opt. Lett. 1985; 10 151-153.
[6] Wong SF, Pun EYB, Chung PS. Er^{3+}-Yb^{3+} codoped phosphate glass waveguide amplifier using Ag^+-Li^+ ion exchange. Photonics Technology Letters, IEEE 2002; 14(1) 80-82.
[7] Doremus RH. Glass Science, 2nd ed. New York, NY, USA: J. Wiley and Sons; 1994.
[8] Jacoby M. New Applications For Glass Emerge. Chemical & Engineering News 2012; 90(25) 34-36.
[9] Donald IW. Review: Method for Improving the Mechanical Properties of Oxide Glasses. J. Mater. Sci. 1989; 24 4177–4208.
[10] Bartholomew RF. Ion-Exchange. In: Ceramics and Glasses, Engineered Materials Handbook, Vol. 4. ASM International. Metals Park, OH, USA, 1991, p460 – 63.
[11] Bartholomew RF and Garfinkel HM. Chemical Strengthening of Glass. In: Uhlmann DR and Kreidl NJ (eds) Glass Science and Technology, Vol. 5. Academic Press, New York, USA, 1980, p217–267.
[12] Varshneya AK, Spinelli IM. High-strength, large-case-depth chemically strengthened lithium aluminosilicate glass. American Ceramic Society Bulletin 2009; 88(5) 27-33.
[13] ASTM standard, C 1356-91, Standard Test Method for Assignment of the Glass Transition Temperatures by Differential Scanning Calorimetry or Differential Thermal Analysis. West Conshohocken, PA, USA: ASTM Book of Standards, ASTM International; 1991.
[14] European standard, UNI EN 1288-1/5, Glass in building - Determination of the bending strength of glass. Milan, Italy: UNI; 2001.
[15] ASTM standard, C 1499-05, Standard Test Method for Monotonic Equibiaxial Flexural Strength of Advanced Ceramics at Ambient Temperature. West Conshohocken, PA, USA: ASTM Book of Standards, ASTM International; 2005.
[16] European standard, UNI EN 572-1/9, Glass in building - Basic soda lime silicate glass products. Milan, Italy: UNI; 2004.
[17] Green DJ., Introduction to Mechanical Properties of Ceramics. Cambridge University Press, Cambridge, U.K., 1998.
[18] Sglavo VM, Green DJ. Indentation Determination of Fatigue Limits in Silicate Glasses. J. Am. Ceram. Soc. 1999; 82(5) 1269–74.
[19] Lawn BR. Fracture of brittle solids. 2nd ed. Cambridge, UK: Cambridge University Press; 1993.
[20] Saunders AE, Kubichan RE. 'Strengthening Glass by Multiple Alkali Ion Exchange. U.S. Patent No. 4 119 760, October 10, 1978.
[21] Mallick KK, Holland D. Strengthening of container glasses by ion-exchange dip coating. J. Non-Cryst. Solids 2005; 351 2524–2536.
[22] Younei J, Linge J. Effect of additives in the salt bath on glass strengthening. J. Non-Cryst. Solids 1986; 80(1–3) 300–306.
[23] Xiangchen Z, Ouli H, Cengzuo X, Yinghuan Z. The effect of impurity ions in molten salt KNO_3 on ion-exchange and strengthening of glass. J. Non-Cryst. Solids 1986; 80(1–3) 313–318.

Chapter 13

Unheated and Heated Batch Methods in Ion Exchange of Clinoptilolite

Tevfik Ünaldı and Selahattin Kadir

Additional information is available at the end of the chapter

http://dx.doi.org/10.5772/51441

1. Introduction

It is well known that natural zeolites consist of aluminia and silica tetrahedra which, bound in a definite way, include crystal structure vacancies, channels and pores [1,2]. About 40 natural zeolites have been identified during the past 200 years; the most widespread are analcime, chabazite, clinoptilolite, erionite, ferrierite, heulandite, laumontite, mordenite, and phillipsite. More than 150 zeolites have been synthesized; the most common are zeolites A, X, Y, and ZMS-5. Clinoptilolite has the structural formula $(Na,K)_6(Al_6Si_{30}O_{72})$. $20H_2O$, characterized by two different rings, which are 8(3,3x4,6 Å) T and 10(3,0x7,6 Å) T on the ab-plane, and channels with rings of 8(2,6x4,7 Å) T on the bc-plane. As shown in Figure 1, there are exchangeable cations of Na^+, K^+, Ca^{2+} and Mg^{2+} through the channels

Because clinoptilolite has rings of two different dimensions, it is used in retention and separation of various gases, as an adsorbent, and as a molecular sieve. Because of its exchangeable cations clinoptilolite, which has the property of ion exchange, is used in producing biological and water filters, in retention of various heavy metal ions and radio-isotopes, in production of fillers in animal feeds, and in the horticulture and agriculture as a soil additive.[4, 5, 6]

Clinoptilolite is a material that is used in both its natural and modified forms. Both forms of clinoptilolite have widespread application but, apart from their general ion-exchange properties and the high quantities of exchangeable cations through their channels, their general properties are different [7, 8].

2. Materials and methods

2.1. Unheated batch method

Ion-exchanged forms of Na-clinoptilolite from the Bigadiç area were produced via the batch method, using solutions of 0.1N, 0.5N and 1N – with salts of Na^+, K^+, Ca^{2+} , Cr^{3+}, Ag^+ and 100

INTECH
open science | open minds

© 2012 Ünaldı and Kadir, licensee InTech. This is a paper distributed under the terms of the Creative Commons Attribution License (http://creativecommons.org/licenses/by/3.0), which permits unrestricted use, distribution, and reproduction in any medium, provided the original work is properly cited.

ml of deionized water. Five grams of 300-mesh clinoptilolite were poured into each solution and stirred gently, and then put in suspension for 72 hours at room temperature. Subsequently, the clinoptilolite-solution suspension system was filtered. After filtration, the clinoptilolite was washed eight times with deionized water at 98°C and dried at 110°C for 16h. [9, 10]

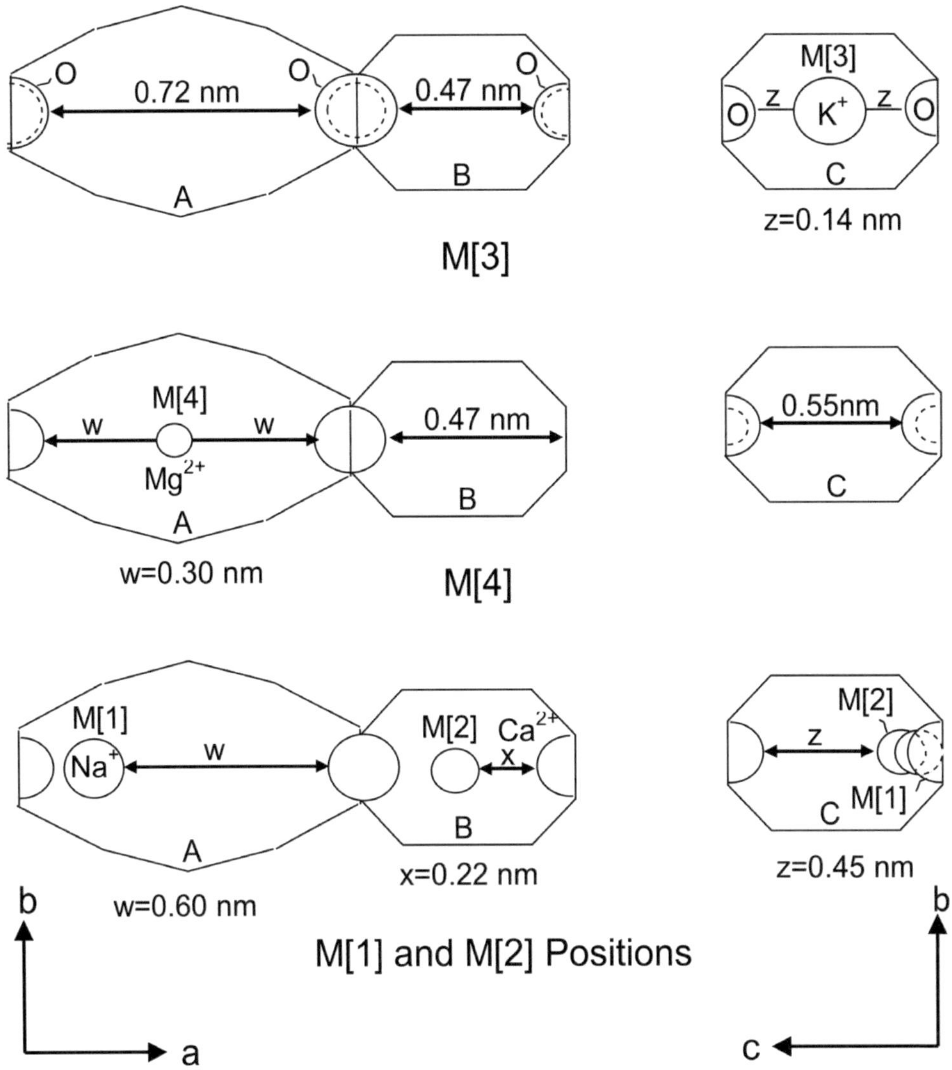

Figure 1. Cation positions in clinoptilolite.[3]

2.2. Heated batch methods

Ion exchange is conducted using a heated/cooled back magnetic stirrer system. A 100 ml solution was prepared by mixing 5 g of 300 mesh clinoptilolite with deionized water kept at 98°C for 2h.

The clinoptilolite-solution suspension system was then filtered. After filtration, the clinoptilolite was washed eight times with deionized water at 98°C and then dried at 110°C for 16 h. [11, 12]

2.3. Ion exchange

The process of ion exchange occurs between the A^{Z_A} cation in solution and B^{Z_B*} cation in the zeolite, and can be formulated as follows:

$$Z_B A^{Z_A} + Z_A B^{Z_B*} \leftrightarrow Z_B A^{Z_A*} + Z_A B^{Z_B}$$

where Z_A and Z_B show the valences of the cations, and A^{Z_B*} and B^{Z_B*} show the cations in the zeolite structure.

Ion-exchange reactions are stoichiometric, graphical representations of equilibrium concentrations of exchangeable ions in both solutions; the structure of zeolite may be ascertained from ion-exchange isotherms.

Before an ion-exchange isotherm may be obtained, equilibrium of ion exchange must be reached. In zeolites A, X, and Y with low framework densities, the equilibrium of exchange between one valence ions (such as Na^+ and K^+) is obtained in approximately one week. In zeolite structures with high framework densities, the equilibrium of exchange among high valence ions is obtained in a few months.

After the time of equilibrium is defined, this procedure could be utilized in order to plot an isotherm. Zeolites react with a solution containing ions of both A^{Z_A} and B^{Z_B}. Although the relative amounts of ions A^{Z_A} and B^{Z_B} might vary, solutions must have a constant total normality (N). According to the condition of equinormality, total ionic intensities of any solution in the system of the zeolite/solution must be constant before and after ion-exchange reactions.

The ionic intensity of any solution is

$$I = (1/2)\sum C_i Z_i^2$$

where C_i are the concentrations of opposing ions in ion-grams per liter, and Z_i are the valences of opposing ions.

Because of the distribution of A^{Z_A} and B^{Z_B} between the phases, the solution and solid phases in equilibrium must be analyzed. Thus, a plot of the equivalent fraction of ion in solution (A_s) versus equivalent fraction of the same ion in zeolite (A_Z^*) isotherm may be obtained.

The ion-exchange isotherm indicates the relative preferences of any ion within the zeolite structure. Besides, the separation factor of ion A within the zeolite structure is

$$\alpha = (A_Z^*/B_Z^*)(m_B/m_A)$$

where A_Z^* ve B_Z^* are equivalent fractions of ions A and B in zeolite, respectively, and m_A and m_B are the concentrations of ions in solutions in mole/liter. The total of the equivalent fractions of A_Z^* and B_Z^* must equal 1. On the basis of ion selectivity, if $\alpha > Z_A/Z_B$, the zeolite prefers A^{Z_A} ions; if $\alpha = Z_A/Z_B$, the zeolite has no preference; and if $\alpha < Z_A/Z_B$, the zeolite prefers B^{Z_B} ions. [13, 14]

2.4. Ion-exchange rate

XRF analyses were conducted on 0.1N-, 0.5N- and 1N-modified forms of solid-phase Na-clinoptilolite, and especially the values of exchangeable and other cations differed greatly from values of the natural form (Table 1). The numbers of atoms in the unit cell were calculated with the knowledge that the unit cell includes 72 oxygen atoms. The numbers of atoms calculated and the following formula were used:

$$X_{form} = [1 - (A_{form} / A_{natural})] \times 100$$

where X_{form} = the ion-exchange rate of the forms and the number of atoms in the unit cell of the same form, and $A_{natural}$ = the number of atoms in the unit cell of the natural form.

2.5. Rate of ion selectivity

The rate of ion selectivity, as termed by us, is different from the ''ion selectivity'' of the ion-exchanged forms; this rate is calculated from the percentages of ions in the structure, and thus is similar to the rate of ion exchange. For Na^+, K^+, Ca^{2+} and Mg^{2+} (exchangeable cations) forms, this quantity may be calculated using

$$\alpha_{form} = [(A_{form}/A_{natural}) - 1] \times 100$$

and for non-exchangeable cations such as Co^{3+}, Cd^{+}, Cr^{3+}, Ag^+ forms using

$$\alpha_{form} = A_{form} \times 100$$

where α_{form} is the ion-selectivity rate of the ionic form, and A_{form} and $A_{natural}$ are the numbers of atoms in the unit cells of the ion-exchanged form and the natural form, respectively.

3. Results and discussion

3.1. Rate of ion exchange

The results of chemical analyses and the numbers of atoms in the unit cells of natural and Na^+, K^+, Ca^{2+}, Mg^{2+}, Co^{3+}, Cd^{2+}, Cr^{3+} and Ag^+ modified forms of clinoptilolite are given in Tables 1 and 2. As shown in the chemical formula of clinoptilolite, the numbers of atoms in the unit cell were calculated with the knowledge that the unit cell includes 72 oxygen atoms (Table 2).

Tables 3 and 4 were derived from data given in Table 2. Ion-exchange rates and the ordering of ion-exchange rate for unheated and heated methods applied to Na-clinoptilolite from Bigadiç-Balıkesir (Turkey) are given in Tables 3 and 4.

Ion-Exchange Method	Ion-Exchange Forms	Molecules												
		SiO_2	Al_2O_3	Fe_2O_3	CaO	MgO	K_2O	Na_2O	Co_2O_3	CdO	Cr_2O_3	Ag_2O	H_2O	Σ
	Natural	69.81	11.92	0.78	3.32	1.33	1.65	0.49	-	-	-	-	10.70	100.00
	0.1N Na^+	68.45	11.72	0.86	2.72	1.24	2.19	0.81	-	-	-	-	11.98	99.97
UHBM	0.5N Na^+	68.44	11.69	0.76	1.80	1.06	4.28	1.13	-	-	-	-	10.85	100.10
	1N Na^+	69.05	11.79	0.83	1.53	0.96	4.22	1.66	-	-	-	-	9.96	100.00
	0.1N Na^+	68.21	11.63	0.83	3.14	1.58	1.92	0.94	-	-	-	-	11.75	100.00
HBM	0.5N Na^+	69.40	11.78	0.85	3.04	1.47	1.79	1.02	-	-	-	-	10.63	100.00
	1N Na^+	68.69	11.67	0.89	1.92	1.14	1.63	3.18	-	-	-	-	10.89	100.01
	0.1N K^+	68.94	11.77	0.85	2.71	1.23	2.72	0.42	-	-	-	-	11.37	100.01
UHBM	0.5N K^+	67.48	11.51	0.70	1.60	0.97	5.82	0.10	-	-	-	-	11.80	99.88
	1N K^+	67.86	11.60	0.69	1.47	0.96	6.64	0.13	-	-	-	-	10.67	100.02
	0.1N K^+	68.91	11.75	0.82	3.57	1.25	1.62	0.46	-	-	-	-	11.63	100.01
HBM	0.5N K^+	68.26	11.61	0.81	2.39	1.35	4.11	0.17	-	-	-	-	11.31	100.01
	1N K^+	67.57	11.54	0.61	0.58	0.55	8.54	0.12	-	-	-	-	10.49	100.00
	0.1N Ca^{2+}	69.49	11.86	0.84	3.38	1.29	1.62	0.57	-	-	-	-	10.95	100.00
UHBM	0.5N Ca^{2+}	70.26	12.02	0.83	3.52	1.31	1.61	0.47	-	-	-	-	9.98	100.00
	1N Ca^{2+}	69.02	11.69	0.55	4.01	1.51	1.72	0.16	-	-	-	-	11.04	99.70
	0.1N Ca^{2+}	70.46	12.00	0.73	3.56	1.22	1.80	0.22	-	-	-	-	10.04	100.03
HBM	0.5N Ca^{2+}	69.49	11.81	0.74	3.67	1.41	1.79	0.18	-	-	-	-	10.91	100.00
	1N Ca^{2+}	69.49	11.87	0.77	3.85	1.24	1.77	0.15	-	-	-	-	10.85	99.99
	0.1N Mg^{2+}	69.71	11.87	0.91	3.15	1.35	1.61	0.52	-	-	-	-	10.88	100.00
UHBM	0.5N Mg^{2+}	68.84	11.74	0.86	3.01	1.45	1.61	0.58	-	-	-	-	11.91	100.00
	1N Mg^{2+}	68.32	11.66	0.94	2.92	1.56	1.58	0.62	-	-	-	-	12.38	99.98
	0.1N Mg^{2+}	67.54	11.42	0.76	3.13	1.27	1.80	0.22	-	-	-	-	13.89	100.03
HBM	0.5N Mg^{2+}	69.62	11.84	0.79	3.15	1.79	1.80	0.22	-	-	-	-	10.79	100.00
	1N Mg^{2+}	67.98	11.56	0.74	2.96	2.26	1.72	0.27	-	-	-	-	12.52	100.01
	0.1N Co^{3+}	68.51	11.71	0.85	2.64	1.20	2.83	0.32	0.19	-	-	-	11.18	100.00
UHBM	0.5N Co^{3+}	67.63	11.56	0.84	2.55	1.15	2.67	0.37	1.69	-	-	-	11.52	99.98
	1N Co^{3+}	66.75	11.36	0.81	2.57	1.14	2.67	0.36	2.37	-	-	-	11.97	100.00
	0.1N Co^{3+}	66.93	11.41	0.55	0.44	0.45	9.09	0.11	0.72	-	-	-	10.32	100.01
HBM	0.5N Co^{3+}	64.75	10.99	0.75	3.15	1.33	1.69	0.19	5.95	-	-	-	11.19	99.99
	1N Co^{3+}	63.74	10.82	0.68	2.81	1.16	1.67	0.20	7.25	-	-	-	11.68	100.07
	0.1N Cd^{2+}	68.66	11.69	0.76	2.61	1.19	2.73	0.39	-	0.13	-	-	11.84	100.00
UHBM	0.5N Cd^{2+}	67.29	11.47	0.73	2.47	1.14	2.79	0.39	-	2.18	-	-	11.54	100.00
	1N Cd^{2+}	65.44	11.10	0.66	2.43	1.07	2.87	0.30	-	3.66	-	-	12.47	100.00

Ion-Exchange Method	Ion-Exchange Forms	Molecules												
		SiO_2	Al_2O_3	Fe_2O_3	CaO	MgO	K_2O	Na_2O	Co_2O_3	CdO	Cr_2O_3	Ag_2O	H_2O	Σ
HBM	0.1N Cd^{2+}	69.21	11.90	0.83	2.87	2.06	1.78	0.19	-	0.61	-	-	10.66	100.11
	0.5N Cd^{2+}	67.65	11.51	0.65	2.85	1.33	1.91	0.20	-	2.30	-	-	11.59	99.99
	1N Cd^{2+}	64.45	10.92	0.52	2.33	1.11	2.08	0.18	-	5.61	-	-	12.81	100.01
UHBM	0.1N Cr^{3+}	68.39	11.65	0.88	3.28	1.27	1.65	0.56	-	-	0.65	-	11.67	100.00
	0.5N Cr^{3+}	68.76	11.74	0.79	3.00	1.25	1.66	0.55	-	-	0.98	-	11.26	99.99
	1N Cr^{3+}	68.70	11.70	0.85	2.81	1.18	2.33	0.33	-	-	0.87	-	11.23	100.00
HBM	0.1N Cr^{3+}	69.47	11.88	0.81	2.92	1.34	1.80	0.31	-	-	0.66	-	10.81	100.00
	0.5N Cr^{3+}	67.79	11.58	0.63	2.77	1.33	3.13	0.19	-	-	1.76	-	10.83	100.01
	1N Cr^{3+}	61.92	10.39	0.27	0.70	0.47	8.36	0.10	-	-	5.64	-	12.15	100.00
UHBM	0.1N Ag^{+}	69.44	11.86	0.74	2.62	1.20	2.82	0.32	-	-	-	0.22	10.79	100.01
	0.5N Ag^{+}	66.25	11.25	0.55	1.85	1.02	2.46	0.33	-	-	-	4.36	11.90	100.00
	1N Ag^{+}	63.17	10.69	0.42	1.57	0.86	2.27	0.23	-	-	-	10.57	10.22	100.00
HBM	0.1N Ag^{+}	68.04	11.51	0.47	2.49	1.26	2.25	0.20	-	-	-	0.36	13.42	100.00
	0.5N Ag^{+}	67.63	11.39	0.59	2.35	1.35	1.67	0.17	-	-	-	1.93	12.91	99.99
	1N Ag^{+}	62.21	10.51	0.16	0.84	0.52	1.34	0.12	-	-	-	10.98	13.31	99.99

UHBM: unheated batch method; HBM: heated batch method

Table 1. Chemical analyses of natural and Na^{+}, K^{+}, Ca^{2+}, Mg^{2+}, Co^{3+}, Cd^{2+}, Cr^{3+}, Ag^{+} modified forms of clinoptilolite.

Ion-Exchange Method	Ion-Exchange Forms	Atoms												
		Si	Al	Fe	Ca	Mg	K	Na	Co	Cd	Cr	Ag	H	Si/Al
	Natural	24.60	4.95	0.21	1.25	0.70	0.75	0.33	-	-	-	-	25.15	4.97
UHBM	0.1N Na^{+}	23.98	4.84	0.23	1.02	0.64	0.98	0.55	-	-	-	--	28.00	4.95
	0.5N Na^{+}	24.40	4.92	0.21	0.69	0.56	1.95	0.78	-	-	-	--	25.80	4.96
	1N Na^{+}	24.80	4.99	0.22	0.59	0.51	1.93	1.16	-	-	-	-	23.86	4.97
HBM	0.1N Na^{+}	23.82	4.79	0.22	1.18	0.82	0.86	0.64	-	-	-	-	28.19	4.97
	0.5N Na^{+}	24.54	4.91	0.23	1.16	0.78	0.81	0.70	-	-	-	-	25.07	5.00
	1N Na^{+}	24.33	4.88	0.24	0.73	0.60	0.74	2.18	-	-	-	-	25.74	4.99
UHBM	0.1N K^{+}	24.28	4.88	0.22	1.02	0.65	1.22	0.29	-	-	-	-	26.71	4.98
	0.5N K^{+}	24.00	4.83	0.19	0.61	0.51	2.64	0.07	-	-	-	-	28.00	4.97
	1N K^{+}	23.73	4.71	0.18	0.54	0.49	2.92	0.09	-	-	-	-	24.53	5.04
HBM	0.1N K^{+}	24.15	4.85	0.22	1.34	0.65	0.78	0.31	-	-	-	-	27.19	4.98
	0.5N K^{+}	24.20	4.85	0.22	0.91	0.71	1.86	0.12	-	-	-	-	26.75	4.99
	1N K^{+}	24.54	4.94	0.17	0.22	0.30	3.96	0.08	-	-	-	-	25.42	4.97

	0.1N Ca^{2+}	24.46	4.92	0.22	1.28	0.62	0.73	0.39	-	-	-	-	25.71	4.97
UHBM	0.5N Ca^{2+}	24.89	5.02	0.22	1.34	0.69	0.73	0.32	-	-	-	-	23.59	4.96
	1N Ca^{2+}	24.33	4.85	0.22	1.51	0.79	0.78	0.11	-	-	-	-	25.96	5.02
	0.1N Ca^{2+}	24.94	5.01	0.20	1.35	0.64	0.81	0.15	-	-	-	-	23.70	4.97
HBM	0.5N Ca^{2+}	24.48	4.90	0.19	1.38	0.74	0.80	0.12	-	-	-	-	25.63	5.00
	1N Ca^{2+}	24.49	4.93	0.20	1.45	0.65	0.80	0.10	-	-	-	-	25.51	4.97
	0.1N Mg^{2+}	24.53	4.92	0.24	1.19	0.71	0.72	0.36	-	-	-	-	25.53	4.99
UHBM	0.5N Mg^{2+}	24.05	4.83	0.23	1.13	0.76	0.72	0.39	-	-	-	-	27.75	4.98
	1N Mg^{2+}	23.80	4.79	0.25	1.09	0.81	0.70	0.42	-	-	-	-	28.77	4.97
	0.1N Mg^{2+}	23.26	4.64	0.20	1.15	0.65	0.79	0.14	-	-	-	-	31.91	5.01
HBM	0.5N Mg^{2+}	24.52	4.91	0.21	1.19	0.94	0.81	0.15	-	-	-	-	25.35	4.93
	1N Mg^{2+}	23.67	4.75	0.19	1.10	1.17	0.77	0.18	-	-	-	-	29.09	4.98
	0.1N Co^{3+}	23.93	4.82	0.22	0.99	0.63	1.26	0.22	0.19	-	-	-	26.05	4.97
UHBM	0.5N Co^{3+}	23.94	4.82	0.23	0.97	0.61	1.20	0.26	0.43	-	-	-	27.20	4.97
	1N Co^{3+}	23.62	4.74	0.22	0.97	0.60	1.20	0.25	0.61	-	-	-	28.25	4.98
	0.1N Co^{3+}	24.48	4.92	0.15	0.17	0.25	4.24	0.07	0.19	-	-	-	25.18	4.98
HBM	0.5N Co^{3+}	23.29	4.66	0.20	1.21	0.71	0.77	0.13	2.33	-	-	-	26.85	5.00
	1N Co^{3+}	22.92	4.59	0.19	1.08	0.62	0.84	0.14	1.89	-	-	-	28.02	4.99
	0.1N Cd^{2+}	24.03	4.87	0.20	0.99	0.63	1.23	0.27	-	0.02	-	-	27.91	4.93
UHBM	0.5N Cd^{2+}	24.02	4.83	0.20	0.94	0.61	1.27	0.27	-	0.37	-	-	27.48	4.97
	1N Cd^{2+}	23.46	4.63	0.18	0.93	0.57	1.31	0.21	-	0.61	-	-	29.82	5.07
	0.1N Cd^{2+}	24.46	4.96	0.22	1.09	1.09	0.80	0.13	-	0.10	-	-	25.13	4.93
HBM	0.5N Cd^{2+}	24.04	4.82	0.18	1.09	0.70	0.87	0.14	-	0.38	-	-	27.48	4.99
	1N Cd^{2+}	23.22	4.64	0.14	0.90	0.60	1.00	0.13	-	0.95	-	-	30.78	5.00
	0.1N Cr^{3+}	24.01	4.82	0.23	1.23	0.66	0.73	0.38	-	-	0.18	-	27.33	4.98
UHBM	0.5N Cr^{3+}	24.20	4.87	0.21	1.13	0.66	0.74	0.38	-	-	0.27	-	26.44	4.97
	1N Cr^{3+}	24.32	4.88	0.22	1.07	0.62	1.05	0.23	-	-	0.08	-	26.52	4.98
	0.1N Cr^{3+}	24.49	4.93	0.22	1.10	0.70	0.81	0.21	-	-	0.18	-	25.42	4.97
HBM	0.5N Cr^{3+}	24.15	4.86	0.17	1.06	0.71	1.42	0.13	-	-	0.50	-	25.74	4.97
	1N Cr^{3+}	22.68	4.48	0.07	0.28	0.26	3.90	0.07	-	-	1.63	-	29.68	5.06
	0.1N Ag^{+}	24.58	4.95	0.20	0.99	0.63	1.27	0.22	-	-	-	0.04	25.48	4.97
UHBM	0.5N Ag^{+}	23.94	4.79	0.15	0.72	0.55	1.13	0.23	-	-	-	0.83	28.69	5.00
	1N Ag^{+}	24.30	4.85	0.12	0.65	0.49	1.11	0.17	-	-	-	2.11	26.23	5.01
	0.1N Ag^{+}	23.55	4.70	0.12	0.92	0.65	0.99	0.13	-	-	-	0.07	30.99	5.01
HBM	0.5N Ag^{+}	23.70	4.71	0.17	0.88	0.71	0.75	0.11	-	-	-	0.35	30.18	5.03
	1N Ag^{+}	23.20	4.62	0.05	0.34	0.29	0.64	0.09	-	-	-	2.13	33.12	5.02

Table 2. Numbers of atoms in the unit cells of natural and Na^{+}, K^{+}, Ca^{2+}, Mg^{2+}, Co^{3+}, Cd^{2+}, Cr^{3+}, Ag^{+} modified forms of clinoptilolite.

Ion-Exchange Method	Ion-Exchange Forms	Ions				
		Na^+	K^+	Ca^{2+}	Mg^{2+}	Fe^{3+}
UHBM	0.1N Na^+	-	-	18.40	8.57	-
	0.5N Na^+	-	-	44.80	20.00	-
	1N Na^+	-	-	52.80	27.14	-
HBM	0.1N Na^+	-	-	5.60	-	-
	0.5N Na^+	-	-	7.20	-	-
	1N Na^+	-	1.33	41.60	14.29	-
UHBM	0.1N K^+	12.12	-	18.40	7.14	-
	0.5N K^+	78.79	-	51.20	27.14	9.52
	1N K^+	72.72	-	56.80	30.00	14.29
HBM	0.1N K^+	6.06	-	-	7.14	-
	0.5N K^+	63.63	-	27.20	-	-
	1N K^+	75.76	-	82.40	57.14	19.05
UHBM	0.1N Ca^{2+}	-	2.67	-	2.67	-
	0.5N Ca^{2+}	3.03	2.67	-	1.43	-
	1N Ca^{2+}	66.67	-	-	-	-
HBM	0.1N Ca^{2+}	54.55	-	-	8.57	4.76
	0.5N Ca^{2+}	63.64	-	-	-	9.52
	1N Ca^{2+}	69.70	-	-	7.14	4.76
UHBM	0.1N Mg^{2+}	-	4.00	4.80	-	-
	0.5N Mg^{2+}	-	4.00	9.60	-	-
	1N Mg^{2+}	-	6.67	12.80	-	-
HBM	0.1N Mg^{2+}	57.58	-	8.00	-	4.76
	0.5N Mg^{2+}	54.55	-	4.80	-	-
	1N Mg^{2+}	45.45	-	12.00	-	9.52
UHBM	0.1N Co^{3+}	33.33	-	20.08	10.00	-
	0.5N Co^{3+}	21.21	-	22.40	12.85	-
	1N Co^{3+}	24.24	-	22.40	14.29	-
HBM	0.1N Co^{3+}	78.79	-	86.40	64.29	28.57
	0.5N Co^{3+}	60.61	-	3.20	-	4.76
	1N Co^{3+}	57.57	-	13.60	11.43	9.52
UHBM	0.1N Cd^{2+}	18.18	-	20.80	10.00	4.76
	0.5N Cd^{2+}	18.18	-	24.80	12.85	4.76
	1N Cd^{2+}	36.36	-	25.60	18.57	14.29
HBM	0.1N Cd^{2+}	60.61	-	12.80	-	-
	0.5N Cd^{2+}	57.58	-	12.80	-	14.29
	1N Cd^{2+}	60.61	-	28.00	14.29	33.33

Ion-Exchange Method	Ion-Exchange Forms	Ions				
		Na^{+}	K^{+}	Ca^{2+}	Mg^{2+}	Fe^{3+}
UHBM	0.1N Cr^{3+}	-	2.67	1.60	5.71	-
	0.5N Cr^{3+}	-	1.33	9.60	5.71	-
	1N Cr^{3+}	30.30	-	14.40	11.43	-
HBM	0.1N Cr^{3+}	36.36	-	12.00	-	-
	0.5N Cr^{3+}	60.61	-	15.20	-	19.05
	1N Cr^{3+}	78.79	-	77.60	62.86	66.67
UHBM	0.1N Ag^{+}	33.33	-	20.80	10.00	4.76
	0.5N Ag^{+}	30.30	-	42.40	21.43	28.57
	1N Ag^{+}	48.48	-	48.00	30.00	42.86
HBM	0.1N Ag^{+}	60.61	-	26.40	7.14	42.86
	0.5N Ag^{+}	66.67	-	29.60	-	19.05
	1N Ag^{+}	72.73	14.67	72.80	58.57	76.19

Table 3. Ion-exchange rates for Na^{+}, K^{+}, Ca^{2+}, Mg^{2+}, Co^{3+}, Cd^{2+}, Cr^{3+}, Ag^{+} modified forms compared to natural clinoptilolite.

Ion-Exchange Method	Ion-Exchange Forms	Ordering of Ion-Exchange Rate
UHBM	Na^{+}	$Ca^{2+}>Mg^{2+}$
HBM	Na^{+}	$Ca^{2+}>Mg^{2+}$
UHBM	K^{+}	$Na^{+}> Ca^{2+}> Mg^{2+} > Fe^{3+}$
HBM	K^{+}	$Na^{+}> Ca^{2+}> Mg^{2+} > Fe^{3+}$
UHBM	Ca^{2+}	$Na^{+}>K^{+}>>Mg^{2+}$
HBM	Ca^{2+}	$Na^{+}>Mg^{2+}>Fe^{3+}$
UHBM	Mg^{2+}	$Ca^{2+}> K^{+}$
HBM	Mg^{2+}	$Na^{+}>Ca^{2+}>Fe^{3+}$
UHBM	Co^{3+}	$Na^{+}> Ca^{2+}>Mg^{2+}$
HBM	Co^{3+}	$Na^{+}> Ca^{2+}>Mg^{2+}>Fe^{3+}$
UHBM	Cd^{2+}	$Ca^{2+}>Na^{+}> Mg^{2+}>Fe^{3+}$
HBM	Cd^{2+}	$Na^{+}> Ca^{2+}>Mg^{2+}>Fe^{3+}$
UHBM	Cr^{3+}	$Na^{+}> Ca^{2+}>Mg^{2+}> K^{+}$
HBM	Cr^{3+}	$Na^{+}> Ca^{2+}> Fe^{3+}>Mg^{2+}$
UHBM	Ag^{+}	$Na^{+}> Ca^{2+}> Fe^{3+}>Mg^{2+}$
HBM	Ag^{+}	$Na^{+}> Ca^{2+}> Fe^{3+}>Mg^{2+}> K^{+}$

Table 4. Ordering of ion-exchange rate for unheated and heated methods for Na-clinoptilolite from Bigadiç-Balıkesir (Turkey)

The order of ion-exchange rate of Na^+ and K^+ forms is constant under both unheated and heated conditions. On the other hand, high ion-exchange rates occur under the effects of heating. The ion-exchange rate order of Ca^{2+}, Mg^{2+}, Co^{3+}, Cd^{2+}, Cr^{3+} and Ag^+ forms changes upon heating, plus Fe^{3+} is leached from the structure. Although iron occurs as Fe^{3+} in the general order, Fe^{2+} was depleted instead of Fe^{3+} because Fe^{3+} cannot be depleted from the structure of clinoptilolite. Generally speaking, the ion-exchange process acts more on the surface of clinoptilolite than on its inner sites.[15]

Generally, the ion-exchange rate of the cations increases with increase in normality via application of both unheated and heated batch methods. Ion-exchange rates increase via the heating method compared to the unheated one. Forced-ion exchange occurs using the heated batch method as compared to natural ion exhange via the unheated method. The ion exchange of cations is controlled by cation valence, cation radius, ionization potential and the location of cations within pores.

The weak connection of the +1 valence of the Na^+ cation to the structure of 10T is related to its low ionization potential, resulting in leaching -with a high ion-exchange rate- from its structure (Figure 1). The Mg^{2+} cation exhibits low depletion due to its having symmetrical binding to the structure of 8T, despite having the lowest ionization potential. Ca^{2+} is more depleted than Mg^{2+} because of the unsymmetrical binding of Ca^{2+} to the structure, although the ionization potential of Ca^{2+} is higher than that of Mg^{2+}. The K^+ cation is strongly bound between the 8T and 10T structures; therefore, it is only slightly depleted.

Exchangeable Cations	Ionic Radii (Å)	Ionization Potential (eV)
Na^+	0.99	5.13
K^+	1.37	4.34
Ca^{2+}	1.00	6.11
Mg^{2+}	0.57	7.64
Cr^{3+}	0.62	6.77
Ag^+	0.79	7.57
Co^{3+}	0.55	7.88
Cd^{2+}	0.78	8.99
Fe^{+2}	0.63	7.90
Fe^{3+}	0.49	7.90

Table 5. Ionic radii and ionization potentials of exchangeable cations. [16]

Ion-Exchange Method	Ion-Exchange Forms	Ions							
		Na^+	K^+	Ca^{2+}	Mg^{2+}	Co^{3+}	Cd^{2+}	Cr^{3+}	Ag^+
UHBM	0.1N Na^+	66.67	-	-	-	-	-	-	-
	0.5N Na^+	136.36	-	-	-	-	-	-	-
	1N Na^+	251.52	-	-	-	-	-	-	-
HBM	0.1N Na^+	93.94	-	-	-	-	-	-	-
	0.5N Na^+	112.12	-	-	-	-	-	-	-
	1N Na^+	560.61	-	-	-	-	-	-	-
UHBM	0.1N K^+	-	62.67	-	-	-	-	-	-
	0.5N K^+	-	252.00	-	-	-	-	-	-
	1N K^+	-	289.33	-	-	-	-	-	-
HBM	0.1N K^+	-	4.00	-	-	-	-	-	-
	0.5N K^+	-	148.00	-	-	-	-	-	-
	1N K^+	-	428.00	-	-	-	-	-	-
UHBM	0.1N Ca^{2+}	-	-	2.40	-	-	-	-	-
	0.5N Ca^{2+}	-	-	7.20	-	-	-	-	-
	1N Ca^{2+}	-	-	20.80	-	-	-	-	-
HBM	0.1N Ca^{2+}	-	-	8.00	-	-	-	-	-
	0.5N Ca^{2+}	-	-	10.40	-	-	-	-	-
	1N Ca^{2+}	-	-	16.00	-	-	-	-	-
UHBM	0.1N Mg^{2+}	-	-	-	1.43	-	-	-	-
	0.5N Mg^{2+}	-	-	-	8.57	-	-	-	-
	1N Mg^{2+}	-	-	-	15.71	-	-	-	-
HBM	0.1N Mg^{2+}	-	-	-	-	-	-	-	-
	0.5N Mg^{2+}	-	-	-	34.29	-	-	-	-
	1N Mg^{2+}	-	-	-	67.14	-	-	-	-
UHBM	0.1N Co^{3+}	-	-	-	-	19	-	-	-
	0.5N Co^{3+}	-	-	-	-	43	-	-	-
	1N Co^{3+}	-	-	-	-	61	-	-	-
HBM	0.1N Co^{3+}	-	-	-	-	19	-	-	-
	0.5N Co^{3+}	-	-	-	-	233	-	-	-
	1N Co^{3+}	-	-	-	-	189	-	-	-
UHBM	0.1N Cd^{2+}	-	-	-	-	-	2	-	-
	0.5N Cd^{2+}	-	-	-	-	-	37	-	-
	1N Cd^{2+}	-	-	-	-	-	61	-	-
HBM	0.1N Cd^{2+}	-	-	-	-	-	10	-	-
	0.5N Cd^{2+}	-	-	-	-	-	38	-	-
	1N Cd^{2+}	-	-	-	-	-	95	-	-
UHBM	0.1N Cr^{3+}	-	-	-	-	-	-	18	-
	0.5N Cr^{3+}	-	-	-	-	-	-	27	-
	1N Cr^{3+}	-	-	-	-	-	-	8	-

Ion-Exchange Method	Ion-Exchange Forms	Ions							
		Na^+	K^+	Ca^{2+}	Mg^{2+}	Co^{3+}	Cd^{2+}	Cr^{3+}	Ag^+
HBM	0.1N Cr^{3+}	-	-	-	-	-	-	18	-
	0.5N Cr^{3+}	-	-	-	-	-	-	50	-
	1N Cr^{3+}	-	-	-	-	-	-	163	-
UHBM	0.1N Ag^+	-	-	-	-	-	-	-	4
	0.5N Ag^+	-	-	-	-	-	-	-	83
	1N Ag^+	-	-	-	-	-	-	-	211
HBM	0.1N Ag^+	-	-	-	-	-	-	-	7
	0.5N Ag^+	-	-	-	-	-	-	-	35
	1N Ag^+	-	-	-	-	-	-	-	213

Table 6. The ion-selectivity rates of Na^+, K^+, Ca^{2+}, Mg^{2+}, Co^{3+}, Cd^{2+}, Cr^{3+} and Ag^+ modified forms of clinoptilolite compared to natural clinoptilolite.

3.2. Rate of ion selectivity

Table 6 shows the ion-selectivity rates of Na^+, K^+, Ca^{2+}, Mg^{2+}, Co^{3+}, Cd^{2+}, Cr^{3+} and Ag^+ modified forms compared to natural clinoptilolite. Entrance of the cation into the structure depends on the ion-selectivity rate, that is, the ion-selectivity coefficient of cations. The rate of ion selectivity is controlled by cation valence, cation radius, ionization potential, pore size and location of the cation within the pore. Entrance of the cation into the structure increases with increasing normality and heating (Table 6). Monovalent cations such as K^+, Na^+ and Ag^+ predominantly enter the structure of Na-clinoptilolite; these cations have relatively low ionization potentials. These cations are followed by other cations, such as Ca^{2+}, Mg^{2+}, Co^{3+}, Cd^{2+} and Cr^{3+}.

Selectivity Order	References
$Pb^{2+}>NH_4^+>Ba^{2+}>Cu^{2+}>Zn^{2+}>Cd^{2+}>Co^{2+}$	[19]
$Pb^{2+}>Cd^{2+}>Cs^{2+}>Cu^{2+}>Co^{2+}>Cr^{3+}>Zn^{2+}>Ni^{2+}>Hg^{2+}$	[20]
$Pb^{2+}>Ag^+>Cd^{2+}>Zn^{2+}>Cu^{2+}>Na^+$	[21]
$Pb^{2+}>Cu^{2+}>Cd^{2+}>Zn^{2+}>Cr^{3+}>Co^{2+}>Ni^{2+}$	[22]

Table 7. Global experimental ion-selectivity order for modified clinoptilolite.

The ordering of ion-selectivity rates via application of unheated and heated methods is as follows:

$$\text{Unheated: } K^+>Na^+>Ag^+>Co^{3+}>Cd^{2+}>Cr^{3+}>Ca^{2+}>Mg^{2+}$$

$$\text{Heated: } Na^+>K^+> Co^{3+}>Ag^+> Cr^{3+}>Cd^{2+}> Mg^{2+}>Ca^{2+}$$

The ordering rate obtained via the unheated batch method is consistent with the results given in the global literature; conversely, this order changes upon heating via application of the heated batch method.

Moreover, the order of the ion-selectivity rates, namely

$$Na^{+} > Ag^{+} > K^{+} \approx NH_4^{+} > Cd^{2+} > Ca^{2+} > Cr^{3+} > Mg^{+2} > Co^{3}$$

was also determined for the Gördes Na-clinoptilolite.[17, 18]. Other results are given in Table 7.

4. Conclusions

1. Rates of ion exchange and ion selectivity increase with increase in normality via application of both unheated and heated methods.
2. Entrance of heavy metals into and depletion of exchangeable cations from the zeolite structure is controlled by cation valence, cation radius, ionization potential, pore size and location of the cation within the pore.
3. The rates of ion exchange and ion selectivity are higher in application of the heated batch method than in the unheated one.
4. Forced ion exchange occurs via application of the heated batch method as compared to natural ion exchange via the unheated method due to the rates of ion exchange and ion selectivity.
5. The order of ion-selectivity rate for the modified forms, except for the exchangeable cations, is $Ag^{+} > Cd^{2+} > Cr^{3+} > Co^{3+}$; this finding is consistent with Ag^{+}, Cd^{2+}, Cr^{3+} and Co^{3+} cation adsorption from wastewater, as determined worldwide.
6. Na-clinoptilolite adsorbs cations such as Na^{+}, K^{+}, Ca^{2+}, Mg^{2+}, Co^{3+}, Cd^{2+}, Cr^{3+} and Ag^{+}. Ion-exchangeable cations such as Na^{+}, K^{+}, Ca^{2+}, Mg^{2+}, 8T and 10T structures and negative structure also exhibit adsorptive capacity with respect to other cations.
7. Heavy-metal cation retention in large volumes of water (such as in rivers) requires application of the unheated batch method. In contrast, heavy-metal cation retention in small volumes of water (such as in the laboratory) over shorter time periods requires application of the heated batch method.

Author details

Tevfik Ünaldı
Department of Physics, Eskişehir Osmangazi University, Eskişehir, Turkey

Selahattin Kadir
Department of Geological Engineering, Eskişehir Osmangazi University, Eskişehir, Turkey

5. References

[1] Ivanova, E., Koumanova, B., J. Hazard. Mat. 2009; 167, 306–312.

[2] Lihareva, N., Dimova, L., Petrov, O., Tzvetanova, Y., Micropor. Mesopor. Mat. 2010; 130, 32–37.

[3] Ackley, M.W., Giese, R.F., Yang, R.T. Zeolites, 1992; 12, 780.

[4] Merkle, A. B. and Slaughter, M., Structure of Heulandite, Amer. Mineral, 1968; 53, 1120-113.

[5] Lam, A., Sierra, L. R., Rojas, G., Rivera, A., Rodriguez-Fuentes, G., Montero, L. A., Theoretical Study of The Physical Adsorption of Aspirin on Natural Clinoptilolite, Microporous and Mesoporous Materials, 1998; 23, 247-25.

[6] Dorfner, K., Ion Exchangers, Properties and Applications, Ann Arbor Science Publishers Inc., Michigan, 1972; p.3.

[7] Orhun, Ö., Ünaldı, T. and Vanturache, C., On the Na^+ Ion Exchanged Natural Clinoptilolite, Analele StiintificeAle Universitatii, "Al.l.Cuza" iasi, Tomul III, s. Chimie, Supliment, 1995; p. 118-122.

[8] Orhun, Ö. and Ünaldı, T., Ion Exchange Properties of Na^+ Forms having some Normalities of Natural Clinoptilolite obtained from Bigadiç-Balıkesir (Turkey) Region, The Fourth National Symposium, Zeolites in Modern Technology, Jassy-Romania, Nov. 1995; p. 34-35.

[9] İnel, O., Yörükoğulları, E., Orhun, Ö., Albayrak, F. Chemica Acta Turcica, 1972; 19, 77-83.

[10] Hulbert, M.H. Clays and Clay Min. 1987; 35, 458-462.

[11] Yücel, H. ve Çulfaz, A. Batı Anadolu Doğal Klinoptilolitlerinin Karakterizasyonu ve Karbondioksit Tutma Özellikleri, TÜBİTAK Araştırma Projesi, MAG 624, 1984; 54 s.

[12] Fraenkel, D., Lazar, R. and Shabtai, J., The Potential of Zeolite Molecular Sieve as Hydrogen Storage Media, Alternative Energy Sources, Hydrogen Energy, 1978; 3771-3882.

[13] Orhun, Ö., Zeolitlerde İyon Değişimi, Anadolu Univ., Eskişehir, 1997; s. 53.

[14] Dyer, A., An Introduction to Zeolite Molekular Sieves, John Wiley & Sons, New York, 1988; p. 148.

[15] Nyembe, D.W., Mamba B.B. and Mulaba-Bafubiandi, A.F., Adsorption Mechanisms of Co^{2+} and Cu^{2+} from Aqueous Solitions using Natural Clinoptilolite: Equilibrium and Kinetic Studies, Journal of Applied Sciences, 2010; 10, 599-610.

[16] Linde, R. D., Handbook of chemistry and physics, 84 Th Edition, CRC press, 2003; 2475 p.

[17] Ünaldı, T., Orhun, Ö. and Kadir, S., Physicochemical characterization of natural and Na^+-, K^+-, Ca^{2+}- and Mg^{2+}-modified clinoptilolite from Gördes (Manisa, Turkey), Adsorption Science & Technology, 2009; 27, 615-631.

[18] Ünaldı, T. and Yıldırım, B., Investigation of NH_4 forms of Gördes region natural zeolites, Anadolu University Journal of Science and Technology, 2009; 10, 2, 485-493.

[19] Blanchard, G., Maunaye, M. and Martin, G., Removal of heavy metals from waters by means of natural zeolites, Water Research, 1984; 18, 1501-1507.

[20] Zamzow, M.J., Eichbaum, R., Sandgren, K.R., and Shanks, D.E., Removal of heavy metals and other cations from wastewater using zeolites. Separation Science Technology, 1990; 25,1555–1569.

[21] Kesraoui-Ouki, S., Cheeseman, C. and Perry, R., Natural zeolite utilization in pollution control: a review of applications to metal's effluents, J. Chem. Technol. Biotechnol., 1994; 59, 121-126.

[22] Ouki, S.K. and Kavanagh, M., Treatment of metals contaminated wastewaters by use of natural zeolites, Water Science and Technology, 1999; 39, 122–155.

Section 4

Ion Exchange Chromatography

Chapter 14

The Role of Ion Exchange Chromatography in Purification and Characterization of Molecules

Hidayat Ullah Khan

Additional information is available at the end of the chapter

http://dx.doi.org/10.5772/52537

1. Introduction

Adsorption chromatography depends upon interactions of different types between solute molecules and ligands immobilized on a chromatography matrix. The first type of interaction to be successfully employed for the separation of macromolecules was that between charged solute molecules and oppositely charged moieties covalently linked to a chromatography matrix. The technique of ion exchange chromatography is based on this interaction.

Ion exchange is probably the most frequently used chromatographic technique for the separation and purification of proteins, polypeptides, nucleic acids, polynucleotides, and other charged biomoleules (1) . The reasons for the success of ion exchange are its widespread applicability, its high resolving power, its high capacity, and the simplicity and controllability of the method.

2. The theory of ion exchange

Purification using ion exchange chromatography depends upon the reversible adsorption of charged solute molecules to immobilized ion exchange groups of opposite charge. Most ion exchange experiments are performed in five main stages. These steps are illustrated schematically.

The first stage is equilibration in which the ion exchanger is brought to a starting state, in terms of pH and ionic strength, which allows the binding of the desired solute molecules. The exchanger groups are associated at this time with exchangeable counter-ions (usually simple anions or cations, such as chloride or sodium). The second stage is sample application and adsorption, in which solute molecules carrying the appropriate charge displace counter-ions and bind reversibly to the gel. Unbound substances can be washed out

INTECH
open science | open minds

© 2012 Khan, licensee InTech. This is a paper distributed under the terms of the Creative Commons Attribution License (http://creativecommons.org/licenses/by/3.0), which permits unrestricted use, distribution, and reproduction in any medium, provided the original work is properly cited.

from the exchanger bed using starting buffer. In the third stage, substances are removed from the column by changing to elution conditions unfavourable for ionic bonding of the solute molecules. This normally involves increasing the ionic strength of the eluting buffer or changing its pH. In Figure 1 desorption is achieved by the introduction of an increasing salt concentration gradient and solute molecules are released from the column in the order of their strengths of binding, the most weakly bound substances being eluted first.

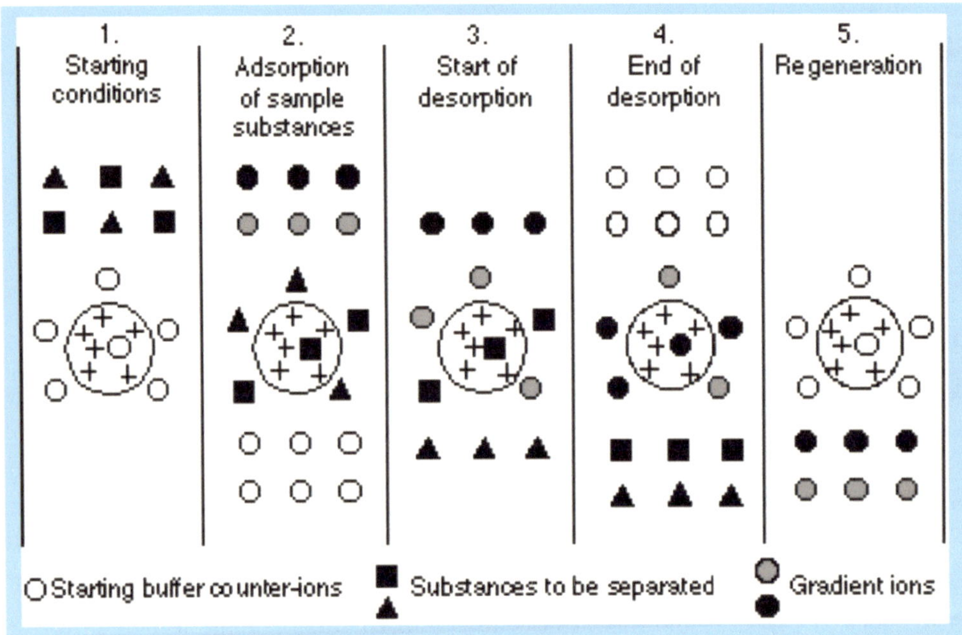

Figure 1. The principle of ion exchange chromatography (salt gradient elution).

The fourth and fifth stages are the removal from the column of substances not eluted under the previous experimental conditions and re-equilibration at the starting conditions for the next purification. Separation is obtained since different substances have different degrees of interaction with the ion exchanger due to differences in their charges, charge densities and distribution of charge on their surfaces. These interactions can be controlled by varying conditions such as ionic strength and pH. The differences in charge properties of biological compounds are often considerable, and since ion exchange chromatography is capable of separating species with very minor differences in properties, e.g. two proteins differing by only one charged amino acid, it is a very powerful separation technique. In ion exchange chromatography one can choose whether to bind the substances of interest and allow the contaminants to pass through the column, or to bind the contaminants and allow the substance of interest to pass through. Generally, the first method is more useful since it allows a greater degree of fractionation and concentrates the substances of interest. In addition to the ion exchange effect, other types of binding may occur. These effects are small and are mainly due to van der Waals forces and non-polar interactions. Ion exchange separations may be carried out in a column, by a batch procedure or by expanded bed adsorption. All three methodologies are performed in the stages of equilibration, sample adsorption etc. described previously.

3. The matrix

An ion exchanger consists of an insoluble matrix to which charged groups have been covalently bound. The charged groups are associated with mobile counter ions. These counter-ions can be reversibly exchanged with other ions of the same charge without altering the matrix. It is possible to have both positively and negatively charged exchangers (Fig. 2). Positively charged exchangers have negatively charged counter-ions (anions) available for exchange and are called anion exchangers. Negatively charged exchangers have positively charged counter-ions (cations) and are termed cation exchangers. The matrix may be based on inorganic compounds, synthetic resins or polysaccharides. The characteristics of the matrix determine its chromatographic properties such as efficiency, capacity and recovery as well as its chemical stability, mechanical strength and flow properties.

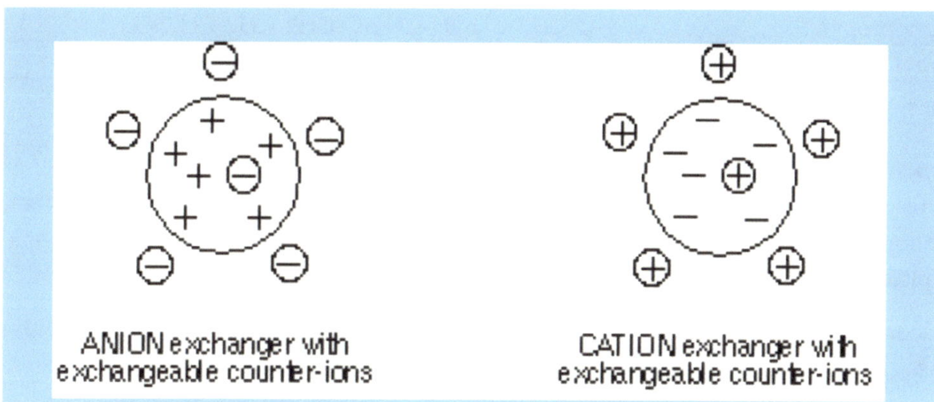

Figure 2. Ion exchanger types.

The nature of the matrix will also affect its behaviour towards biological substances and the maintenance of biological activity. The first ion exchangers were synthetic resins designed for applications such as demineralisation, water treatment, and recovery of ions from wastes. Such ion exchangers consist of hydrophobic polymer matrices highly substituted with ionic groups, and have very high capacities for small ions. Due to their low permeability these matrices have low capacities for proteins and other macromolecules. In addition, the extremely high charge density gives very strong binding and the hydrophobic matrix tends to denature labile biological materials. Thus despite their excellent flow properties and capacities for small ions, these types of ion exchanger are unsuitable for use with biological samples. The first ion exchangers designed for use with biological substances were the cellulose ion exchangers (2) . Because of the hydrophilic nature of cellulose, these exchangers had little tendency to denature proteins. Unfortunately, many cellulose ion exchangers had low capacities (otherwise the cellulose became soluble in water) and had poor flow properties due to their irregular shape. Ion exchangers based on dextran (Sephadex), followed by those based on agarose (Sepharose CL-6B) and cross-linked cellulose (DEAE Sephacel) were the first ion exchange matrices to combine a spherical form with high porosity, leading to improved flow properties and high capacities for macromolecules.

The presence of charged groups is a fundamental property of an ion exchanger. The type of group determines the type and strength of the ion exchanger; their total number and availability determines the capacity. There is a variety of groups which have been chosen for use in ion exchangers (3); some of these are shown in Table 1

Anion exchangers	**Functional group**
Diethylaminoethyl (DEAE)	-O-CH2-CH2-N+H(CH2CH3)2
Quaternary aminoethyl (QAE)	-O-CH2-CH2-N+(C2H5)2-CH2-CHOH-CH3
Quaternary ammonium (Q)	-O-CH2-CHOH-CH2-O-CH2-CHOH-CH2-N+(CH3)3
Cation exchangers	**Functional group**
Carboxymethyl (CM) -	O-CH2-COO
Sulphopropyl (SP)	-O-CH2-CHOH-CH2-O-CH2-CH2-CH2SO3
Methyl sulphonate (S)	-O-CH2-CHOH-CH2-O-CH2-CHOH-CH2SO3

Table 1. Functional groups used on ion exchangers.

Sulphonic and quaternary amino groups are used to form strong ion exchangers; the other groups form weak ion exchangers. The terms strong and weak refer to the extent of variation of ionization with pH and not the strength of binding. Strong ion exchangers are completely ionized over a wide pH range. Whereas with weak ion exchangers,

the degree of dissociation and thus exchange capacity varies much more markedly with pH. Some properties of strong ion exchangers are:

- Sample loading capacity does not decrease at high or low pH values due to loss of charge from the ion exchanger.
- A very simple mechanism of interaction exists between the ion exchanger and the solute.
- Ion exchange experiments are more controllable since the charge characteristics of the media do not change with changes in pH.

4. Resolution in ion exchange chromatography

The result of an ion exchange experiment, as with any other chromatographic separation, is often expressed as the resolution between the peaks of interest. The resolution is defined as the distance between peak maxima compared with average base width of the two peaks. Elution volumes and peak widths should be measured with the same units to give a dimensionless value to the resolution (4, 5). The resolution (Rs) is determined from the chromatogram as shown in Figure 3.

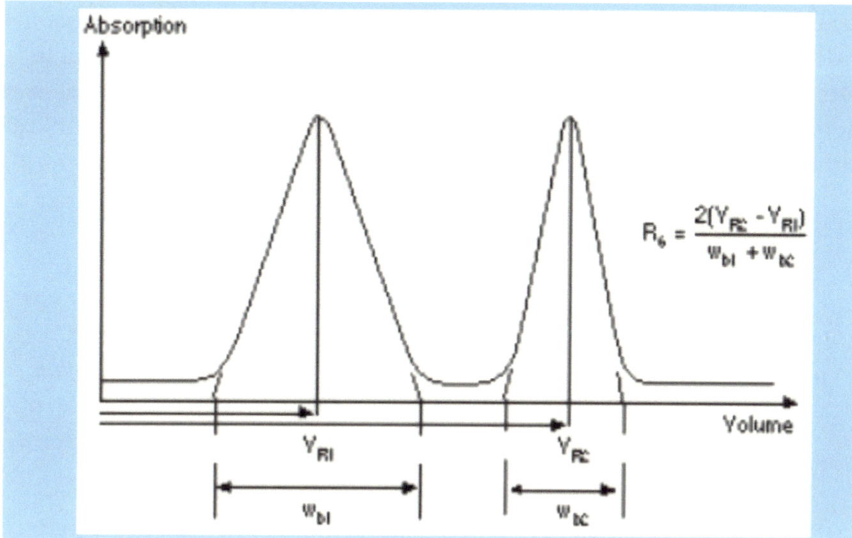

Figure 3. Determination of the resolution (Rs) between two peaks.

Rs is a measure of the relative separation between two peaks and can be used to determine if further optimization of the chromatographic procedure is necessary.

4.1. Capacity

The capacity of an ion exchanger is a quantitative measure of its ability to take up exchangeable counter-ions and is therefore of major importance. The capacity may be expressed as total ionic capacity, available capacity or dynamic capacity. The total ionic capacity is the number of charged substituent groups per gram dry ion exchanger or per ml swollen gel. Total capacity can be measured by titration with a strong acid or base. The actual amount of protein which can be bound to an ion exchanger, under defined experimental conditions, is referred to as the available capacity for the gel. If the defined conditions include the flow rate at which the gel was operated, the amount bound is referred to as the dynamic capacity for the ion exchanger. Available and dynamic capacities depend upon: The properties of the protein. The properties of the ion exchanger. The chosen experimental conditions. The properties of the protein which determine the available or dynamic capacity on a particular ion exchange matrix are its molecular size and its charge/pH relationship. The capacity of an ion exchanger is thus different for different protein.

5. Choice of exchanger group

Substances are bound to ion exchangers when they carry a net charge opposite to that of the ion exchanger. This binding is electrostatic and reversible. In the case of substances which carry only one type of charged group the choice of ion exchanger is clear-cut. Substances which carry both positively and negatively charged groups, however, are termed amphoteric and the net charge which they carry depends on pH (Fig. 4). Consequently at a certain pH value an amphoteric substance will have zero net charge. This value is termed the isoelectric point (pI) and at this point substances will bind to neither anion or cation exchangers (6). The pH ranges in which the protein is bound to anion or cation exchangers and an arbitrary range of stability are shown in Figure 4.

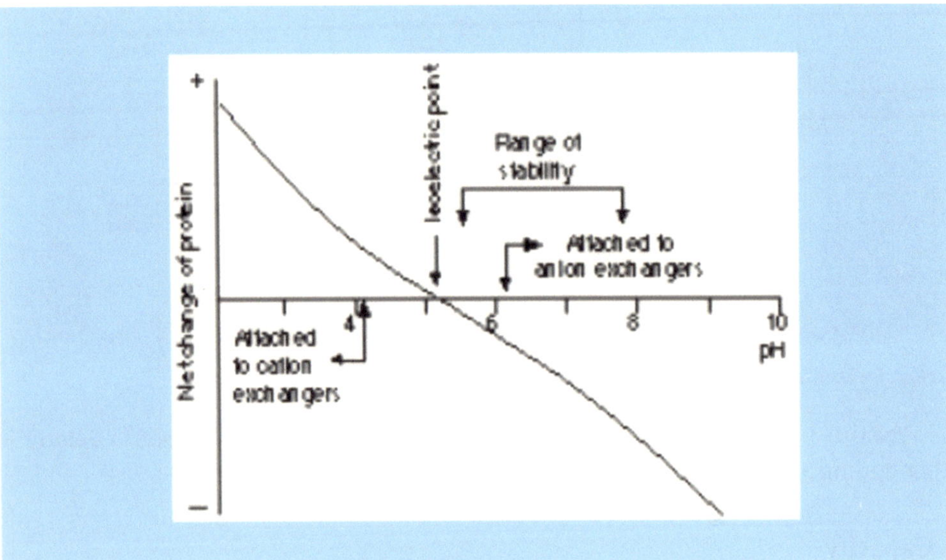

Figure 4. The net charge of protein as a function of pH.

The pH of the buffer thus determines the charge on amphoteric molecules during the experiment. In principle therefore, one could use either an anion or a cation exchanger to bind amphoteric samples by selecting the appropriate pH. In practice however, the choice is based on which exchanger type and pH give the best separation of the molecules of interest, within the constraints of their pH stability (7, 8).

Many biological macromolecules become denatured or lose activity outside a certain pH range and thus the choice of ion exchanger may be limited by the stability of the sample. This is illustrated in Figure 4. Below its isoelectric point a protein has a net positive charge and can therefore adsorb to cation exchangers. Above its pI the protein has a net negative charge and can be adsorbed to anion exchangers. However, it is only stable in the range pH 5-8 and so an anion exchanger has to be used.

5.1. In summary

1. If the sample components are most stable below their pI's, a cation exchanger should be used.
2. If they are most stable above their pI's, an anion exchanger should be used.
3. If stability is high over a wide pH range on both sides of pI, either type of ion exchanger can be used.

6. Determination of starting conditions

6.1. The isoelectric point

The starting buffer pH is chosen so that substances to be bound to the exchanger are charged. The starting pH should be at least 1 pH unit above the isoelectric point for anion exchangers or at least 1 pH unit below the isoelectric point for cation exchangers to facilitate adequate binding (6). Substances begin to dissociate from ion exchangers about 0.5 pH units from their isoelectric points at ionic strength 0.1 M (15). If the isoelectric point of the sample is unknown, a simple test can be performed to determine which starting pH can be used. range of pH 5-9 for anion and pH 4-8 for cation exchangers, with 0.5 pH unit intervals between tubes.

6.2. Choice of buffer

As with the choice of ion exchanger, there are a number of variables which have to be considered. These include:

1. The choice of buffer pH and ionic strength.
2. The choice of buffering substance.

It should be pointed out, however, that in many applications the optimum separation may be achieved by choosing conditions so that major and troublesome contaminants are bound to the exchanger while the substance of interest is eluted during the wash phase (9). This procedure is sometimes referred to as "starting state elution".

Note: Concentration of sample does not occur with starting state elution.

The highest ionic strength which permits binding of the selected substances and the lowest ionic strength that causes their elution should normally be used as the starting and final ionic strengths in subsequent column experiments (i.e. the starting and limiting buffers for gradient elution). A third and higher ionic strength buffer is frequently employed as a wash step before column regeneration and re-use.

The required concentration of the start buffer will vary depending on the nature of the buffering substance. In the majority of cases a starting ionic strength of at least 10 mM is required to ensure adequate buffering capacity. Salts also play a role in stabilizing protein structures in solution and so it is important that the ionic strength should not be so low that protein denaturation or precipitation occurs.

6.3. Choice of buffer substance

If the buffering ions carry a charge opposite to that of the functional groups of the ion exchanger they will take part in the ion exchange process and cause local disturbances in pH. It is preferable, therefore, to use buffering ions with the same charge sign as the substituent groups on the ion exchanger. There are of course exceptions to this rule as illustrated by the frequency with which phosphate buffers are cited in the literature in connection with anion exchangers. In those instances when a buffering ion which interacts with the ionic groups on the matrixnis used, extra care must be taken to ensure that the system has come to equilibrium before application of sample.

7. Column chromatography

Good results in column chromatography are not solely dependent on the correct choice of gel media. The design of the column and good packing technique are also important in realising the full separation potential of any gel.

The material used in the construction of the column should be chosen to prevent destruction of labile biological substances and minimize non-specific binding to exposed surfaces. The bed support should be designed so it is easily exchangeable to restore column performance whenever contamination and/or blockage in the column occur. Bed supports made from coarse sintered glass or glass wool cannot be recommended because they soon become clogged, are difficult to clean and cause artifacts (10).

The pressure specifications of the column have to match the back-pressure generated in the packed bed when run at optimal flow rate. This is particularly important when using high performance media with small bead size. All are easy to dismantle and reassemble to allow thorough cleaning, which is a particularly important aspect when handling biological samples.

As for most adsorptive, high selectivity techniques, ion exchange chromatography is normally carried out in short columns. A typical ion exchange column is packed to a bed height of 5-15 cm. Once the separation parameters have been determined, scale-up is easily achieved by increasing the column diameter.

8. Quantity and preparation of ion exchanger

The amount of ion exchanger required for a given experiment depends on the amount of sample to be chromatographed and on the available or dynamic capacity of the ion exchanger for the sample substances. For the best resolution in ion exchange chromatography, it is not usually advisable to use more than 10-20% of this capacity, although this value can be exceeded if resolution is adequate. Preparation of the ion exchanger Having chosen the appropriate ion exchanger and starting buffer it is essential that the exchanger is brought to equilibrium with start buffer before sample application.

To prepare the gel, the supernatant is decanted and replaced with starting buffer to a ratio of approximately 75% settled gel to 25% buffer. If large amounts of ion exchangers are to be equilibrated with a weak buffer, the ion exchanger should first be equilibrated with a 10 times concentrated buffer solution at the correct pH, and then with a few volumes of starting buffer.

9. Sample preparation

The amount of sample which can be applied to a column depends on the dynamic capacity of the ion exchanger and the degree of resolution required. For the best resolution it is not usually advisable to use more than 10-20% of this capacity(11). Information on the available capacities for the different exchangers is given in the relevant product sections.

The ionic composition should be the same as that of the starting buffer. If it is not, it can be changed by gel filtration on Sephadex G-25 using.Desalting Columns, dialysis, diafiltration or possibly by addition of concentrated start buffer.

If the ion exchanger is to be developed with the starting buffer (isocratic elution), the sample volume is important and should be limited to between 1 and 5% of the bed volume. If however, the ion exchanger is to be developed with a gradient, starting conditions are normally chosen so that all important substances are adsorbed at the top of the bed. In this case, the sample mass applied is of far greater importance than the sample volume. This means that large volumes of dilute solutions, such as pooled fractions from a preceding gel filtration step or a cell culture supernatant can be applied directly to the ion exchanger without prior concentration.Ion exchange thus serves as a useful means of concentrating a sample in addition to fractionating it. If contaminants are to be adsorbed, and the component of interest is allowed to pass straight through, then the sample volume is less important than the amount of contaminant which is present. Under these conditions there will be no concentration of the purified component, rather some degree of dilution due to diffusion.

The viscosity may limit the quantity of sample that can be applied to a column. A high sample viscosity causes instability of the zone and an irregular flow pattern. The critical variable is the viscosity of the sample relative to the eluent. This corresponds to a protein concentration of approximately 5%. Approximate relative viscosities can be quickly estimated by comparing emptying times from a pipette.

10. Sample load

When the selectivity parameters have been defined to achieve the most optimal balance between resolution, capacity, speed and recovery, in ion exchange chromatography, as for most other adsorption techniques, there are then basically two alternative routes to follow for optimization of sample load and flow rate to achieve highest possible productivity in the system. I. In a typical capture situation the sample will be applied to the column, nonbound substances will be washed out from the column and the compound of interest will be eluted from the column with a simple step elution procedure. The difference in eluting strength, between the different steps will usually be large, i.e. it will be possible to elute one group of compounds while the others are still retained on the column. In this mode, the entire bed volume can be utilized for sample bin ding and the prime consideration when optimising for highest possible productivity is to define the highest possible sample load over the shortest possible sample application time with acceptable loss in yield.

11. Flow rate

The maximum flow rate that can be applied in any particular ion exchange chromatography step will differ between different parts of the chromatographic cycle. Since low molecular weight substances show high diffusion rates, i.e. are transported rapidly between the mobile phase and stationary phase, the flow rate during equilibration, washing and regeneration procedures is limited primarily by the rigidity of the chromatography media and by system constraints regarding pressure specification. Larger molecules, i.e. the substances to be separated during the

Chromatographic run, show a lower diffusion rate which will limit the flow rate that can be applied during sample adsorption and desorption. In a typical capture situation, the flow rate during sample application has to be controlled so that the residence time in the column allows for a complete binding without leakage in the flow through fraction. Maximum flow rate is defined by running the frontal analysis test (break-through) referred to above at a number of different flow rates. Optimal conditions will depend on the requirements for speed and capacity in the system. If speed, i.e. sample application time, is critical due to proteolysis or other detrimental effects in the feed material, a higher flow rate may have to be used on the expense of the binding capacity in terms of amount of sample that can be applied per volume of media. If speed is not a big issue, binding capacity can be increased on the expense of flow rate which will reduce the scale of work in the final production process (12).

12. Elution

If starting conditions are chosen such that only unwanted substances in the sample are adsorbed, then no change in elution conditions is required since the substance of interest passes straight through the column. Similarly no changes are required if sample components are differentially retarded and separated under starting conditions. This procedure is termed isocratic elution, and the column is said to be developed under starting conditions. Isocratic elution can be useful since no gradient apparatus is required for the run and, if all retarded substances elute, regeneration is not required. Normally, however, separation and elution are achieved by selectively decreasing the affinity of the solute molecules for the charged groups on the gel by continuously changing either buffer pH or ionic strength or possibly both. This procedure is termed gradient elution.

13. Change of pH and ionic strength

The net charge on a molecule depends on pH. Thus altering the pH towards the isoelectric point of a substance causes it to lose its net charge, desorb, and elute from the ion exchanger (13).

At low ionic strengths, competition for charged groups on the ion exchanger is at a minimum and substances are bound strongly. Increasing the ionic strength increases competition and reduces the interaction between the ion exchanger and the sample substances, resulting in their elution.

14. Regeneration

After each cycle, bound substances must be washed out from the column to restore the original function of the media. Ion exchange adsorbents can normally be regenerated after each run by washing with a salt solution until an ionic strength of about 2 M has been reached. This should remove any substances bound by ionic forces. The salt should contain the counter-ion to the ion exchanger to facilitate equilibration. To prevent a slow build up of contaminants on the column over time, more rigorous cleaning protocols may have to be applied on a regular basis

15. Applications

Ion exchange has proven to be one of the major methods of fractionation of labile biological substances. In the development of modern high performance media for purification, ion exchange chromatography has played a major role in the separation and purification of biomolecules and contributed sgnificantly to our understanding of biological processes. Analytical and preparative applications from the research laboratory. Ion exchange chromatography has played a role in the purificationof thousands of enzymes, and using modern matrices with optimized separation conditions gives extremely high recoveries. Normally the isoforms of an enzyme have approximately the same molecular weight. This makes their separation impossible by gel filtration. However, the small differences in charge properties resulting from altered amino acid composition enable the separation of isoenzymes using ion exchange chromatography. Ion exchange is frequently used for the purification of immunoglobulins.

Ion exchange chromatography also has many important applications in the field of industrial and pilot scale preparations. Many blood products such as albumin and IgG as well as the products of recombinant DNA technology, such as growth factors and pharmaceutically important enzymes are purified using this technique. Analytical applications of ion exchange chromatography are to be found in diverse areas such as quality control of purified products or process monitoring in biotechnology.

Other areas of application include food research where FPLC ion exchange can be used in the and in clinical research where ion exchange chromatography has been used in studies such as the relationship between post-partum depression and β-endorphin secretion and the correlation of a chromatogram of the urine from patients exhibiting tubular proteinuria, due to acute pyelonephritis, severe burns or renal transplants, shows distinct peaks corresponding to β2-microglobulin, retinol binding protein and α1-acid glycoprotein.

Author details

Hidayat Ullah Khan
Department of Biotechnology, University of Science and Technology Bannu, Khyber Pakhtunkhwa Pakistan

16. References

[1] Bonnerjera, J., Oh, S., Hoare, M., Dunhill, P *(1986)*.The right step at the right time. *Bio/Technology, 4, 954-958*

[2] Peterson, E.A., Sober, H.A (*(1956)*.Chromatography of Proteins. I. Cellulose ion exchange adsorbents. *J. Amer. Chem. Soc. 78; 751 755.*

[3] Himmelhoch, S.R. *(1971)*. Chromatography of proteins on ion-exchange adsorbents. *Meth. Enzymol. 22 273—286.*

[4] Heftman, E. (Ed.), Van Noostrand Rheinhold Co., New York (1975). Chromatography: a laboratory handbook of chromatographic and electrophoretic techniques.

[5] Giddings, J.C., Keller, R.A. (Eds.), Marcel Dekker Inc., New York (1965). Dynamics of chromatography, Part 1., Principles and theory.

[6] G.P., Tytell, A.A *(1965)*. A simple method for estimating isoelectric points. *Anal. Biochem. 11374—377.*

[7] P.G., Caravaggio, T*(1976)*.Isoelectric points and molecular weights of proteins: a table. *J. Chromatogr.127 1—28.*

[8] Malamud, D., Drysdale, J.W. *(1978)*.Isoelectric points of proteins: a table. *AnaL Biochem. 86; 620—647.*

[9] Stone-Wolff, D.S., Yip, Y.K., Kelker, H.C. et al*(1984)*.Interrelationships of human-interferon gamma with lymphotoxin and monocyte cytotoxin. *J. Exp. Med. 159; 824-843,*

[10] Schwartz, D.P*(1978)*.Glass wool as a potential source of artifacts in chromatography. *J. Chromatogr. 152; 514—516.*

[11] Karlsson, E., Ryden, L., Brewer, J(1989). Ion Exchange Chromatography. Protein Purification, Principles, High resolution methods and Applications, Janson, J.C., Ryden, L. (Eds) VCH, Publishers Inc. New York. 107-148.

[12] Gel Filtration Chromatography. L. Fischer. Elsevier, Amsterdam (1980)

[13] Lamy, J., Lamy, J., Weill, J(1979). Arthropod hemocyanin structure: isolation of eight subunits in the scorpion. Arch. Biochem.Biophys. 193; 140—149

Chapter 15

Nitrogen Isotope Separation by Ion Exchange Chromatography

Xingcheng Ding and Xunyue Liu

Additional information is available at the end of the chapter

http://dx.doi.org/10.5772/51311

1. Introduction

Enriched stable isotopes of many elements have been widely used in many aspects and such isotopes have been primarily used as the tracer for agricultural and biochemical studies and the availability of the isotopically labeled compounds have been now extensively increasing. Chemical exchange separation of isotopes is based on the equilibrium fractionation of isotopes between two phases, ion exchange isotope separation which is one of the chemical exchange methods, is based on the chemical equilibrium between a stationary phase and a mobile fluid phase. Displacement always results in a sharp boundary between the bands of eluted solute and displacing solute in this method. Displacement chromatographic ion exchange technology for isotope separation is described in this chapter. This chapter mainly divided into two sections. Ion exchange system, ion exchange separation and analysis are introduced in the former section as the basic theory and concept for ion exchange separation. The second section is about the experiment parts of nitrogen isotope separation which include nitrogen isotope application, nitrogen isotope separation process, nitrogen isotopic analysis with mass spectrometry and the obtained results and discussion which were performed by our recent experiments.

2. Ion exchange system

Enriched stable isotopes of many elements have been widely used in agriculture, medicine, biochemistry and industry, especially in nuclear industry. Such isotopes have been primarily used as the tracer for agricultural and biochemical studies and the availability of the isotopically labeled compounds have been now extensively increasing. In order to meet the demand, the need of developing a practical cost-effective process has been enlarged. The most heavy stable isotopes are produced in relatively small quantities, many light isotopes

INTECH
open science | open minds

© 2012 Ding and Liu, licensee InTech. This is a paper distributed under the terms of the Creative Commons Attribution License (http://creativecommons.org/licenses/by/3.0), which permits unrestricted use, distribution, and reproduction in any medium, provided the original work is properly cited.

are used in large quantities. Distillation is used in industry to separation the isotopes of hydrogen, carbon, boron, nitrogen and oxygen. Chemical exchange is the satisfactory technique for separating isotopes of light elements because of their relatively high simple process factors (separation factors).

Chemical exchange separation of isotopes is based on the equilibrium fractionation of isotopes between two phases, i.e. one of the isotopes is concentrated more abundantly in one of the phases than in the other phase if two chemical species are distributed in each phase. In this method, the enrichment of the isotopes concerned is achieved when the simple process of isotope separation is multiplied, by arranging the countercurrent contacting of two substances. The countercurrent contacting of the two different chemical substances is most easily realized in vapor-liquid or liquid-liquid system. The contacting takes place in conventional systems such as packed columns, mixer-settlers, etc. In this process it is necessary to provide reflux at the product end of a chemical exchange plant; the end which the desired isotope being enriched. In general, chemical refluxing is used to complete the conversion of one chemical species into the other at the end of the multi-stage contactor system.

Ion exchange isotope separation, which is one of the chemical exchange methods, is based on the chemical equilibrium between a stationary phase and a mobile fluid phase. In the past, many researchers have studied isotope fractionation on the isotopes both for light and heavy elements in this liquid solid system[1-5] and several isotopes especially for nitrogen isotope have been successfully enriched in laboratory scale by using displacement chromatographic technique [6, 7]. In this process a band of adsorbate is eluted through the column by a displacing solution, which the rate of movement of the band is determined by the equilibrium between solution and adsorbent, not of the material of the band but of the displacing adsorbate. Displacement always results in a sharp boundary between the bands of eluted solution and displacing solution. Thus, the adsorption band moves with keeping a constant band length. During the migration the isotopes in the band are rearranged in order of the distribution coefficients.

3. Ion exchange separation and analysis

3.1. Ion exchange column

3.1.1. Batch and column operation

Ion exchanges are generally employed in the two different modes of operations: batch and column operation. The batchwise operation consists of contacting the whole of an electrolyte solution to be treated with a mass of exchanger and then separating the two phases by means of filtration, decantation, etc. It is quite obvious that for those exchange reactions that do not approach completion, a batchwise operation must be repeated many times before complete transformation is realized. The more unfavorable the equilibrium, the larger the number of stages are necessary for a given degree of exchange or separation.

The operation of an ion-exchange reaction by leaching the exchanger in a columnwise manner achieves most efficiency the same goal that would take many stepwise stages in a batch method. The column operation is essentially an elaborate multiple batchwise technique in which the uppermost portion of the column is constantly contacting fresh electrolyte whereas the lower portions contact the electrolyte not adsorbed by the upper exchanger. This procedure permits the exchanger bed to become fully exhausted at the top and then gradually downward. Ion exchange chromatography is a technique in which resolution of a mixture is achieved by virtue of differences in migration rates of the components in a column packed with an ion exchanger.

3.1.2. Chromatographic column operation

Chromatography was first applied to colored substances where "bands" of different colors can be seen while they move down the column. The chromatographic column operations are generally conducted in the manners of elution development, displacement development, and frontal analysis [8-10]. In the development techniques, a certain amount of the mixture is introduced at the top of the column and is then "developed" or "eluted" by a suitable agent. In frontal analysis, the mixture is continuously fed to the column. The front boundaries of the various components emerge at different times due to the differences in the migration rates.

3.1.3. Preparation of ion exchange column

In the analytical applications of ion exchange, it is necessary to utilize uniform particle size ion exchangers in order to avoid irregular flow due to the distribution of particle size. For this purpose, the wet resins in the form of H^+ were screened prior to the packing. The stainless steel screens of 42, 80, 200 and 400 meshes are used commonly. Then the ion exchanger with water in a beaker was stirred, and poured the slurry of resin into the column.

Under no circumstances must bubble or air be allowed to form in the bed of the ion exchanger beads. In operation, therefore, the bed must never be allowed to drain; the liquid level must not fall below the top of the bed. The water level was kept at 1-2 cm above the top of the bed during the operation.

The fresh ion exchanger provided by the supplier usually contains impurities, such as low molecular weight organic substances, and metallic ions, such as ferric ions. The new ion exchanger bed was, thus, washed in order to remove these impurities and metallic ions by following manner:

1. The bed is first washed with a ten fold column volume of 2.0 M hydrochloric acid.
2. The same amount of 10 M sodium chloride is then fed into the column.
3. Repeat the process (1) and (2) for three times.
4. Repeat the process (1).
5. The bed is finally rinsed with distilled water until no leakage of H^+ is recognized.

3.1.4. *Apparatus*

In the ion exchange chromatography, high pressure resistant pyrex glass columns with a water jacket were used in the studies of isotope separation. The column of which the inside diameters of 8, 10 and 20 mm were prepared so that the operations are able to be conducted up to the column pressure of 50, 30 and 20 kg/cm^2 in each column. To feed solutions, the microprocessor controlled, double plunger type titanium high pressurized pumps were used. The Teflon tubing of 1.0 mm i.d. was employed to connect column and pump. To monitor the column pressure, a pressure gauge with a safety device was placed between the column and the pump. In addition, an air damper was used in order to prevent the piston flow, especially at high flow rate runs, when loading solutions into columns. A schematic diagram of the apparatus used for the ion exchange separation is shown in Figure1. Water jacket was employed to remove the heat of reaction and maintain the column temperature throughout the experiments by circulation the thermostated water.

4. Nitrogen isotope application

Enriched isotope of ^{15}N has been a very valuable commercial product for which there is presently a growing demand in various scientific research and industry applications. In the agricultural field, ^{15}N has been extensively used to study nitrogen cycling in soil-plant relationships. The fertilizer efficiency, distribution, assimilation and metabolic plants in the plant, symbiotic nitrogen fixation and the behavior of fertilizer in soils have been studied with use of ^{15}N. Adriana et al. evaluated the nitrogen fixing capacity of a range of commercial cultivars of maize (*Zea mays L.*) by the ^{15}N isotope-dilution method. Biological nitrogen fixation expressed as percent nitrogen derived from air ranged from 12 to 33 regardless of nitrogen fertilization [11]. Their results demonstrated that maize cultivars obtain significant nitrogen from biological nitrogen fixation, varying by maize cultivar and nitrogen fertilization level. In order to improve yields of crop production in many areas, one strategy is to choose crops with high nitrogen use efficiency that can produce economic yields under limited water supply. Sarr et al. performed the pearl millet (*Pennisetum glaucum* L.R.Br.) and cowpea (*Vigna unguiculata* L. Walp.) in sole crops and intercrops systems for the nitrogen use efficiency of applied fertilizers [12]. ^{15}N-labeled urea at rates of 20 kg ha^{-1}(sole and intercrop cowpea) and 41 kg ha^{-1}(sole millet and intercrop millet) was derived from the nitrogen fertilizer and 84.70% from nitrogen mineralized in soil. In addition, many other publications reported the ^{15}N application in agroecological system [13-17]. In the industrial application, growing attention has been recently placed on the use of ^{15}N isotope for the nitride fuels of FBRs (Fast Breeders Reactors) because of their desirable properties of large thermal conductivities and large breeding ratio [18-20].

4.1. Nitrogen isotope separation process

The separation of the nitrogen isotope has been studied since Urey et al. first succeeded in concentration of the nitrogen isotope of atomic weight 15 by using chemical exchange between ammonia gas and aqueous solutions of ammonium salts [21]. In their work, the

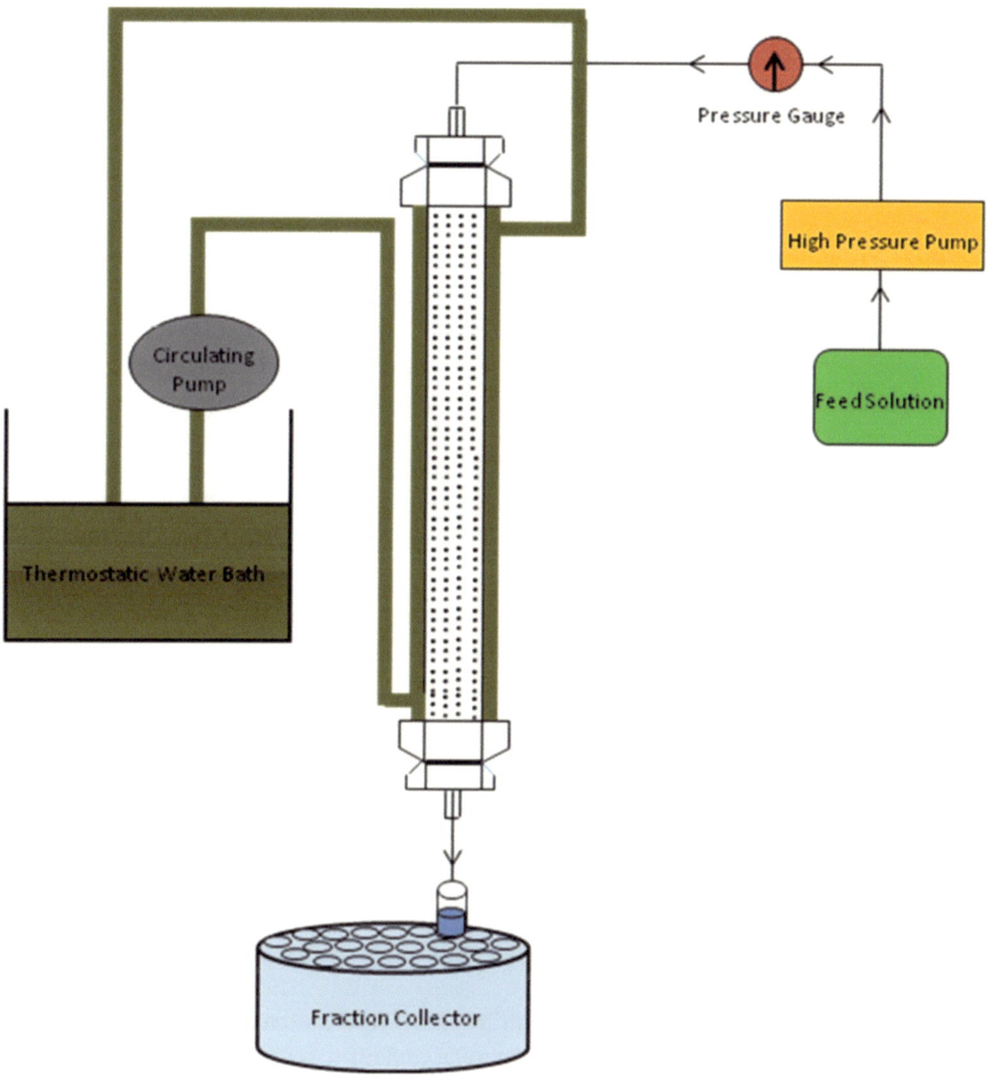

Figure 1. Schematic diagram of the apparatus for ion exchange separation

considerable quantities of material containing up to 75% atom of ^{15}N was obtained by exchange between ammonia and aqueous ammonium nitrate solutions. Later, in 1955, Spinder and Taylor [22] developed the chemical exchange method involving the exchange between nitric acid and nitric oxide as Equation 1:

$$^{15}NO + H^{14}NO_3 = {}^{14}NO + H^{15}NO_3 \tag{1}$$

The fractionation factor of the exchange reaction obtained at 1.0 M nitric acid 1.062 is extremely favorable compared with the separation factors of most other possible separation methods of nitrogen isotopes by chemical system. In this process, the ^{15}N concentrates in the nitric acid and thus ^{15}N enriches in the lower part of the exchange column. In Japan, ^{15}N has been commercially produced using this technique, which is called NITROX process. In Romania, Axente et al. have studied this process for thirty years using a laboratory-scale experimental plant [23]. Some other isotope separation methods such as distillation and thermal diffusion have also proved to be satisfactory for ^{15}N separation [24-27].

Ion exchange process on nitrogen isotope separation has been studied since Spedding et al. first succeeded in the enrichment of ^{15}N by cation exchange chromatography [6]. In their experiment, the separation factor of the isotopic exchange between the two phases of the dilute ammonium hydroxide solution and the cation exchange resin (Dowex 50-x12) in ammonium form was measured as 1.0257±0.0002. A few years later, the studies on the nitrogen isotope separation by ion exchange were reported by two Japanese research groups Ishimori[28] and Kakihana [29]. Ishimori investigated the influence of operating temperature and concentration of ammonium solution on the isotope effects in a batch equilibrium system. Kakihana measured the separation factors of nitrogen isotopes for NH_3 and NH_4Cl in acetone water and ethanol water mixture systems. Afterwards, in 1980s, Park and Michaels studied the separation process developed by Spedding et al. under various operating conditions using columns packed with sulfonated polystyrene-divinylbenzene copolymers [30]. In the recent years, Krugrov et al. studied NH_3-NH_4^+ system using SMB (simulated moving bed) process for various flow rates under total reflux [31] and the theoretical analysis of separating nitrogen isotopes by ion exchange was reported by Aoki et al [32].

4.2. Nitrogen isotopic analysis with mass spectrometry

If an aqueous solution of ammonium hydroxide is placed in contact with a cation exchanger, the following exchange reaction occurs:

$$^{14}NH_4^+ + {}^{15}NH_4OH = {}^{15}NH_4^+ + {}^{14}NH_4OH \qquad (2)$$

the result of which, under equilibrium conditions, is that the heavy nitrogen isotope, by means of the ammonium ion, is distributed more abundantly in the solid phase, which is the cation exchanger. In order to multiply the elementary effect for separating ^{15}N, the following chromatographic separation process developed by Spedding was employed [6]: the ion exchanger bed was first converted to the hydrogen form by passing 2.0 M hydrochloric acid through the column until the bed was saturated. The bed was then rinsed with distilled water and an ammonium hydroxide solution of certain concentration was added. At the lower end of an ammonium band, the following reaction takes place:

$$H - R + NH_4OH \rightarrow NH_4 - R + H_2O \qquad (3)$$

Where -R denotes the fixed anion of cation exchanger. Since the equilibrium constant of this exchange reaction is on the order of 10^9, the reaction is far to the right, resulting in an extremely sharp boundary. The ammonium ion comes to chemical equilibrium within the system very rapidly and isotopic equilibrium tends to be approached as the solution passes over the cation exchanger. In the frontal analysis, the effluent is collected in fractions at the bottom of the column when the ammonia solution of which the front part of the ammonium band is emerged. Prior to the collection, certain volumes of hydrochloric acid are initially added in the sampling tubes to prevent the ammonia from degassing. Since the ^{15}N isotope is depleted in the front of the band, ^{14}N isotope is enriched in the effluent.

In the displacement band chromatographic process, an ammonia solution is fed into the column until an ammonium adsorption band of the desired length is formed. The concentration of the fed ammonia solution was adjusted approximately at the concentration of the eluent in each operation. A caustic solution (NaOH or LiOH) was then fed into the column to elute the ammonium band in the reverse breakthrough manner. At the top of the ammonium band, the following reaction takes place:

$$NH_4 - R + NaOH(LiOH) \rightarrow Na(Li) - R + NH_4OH \tag{4}$$

which, having an equilibrium constant of about 10^5, in turn guarantees a very sharp boundary at the rear end of the band. The ammonium ions are continuously released from the cation exchanger and converted to ammonium hydroxide by an equivalent of sodium (or lithium) ions which will be deposited at the rear edge of the band. The ammonium hydroxide solution moves down along the band until it reaches the front edge of the band. The diagram of the displacement band chromatographic process for separating nitrogen isotopes with the use of a cation exchanger is shown in Figure 2. The countercurrent movement of NH_3-NH_4^+ ions in two phases develops a longitudinal isotopic distribution profile within the band. The ^{15}N isotope is enriched in the rear end of the band. The effluent of the rear part of the band was collected in fractions at the bottom of the column.

The volume of the fraction was determined by gravimetry, measuring the weight of the sample solution. The concentrations of ammonium and sodium (lithium) ions in the fractions were determined by an ion chromatography analyzer. The collected samples in the form of ammonium chloride were converted to nitrogen gas by adding a solution of potassium hypobromide(KBrO). The reaction involved is as follows:

$$2NH_4Cl + 3KBrO + 2NaOH = N_2 + 3KBr + 2NaCl + 5H_2O \tag{5}$$

The solution of potassium hypobromide was prepared by following process:

1. Add 36.5 g of sodium hydroxide and 2.2 g of potassium iodide into 170 g of distilled water.
2. Cool the solution 1) by ice and then 25 g of bromine is added with agitation.
3. The yellow colored solution obtained is left standing in a refrigerator for overnight.

The KBrO solution was stored in a refrigerator since potassium hypobromide is easily decomposed to potassium bromide and oxygen in light or by elevation of temperature. The Rittenberg was employed in order to prepare nitrogen gas samples for isotopic analysis and the apparatus of the nitrogen conversion system is shown in Figure 3. The preparation procedure of nitrogen gas samples is described below:

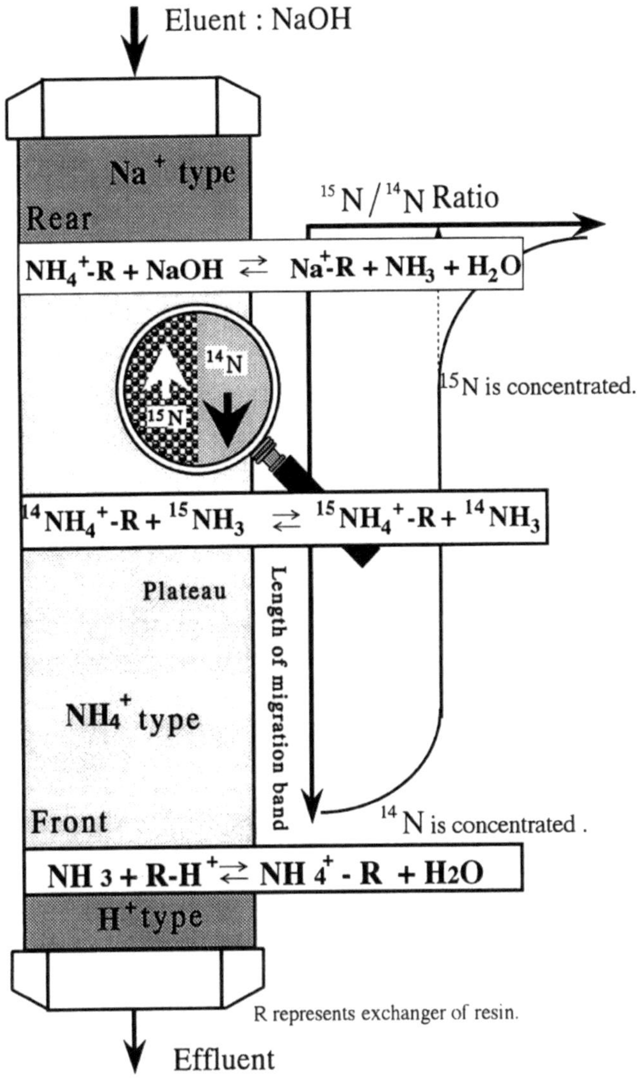

Figure 2. The diagram of the displacement band chromatographic process for nitrogen isotope separation

1. Put a certain volume of ammonium chloride sample (0.1-1.0 cm^3) in one side and a stoichiometrically excess amount of potassium hypobromide in the other side of a Rittenberg glass tube A.
2. A cold trap D is cooled with liquid nitrogen to prevent contamination of water in the vacuum system during evacuation.
3. Freeze the solutions in the Rittenberg galss tube with liquid nitrogen.
4. Open the glass stopcocks B and C which are used to control the vacuum manipulations to pump out the air by an oil rotary pump. After the evacuation, the stopcock B is closed.
5. The liquid nitrogen vessel is then removed from the Rittenberg tube to admit the solutions to be melted.

6. Freeze the solutions again and open the stopcock B so that the gas generated during the melting is evacuated.
7. After the evacuation, both stopcocks B and C are closed and Rittenberg tube is taken off from the conversion system.
8. Finally, nitrogen gas is obtained by mixing the solutions of ammonium chloride sample and potassium hypobromide in the Rittenberg tube.
9. The nitrogen gas sample is then introduced to the mass spectrometer for the isotopic analysis.

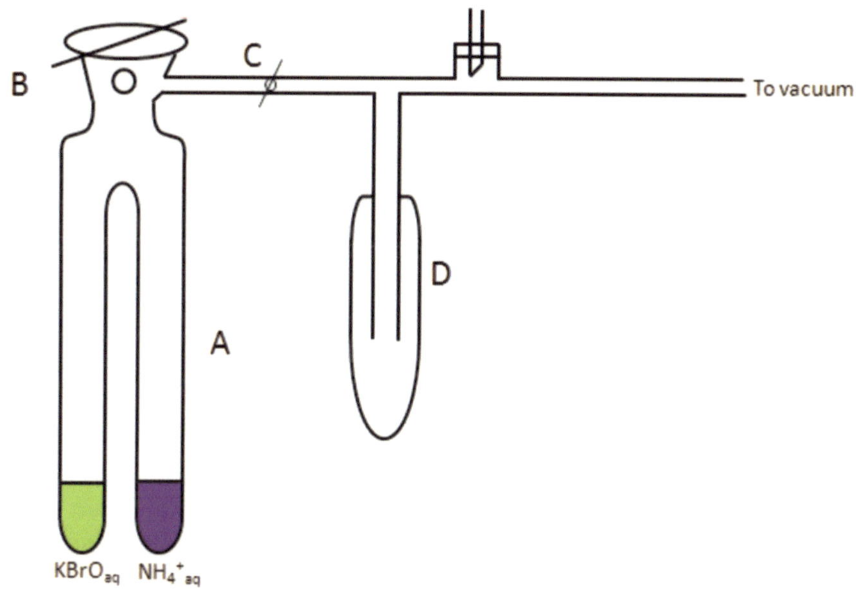

Figure 3. Schematic apparatus of the nitrogen gas conversion system

5. The effect of migration distance to the nitrogen isotope separation

The sulfuric strong acid cation exchange resin SQS-6(high porous type, cross linking 8%, particle size 60-90 um in the form of NH_4^+, provided by Asahi Chemical Industry Co., Japan) was packed uniformly in a pressurized glass column with a water jacket. The microscope view of H^+ type SQS-6 resin was shown in Figure 4. The ion exchange resin in the column preliminarily conditioned to the H^+ form with 2.0 M HCl solution, then pure water was feed into the top of the column to remove the free H^+ in the resins. 0.2M NH_4OH solution was feeded into the column to form 1.0 m ammonium adsorption band and the band was eluted by the displacing solution of 0.2M NaOH in the reverse break-through manner. The flow rate and band velocity were controlled by feeding pump and the column pressure was monitored by high pressure pump. The temperature of the column was kept constant throughout the experiments by passing the thermostated water through the water jacket. The effluent, which emerged from the column, was collected using a fraction collector. In order to prevent the NH_4OH samples from de-gassing, certain volumes of excess HCl solutions were added in the collection tubes prior to the samplings. Considering the long migration distance for the purpose of large scale production, it is necessary to regeneration column after the ammonium adsorption band pass through certain column. When the rear boundary of ammonium adsorption band enter into the next column, the first column was separated from the connection system and eluted by 2.0M HCl for regeneration, the rinse volume was at least five times larger than that of the resin's volume. The concentrations of NH_4^+ in the fractions were determined by ion chromatography analyzer and the mass peaks of $^{14}N^{15}N$ and $^{14}N^{14}N$ of the samples were measured with the mass spectrometer and the isotopic abundance of ^{15}N was calculated from the ratio of the peak height.

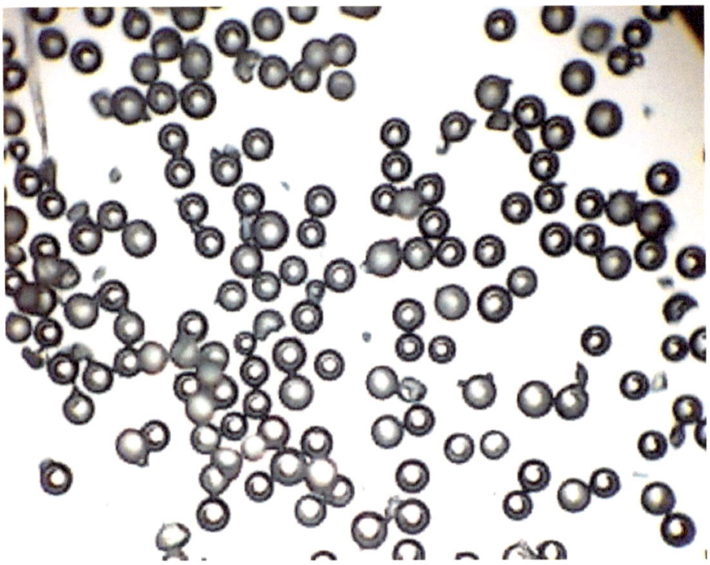

Figure 4. Microscope view of H^+ type SQS-6 resin

The separation of nitrogen isotopes by means of ion exchange resin is based on the isotopic fraction between ammonia in aqueous solution and ammonium ion in the ion-exchange resin as shown in the Equation 6. The single stage separation factor α or the separation coefficient ε is defined by:

$$\alpha = 1 + \varepsilon = \overline{[15N]}[14N] / [15N]\overline{[14N]} \tag{6}$$

Where $[\ ]$ denotes the concentration of isotopes in the aqueous phase and $\overline{[\ ]}$ the concentration of the isotopes in the resin phase.

In order to evaluate the chromatographic efficiency of nitrogen isotope enrichment, HETP (height equivalent to a theoretical plate) is introduced. HETP is usually calculated using the slope of the isotopic distribution curve in the steady-state of isotope separation after migration and is calculated by the following Equation 7:

$$H = \frac{\varepsilon}{k} + \frac{1}{k^2 L} \tag{7}$$

Where H is the HETP and L is the migration length. The slope k is experimentally determined by using the following Equation 8:

$$Ln(r - r_0) = k(L - x) \tag{8}$$

Where r is the isotopic ratio of nitrogen sample and r_o is the isotope ratio of feeding material. The term (L-x) is the distance between inner band location x and the rear or front boundary of which the migration length is L. Equation 8 indicates that plotting $\ln(r-r_o)$ against (L-x) for the experimental data produces a linear line with the slope of k.

Nitrogen isotope separations with different migration distance were performed and the experimental conditions were mentioned in Table 1.

Run	Migration distance (m)	Column ϕ (mm)	Eluent NaOH Conc.(M)	Flow rate (cm^3/min)	Band velocity (m/d)
1	3	10	0.22	0.98	2.3
2	4	10	0.21	0.99	2.3
3	10	10	0.23	1.0	2.3
4	30	10	0.25	0.98	2.3
5	30	8	0.20	1.5	4.8

Table 1. Experimental conditions of chromatographic nitrogen isotope separation with different migration distance

The chromatographic elution curve for Run 4 as an example is given in Figure 5. It is seen that ideal displacement chromatograms are obtained and a sharp band boundary between the ammonium band and sodium band is maintained during the long migration at the current experimental conditions. The maintenance of the sharp band boundary is most important to obtain highly enriched isotope at the boundary region. Light isotope of ^{14}N was enriched at the front boundary and the heavy isotope ^{15}N was enriched at the rear boundary.

The two parameters of separation coefficient and HETP can be calculated from both front and rear boundary regions and the values were listed in Table 2.The observed ε at front and rear boundary should be coincident each other. As shown in Table 2, the observed separation coefficients at front and rear boundary are in good agreement in each Run. The separation coefficient is an equilibrium factor and should be constant, being independent of the migration distance. It is confirmed from the results listed in Table 2 in the cases other than Run1. The small values of separation coefficient of Run1 are probably due to the feed material of $(NH_4)_2SO_4$ solution and one in NH_4SO_3-R in the ion exchange resin are resemble each other. In such case, isotope effects do not occur between the two chemical species. Due to this fact, the effective length of migration is reduced by approximately 20%. This effect is reflected in the decrease in the separation coefficient in Run1.

It is also confirmed that the values of ε obtained from Run 4 are slightly reduced. Probably this is due to the remixing in the middle of the adsorption band. Figure 5 showed the chromatographic profile of Run 4. From the figure it can be seen that the isotopic plateau region is not seen, the enriched zone and the depleted zone are directly contacted because of the long migration. It should be noted that the values of ε obtained from the experiments using natural nitrogen (Run 3) are practically the same as those observed by using enriched nitrogen (Run 5);

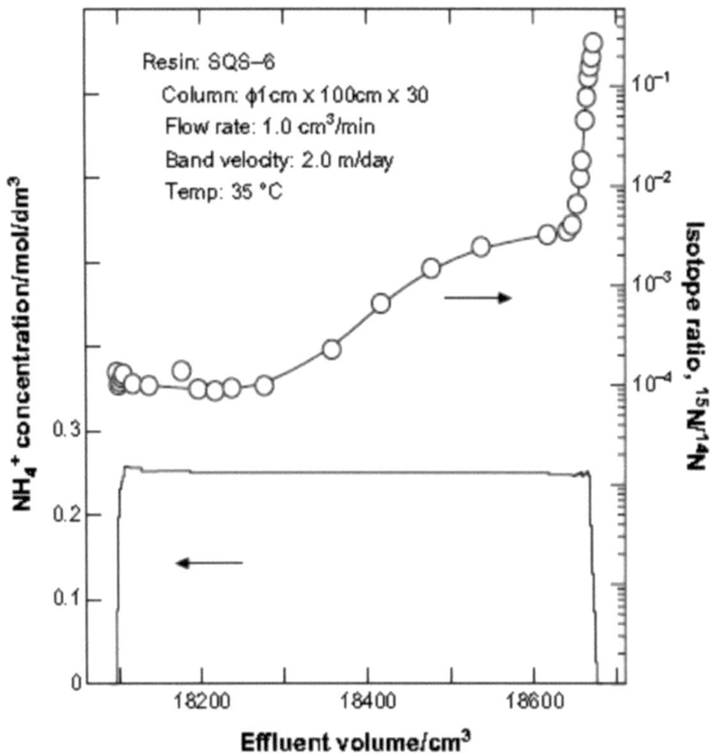

Figure 5. Chromatographic profiles of ammonium ion concentration and isotopic enrichment of 30 m migration distances.

Run	Max. enrichment(%)		$\beta_{rear} = r_{max}/r_o$	Separation coefficient		HETP (mm)	
	Front ^{14}N	Rear ^{15}N		Front	Rear	Front	Rear
1	99.956	2.29	6.4	0.016	0.018	0.37	0.21
2	99.987	3.91	11.1	0.025	0.025	0.74	0.19
3	99.994	9.54	28.7	0.024	0.024	1.22	0.16
4	99.990	21.55	78.9	0.021	0.019	2.6	0.16
5[a]	99.143	99.678	77.4	0.023	0.022	0.39	1.58

Run 5 used 80% ^{15}N as feed solution, of which r_o =4.

Table 2. Experimental results of chromatographic nitrogen isotope separation with different migration distance

The maximum enrichment β values are defined as the local enrichment factor of ^{15}N at the boundary as follows:

$$\beta_{rear} = r_{\max} / r_o \tag{9}$$

Where r is the isotopic abundance ratio of ^{15}N against ^{14}N. The values of β calculated are listed in Table 2 and plotted as a function of migration distance in Figure 6. The slope of plots is unity, which means that the maximum enrichment β values are proportional to the migration distance up to 30 m. This fact suggests that the enrichment proceeds, forming an ideal shape of exponential enrichment curve at the band boundary region. In addition, it is quite interesting that β value of Run 4 using natural nitrogen is in very good agreement with that of Run 5 where 80% enriched nitrogen was used as feed material. This information is important and useful for designing the enrichment plant based on the present method of ion exchange chromatography.

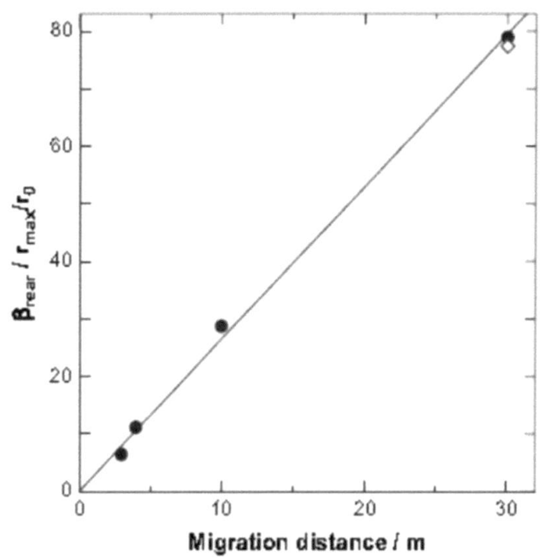

Figure 6. Correlation between migration distance and the maximum enrichment of ^{15}N

In order to evaluate the efficiency of nitrogen enrichment in chromatographic migration, HETP is introduced and the calculated values of HETP are listed in Table 2. HETP values are small enough in the rear band region of Runs 1-4 where the enrichment of ^{15}N is steadily proceeding. Similar phenomenon is observed in the front ^{14}N enrichment zone of Run 5. In general, the HETP values are small when the enrichment process does not reach the high level. On the other hand, saturation of isotope enrichment gradually takes place and slope of the enrichment curve sharply decreases when high enrichment is attained. This is the case for the enrichment of ^{14}N in Runs 1-4 and the enrichment of ^{15}N in Run 5.

The results of Run 1-4 clearly indicate that the practical limit of ^{14}N enrichment by the ion exchange packed column is 99.99%. The reason for this limit has not yet been elucidated, but it is estimated that this limit of 99.99% may be applicable for the enrichment of ^{15}N as well. So far, the target of ^{15}N enrichment for nitride fuels is 99.9%. The present work realized the enrichment of 99.678% ^{15}N in Run 5. The results suggest that the 99.9% ^{15}N is attainable by ion exchange enrichment.

6. The effect of cross linking to the nitrogen isotope separation

Separation and concentration of a stable isotope from an isotopic mixture with natural occurrence is a very complex problem and usually, the isotopic separation coefficients of natural abundance are very low. In general, cation exchange process is a promising technique to produce highly enriched isotope due to the nature of small HETP value. Among the operating parameters in cation exchange process, cross linking is known as a decisive factor on the process of nitrogen isotope separation. It is interesting to evaluation the chromatographic performance of ^{15}N isotope separation by ion exchange resin with different percentage of cross linking.

The cation exchange resins with different cross linking were synthesized from the law material of styrene and the synthetic method was given in Figure 7. When different cross

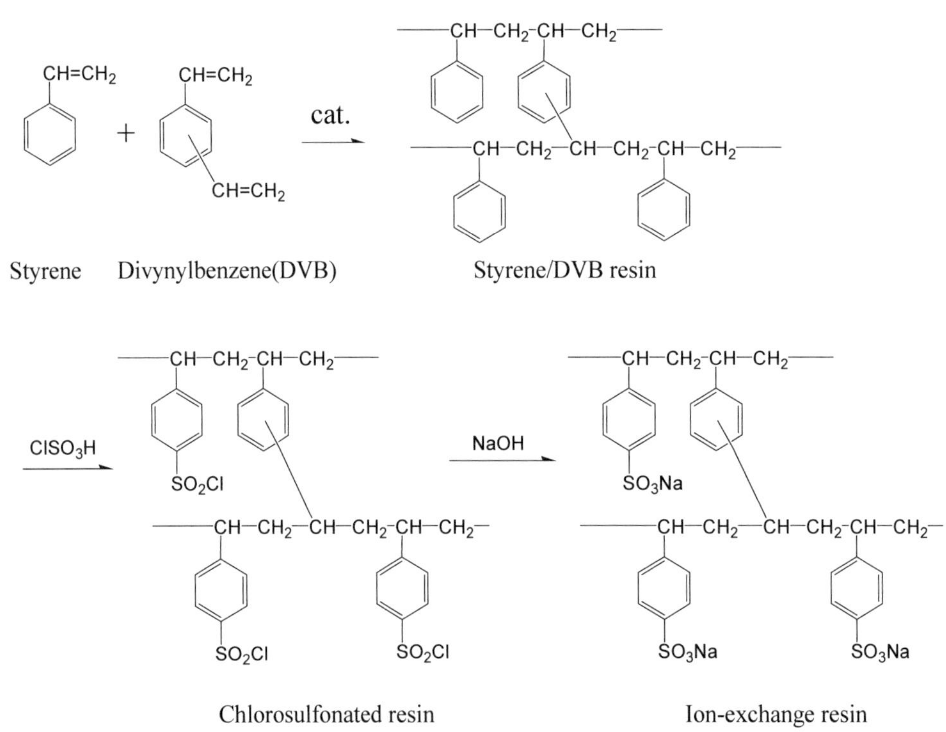

Figure 7. Synthetic method of the cation exchange resin with different cross linking percentage

linking was synthesized, high cross linking means high divynybenzene (DVB) materials contained in the resin structure and this means the percentage of ion exchange functional group was decreased in the same amount resin. The relationship between ^{15}N enrichment percentage and cross linking with 2.0 m migration was given in Figure 8. Enriched ^{15}N isotopes were decreased from 0.93 to 0.68 when compared with the cross linking of twenty and forty percentage. Since high cross linking means low exchange capacity, it is reasonable that the ^{15}N enrichment percentage is reduced with the increase of cross linking at the same given migration distances. SQS-6 resin has 8% cross linking and ^{15}N enrichment percentage was 1.56% when SQS-6 resin was performed under the same condition, and low cross linking resin has much higher enrichment ability than high cross linking resin.

The observed HETP values of different cross linking were plotted in Figure 9. It is seen in the figure that cross linking can affect the HETP value obviously. HETP value proportionally increase with the cross linking at the present work and low cross linking has much advantage for HETP. HETP value of 0.036 cm was obtained at the present system by using 20% cross linking and the present resin can be used for the large scale of nitrogen isotope industrial production.

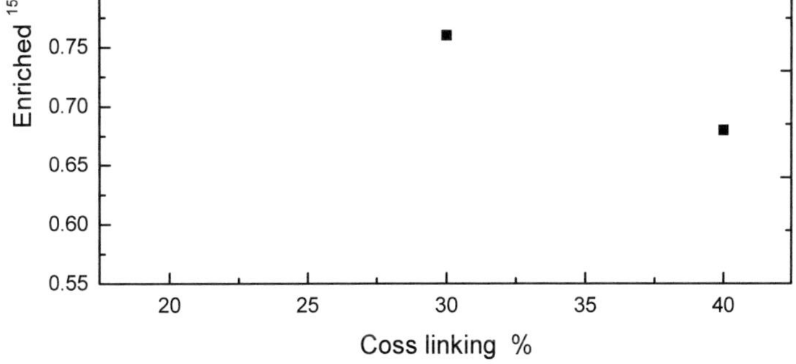

Figure 8. Relationship between enriched ^{15}N and cross linking

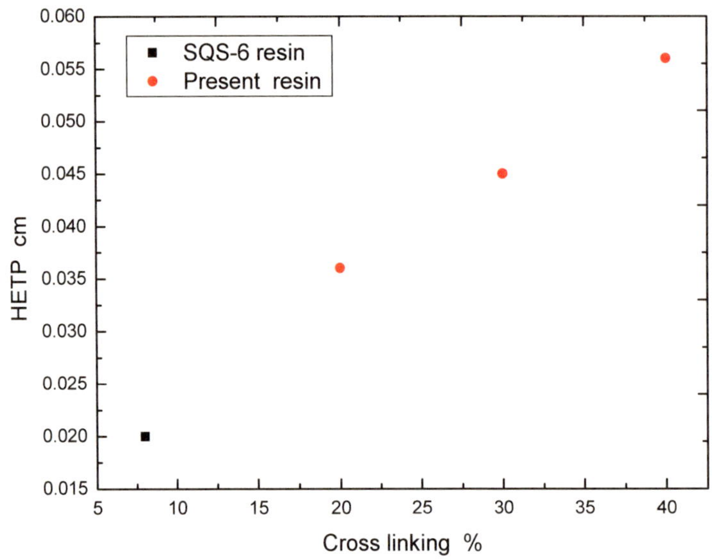

(Square dot was the result of SQS-6 resin)

Figure 9. Observed HETP values and cross linking

7. Preparation of highly enriched ^{15}N by chemical exchange

The migration distance and cross linking can both affect the nitrogen isotope separation. It has relatively low separation coefficient by ion exchange chromatographic method when compared with that of NITROX method, it is very difficult to prepare 99.9% ^{15}N directly from natural abundance of 0.366%. At the present stage of industrial factory, NITROX was used widely to produce various percentages of ^{15}N products. It is very difficult to get highly enriched ^{15}N by this method. In order to improve feasibility of this method, one good idea is to combine NITROX method and ion exchange chromatography. Starting from the natural abundance, NITROX has advantage to enrich ^{15}N because of large separation coefficient; while when it reaches to relative highly enriched ^{15}N, due to the good merit of very small HETP value, the ion exchange chromatographic method is performed to get highly enriched ^{15}N.

In order to simulate the industry scale, large diameter column (ϕ=3.0 cm) and high speed band velocity were used in this stage. 100 g commercial NH_4Cl (^{15}N-80%) were feed into 3.0 cm diameter ion exchange columns. In order to shorten the ammonium adsorption time, the concentration of NH_4Cl was adjusted to about 0.5 mol/dm^3 and the length of ammonium adsorption band was 138 cm. Then 0.22 mol/dm^3 NaOH was feed to the rear part of ammonium adsorption band to develop the adsorption band. The flow rate of feeding solution is 50 ml/min and the band velocity is 12 m/d. When the front boundary of ammonium adsorption band reached to twenty five meters, the three way valve was connected to the column for portion sampling. The sampling volume for each sample was about 0.5 ml for every seven minute and it's corresponding to 5.0 cm of adsorption band for each sampling. These samples were analyzed by mass spectrometer for determine the nitrogen isotopic ratio as the result of the migration distance equal to 25 meters. When the rear boundary of ammonium adsorption band passes through the three way valve, the sampling was stopped. In order to keep the length of ammonium adsorption band constant and to avoid band diffusion, total amount of 3.0 L pure water with same flow rate was feed to the top of the column to develop the inside part of free ammonium molecular of adsorption band, the free ammonium molecular was move to the front boundary and re-adsorb again at the front boundary. The adsorption band moved forward 22.5 cm after 3.0 L pure water washing. Five days later, NaOH solution was feed again to develop the migration distance until fifty meters. The operation condition was the same with the first stage. When the front boundary of ammonium adsorption band was reached to fifty meters, three way valve was connected to the column for portion sampling. The sampling volume for each sample was about 0.5 ml for every five minute and it's corresponding to 4.0 cm of adsorption band for each sampling. Before the rear boundary of ammonium adsorption band reach to three way valve, about 10 cm left the boundary, sampling was changed to whole fraction sampling and the volume of each sampling was 30 ml and each corresponds to 1.0 cm of adsorption band, the sampling time was 45 second. Before sampling, excess amount of HCl was added to each bottle for neutralization. After the rear boundary pass through three way valve, 3.0 L pure water was feed to the top of column.

The large scale experimental conditions and results were listed in Table 3 and the chromatographic isotopic distribution curve of 25 m and 50 m migration distance were given in Figure 10 and Figure 11. During the long chromatographic operation, small amount ammonium ion exists in the tailing and may cause to the decrease the final isotope separation effect; this is the disadvantage of the long chromatographic operation. From the isotopic ratio curve of Figure 10 and Figure 11, the front boundary of ammonium adsorption band, the ^{14}N isotopes could be continually enriched to over 99% which starting from 20%. On the contrary, ^{15}N isotope was enriched in the rear boundary regions steadily. Compare the results by Run1 and Run2 in Table 3, both are starting from 80%-^{15}N NH_4Cl, the main differences are column diameter and flow rate. In Run3, high flow rate (50 ml/min) and 3.0 cm diameter column was used. The results show that at the rear band, both results have nearly the same values of separation coefficient (0.022) and HETP (0.16cm). At Run3, the highest percentage of enriched ^{15}N was 99.756%, higher than the value of 99.672% by Run2, even if the migration distance in Run3 is shorten than Run2. This means the speed of ion exchange between sodium and ammonium is very fast, and high flow rate does not affect to the HETP and separation coefficient in this operation systems. It is obviously that high flow rate has advantage to get highly enriched ^{15}N and is much suitable to industrial operation. Run 3 and Run 4 were within the same experiment in a whole experiment; here it was divided into two parts. From the results of Run3 and Run4, ^{15}N was steadily enriched with

Run No.	Run 1		Run 2		Run 3	
Resin	SQS-6 high porous cation exchange resin					
Feed solution	0.19M NH_4Cl (^{15}N-80%)		0.50M NH_4Cl (^{15}N-80%)		0.50M NH_4Cl (^{15}N-80%)	
Effluent solution	0.20M NaOH		0.23M NaOH		0.24M NaOH	
Temp. K	308					
Migration distance m	30		25		50	
Column diameter mm	8		30		30	
Ammonium adsorption band m	1.0		1.38		1.38	
Flow rate ml/min	1.5		50		50	
Band velocity m/d	4.8		11.1		14.4	
Adsorption capacity Q mmol/m	74.3		1230		1250	
Boundary	Front ^{14}N	Rear ^{15}N	Front ^{14}N	Rear ^{15}N	Front ^{14}N	Rear ^{15}N
Nitrogen %	99.143	99.672	96.617	99.756	99.710	99.859
Separation coefficient	0.023	0.022	0.022	0.022	0.023	0.023
Slope coefficient Ks	0.6104	0.1541	0.5218	0.1569	0.4919	0.1599
HETP cm	0.039	0.158	0.044	0.157	0.048	0.145

Table 3. The experimental conditions and results for nitrogen isotope separation by using 80% $^{15}NH_4Cl$

the increase of the long migration distance, but the increasing trend becomes very low. For the first twenty five meter's migration, ^{15}N has been enriched to 99.756%, after another twenty five meter's operation, the final maximum enrichment percentage reached to 99.859%. Enriched ^{15}N is only increased 0.1% by another twenty five meter's migration, although the highly enriched regions became broaden and we can get much volume of ^{15}N which the percentage is over 99.8%. The separation coefficient of Run3 and Run4 is same because of the same operation condition; Run4 has relatively smaller HETP value than the value of Run 3. It indicates that long chromatographic operation has advantage and can steadily enrich ^{15}N to very high percentage. The reason why the enriched ^{15}N could not enrich to higher percentage in the present operation system may come from the stopping during the operation between Run3 and Run4. Because of the stopping by one week, the inner ammonium adsorption band may take remixing within the middle adsorption band. Highly enriched ^{15}N at rear band will mix with middle level enriched ^{15}N which was located in the middle band. If this operation restarts again, it will take long distance chromatographic operation to compensate the mixing and reach to the same percentage of ^{15}N with twenty five meter's operation. From this result, the effective migration distance may be less than fifty meters. This can be confirmed by the analysis of the isotopic curve of Run 3 and Run 4 in Figure 10 and Figure 11. In Figure 10, there is a flat regions existing in the middle ammonium adsorption band, in this region, the ratio of $^{15}N/^{14}N$ is the same with

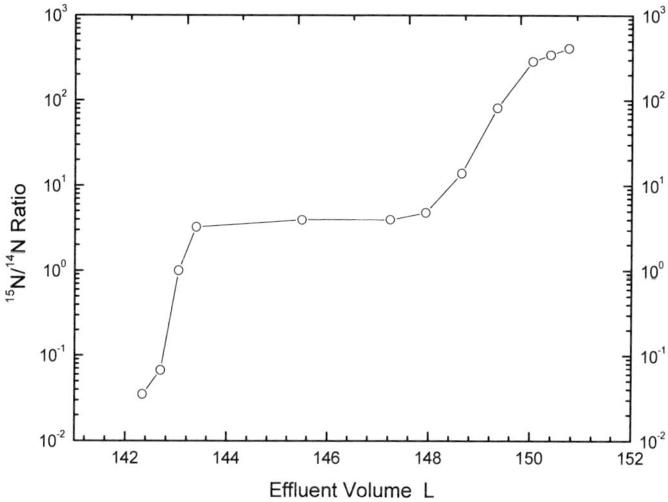

Figure 10. $^{15}N/^{14}N$ isotopic ratio (25 meter)

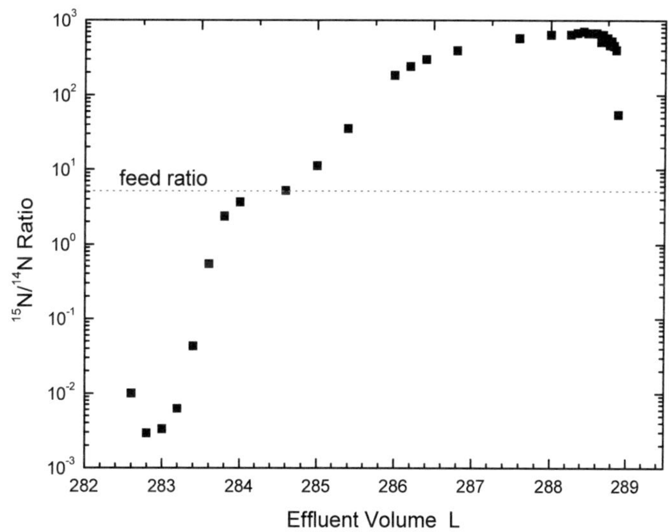

Figure 11. $^{15}N/^{14}N$ isotopic ratio (50 meter)

the original feeding solution, 80%. At the rear boundary of the isotopic distribution curve, the slope of isotopic curve is still kept sharpness. In Figure 11, it was found that this flat region almost disappears and the isotopic ratio of $^{15}N/^{14}N$ was steadily increasing in the whole ammonium adsorption band. At the rear boundary region of the isotopic curve, there is a large amount of highly enriched ^{15}N and the slope coefficient of isotopic distribution curve becomes flat. This is the evidence that by long chromatographic operation, highly enriched ^{15}N was remixing again with the middle level enriched ^{15}N in the middle position of ammonium adsorption band. The percentage and gram distribution of enriched ^{15}N which along with ammonium adsorption band were listed in Table 4. ^{15}N was mainly concentrated in the adsorption band which range from 64.5-152.03 cm in the adsorption band and 4.6 gram of highly enriched ^{15}N(>99.82%) were successfully obtained.

Adsorption band length cm	Percentage distribution of ^{15}N %	^{15}N g
0～28.5	＜4.118	0.0566
28.5～ 33	4.118 ～ 35.32	0.3539
33 ～ 46.5	35.32 ～ 80.39	2.2598
46.5 ～ 64.5	80.39 ～91.81	3.2737
64.5 ～ 82.5	91.81 ～ 98.73	3.5679
82.5～ 123	98.73～ 99.79	8.1944
123 ～147.3	99.79 ～99.86	4.6148
147.3～ 152.03	99.86 ～ 99.69	0.8185

Table 4. Percentage and gram distribution of enriched ^{15}N along ammonium adsorption band by 50 m chromatography

8. Conclusion of nitrogen isotope separation by ion exchange method and the comparison with NITROX separation method

Long chromatographic operations were studied to nitrogen isotope separation by using 80% enriched ^{15}N-NH_4Cl. The main purpose was to get very highly enriched ^{15}N isotope and to study how the flow rate and column diameter affect to nitrogen isotope separation. From the results obtained by each runs, it was confirmed that front and rear boundary could enrich ^{14}N and ^{15}N steadily to very high percentage. Since the feeding and elution solution are same for Run 1, Run 2 and Run 3, the same separation coefficients should be obtained no matter the migration distance. Due to the small amount of ammonium remained on the column, separation effect became slightly decrease with long chromatographic migration distance, especially by large diameter columns. When compared with the results of Run 2 and Run 3, the HETP value of Run 3 is smaller than Run2.

During the long operation, ^{15}N can be steadily enriched to very high purity. Because the remixing phenomena take place, the highly enriched ^{15}N will penetrated into the middle band, it decreased the maximum enrichment percentage of ^{15}N, but the highly enriched zones became broadened with long migration distance. The maximum percentage of enriched ^{15}N was 99.859% and 4.6 gram of highly enriched ^{15}N (99.82%) were successfully obtained by fifty meter's operation.

Chemical exchange of nitrogen oxide system has been discussed for nitrogen isotope separation deeply in early times. Fifty years ago, W.Spindel and T. Taylor performed nitrogen isotope separation by chemical exchange (NITROX) with the condition of 15 ml/min of 10 mol/dm^3 nitric acid, containing the natural ^{15}N abundance [7]. They produced material containing 6.0 g of nitrogen analyzing 99% ^{15}N by twenty-five days continuously columns operation. Because of the large separation factor (1.055) by using NITROX method, most of industrial scale plants in the world are adopted with this method to produce nitrogen isotopes. As mentioned above, ion exchange method has the advantage of very small HETP, it can obtain very highly enriched nitrogen isotopes and no large amount of waste gas and liquid occurred, it can be widely used if combined with NITROX method when ^{15}N has been enriched to the percentage of, for example, 80%. One problem come from ion exchange method is the capacity of resin. The ion exchange NH_4-R/NH_3aq on cation exchange resin being inefficient for large scale production, according to the small flow rate accepted in the separation columns. That involves utilization of too large diameter separation columns with prohibitive amount of cation exchange resin.

Taking into account that the world market is about 20 to 40 Kg ^{15}N annually, the supply of that isotope for nitride fuel production for nuclear power reactors (NPR) and accelerator driven system (ADS) would therefore demand an increase in production capacity by a factor of 1000. For an industrial plant producing 100 t/y ^{15}N, using present technology of isotopic exchange in NITROX system, the first separation stage of the cascade would be feed with 10M HNO_3 solution of 600 ml/h flow rate. If conversion of HNO_3 into NO, NO_2, at the enriching end of the columns, would be done with gaseous SO_2, for a production plant of 100 t/y ^{15}N, a consumption of 4 million ton SO_2/y and a production of 70% H_2SO_4 waste solution of 4.5 million ml/y are estimated. The reconversion of H_2SO_4 into SO_2 in order to recycle of SO_2 is a problem to be solved to compensate the cost of SO_2, and to diminish the amount of H_2SO_4 waste solution. If it's considered that all ADS installations for minor actinides transmutation utilize nitride fuel with ^{15}N, the need of that isotope is about 4 t/y, with a cost of 400 million USD (the price of ^{15}N was considered 100 USD). It should be taken into consideration an important price reduction of ^{15}N in order to make possible its utilization for large scale production of nitride fuel for NPRs and ADSs.

Author details

Xingcheng Ding* and Xunyue Liu
Institute of Nuclear Agricultural Science, Zhejiang University, Hangzhou, P.R.China

Acknowledgement

The authors would like to express best thanks to Professor Yasuhiko Fujii, Dr. Tatsuya Suzuki and Dr. Masao Nomura, Research Laboratory for Nuclear Reactors, Tokyo Institute of Technology, Japan for their kindly guidance in technical and material supporting.

9. References

[1] Ban Y, Nomura M, Fujii Y (2002)Chromatographic Separation of Lithium Isotopes With Silica Based Monobenzo-15-crown-5 Resin. J.nucl.sci.technol.39(3): 279-281.

[2] Ban Y, Nomura M, Fujii Y(2001)Isotope Effects of Strontium in Crown Ether Chromatography. Sep.sci.technol. 36(10):2165-2180.

[3] Ismail I, Fukami A, Nomura M, Fujii Y(2000)Anomaly of ^{155}Gd and ^{157}Gd Isotope Effects in Ligand Exchange Reactions Observed by Ion Exchange Chromatography. Anal. chem.72: 2841-2845.

[4] Matin M, Ismail I, Nomura M, Fujii Y(2002) Isotope Effects of Copper in Cu(II) Malate Ligand Exchange System Studied by Using Ion Exchange Displacement Chromatography, Sep.sci.technol. 37(9): 2129-2142.

[5] Nomura M, Higuchi N, Fujii Y(1996)Mass Dependence of Uranium Isotope Effects in the U(IV)-U(VI)Exchange Reaction. J.am.chem.soc. 118: 9127-9130.

[6] Spedding F, Powell J, Svec H(1955) A Laboratory Method for Separating Nitrogen Isotopes by Ion Exchange. J.am.chem.soc. 77(23): 6125-6132.

[7] Spindel W, Taylor T(1956)Preparation of 99.8% Nitrogen-15 by Chemical Exchange. J.chem.phys. 24: 626-627.

[8] Fujii Y, Aida M, Okamoto M, Oi T(1985)A Theoretical Study of Isotope Separation by Displacement Chromatography. Sep.sci.technol. 20(5&6): 377- 392.

[9] Ishida T(2002)Isotope Effect and Isotope Separation: A Chemist's View. J.nucl.sci.technol.39(4): 407-412.

[10] Ishida T, Ono Y(2006)Early History of Chemical Exchange Isotope Enrichment and Lessons We Learn. J.nucl.sci.technol. 43(4): 391-399.

[11] Montanez A, Abreu C, Gill P, Hardarson G, Sicardi M(2009)Biological Nitrogen Fixation in Maize(Zea mays L.) by ^{15}N Isotope-Dilution and Identification of Associated Culturable Diazotrophs. Biol fertil soils.45: 253-263.

[12] Sarr P, Khouma M, Sene M, Guisse A, Badlane A, Yamakawa T(2008)Effect of Pearl Millet-Cowpea Cropping Systems on Nitrogen Recovery, Nitrogen Use Efficiency and Biological Fixation Using the ^{15}N Tracer Technique. Soil. sci .plant. nutr. 54: 142-147.

[13] Bosshard C, Sorensen P, Frossard E, Dubois D, Mader P, Nanzer S, Oberson A(2009)Nitrogen Use Efficiency of ^{15}N-labelled Sheep Manure and Mineral Fertilizer Applied to Microplots in Llong-term Organic and Conventional Cropping Systems. Nutr. cycl.agroecosyst.83:271-287.

[14] Hood R(2001)Evaluation of A New Approach to the Nitrogen-15 Isotope Dilution Technique, To Estimate Crop N Uptake From Organic Residues in The Field. Biol.fert.soils. 34(3): 156-161.

[15] Sorensen P, Thomsen I(2006)Separation of Pig Slurry and Plant Utilization and Loss of Nitrogen-15-labeled Slurry Nitrogen. Soil. sci. soc. am. j. 69: 1644-1651.

[16] Thomsen I, Kjellerup V, Jensen B(1997)Crop Uptake and Leaching of ^{15}N Applied in Ruminant Slurry With Selectively Labeled Faeces And Urine Fractions. Plant soil.197: 233-239.

[17] Gutser R, Ebertseder Th, Weber A, Schraml M, Schmidhalter U(2005)Short-term and Residual Availability of Nitrogen After Long-term Application of Organic Fertilizers On Arable Land. J. plant. nutr. soil sci. 168(4):439-446.

[18] Ding X, Suzuki T, Nomura M, Aida M, Fujii Y(2005)Nitrogen Isotope Enrichment For Nitride Fuel by Using Hybrid Chemical Exchange Process. Prog.nucl.Energy. 47: 420-425.

[19] Ding X, Kaneshiki T, Arima M, Nomura M, Suzuki T, Fujii Y(2008) High Enrichment of ^{15}N Isotope by Ion Exchange For Nitride Fuel Development. Prog. nucl. Energy. 50: 504-509.

[20] Ding X, Nomura M, Suzuki T, Fujii Y(2008)High Enrichment of ^{15}N by Chromatographic Chemical Process.J.Chroms.A 1201: 65-68.

[21] Urey H, Huffman J, Thode H, Fox M(1937)Concentration of ^{15}N by Chemical Methods. J. chem. phys. 5: 856-867.

[22] Spindel W, Taylor T(1956) Preparation of 99.8% Nitrogen-15 by Chemical Exchange. J. chem. phys. 24: 626-627.

[23] Axente D, Baldea A, Abrudean M(1992) Isotope Separation by Chemical Exchange, Proceedings of the International Symposium on Isotope Separation and Chemical Exchange Uranium Enrichment. In Fujii Y, Ishida T, Takeuchi K, editors. Bulletin of the Research Laboratory for Nuclear Reactors, Tokyo Institute of Technology. pp 357-367.

[24] Mills T, Garcia M, Vandervoort R, McInteer B(1989) A Chemical Exchange System for Isotopic Feed to a Nitrogen and Oxygen Isotope Separation Plant. Sep. sci.technol.24:415-428.

[25] Kendall J(1942)Separation of Isotopes and Thermal Diffusion. Nature .150:136-140.

[26] Majumdar S(1951)The Theory of The Separation of Isotopes by Thermal Diffusion. Phys. review, 81(5): 844-848.

[27] Spindel W, Taylor T (1955) Concentration of Nitrogen-15 by Chemical Exchange in a Thermal Diffusion Column.J. chem. phys. 23(7): 1318-1322.

[28] Tatsukiro I(1960)The Nitrogen Isotopic Equilibrium Between Ammonia and Ammonium Ion. Bull.chem.soc.jpn. 33: 516-519.

[29] Kakihana H(1963)A Fundamental Study On the Ion Exchange Separation of Lithium, Nitrogen and Uranium Isotopes. J. DE chimie physique ET DE physico-chimie biologique 60(1-2): 81-88.

[30] Park W, Michaels E(1988)Separation of Nitrogen Isotopes by Displacement Band Chromatography. Sep. sci. technol. 23: 1875-1889.

[31] Kruglov A, Andreev B, Pojidaev Y(1996)Continuous Isotope Separation in Systems With Solid Phase. II. Separation of Nitrogen Isotopes With Use of Ion-exchange Resin. Sep. sci. technol.31: 471-490.